A Course in Advanced Calculus

Robert S. Borden

DOVER PUBLICATIONS, INC.
Mineola, New York

Copyright

Published in Canada by General Publishing Company, Ltd., 30 Lesmill Road, Don Mills, Toronto, Ontario.
Published in the United Kingdom by Constable and Company, Ltd., 3 The Lanchesters, 162–164 Fulham Palace Road, London W6 9ER.

Bibliographical Note

This Dover edition, first published in 1998, contains the unabridged and slightly corrected text of the work first published by North Holland/Elsevier Science Publishing Co., Inc., New York, in 1983. A new chapter, "Tips and Solutions for Selected Problems," has been prepared specially for this edition.

Library of Congress Cataloging-in-Publication Data

Borden, Robert S.
A course in advanced calculus / Robert S. Borden.
p. cm.
"Contains the unabridged and slightly corrected text of the work first published by North Holland/Elsevier Science Publishing Co., Inc., New York, in 1983. A new chapter, 'Tips and solutions for selected problems,' has been prepared specially for this edition"—T.p. verso.
Includes bibliographical references and index.
ISBN 0-486-67290-5 (pbk.)
1. Calculus. I. Title.
QA303.B73 1997
515—dc21 97-43366
CIP

Manufactured in the United States of America
Dover Publications, Inc., 31 East 2nd Street, Mineola, N.Y. 11501

To Mary

CONTENTS

PREFACE

This book has been a long time in developing, but the original goal of the author has remained unchanged: to present a course in calculus that would unify the concepts of integration in Euclidean space while at the same time giving the student a feeling for some of the other areas of mathematics which are intimately related to mathematical analysis. For several years at Knox College notes that might be called the original form of this book were used as a text for the honors section of advanced calculus, so in a sense this book has been classroom tested. It should be mentioned, however, that the students had all been through Tom Apostol's classic two-volume text of calculus with linear algebra, and most were graduate-school bound. Thus, it is only fair to say that although the book was written with the undergraduate in mind, it might be a bit heavy going for students with limited background. On the other hand, it would be presumptuous to say that this is a reference book or a treatise; it is simply a survey of a number of topics the author feels serious mathematics students should know something about, topics that are at the core of undergraduate analysis.

Anyone who studies advanced calculus must be strongly motivated by a genuine love of mathematics. By its very nature the subject is hard, and perhaps not so glamorous. To spend a year at it requires dedication. Therefore the author has made a real effort to "break the monotony," so to speak, by shifting abruptly from one topic to another. Of course, the topics are related; topology, linear algebra, and inequalities all fit into the grand scheme of advanced calculus. To the student, however, they will probably be as distinct as night and day. If the author's timing has been good, just when one has had all the topology one can stand, inner-product spaces present themselves in all their pristine glory, Fourier series for the most striking curves unfold, and the surprising secret of Pythagoras is revealed.

Most books on advanced calculus are good; so is this one. It is the sincere hope of the author that this book will, by stimulating and inspiring the readers, help develop some future mathematicians. The author would certainly never have started this project had not others had a marked influence on him through their own texts, most notably Tom Apostol, Creighton Buck, Serge Lang, Angus Taylor, and in particular, Harold Edwards, who first brought to the attention of this author the idea of interpreting the indeterminant dx as a functional.

If this work is successful, much credit should be given to the author's friend and dissertation advisor, Fernando Bertolini; his teaching was creative and inspirational, never ordinary. One pearl of wisdom of his should be passed on. Mathematics is beautiful, but it is crystal clear only to the genius. Therefore, dear reader, do not expect too much of this author; he cannot make the assimilation of the material in this text an easy task. He can only do his best to make it all beautiful.

I owe a great debt of gratitude to Professors J. Diestel, J. A. Seebach, R. Stern, and J. J. Uhl, who made numerous valuable suggestions for improving this work. Needless to say, my colleagues Bryan, Schneider, and Steeg have done far more than they realize by helping me iron out difficulties in the text as they arose, and I am exceedingly grateful for their efforts. My sincere thanks also to John Haber, Robert Hilbert, and all the people at Elsevier who have contributed to this project.

Finally, this book would never have been written had it not been for my wife, Mary, who did all the typing and managed to maintain her good humor throughout the entire ordeal. When things went badly, she was always able to provide the necessary encouragement, the timely suggestions, and the inspiration that made all turn out well, and during those periods when my enthusiasm ran high, she listened with remarkable patience as I rambled on and on about the beauty of the calculus. It is to Mary that this book is dedicated.

Most of the symbols used in this text are standard; in almost every case they are defined the first time they are used, unless their meanings are obvious. We use the ordinary symbols

$\forall$	for each, for every
$\exists$	there exists
$\Rightarrow$	implies
iff	if and only if
$\Leftrightarrow$	iff, is equivalent to
s.t.	such that

CHAPTER

1

SETS AND STRUCTURES

1.1 SETS

It is fitting that we begin this course in advanced calculus by introducing the concept of a set, perhaps the most fundamental idea in all mathematics. Having done this, we can proceed in any number of directions, building different kinds of mathematics by imposing various structures on sets. The definition that we are about to give is a naive one, but it will certainly suffice for our purposes.

Definition 1.1.1. A *set* is a collection of objects, called *elements* of the set, for which the following restrictive axioms must hold:

S1. If x is an arbitrary object, then exactly one of the two possibilities is the case: x is an element of the set, or x is not an element of the set.
S2. An element of a set must be distinguishable from another element of the set.
S3. No set is an element of itself.

Axiom S1 implies that with every set there must be an underlying rule which tells unambiguously whether an arbitrary thing does or does not belong to the set. Axiom S2 simply requires that a set be a collection of distinct objects, distinguishable from one another in the light of the underlying rule which characterizes membership in the set. The third axiom is, in a way, an afterthought, inserted to avoid a paradoxical situation. Suppose we had omitted S3. Having defined what we mean by a set, we could then consider the collection of all sets; this collection would itself be a set and hence contain itself as an element. Clearly, such a set is larger than any other set imaginable, yet once we have it, we can easily conceive of the set of all its subsets, which is an even larger set as we shall see later. Axiom S3 avoids this paradox; the collection of all sets is not a set but only a collection. There is no largest set.

A set A cannot contain itself as an element, but it does contain itself as a subset. Symbolically, this is to say $A \notin A$, but $A \subset A$. One particularly noteworthy set is the *empty set*, the set containing no elements, and we denote it by the Scandinavian letter $\varnothing$. (How *does* one pronounce this?) Although $\varnothing$ is empty, it is true that $\varnothing \subset \varnothing$. In fact, if A is any set, it is always true that $\varnothing \subset A$. If a set B is such that every element of B is also an element of A, we write $A \supset B$ or $B \subset A$. In particular, if $B \neq \varnothing$ and A contains an element not in B, we call B a *proper subset* of A.

If $x \in A$, we denote by $\{x\}$ the subset of A consisting of x alone. Such a set is called a *singleton*. We may write $x \in \{x\} \subset A$, distinguishing between the element x and the singleton subset $\{x\}$. Associated with any set A is a particular collection of sets, the collection of all subsets of A. We denote this collection $P(A)$, and it is indeed a set. $P(A)$ is called the *power set* of A; occasionally it is denoted by the strange symbol 2^A, the significance of which will be clearer very shortly.

If $\mathbb{S}$ is a class, or collection, of sets, there are several fundamental binary operations we can perform over $\mathbb{S}$. We list them for the record. For any two sets $A, B \in \mathbb{S}$:

$$A \cup B = \{x : x \in A \text{ or } x \in B\}$$

(the *union* of A and B),

$$A \cap B = \{x : x \in A \text{ and } x \in B\}$$

(the *intersection* of A and B),

$$A \setminus B = \{x : x \in A \text{ and } x \notin B\}$$

(the *difference* of A and B),

$$A \,\triangle\, B = \{x : x \in A \text{ or } B \text{ but } x \notin A \cap B\}$$

(the *symmetric difference* of A and B),

$$A \times B = \{(x, y) : x \in A \text{ and } y \in B\}$$

(the *Cartesian product* of A and B), and

$$A^B = \{x : x \text{ is a function from } B \text{ into } A\}$$

(the *exponentiation* of B by A).

A set A has no algebraic structure, but the power set $P(A)$, or any nonempty collection of sets for that matter, does have an algebraic structure once we have introduced the first four binary set operations listed immediately above and noted that two sets are equal if and only if each element of one is also an element of the other.

However, each set has one intrinsic property that is arithmetic in nature; we refer to its *cardinal number*. This number simply expresses "how many" elements the set contains. The cardinal number of $\varnothing$ is zero, and that of the set of the reader's toes is (probably) 10. We say that two sets A and B are *equivalent*, or *set-isomorphic*, if they have the same cardinal number. In order to show that two sets are equivalent, one must show that there exists a one-to-one correspondence between them. This is to say, for a set A to be equivalent to a set B, there must exist a *bijectivity* from A to B; i.e., there must be a function, or mapping, having A as domain which maps each $x \in A$ to a unique $y \in B$ with the additional characterizing property that each $y \in B$ is the image of a unique $x \in A$.

A set is said to be *finite* if it is empty or equivalent to the set of positive integers $\{1,2,3,\ldots,n\}$ for some particular integer $n>0$. A set which is equivalent to the set $\mathbb{Z}^+$ of all positive integers is said to be *countable* or *denumerable*. Most mathematicians include the finite sets among the countable ones, and we shall also. Common countably infinite sets are $\mathbb{Z}$, the set of integers; $\mathbb{Q}$, the set of rational numbers; $\mathbb{A}$, the set of algebraic numbers (an algebraic number is one which is a zero of some polynomial in x having rational coefficients); and $\mathbb{P}$, the set of prime integers.

An infinite set which is not countable is called *uncountable* or *nondenumerable*. Examples of such sets are $\mathbb{R}$, the set of real numbers; $\mathbb{C}$, the set of complex numbers; and $\mathbb{F}$, the set of real-valued functions of a real variable. If A is an infinite set, it is often convenient to imagine that we have *indexed* each element of A by its own distinct index α from some suitably large index set $\mathcal{Q}$ and to write $A=\{x_\alpha\}_{\alpha\in\mathcal{Q}}$. We think of α as running through the index set $\mathcal{Q}$. If A is a finite set, with card $A=n$, we could write $A=\{x_i\}_{i=1}^n$; in this case i is the index and $\{1,2,\ldots,n\}$ the index set.

If A is a finite set, then card A is a nonnegative integer. But if A is an infinite set, what symbol should we use to denote the cardinality of A? The symbol ∞ is a common symbol for infinity, but unless all infinite sets are equivalent, we may need many different symbols to represent the different transfinite cardinal numbers. Our first theorem shows that we do indeed need infinitely many such symbols.

Theorem 1.1.2. *For any set* A, card $A<$ card $P(A)$.

PROOF: If $A=\varnothing$, card $A=0<$ card $P(A)=1$. If $A\neq\varnothing$, but A is finite, then card $A=n$ for some positive integer $n>0$. Among n distinct things there are "binomial n,k,"

$$\binom{n}{k}=\frac{n!}{k!(n-k)!},$$

distinct k-combinations, or subsets of cardinality k. Hence the total number of subsets of a set of n elements is the number

$$\sum_{k=0}^{n}\binom{n}{k}=\binom{n}{0}+\binom{n}{1}+\cdots+\binom{n}{k}+\cdots+\binom{n}{n}.$$

But this number is precisely $(1+1)^n = 2^n$ by the binomial formula. Hence, if card $A = n$, card $P(A) = 2^n$. That $n < 2^n$ for all integers $n \geqslant 0$ is easy to prove by induction. It is certainly true if $n = 0$. Suppose for some $k \geqslant 0$ we have $k < 2^k$. Then $k+1 < 2^k + 1 \leqslant 2^k + 2^k = 2^{k+1}$. That does it.

If A is infinite, we need a more elaborate proof, and here it is. This is a proof by contradiction. Now it is quite clear that card $A \not> \text{card } P(A)$ since there is a clear isomorphism between A and the set of singleton subsets of A, so we can be sure that card $A \leqslant \text{card } P(A)$. Suppose equality holds. Then there must exist a bijective mapping (see Section 1.3) of A onto $P(A)$; call it f. For each $x \in A$, $f(x)$ is a subset of A, and if x_1, x_2 are distinct elements of A, then $f(x_1)$, $f(x_2)$ are distinct subsets of A. Moreover, if B is any subset of A, there is a unique $y \in A$ such that $f(y) = B$. Specifically, let B be the subset of A unambiguously defined by $B = \{x \in A : x \notin f(x)\}$. If $y \in A$ is such that $f(y) = B$, where is y? It must be either in B or in $A \setminus B$. If $y \in B$, then by the definition immediately above $y \notin B = f(y)$. If $y \in A \setminus B$, then $y \notin f(y) = B$, so $y \in B$. Clearly, we have an impossible situation here; we can only conclude that there is no element $y \in A$ such that $f(y) = B$. This is to say, no mapping $f: A \to P(A)$ can be bijective. We conclude that card $A <$ card $P(A)$ must hold, and the proof is complete. □

The reader should be aware of the fact that the latter half of the proof is valid for any nonempty set A, finite or infinite.

This theorem implies that if A is an infinite set, we can form an infinite sequence of sets $A, P(A), P(P(A)), \ldots$, giving rise to a strictly increasing sequence of transfinite cardinal numbers. There ought to be a smallest infinite cardinal, and indeed there is; we denote it $\aleph_0$ (aleph null).

Theorem 1.1.3. *Every infinite set contains a countable subset. Hence the cardinal number of* $\mathbb{Z}^+$, $\aleph_0$, *is the smallest infinite cardinal.*

PROOF: Let A be an infinite set. Choose an element $a_1 \in A$, and let $A_1 = A \setminus \{a_1\}$. From A_1 choose an element a_2, and let $A_2 = A_1 \setminus \{a_2\} = A \setminus \{a_1, a_2\}$. Continue in this way inductively; having chosen $\{a_1, a_2, \ldots, a_n\}$, choose a_{n+1} from $A \setminus \{a_1, \ldots, a_n\} = A_n$. Since A is infinite, we cannot exhaust A after a finite number of choices, and from the way we have set things up it is clear that all the chosen elements are distinct. The set $A_0 = \{a_1, a_2, \ldots\}$ is evidently a countably infinite set which consists of elements of A. Hence $A_0 \subset A$, and the assertion of the theorem is proved. □

The next theorem provides a characterization of infinite sets.

Theorem 1.1.4. *A set A is infinite if and only if A is equivalent to a proper subset of itself.*

PROOF: If A is finite, it is evident that A is not equivalent to a proper subset of itself. If A is infinite, it contains a countably infinite subset A_0. Suppose the elements of A_0 have been indexed, so that $A_0=\{a_1, a_2, a_3,\dots\}\subset A$. Let $A_1=\{a_2, a_3, a_4,\dots\}$. A_0 and A_1 are clearly equivalent by the isomorphism that maps each $a_n\in A_0$ to $a_{n+1}\in A_1$. Let $A'=A\setminus\{a_1\}$. A' is a proper subset of A, and A' and A are isomorphic; the isomorphism is the mapping $f\colon A\to A'$ defined by

$$f(x)=x, \qquad \forall x\in A\setminus A_0,$$
$$f(a_n)=a_{n+1}, \qquad \forall a_n\in A_0.$$

This completes the proof. □

The previous three theorems give some insight into the nature of infinite sets and their corresponding cardinal numbers. Evidently, the arithmetic of transfinite numbers is by no means the same as that of real numbers. We leave it to the reader to demonstrate that the cardinalities of $\mathbb{Z}^+$, the set of odd integers, the set of even integers, and $\mathbb{Z}$ are all equal, which implies that $\aleph_0+\aleph_0=\aleph_0$. The last theorem suggests that one method for showing that two infinite sets are equivalent and hence have the same cardinal number is to show that each is isomorphic to a subset of the other. On the other hand, if one set is equivalent to a subset of a second set, but the latter is equivalent to no subset of the former, we can conclude that the cardinal number of the first is less than the cardinal number of the second.

Theorem 1.1.5

$$\operatorname{card}\mathbb{Q}=\aleph_0.$$

PROOF: We first establish a one-to-one correspondence between the set F^+ of positive fractions i/j, $i, j\in\mathbb{Z}^+$, and the set of positive integers. Think of F^+ as the countable union of the disjoint countable sets $A_1, A_2, A_3,\dots,$ where

$$\begin{array}{llllll}
A_1=\frac{1}{1} & \frac{1}{2} & \frac{1}{3} & \frac{1}{4} & \frac{1}{5} & \cdots\\
A_2=\frac{2}{1} & \frac{2}{2} & \frac{2}{3} & \frac{2}{4} & \frac{2}{5} & \cdots\\
A_3=\frac{3}{1} & \frac{3}{2} & \frac{3}{3} & \frac{3}{4} & \frac{3}{5} & \cdots\\
A_4=\frac{4}{1} & \frac{4}{2} & \frac{4}{3} & \frac{4}{4} & \frac{4}{5} & \cdots\\
A_5=\frac{5}{1} & \frac{5}{2} & \frac{5}{3} & \frac{5}{4} & \frac{5}{5} & \cdots\\
\quad\vdots & & & & &
\end{array}$$

This doubly infinite square array on the right-hand side of the equal signs contains precisely the elements of F^+. Count these elements by counting down along the diagonals from upper right to lower left. Note that on each such diagonal, the sum of numerator and denominator for each fraction is the same; that sum is $k+1$ on the kth diagonal from the upper left-hand corner, this corner being considered as the first diagonal.

The fraction i/j therefore is on the $(i+j-1)$th diagonal. Above this diagonal are $1+2+\cdots+(i+j-2)=\frac{1}{2}(i+j-2)(i+j-1)$ fractions, so if we start counting fractions with the upper left-hand corner, counting down the diagonals as we move left to right, the fraction i/j will be the nth fraction in the array, where $n=\frac{1}{2}(i+j-2)(i+j-1)+i$. It is reasonably apparent because of the counting scheme we have set up that to each fraction $i/j \in F^+$ there corresponds a unique positive integer

$$n=\tfrac{1}{2}(i+j-2)(i+j-1)+i,$$

and to each positive integer n there corresponds a unique fraction of F^+. Hence card $F^+ = \aleph_0$.

There is an obvious bijectivity that exists between $\mathbb{Z}^+$ and a proper subset of F^+, namely, the first column of the matrix of the elements of F^+; denote this column A^1. Since $\mathbb{Q}^+$ is a proper subset of F^+, and $A^1 \subset \mathbb{Q}^+ \subset F^+$, we can conclude that card $\mathbb{Q}^+ = \aleph_0$. We leave it to the reader to complete the proof that card $\mathbb{Q} = \aleph_0$. □

The following rather pleasant result is an immediate consequence of what we have just done.

Theorem 1.1.6. *If each set A_i of a countable collection of sets $\{A_i\}_{i=1}^{\infty}$ is at most countable, then the union of all the sets, $\bigcup_{i=1}^{\infty} A_i$, is at most a countable set.*

PROOF: Assume the extreme situation, that the A_i are all mutually disjoint so that $\bigcup_{i=1}^{\infty} A_i$ is of maximal cardinality, and that each A_i is countable. Index the elements of each A_i precisely as we did in the proof of the preceding theorem; e.g.,

$$A_i = \frac{i}{1} \quad \frac{i}{2} \quad \frac{i}{3} \quad \cdots$$

so that $\bigcup_{i=1}^{\infty} A_i$ and the set F^+ are equivalent. □

This theorem implies the arithmetic result that $\aleph_0^2 = \aleph_0$, and indeed it can be shown that for any positive integer n

$$\aleph_0^n = \aleph_0.$$

Having shown that there are just as many integers as rational numbers, we now show that the number of real numbers is a larger transfinite number.

Theorem 1.1.7

$$\aleph_0 < \operatorname{card} \mathbb{R}$$

PROOF: The mapping $x \mapsto \tan(\pi x - \pi/2)$ is a bijectivity from the interval $(0,1)$ onto $(-\infty, \infty)$, so these two intervals of real numbers are isomorphic. Thus it suffices to show that the open interval $(0,1)$ is uncountable. Since $\mathbb{Z}^+$ is a proper subset of $\mathbb{R}$, we need only to show that $\mathbb{R}$ is not equivalent to any subset of $\mathbb{Z}^+$. Let us accept as a fact for the moment that each real number $x \in (0,1)$ has a unique decimal (binary) expansion of the form

$$x = .x_1 x_2 x_3 \ldots,$$

provided that we agree that if any such expansion is such that for some index N all the digits x_n with $n > N$ are nines (ones) but x_N is not a nine (one), we shall replace those digits with zeros and replace x_N with $x_N + 1$.

Suppose $(0,1)$ were equivalent to $\mathbb{Z}^+$. Then we could list all the real numbers in $(0,1)$ in decimal form in a countably infinite column, like so.

$$\begin{aligned} x &= .x_1 x_2 x_3 \ldots, \\ y &= .y_1 y_2 y_3 \ldots, \\ z &= .z_1 z_2 z_3 \ldots, \\ &\vdots \end{aligned}$$

Now form the real number $\alpha = .a_1 a_2 a_3 \ldots$, where the digit $a_1 = x_1 + 1$ if $0 \leqslant x_1 < 5$, or $a_1 = x_1 - 1$ if $5 \leqslant x_1 \leqslant 9$, and $a_2 = y_2 + 1$ if $0 \leqslant y_2 < 5$, or $a_2 = y_2 - 1$ if $5 \leqslant y_2 \leqslant 9$, and so forth. The number α has no zeros or nines in its expansion, so the expansion is an acceptable one, and $0 < \alpha < 1$, but the expansion for α differs from each of the listed expansions in at least one place. Hence the assumption that we can list all the real numbers in $(0,1)$ in a countable list is false, so we conclude that no one-to-one correspondence can exist between $(0,1)$ and $\mathbb{Z}^+$. It follows that $\aleph_0 < \operatorname{card} \mathbb{R}$, and the proof is complete. This proof, incidentally, is generally attributed to Georg Cantor, one of the pioneers of set theory. □

Theorem 1.1.8

$$\operatorname{card} \mathbb{R} = \operatorname{card} P(\mathbb{Z}^+).$$

PROOF: In the proof of Theorem 1.1.7 we stated that every real number in $(0,1)$ has a unique binary expansion, with the particular proviso described. Now consider the set

$$D = \{1_1, 1_2, 1_3, \ldots, 1_n, \ldots\}.$$

This set is evidently isomorphic to $\mathbb{Z}^+$. The binary expansion for an arbitrary $x \in (0,1)$ consists of a sequence of zeros and ones; in a natural way we can make each such binary expansion correspond to a unique nonempty subset of D. For example, $x = .0100100\ldots$ corresponds uniquely to the subset $D_1 = \{1_2, 1_5\}$ and $.1\overline{01}$ corresponds uniquely to $D_2 = \{1_1, 1_3, \ldots, 1_{2k+1}, \ldots\}$. Note that the subset $D_3 = \{1_1, 1_7, 1_8, 1_9, \ldots\}$ corresponds to no acceptable binary expansion, since we have agreed to write $.10000100\ldots$ for the expansion $.100000111\ldots$. Hence we conclude that the set $(0,1)$ is equivalent to a proper subset of $P(D)$.

It is not hard to see that those subsets of D which do not correspond to a real number $x \in (0,1)$ in the way we have described are those whose complements in D are finite (the cofinite sets of D). Suppose $C \subset D$ is a proper cofinite set; let $N > 1$ be the smallest integer such that for all $n \geqslant N$, $1_n \in C$. Make C correspond to the real number, in binary notation, whose integral part is N and whose fractional part is the acceptable binary representation for the expansion determined by C. For example, $D_3 = \{1_1, 1_7, 1_8, \ldots\}$ would correspond to the binary number 111.100001. Let D itself correspond to 1, and let $\varnothing$ correspond to 0. It is apparent that we have here a one-to-one correspondence between $P(D)$ and a subset of real numbers.

Since $(0,1)$ and $\mathbb{R}$ are equivalent and D and $\mathbb{Z}^+$ are equivalent, we conclude that $\operatorname{card} \mathbb{R} = \operatorname{card} P(\mathbb{Z}^+)$, and the theorem is proved. □

We have stated that $\aleph_0 = \operatorname{card} \mathbb{Z}^+$ is the smallest transfinite number and that there must be infinitely many strictly larger transfinite numbers. By $\aleph_1$ we mean the next larger one, followed by $\aleph_2$, and so on. The assumption that $\operatorname{card} \mathbb{R}$ is indeed equal to the second transfinite number $\aleph_1$ has long been called the *continuum hypothesis*. For this course we assume a *generalized continuum hypothesis*; which is to say, we assume that if A is a set and $\operatorname{card} A = \aleph_n$, then $\operatorname{card} P(A) = \aleph_{n+1}$ for $n = 0, 1, 2, \ldots$. We have seen that if $\operatorname{card} A = n$, $\operatorname{card} P(A) = 2^n$, so we shall formally define $2^{\aleph_n} = \aleph_{n+1}$, $n = 0, 1, 2, \ldots$, and state that, for any integer $n > 1$, $n < 2^n < \aleph_0 = \aleph_0^n < 2^{\aleph_0} = \aleph_0^{\aleph_0} = \aleph_1 = \aleph_1^{\aleph_0} < 2^{\aleph_1} = \aleph_1^{\aleph_1} = \aleph_2 < 2^{\aleph_2} = \aleph_3$.

How many sequences of rational numbers are there? There are $\aleph_0$ terms in a sequence, and for each term there is a choice of $\aleph_0$ rational numbers, so it is plausible to conclude that there are $\aleph_0^{\aleph_0} = \aleph_1$ such sequences. Analogously, the number of real sequences is $\aleph_1^{\aleph_0} = \aleph_1$. However, the number of real-valued functions of a real variable is computed in this way: for each real number x there are $\aleph_1$ choices of values for $f(x)$, and there are $\aleph_1$ real numbers, so $\operatorname{card} \mathbb{F} = \aleph_1^{\aleph_1} = \aleph_2$. Hence the cardinality of $P(\mathbb{F}) = \aleph_3$.

Suppose $\mathbb{S}$ is a collection of sets and the cardinality of $\mathbb{S}$ may be arbitrarily large. It is convenient to introduce an index set $\mathcal{Q}$, having the same cardinality as $\mathbb{S}$, and to suppose the elements of $\mathbb{S}$ have been indexed by $\mathcal{Q}$. This is to say, we write $\mathbb{S} = \{A_\alpha\}_{\alpha \in \mathcal{Q}}$. With this indexing notation, we can represent arbitrary unions, intersections, and Cartesian products of sets of $\mathbb{S}$.

$$\bigcup_\alpha A_\alpha = \{x : x \in A_\alpha \text{ for at least one } \alpha \in \mathcal{Q}\},$$

$$\bigcap_\alpha A_\alpha = \{x : x \in A_\alpha \text{ for each } \alpha \in \mathcal{Q}\},$$

$$\times_\alpha A_\alpha = \text{the set of all functions having the index set } \mathcal{Q} \text{ as domain and whose values at } \alpha \text{ are in } A_\alpha.$$

This last Cartesian product is a real puzzler for most students, but consider for a moment. A function is a special kind of correspondence that exists between two sets, one called the domain, the other the range; the correspondence is special in that it goes one way, from the domain to the range. Moreover, to each element of the domain must correspond exactly one element in the range under the correspondence. Now a function is characterized by its domain and its values. A typical element of $\times_\alpha A_\alpha$ is a "generalized sequence" of elements (values); corresponding to each $\alpha \in \mathcal{Q}$ is a unique term of the sequence, $x_\alpha \in A_\alpha$. This describes a function. Changing just one x_α in the generalized sequence gives rise to a different function. Note that if all the sets A_α are one and the same set A, we could write this Cartesian product in the form $A^{\mathcal{Q}}$.

The following theorem is fundamental in set theory; it is easy to prove, so we leave the proof as an exercise.

Theorem 1.1.9 (De Morgan). *Let S be a set and $\{A_\alpha\}$ a collection of sets. Then*

$$\bigcup_\alpha (S \setminus A_\alpha) = S \setminus \bigcap_\alpha A_\alpha,$$

$$\bigcap_\alpha (S \setminus A_\alpha) = S \setminus \bigcup_\alpha A_\alpha.$$

The theorem states that *the union of the complements equals the complement of the intersection* and *the intersection of the complements equals the complement of the union*. If S is some clearly understood "universal set," we sometimes write $\tilde{A}_\alpha$ for $S \setminus A_\alpha$, the complement of A_α. If $S \supset \bigcup A_\alpha$, we can more simply write

$$\bigcup_\alpha \tilde{A}_\alpha = \widetilde{\bigcap_\alpha A_\alpha} \qquad \text{and} \qquad \bigcap_\alpha \tilde{A}_\alpha = \widetilde{\bigcup_\alpha A_\alpha}.$$

To prove that two sets are equal, one usually shows each to be a subset of the other. We shall use this technique extensively in the proof of the following theorem. Before stating the theorem we should recall to the reader what we mean by the image and the preimage of a mapping. If f is a mapping from a set X into a set Y, Y is the range of f, $f(X)$ is the image of f, and X is the domain of f. If $B \subset Y$ is a set, we define the preimage of B, under f, to be the set of all elements $x \in X$ such that $f(x) \in B$, and we denote this preimage by $f^{-1}(B)$. This is to say

$$f^{-1}(B) = \{x \in X: f(x) \in B\}.$$

If $A \subset X$ is a subset, the image of A under f is simply the set

$$f(A) = \{y \in Y: y = f(x) \text{ for at least one } x \in A\}.$$

Finally, we shall agree that $f(\varnothing) = \varnothing$ and $f^{-1}(\varnothing) = \varnothing$.

Theorem 1.1.10. *Let X, Y be nonempty sets and $f: X \to Y$ a mapping (function). Let $\{A_\alpha\}, \{B_\beta\}$ be collections of subsets of X, Y, respectively. Then*

a. $f(\bigcup A_\alpha) = \bigcup f(A_\alpha)$,
b. $f^{-1}(\bigcup B_\beta) = \bigcup f^{-1}(B_\beta)$,
c. $f(\bigcap A_\alpha) \subset \bigcap f(A_\alpha)$,
d. $f^{-1}(\bigcap B_\beta) = \bigcap f^{-1}(B_\beta)$,
e. $A_\alpha \subset f^{-1} f(A_\alpha)$,
f. $f f^{-1}(B_\beta) \subset B_\beta$,
g. $f(X \setminus A_\alpha) \subset Y$,
h. $f^{-1}(Y \setminus B_\beta) = X \setminus f^{-1}(B_\beta)$.

PROOF: (a) $y \in f(\bigcup A_\alpha) \Rightarrow \exists x \in \bigcup A_\alpha$ s.t. (read "such that") $f(x) = y \Rightarrow \exists \alpha_0$ s.t. $x \in A_{\alpha_0}$, $f(x) \in f(A_{\alpha_0}) \Rightarrow y = f(x) \in \bigcup f(A_\alpha)$. Since y was arbitrary, we have $f(\bigcup A_\alpha) \subset \bigcup f(A_\alpha)$. Conversely, $y \in \bigcup f(A_\alpha) \Rightarrow \exists \alpha_0$ s.t. $y \in f(A_{\alpha_0}) \Rightarrow \exists x \in A_{\alpha_0}$ s.t. $y = f(x) \Rightarrow x \in \bigcup A_\alpha \Rightarrow y = f(x) \in f(\bigcup A_\alpha)$. Since $y \in \bigcup f(A_\alpha)$ was arbitrary, we have $\bigcup f(A_\alpha) \subset f(\bigcup A_\alpha)$. The two inclusions imply equality.

(b) $x \in f^{-1}(\bigcup B_\beta) \Rightarrow y = f(x) \in \bigcup B_\beta \Rightarrow y \in B_{\beta_0}$ for some $\beta_0 \Rightarrow x \in f^{-1}(B_{\beta_0}) \Rightarrow x \in \bigcup f^{-1}(B_\beta)$. Since x was arbitrary, we conclude $f^{-1}(\bigcup B_\beta) \subset \bigcup f^{-1}(B_\beta)$. Conversely, $x \in \bigcup f^{-1}(B_\beta) \Rightarrow x \in f^{-1}(B_{\beta_0})$ for some $\beta_0 \Rightarrow y = f(x) \in B_{\beta_0} \Rightarrow y \in \bigcup B_\beta \Rightarrow x \in f^{-1}(\bigcup B_\beta)$. Since $x \in \bigcup f^{-1}(B_\beta)$ was arbitrary, we have $\bigcup f^{-1}(B_\beta) \subset f^{-1}(\bigcup B_\beta)$, and the two inclusions imply equality.

(c) $y \in f(\bigcap A_\alpha) \Rightarrow \exists x \in \bigcap A_\alpha$ s.t. $y = f(x)$. This x belongs to A_α for each α; hence $y \in f(A_\alpha)$ for each α, so $y \in \bigcap f(A_\alpha)$. We conclude that $f(\bigcap A_\alpha) \subset \bigcap f(A_\alpha)$. In the event that $\bigcap A_\alpha = \varnothing$, the inclusion holds trivially. To show that we cannot get the reverse inclusion in general, we give a counterexample. Let A_1 be the interval $[-1, 0]$ and A_2 be the interval $[0, 1]$, and suppose $f: \mathbb{R} \to \mathbb{R}$ is defined by $f(x) = x^2$. In this case $f(A_1 \cap A_2) = f(\{0\}) = \{0\} \neq \bigcap_{i=1}^2 (f(A_i)) = [0, 1]$.

(d) $x \in f^{-1}(\bigcap B_\beta) \Leftrightarrow y = f(x) \in \bigcap B_\beta \Leftrightarrow y = f(x) \in B_\beta$ for each $\beta \Leftrightarrow x \in f^{-1}(B_\beta)$ for each $\beta \Leftrightarrow x \in \bigcap f^{-1}(B_\beta)$. Since x was arbitrary, we conclude that $f^{-1}(\bigcap B_\beta) = \bigcap f^{-1}(B_\beta)$.

(e) $x\in A_\alpha \Rightarrow f(x)\in f(A_\alpha)\Rightarrow x\in f^{-1}f(A_\alpha)$. Hence $A_\alpha \subset f^{-1}f(A_\alpha)$. The counterexample in (c) proves that in general we cannot have equality.

(f) $y\in ff^{-1}(B_\beta)\Rightarrow \exists x\in f^{-1}(B_\beta)$, s.t. $y=f(x)$ and $y=f(x)\in B_\beta$. Hence $ff^{-1}(B_\beta)\subset B_\beta$. That the reverse inclusion does not in general hold is easy to see. Let $f:\mathbb{R}\to\mathbb{R}$ be given by $f(x)=x^2$, and let $B_\alpha=[-1,1]$. Then $f^{-1}(B_\alpha)=[-1,1]$, and $ff^{-1}(B_\alpha)=[0,1]$, a proper subset of B_α.

(g) Obvious. f maps all of X into Y.

(h) $x\in f^{-1}(Y\setminus B_\alpha)\Leftrightarrow f(x)\in Y\setminus B_\beta \Leftrightarrow f(x)\notin B_\beta \Leftrightarrow x\notin f^{-1}(B_\beta)\Leftrightarrow x\in X\setminus f^{-1}(B_\beta)$. This establishes the equality. The proof of the theorem is complete. □

A very careful reader, or a rather experienced one, may notice that we did not specify in the statement of the above theorem that the collections $\{A_\alpha\},\{B_\beta\}$ of subsets of X, Y, respectively, were nonempty collections. This was done intentionally, because we wanted to introduce a strange but useful convention. Suppose $\{S_\alpha\}$ is a collection of sets. Some of the sets of this collection may be empty, but this causes no concern. However, what if the collection itself is empty? How do we define $\bigcap S_\alpha$ and $\bigcup S_\alpha$ in case the collection $\{S_\alpha\}$ is empty? We agree to do the following: if $\{S_\alpha\}$ is empty, we define $\bigcup S_\alpha=\varnothing$ and $\bigcap S_\alpha=X$, where X is understood to be a universal set which contains all the sets we might possibly be considering.

To help the reader swallow this, we suggest the following bit of logic. Suppose X is an understood universal set and $\{S_\alpha\}$ is a collection of subsets of X. If we increase the number of sets in the collection, the union of all the sets in the collection does not get smaller and the intersection of all the sets does not get larger. On the other hand, decreasing the number of sets in the collection has just the opposite effects. Hence it is quite reasonable that the limiting sets for the union and intersection of the sets of a "decreasing to $\varnothing$" collection of sets should be $\varnothing$ and X, respectively.

Let A be a nonempty set and $\{A_k\}$ a finite collection of nonempty subsets of A such that $\bigcup_{k=1}^n A_k=A$ and $A_i\cap A_j=\varnothing$ whenever $i\neq j$. In this case we say that the collection $\{A_k\}$ is a *partition* of A. If we remove the restriction that the collection be finite, we shall call $\{A_k\}$ a *generalized partition* of A. Now suppose $\{A_k\}$ is a partition, perhaps a generalized partition, of a set A. We can define a binary relation R over A by means of this partition in the following way. For any two not necessarily distinct elements of A, say x, y, we shall say x is related to y and write xRy iff x is in the same A_k as y. Note that this binary relation over A satisfies the following three axioms.

E1. For each $x\in A$, xRx (reflexivity).
E2. For any $x, y\in A$, $xRy\Rightarrow yRx$ (symmetry).
E3. For any $x, y, z\in A$, if xRy and yRz, then xRz (transitivity).

Conversely, if we have a set A and a binary relation R defined over A which satisfies the above three axioms, it is not hard to see that if we assemble into subsets all the elements of A which are related to one another, we have effected a partition, perhaps a generalized partition, of A. We are led to make the following definition.

Definition 1.1.11. If a binary relation R over a set A satisfies the three axioms E1, E2, and E3, we shall call R an *equivalence relation* over A. The subsets A_k that are comprised by the partition $\{A_k\}$ arising from the equivalence relation R over A will be called *equivalence classes* of A, relative to R.

We shall meet a number of binary relations defined over sets; some will be equivalence relations, others will not. Isomorphism is a binary relation defined over the category of sets, obviously an equivalence relation. On the other hand, set inclusion, a binary relation defined over the category of sets, is not an equivalence relation. Equality is a very strong equivalence relation; when we say two things are equal, we mean that in some respect they are indistinguishable from each other or identical.

1.2 ALGEBRAIC STRUCTURES

We now move on to some new ideas. Having fixed the basic notion of set, we can now add various algebraic structures to sets. Let S be a nonempty set. A binary operation $\circ$ defined over S is simply a rule which assigns to each pair of not necessarily distinct elements of S a unique object. This is to say, the operation combines two elements, in the particular order given, into an object which may or may not be an element of S. It may be that x combined with y is a different object than y combined with x; if the order of combination never makes any difference, the operation is said to be commutative. Perhaps a more sophisticated way of describing a binary operation $\circ$ on S would be to call it a mapping of $S \times S$ into some image set or range.

Suppose we have a nonempty set S, together with a binary operation $\oplus$ on S and a well-defined notion of equality; we regard all this as an algebraic structure and denote it by $(S, \oplus, =)$. Consider the following five axioms.

A1. $\forall x, y \in S, x \oplus y \in S$	(closure).
A2. $\forall x, y, z \in S, x \oplus (y \oplus z) = (x \oplus y) \oplus z$	(associativity).
A3. $\exists o \in S$ s.t. $\forall x \in S, o \oplus x = x = x \oplus o$	(identity element).
A4. $\forall x \in S, \exists {}^-x$ s.t. $x \oplus {}^-x = o = {}^-x \oplus x$	(inverse elements).
A5. $\forall x, y \in S, x \oplus y = y \oplus x$	(commutativity).

Definition 1.2.1. If $(S, \oplus, =)$ satisfies

axioms	$(S, \oplus, =)$ is a
A1,2,	semigroup.

A1,2,3,	monoid.
A1,2,3,4,	group.
A1,2,3,4,5,	abelian group.

If A is a nonempty set, then $(P(A), \cup, =)$, where $=$ is ordinary set equality and $\cup$ is set union, is a monoid, or semigroup with identity, the empty set being the identity element. Similarly, $(P(A), \cap, =)$ is also a monoid, with A the identity element. The set V of all vectors in Euclidean 3-space, with ordinary vector addition and equality, becomes an abelian, or commutative, group.

Now suppose we have an abelian group $(S, \oplus, o, =)$, where we have put in view the identity element o, relative to the operation $\oplus$. Suppose there is a secondary binary operation $\circ$ defined on S (for which the notion of equality remains the same). Consider the following set of axioms:

M1. $\forall x, y \in S, x \circ y \in S$	(closure).
M2. $\forall x, y, z \in S, x \circ (y \circ z) = (x \circ y) \circ z$	(associativity).
M3. $\exists e \in S$ s.t. $\forall x \in S, e \circ x = x = x \circ e$	(identity).
M4. $\forall x \in S, x \neq o, \exists x^-$ s.t. $x^- \circ x = e = x \circ x^-$	(inverse).
M5. $\forall x, y \in S, x \circ y = y \circ x$	(commutativity).
D1. $\forall x, y, z \in S, x \circ (y \oplus z) = (x \circ y) \oplus (x \circ z)$, $(y \oplus z) \circ x = (y \circ x) \oplus (z \circ x)$	(two-sided distributivity of the secondary operation with respect to the primary operation).

Definition 1.2.2. If $(S, \oplus, \circ, =)$ satisfies

axioms	$(S, \oplus, \circ, =)$ is a
A1,2,3,4,5, M1,2,3,4,5, D1;	field.
A1,2,3,4,5, M1,2,3,4, D1;	division ring.
A1,2,3,4,5, M1,2,3,5, D1;	commutative ring with unity.
A1,2,3,4,5, M1,2,3, D1;	ring with unity.

A1,2,3,4,5, nonassociative ring.
M1,
D1;

Needless to say, there are more possibilities for structures on a set utilizing one or both of the binary operations $\oplus$ and $\circ$. For example, let $\mathbb{S}$ be a set of sets, with $\cup$ as the primary binary operation on $\mathbb{S}$ and $\setminus$ as the secondary operation, where equality is ordinary set equality. Suppose further that $\mathbb{S}$ is closed with respect to the operations $\cup$ and $\setminus$. Then $(\mathbb{S}, \cup, =)$ is an abelian group, with $\varnothing$ as the primary identity. The structure $(\mathbb{S}, \cup, \setminus, =)$ evidently satisfies all the A-axioms, but only M1 of the M-axioms, and not axiom D1. Hence this structure is not a ring.

However, notice what the fact that axioms A1 and M1 hold leads to. For any sets $A, B \in \mathbb{S}$, $A \triangle B \in \mathbb{S}$ since

$$A \triangle B = (A \setminus B) \cup (B \setminus A) \in \mathbb{S}.$$

Moreover, $A \cap B \in \mathbb{S}$ since $A \cap B = (A \cup B) \setminus (A \triangle B)$. We leave it to the reader to verify that $(\mathbb{S}, \triangle, \cap, =)$ is indeed a ring, and we give the following definitions.

Definition 1.2.3. A nonempty collection of sets $\mathbb{S}$ which is closed with respect to finite unions and differences is called a *ring* of sets. Moreover, if $\mathbb{S}$ contains a universal set S, i.e., a set which contains as a subset every set of the collection $\mathbb{S}$, so that $\mathbb{S}$ is also closed with respect to complementation, $\mathbb{S}$ is called an *algebra* of sets. It is understood that $\triangle$ is the primary operation, and $\cap$ the secondary operation in the ring $\mathbb{S}$.

If $\mathbb{S}$ is a ring (algebra) of sets such that for every countable collection of sets $\{A_n\}_{n=1}^{\infty} \subset \mathbb{S}$, we have $\bigcup_{n=1}^{\infty} A_n \in \mathbb{S}$, we call $\mathbb{S}$ a σ-ring (σ-algebra); σ is pronounced sigma.

Definition 1.2.4. Suppose we have a commutative group $(V, \oplus, =)$ and a field $(F, +, \cdot, =)$. Let $\circ$ be an operation which maps $F \times V$ into V, as well as $V \times F$ into V, such that for each $\alpha \in F$, $\mathbf{v} \in V$, we have $\alpha \circ \mathbf{v} = \mathbf{v} \circ \alpha \in V$. Then the structure $(V, \oplus, F, +, \cdot, \circ, =)$ is called a *linear space*, or a *vector space*, provided the following axioms hold:

V1. $\forall \mathbf{x}, \mathbf{y} \in V$, $\alpha \in F$, $\alpha \circ (\mathbf{x} \oplus \mathbf{y}) = (\alpha \circ \mathbf{x}) \oplus (\alpha \circ \mathbf{y})$.
V2. $\forall \alpha, \beta \in F$, $\mathbf{x} \in V$, $(\alpha + \beta) \circ \mathbf{x} = (\alpha \circ \mathbf{x}) \oplus (\beta \circ \mathbf{x})$.
V3. $\forall \mathbf{x} \in V$, $0 \circ \mathbf{x} = \mathbf{0}$, where the 0 on the left is the primary identity in F and the $\mathbf{0}$ on the right is the identity in V.
V4. $\forall \mathbf{x} \in V$, $e \circ \mathbf{x} = \mathbf{x}$, where e is the secondary identity in F.
V5. $\forall \alpha, \beta \in F$, $\mathbf{x} \in V$, $\alpha \circ (\beta \circ \mathbf{x}) = (\alpha \cdot \beta) \circ \mathbf{x}$.

We usually refer to V as the vector space and to F as the underlying scalar field. In our applications, the scalar field will almost always be the field $\mathbb{R}$ of real numbers or the field $\mathbb{C}$ of complex numbers.

If we replace the field F in the above structure by a ring A with unity and keep the same set of axioms, we call this structure a *linear set* (over A), or an *A-module*. If $V(A)$ is an arbitrary linear set over a ring A, and B, C are subsets of V, we define sets

$$B \pm C = \{\mathbf{x} \in V : \mathbf{x} \text{ is of the form } \mathbf{b} \pm \mathbf{c},\ \mathbf{b} \in B,\ \mathbf{c} \in C\},$$

and if $\lambda \in A$,

$$\lambda B = \{\mathbf{x} \in V : \mathbf{x} \text{ is of the form } \lambda\mathbf{b},\ \mathbf{b} \in B\}.$$

Note the distinction between the sets $B - C$ and $B \backslash C$.

If $\{\mathbf{x}_i\}_{i=1}^n \subset V$ and $\{\lambda_i\}_{i=1}^n \subset A$, we call the element $\mathbf{x} = \Sigma_{i=1}^n \lambda_i \mathbf{x}_i$ in the linear set $V(A)$ a *linear combination* of the $\mathbf{x}_i$; a linear combination consists of a finite number of terms. The set of all possible linear combinations of the $\mathbf{x}_i$ is called the *linear span* of the $\mathbf{x}_i$ in $V(A)$, and we denote this $\mathrm{sp}\{\mathbf{x}_i\}$. It is not hard to show that $\mathrm{sp}\{\mathbf{x}_i\}$ is a linear subset of $V(A)$, and is in fact the smallest linear subset which contains all the $\mathbf{x}_i$.

In case we have an infinite set of vectors $\{\mathbf{x}_\alpha\} \subset V(A)$, we define $\mathrm{sp}\{\mathbf{x}_\alpha\}$ to be the set of all possible linear combinations that can be formed from the finite subsets of $\{\mathbf{x}_\alpha\}$.

Consider for a moment the two special classes of linear sets $V(\mathbb{R})$ and $V(\mathbb{C})$, the real linear spaces and the complex linear spaces. A subset S of one of these spaces is called *convex* if for each $\mathbf{x}, \mathbf{y} \in S$ the point

$$\mathbf{z} = (1-\lambda)\mathbf{x} + \lambda\mathbf{y},$$

where λ is any real number satisfying $0 \leqslant \lambda \leqslant 1$, is also in S. If $\{\mathbf{x}_\alpha\}$ is a collection of points of one of these spaces, we call the smallest convex set which contains all the $\mathbf{x}_\alpha$ the *convex hull* of the $\mathbf{x}_\alpha$. We remark that the convex hull of $\{\mathbf{x}_\alpha\}$ is the intersection of all convex sets which contain all the $\mathbf{x}_\alpha$; if $\{\mathbf{x}_i\}_{i=1}^n$ is a finite set of points, then the convex hull of this set is the set of all linear combinations of the $\mathbf{x}_i$ one can get of the form $\Sigma_{i=1}^n \lambda_i \mathbf{x}_i$, where each $\lambda_i \geqslant 0$, and $\Sigma_{i=1}^n \lambda_i = 1$. Hence the convex hull of $\{\mathbf{x}_\alpha\}$ is the set of all linear combinations one can form from the finite subsets of $\{\mathbf{x}_\alpha\}$ with the proviso that any such linear combination, for each $n = 1, 2, 3, \ldots$, is of the form immediately above. In case there is any doubt, a singleton is convex, and we shall always consider the empty set to be convex.

The last algebraic structure we are going to introduce at this time is one known as a linear algebra, or more commonly, an algebra.

Definition 1.2.5. A structure $V(A)$ is called an *algebra* over A if, first, $V(A)$ is a linear set over A; second, V is itself a ring with respect to a secondary operation $\otimes$; and, third, the "multiplication by a scalar" $\circ$ commutes with the "ring multiplication" $\otimes$. This last is to say that $\forall \mathbf{x}, \mathbf{y} \in V$, $\lambda \in A$,

$$(\lambda \circ \mathbf{x}) \otimes \mathbf{y} = \mathbf{x} \otimes (\lambda \circ \mathbf{y}) = \lambda \circ (\mathbf{x} \otimes \mathbf{y}).$$

If V is a commutative ring, an algebra $V(A)$ is called a commutative algebra. It is usually the case that the ring V has a unity, so we shall always assume that an algebra has a unity, unless the contrary is specifically stated.

The standard model for a (nonabelian) linear algebra is the algebra of matrices of size $n \times n$ with entries from $\mathbb{R}$; the unity is the $n \times n$ identity matrix.

We bring this brief introduction to structures to a close by giving a concrete example of a linear set over a ring with unity. Consider the set of mathematical vectors in E^n, n-dimensional Euclidean space; these are simply those vectors in E^n which initiate at the origin. Let the symbol dx_i stand for the mapping which assigns to each such vector in E^n the "signed length" of its projection on the ith coordinate axis. This is to say, if $\mathbf{v}$ is an arbitrary mathematical vector in E^n, $dx_i(\mathbf{v})$ is the ith coordinate of the projection of $\mathbf{v}$ on the ith coordinate axis. It is easy to see that dx_i is a linear mapping in that for any vectors $\mathbf{v}, \mathbf{w}$ and any real numbers α, β, $dx_i(\alpha\mathbf{v}+\beta\mathbf{w}) = \alpha\, dx_i(\mathbf{v}) + \beta\, dx_i(\mathbf{w})$.

Now consider the set of n symbols $\{dx_1, \ldots, dx_n\}$, and let V be the additive abelian group generated by these n symbols. Let $\mathcal{C}$ be the ring of continuous real-valued functions defined on E^n, whose unity is the function identically equal to 1 on E^n. Then $V(\mathcal{C})$ is a linear set, and a typical element $\omega \in V(\mathcal{C})$ is of the form $\omega = \sum_{i=1}^{n} f_i\, dx_i$, where each $f_i \in \mathcal{C}$. In later chapters we refer to this linear set $V(\mathcal{C})$ as the set of "1-forms on E^n."

1.3 MORPHISMS

If A and B are sets, we call f a *mapping* from A into B if f maps each element of A to exactly one element of B; A is called the *domain* of f, and B the *range* of f. Function is a synonym for mapping. If f is a mapping, then $x = y \Rightarrow f(x) = f(y)$, for any $x, y \in A$. For the domain of f we sometimes write dom(f), and im(f) represents the image of A under f; $\text{im}(f) \subset B$. A mapping $f: A \to B$ is *injective* if for any $x, y \in A$, $f(x) = f(y) \Rightarrow x = y$. A mapping $f: A \to B$ is called *surjective* if $\text{im}(f) = B$; a *bijective* mapping is one that is both injective and surjective. Injective mappings are often called *one-to-one*, surjective ones *onto*, and bijective ones *one-to-one onto*.

In Section 1.1 we said that two sets A and B are equivalent, or isomorphic, if there exists a bijective mapping $f: A \to B$, and we spoke of f as the isomorphism. The closest thing that sets have to an algebraic structure is a cardinal number, so if f is an isomorphism from A to B, we have card $A =$ card $f(A)$. This is to say, f preserves the algebraic structure of its domain. This leads us to define what we mean by an isomorphism between structures.

Definition 1.3.1. Suppose $(A, +, =), (B, \oplus, \equiv)$ are the two structures with their respective operations $+, \oplus$ and their respective notions of equality.

Let f be a bijectivity from A to B. Then f is an *isomorphism* if for each $x, y \in A$,

$$f(x+y) \equiv f(x) \oplus f(y).$$

If several operations are involved, for example, if we have $(A, +, \times, \cdot, =)$ and $(B, \oplus, \otimes, \circ, \equiv)$, an isomorphism from A to B is a bijective mapping such that for each $x, y \in A$

$$f(x+y) \equiv f(x) \oplus f(y),$$
$$f(x \times y) \equiv f(x) \otimes f(y),$$
$$f(x \cdot y) \equiv f(x) \circ f(y).$$

More simply, an isomorphism is a bijectivity from one structure to another which preserves the characterizing properties of the structures. For example, a linear isomorphism between linear sets with a common scalar ring is a bijective map f such that for all $\mathbf{x}, \mathbf{y}$ in dom(f) and all λ, μ in the scalar ring, $f(\lambda \mathbf{x} + \mu \mathbf{y}) = \lambda f(\mathbf{x}) + \mu f(\mathbf{y})$.

It may be that two rings A and B are isomorphic as groups but not as rings. For example, let A be the ring of even integers with ordinary addition and multiplication, and let B be the additive group of even integers with a multiplication defined by $x \cdot y = 0$ for all $x, y \in B$. Consider the two fields $\mathbb{R}$ and $\mathbb{C}$; as sets they are isomorphic, but they are not even ring-isomorphic.

Note that isomorphism over a class of structures is an equivalence relation.

Theorem 1.3.2. *If A and B are both n-dimensional linear sets with a common scalar ring R, they are linearly isomorphic.*

PROOF: We say that a linear set V has dimension n if and only if there is a set of n vectors $\{\mathbf{e}_1, \ldots, \mathbf{e}_n\} \subset V$ such that each vector $\mathbf{v} \in V$ has a unique representation of the form $\mathbf{v} = \sum_{i=1}^{n} v_i \mathbf{e}_i$, where the v_i are elements of the scalar ring. The set $\{\mathbf{e}_i\}_{i=1}^{n}$ is called a *basis* for V, and the cardinality of the basis is the *dimension* of V. If R is a ring (usually a field), we denote by R^n the Cartesian n-space associated with R; i.e., the set of ordered n-tuples $(r_1, \ldots, r_n)$, where each $r_i \in R$. R^n is itself a linear set over R if we define for any $\lambda \in R$ and any $(a_1, \ldots, a_n), (b_1, \ldots, b_n) \in R^n$, $(a_1, \ldots, a_n) + (b_1, \ldots, b_n) = (a_1 + b_1, \ldots, a_n + b_n)$, $\lambda(a_1, \ldots, a_n) = (\lambda a_1, \ldots, \lambda a_n)$. Since isomorphism is an equivalence relation over the category of linear sets of dimension n with a common scalar ring, it is enough to show that an arbitrary n-dimensional linear set over a ring R, $V(R)$, is isomorphic to R^n. Let $\{\mathbf{e}_1, \ldots, \mathbf{e}_n\} \subset V(R)$ be a basis; then each $\mathbf{v} \in V$ has the unique representation in terms of this basis: $\mathbf{v} = \sum_{i=1}^{n} v_i \mathbf{e}_i$, where $v_i \in R$, $i = 1, \ldots, n$. (These v_i are called the *coordinates* or *components* of the vector $\mathbf{v}$.) The mapping $\varphi: V(R) \to R^n$ defined by $\varphi(\mathbf{v}) = (v_1, \ldots, v_n)$, $\forall \mathbf{v} = \sum_{i=1}^{n} v_i \mathbf{e}_i \in V(R)$, is evidently bijective. For any

$\mathbf{v},\mathbf{w}\in V(R)$, $\mathbf{v}\neq\mathbf{w}$ implies that $\varphi(\mathbf{v})\neq\varphi(\mathbf{w})$, and for each $(x_1,\ldots,x_n)\in R^n$, there is a vector $\mathbf{x}\in V(R)$ such that $\varphi(\mathbf{x})=(x_1,\ldots,x_n)$.

If $\mathbf{v},\mathbf{w}\in V(R)$ and $\lambda,\mu\in R$, $\lambda\mathbf{v}+\mu\mathbf{w}$ is a linear combination of $\mathbf{v}$ and $\mathbf{w}$, and hence is a vector in $V(R)$. But $\lambda\mathbf{v}+\mu\mathbf{w}=\lambda(v_1\mathbf{e}_1+\cdots+v_n\mathbf{e}_n)+\mu(w_1\mathbf{e}_1+\cdots+w_n\mathbf{e}_n)$, and, as a consequence of the way we define addition and multiplication by a scalar in a linear set, this is equal to $(\lambda v_1+\mu w_1)\mathbf{e}_1+\cdots+(\lambda v_n+\mu w_n)\mathbf{e}_n$. Because of the linear structure with which R^n is endowed, it is apparent that $\varphi(\lambda\mathbf{v}+\mu\mathbf{w})=\lambda\varphi(\mathbf{v})+\mu\varphi(\mathbf{w})$. Hence φ is a linear isomorphism. We can conclude that the linear sets A and B in the statement of the theorem are linearly isomorphic. □

This theorem, although quite elementary, is important; it establishes the fact that all finite-dimensional linear sets having the same dimension and scalar ring are algebraically indistinguishable from one another.

We bring this section to a close by giving a comprehensive definition of various morphisms the reader may someday encounter.

Definition 1.3.3. Let V and W be sets having algebraic structure (i.e., certain laws of composition, together with a set of axioms satisfied by these operations) of the same type, and let $\varphi: V\to W$ be a mapping.

a. If φ preserves the given operations, i.e., for each pair of corresponding operations $\cdot,\circ$ in V, W, respectively, and all $x, y\in V$, $\varphi(x\cdot y)=\varphi(x)\circ\varphi(y)$, φ is called a *homomorphism*.
b. If φ is an injective (one-to-one) homomorphism, φ is called a *monomorphism*.
c. If φ is a surjective (onto) monomorphism ($\varphi(V)=W$), φ is called an *isomorphism*.
d. If φ is a isomorphism and $W=V$, φ is called an *automorphism*.
e. If φ is a surjective (onto) homomorphism ($\varphi(V)=W$), φ is called an *epimorphism*.
f. If φ is a homomorphism and $W=V$, φ is called an *endomorphism*.

We remark that if the structure of V and W is that of a linear set, one invariably uses the term linear mapping for homomorphism. The term linear transformation is used interchangeably with the term linear mapping, although it used to be reserved for automorphisms on linear sets.

1.4 ORDER STRUCTURES

In Section 1.1 we introduced the notion of an equivalence relation on a set S, a particular binary relation on S that satisfied three characterizing axioms. In this section we are going to consider another important kind of binary relation on a set P characterized by a slightly different trio of axioms.

Definition 1.4.1. Let P be a nonempty set of elements, and suppose there is a binary relation on P, denoted $\prec$, for which the following "partial-order" axioms hold for all $a, b, c \in P$:

P1. If $a \prec b$ and $b \prec c$, then $a \prec c$ (transitivity).
P2. $a \prec a$ (reflexivity).
P3. If $a \prec b$ and $b \prec a$, then $a = b$.

Then $(P, \prec)$ is called a *partial order*. (For the symbol $\prec$ read "precedes.")

The standard model for a partial order is the power set of a set A, where set inclusion $\subset$ is the specific order relation which partially orders $P(A)$. Note that in a partial order $(P, \prec)$ there may be many pairs $a, b \in P$ where we have neither $a \prec b$ nor $b \prec a$, in which case we say a and b are noncomparable. On the other hand, it may be that for every $a, b \in P$ we have either $a \prec b$ or $b \prec a$, which is to say each element of P is comparable with every other element of P. In this case we call P a *chain* or *total order*. The set $\mathbb{R}$ with its natural ordering is a chain. A sequence $\{a_n\}$ can be totally ordered if we define an ordering by $a_i \prec a_j \Leftrightarrow i \leqslant j$. Sometimes we order the complex number field by an order $\prec$ defined in this way: $\forall z_1 = x_1 + iy_1, z_2 = x_2 + iy_2 \in \mathbb{C}$,

$$z_1 \prec z_2 \quad \Leftrightarrow \quad x_1 < x_2 \quad \text{or} \quad x_1 = x_2,\ y_1 \leqslant y_2.$$

$\mathbb{C}$ with this ordering is a chain.

We are going to present at this time a multitude of definitions so that the reader will have a basic vocabulary of the terms commonly used to describe partial orders. If S is a subset of a partial order P, an element $m \in P$ is called an *upper bound* of S if $x \prec m$ for every $x \in S$. A lower bound of S is defined analogously. If $\mu \in P$ is an upper bound for S and if $\mu \prec m$ for every upper bound m of S, we call μ the *supremum* (least upper bound) of S and write $\mu = \sup S$ ($\mu = \operatorname{lub} S$). The *infimum* (greatest lower bound) of S is defined analogously and is written $\inf S$ ($\operatorname{glb} S$). Note that it is possible for a set S to have no upper bound, or more than one upper bound, and these several upper bounds may not be comparable to one another. Moreover, S may have many upper bounds but no supremum.

If $m \in S$ is such that for every $x \in S$ which is comparable with m we have $x \prec m$, then m is said to be a *maximal element* of S. A minimal element of S is defined analogously. Note that if a set has more than one maximal element, no two of them are comparable. The same is true for distinct minimal elements. Hence a chain can have at most one maximal element and one minimal element.

If a partial order P is such that for each finite subset $S \subset P$, $\sup S \in P$ and $\inf S \in P$, we call P a *lattice*. If the same can be said for each bounded countable set S, P is called a conditionally σ-complete lattice. If $\sup S$, $\inf S$ exist in P for every bounded infinite set S, P is a conditionally complete lattice. Finally, if $\sup S$, $\inf S$ exist in P for each (countably) infinite set S, P is said to be a (σ-complete) complete lattice.

The following notation is convenient. If $\{x_\alpha\}$ is a subset of a partial order or a lattice, we frequently denote the supremum and infimum of that subset by

$$\sup\{x_\alpha\} = \bigvee_\alpha x_\alpha, \qquad \inf\{x_\alpha\} = \bigwedge_\alpha x_\alpha.$$

There is another specialization of partial order that we want to mention, the directed set. If P is a partial order with the property that for any $\alpha, \beta \in P$, $\exists \gamma \in P$ s.t. $\alpha \prec \gamma$ and $\beta \prec \gamma$, we say that P is a *directed set*, or more precisely, an upward directed set. A downward directed set is defined analogously. A lattice is a directed set (both directions), but the converse may not be true. Directed sets appear extensively as index sets.

Suppose a partial order P has this very restrictive property: for each subset $S \subset P$, $\inf S$ exists and is contained in S. We say that P is *well ordered*. The classic model of a well-ordered set is the set $\mathbb{Z}^+$. Evidently, if a set is well ordered, it is already a chain, but the converse is not true; consider the chain $\mathbb{R}$, the subset $(0,1)$, the open unit interval, has no least element in it.

If a set S has an order structure, it is reasonable to expect that we may have occasions to compare it to other order structures. If $(A, \prec)$ and $(B, <)$ are order structures, we say A and B are *order-isomorphic* if there is a bijective map $\varphi: A \to B$ such that $\forall x, y \in A$, $x \prec y \Leftrightarrow \varphi(x) < \varphi(y)$. We remark that if φ is a mapping from $(A, \prec)$ to $(B, <)$, we say that φ is *isotonic* if $\varphi(x) < \varphi(y)$ whenever $x \prec y$. Hence an order isomorphism is a bi-isotonic bijectivity.

It is clear that a set S may have an algebraic structure and an order structure as well. There is no reason to believe that the two structures should in any way be related, but it turns out to be convenient if they are. We are going to present a standard way of putting a partial order on a group. Let $(G, \oplus)$ be such a structure, with o its identity element. Let $P \subset G$ be a nonempty subset (such a subset we are about to describe may not exist in G) such that $o \notin P$ and if $x, y \in P$, then $x \oplus y \in P$. If such a subset exists, it is clear that $x \in P \Rightarrow {}^-x \notin P$ since $o \notin P$. Hence, if each element of G is its own inverse, no such set P exists. Now define $\forall x, y \in G$, $x \neq y$, $x \prec y \Leftrightarrow y \oplus {}^-x \in P$. Note that if $x \in P$, then $o \prec x$, and conversely. Also, if $x \in P$, then ${}^-x \prec x$. If we can find such a set $P \subset G$, we call P the *positive cone* in G. We would naturally call ${}^-P = \{x \in G : {}^-x \in P\}$ the negative cone in G. Evidently, if $G = {}^-P \cup \{o\} \cup P$, then G has been chain ordered.

Suppose G is a group and we have found a positive cone P (of course, there may be many suitable sets P, each giving rise to its own partial ordering of G). If P has the property that for each pair of elements $x, y \in P$, there is a positive integer n such that $x \prec y \oplus \cdots \oplus y$, the "sum" containing n summands, we say that the ordering of G induced by P is *Archimedean*.

Now it may well be that a group G is also a partial order. If such is the case and it turns out that the set $P=\{x\in G: x\neq o, o\prec x\}$ is such that $\forall x\in P$, $^{-}x\notin P$, $\forall x, y\in P$, $^{-}x\prec y$, and $\forall x, y\in G$, $x\neq y$, $x\prec y \Leftrightarrow y\oplus{}^{-}x\in P$, we say that the partial ordering and the group structure on G are compatible.

Now suppose we have a ring with unity, where o is the primary identity element and e is the secondary identity. If $(R, \oplus, \circ, =)$ is not already a partial order, we can try to induce an ordering which will be compatible with the ring structure in the following sense: If x, y are "positive" elements of the ring, then $x\oplus y$ will be "positive" and $x\circ y$ will be positive or o. This is to say, we look for a set $P\subset R$ such that $\forall x, y\in P$, $x\oplus y\in P$ and $x\circ y\in P\cup\{o\}$, $o\notin P$. If such a P exists, we call P the positive cone in R, and we then define $\forall x, y\in R, x\neq y: x\prec y\Leftrightarrow y\oplus{}^{-}x\in P$. Clearly, this is just what we did for a group; the only difference is that in the case of the ring we insist that the set $P\cup\{o\}$ be closed with respect to the secondary operation. If $(R, \oplus, \circ, =)$ is already partially ordered, we say that the order is compatible with the ring structure if the set $P=\{x\in R: x\neq o, o\prec x\}$ is such that $\forall x, y\in P$, $x\oplus y\in P$, $x\circ y\in P\cup\{o\}$, $^{-}x\notin P$, and $^{-}x\prec y$ and that $\forall x, y\in R$, $x\neq y: x\prec y\Leftrightarrow y\oplus{}^{-}x\in P$. Note that if the ordering is compatible with the ring structure, the element $^{-}e\notin P$ since the ring axioms imply that $^{-}e\circ{}^{-}e=e$ and we cannot have both ^{-}e and e in P. Hence, if e and o are comparable, we must have $o\prec e$. Moreover, in such a ring order, the squares of negative elements are positive. Finally, we remark that the order is Archimedean if it is Archimedean as a group order.

Needless to say, we can extend this idea of ordering to linear spaces and algebras; the only thing we need to add is that the underlying scalar set must be ordered in a way consistent with the group or ring ordering. For example, if x is a positive element of a linear set, then all "positive" scalar multiples of x must also be positive.

The set $\mathbb{R}$ is a chain-ordered field (the natural ordering is compatible with the field structure), conditionally complete as a lattice (this fact is generally taken as one of the basic axioms characterizing the field $\mathbb{R}$), and the chain order is Archimedean. The field $\mathbb{C}$ can be totally ordered so as to be a conditionally complete lattice, but it will not then be Archimedean. On the other hand, $\mathbb{C}$ can be partially ordered so as to be a conditionally complete lattice, and Archimedean as well, but it will not be a chain. $\mathbb{Q}$, with its natural ordering inherited from $\mathbb{R}$, is an Archimedean chain-ordered field, but it is not even a conditionally σ-complete lattice. A famous theorem of algebra asserts that the structure $\mathbb{R}$ is very special indeed; it states that every Archimedean totally ordered conditionally complete field is order-field-isomorphic with $\mathbb{R}$.

Consider the set of all real-valued functions having Euclidean n-space as domain. This is simply the collection of all mappings f: $E^n\to\mathbb{R}$; denote this set by $\mathbb{F}$. We make $\mathbb{F}$ a linear algebra over $\mathbb{R}$ by defining $\forall f, g\in\mathbb{F}$,

$\forall\lambda,\mu\in\mathbb{R}$, $\lambda f+\mu g$ is the mapping defined by the formula $\forall\mathbf{x}\in E^n$, $(\lambda f+\mu g)(\mathbf{x})=\lambda f(\mathbf{x})+\mu g(\mathbf{x})$, and the mapping $f\cdot g$ is defined by $\forall\mathbf{x}\in E^n$, $(f\cdot g)(\mathbf{x})=f(\mathbf{x})\cdot g(\mathbf{x})$. It is easy to see that $(\mathbb{F},+,\cdot,=)$ satisfies the axioms for a linear algebra over $\mathbb{R}$; the function identically equal to zero on E^n is the o of $\mathbb{F}$, and the function identically equal to one on E^n is the ring unity of $\mathbb{F}$. We partially order $\mathbb{F}$ in this way. Define $\forall f\in\mathbb{F}$, $o\leqslant f\Leftrightarrow\forall\mathbf{x}\in E^n$, $f(\mathbf{x})\geqslant 0$. If $P\subset\mathbb{F}$ is to be the positive cone, we define $P=\{f\in\mathbb{F}: f\neq o, o\leqslant f\}$, and it is clear that the function $1\in P$. Now we define $\forall f,g\in\mathbb{F}$, $f\neq g$, $f<g\Leftrightarrow g-f\in P$. Clearly this ordering is neither total nor Archimedean. The fact that $\mathbb{R}$ is itself a conditionally complete lattice leads to the conclusion that $\mathbb{F}$ is also conditionally complete as a lattice (that it is a lattice is quite apparent), and it is easy to see that this ordering is compatible with the algebraic structure of $\mathbb{F}$.

We now present one of the most famous lemmas in all analysis. We then use it to prove a theorem about vector spaces. To prove the lemma we must make use of a fundamental axiom of set theory known as the axiom of choice, which we state in the following form:

Axiom of Choice. Given any nonempty collection of nonempty sets $\{S_\alpha\}_{\alpha\in\mathcal{Q}}$, one can choose from each S_α an element x_α and form the collection $S=\{x_\alpha\}_{\alpha\in\mathcal{Q}}$, and the collection S is a set.

The reader should realize that if the collection of sets $\{S_\alpha\}$ is a finite collection, there is no question that we can choose an element x_α from each S_α and form a finite set $\{x_\alpha\}$. The point of the axiom is that we can choose, perhaps simultaneously, one element from each set of an infinitely large collection of sets, and the resulting collection $\{x_\alpha\}$ is itself a set. An alternative way of stating the axiom is as follows: If $\{S_\alpha\}_{\alpha\in\mathcal{Q}}$ is a nonempty collection of nonempty sets, then there exists a function $f:\mathcal{Q}\to\bigcup_\alpha S_\alpha$ such that $f(\alpha)\in S_\alpha$ for each $\alpha\in\mathcal{Q}$. Some mathematicians are uncomfortable accepting this axiom, but we shall take it as a basic part of our mathematical faith.

Lemma 1.4.2 (Zorn's Lemma). *Let $\mathbb{P}$ be a nonempty partially ordered set such that every chain in $\mathbb{P}$ has an upper bound in $\mathbb{P}$. Then $\mathbb{P}$ contains at least one maximal element.*

PROOF: Let $C\subset\mathbb{P}$ be an arbitrary chain in $\mathbb{P}$, and $\gamma\in\mathbb{P}$ be an upper bound for C. Then $C\cup\{\gamma\}$ is also a chain in $\mathbb{P}$. Denote $\mathcal{C}$ the collection of all chains in $\mathbb{P}$, and keep in mind that $\mathcal{C}$ can be partially ordered by inclusion. This is to say, if C_1, C_2 are two members of $\mathcal{C}$, we say $C_1\prec C_2$ if as sets $C_1\subset C_2$.

With each $C\subset\mathbb{P}$ associate an upper bound $\gamma_C\in\mathbb{P}$ and a set T_C defined by $T_C=\{x\in\mathbb{P}:\gamma_C\prec x, x\neq\gamma_C\}$. Assume that each T_C, $C\in\mathcal{C}$, is nonempty, so that the axiom of choice guarantees the existence of a "choice function"

$f\colon \mathcal{C} \to \bigcup\{T_C : C \in \mathcal{C}\}$ such that $\forall C \in \mathcal{C}, f(C) \in T_C$. This is to say, $f(C)$ is an upper bound for C, and $f(C) \notin C$.

Note that the collection $\mathcal{C}$ enjoys the following two properties.

a. If $\{C_\beta\}$ is any totally ordered (by inclusion) subcollection of elements of the collection, then $\bigcup_\beta C_\beta$ is an element of the collection.
b. If C belongs to the collection, $C \cup \{f(C)\}$ belongs to the collection.

Suppose C_0 is a particular chain in $\mathbb{P}$. If $\mathcal{F}$ is the family of all subcollections $\mathcal{C}'$ of $\mathcal{C}$ which contain the chain C_0 and satisfy properties a and b, then the intersection of all these subcollections is a subcollection $\mathfrak{M}_0 \subset \mathcal{C}$ satisfying the following.

a. If $\{M_\beta\}$ is any totally ordered collection of elements of $\mathfrak{M}_0$, then $\bigcup_\beta M_\beta \in \mathfrak{M}_0$.
b. If $M \in \mathfrak{M}_0$, then $M \cup \{f(M)\} \in \mathfrak{M}_0$.
c. $C_0 \in \mathfrak{M}_0$.

It follows that $\mathfrak{M}_0$ is a minimal subset of $\mathcal{C}$ which contains C_0 and is closed with respect to properties a and b. Note these important facts: one subcollection of $\mathcal{C}$ belonging to the family $\mathcal{F}$ consists of C_0 and chains of chains containing C_0; hence it must be true that for each $M \in \mathfrak{M}_0$, $C_0 \subset M$. Furthermore, if $M \in \mathfrak{M}_0$ is distinct from C_0 and we form the subcollection $\mathfrak{M}_M \subset \mathcal{C}$ in the same way that we formed $\mathfrak{M}_0$, except that now M is the "first" element of $\mathfrak{M}_M$, it is apparent from the minimalities of $\mathfrak{M}_0$ and $\mathfrak{M}_M$ that $\mathfrak{M}_M \subset \mathfrak{M}_0$ and that $(\mathfrak{M}_0 \backslash \mathfrak{M}_M) \cup \{M\}$ consists of precisely those elements of $\mathfrak{M}_0$ which comprise a chain starting with C_0 and terminating at M. This is to say, $\mathfrak{M}_0$ must be totally ordered (by inclusion).

Define $A = \bigcup\{M : M \in \mathfrak{M}_0\}$; since $\mathfrak{M}_0$ is a totally ordered subcollection of $\mathcal{C}$, we have that $A \in \mathfrak{M}_0$ and $A \cup \{f(A)\} \in \mathfrak{M}_0$, so we conclude that $f(A) \in A$. Here is a contradiction: $f(A)$ cannot belong to A. Thus γ_A is a maximal element of $\mathbb{P}$ since the set T_A must be empty. □

If one takes this lemma as an axiom, it is possible to prove that the axiom of choice follows as a consequence; logically, the two propositions are equivalent since the truth of each implies the truth of the other. The reader may rest assured that Zorn's lemma is an extremely useful tool in analysis. To show how beautifully this lemma may be applied, we shall prove with consummate grace and ease a hard theorem about linear spaces. Suppose V is a nontrivial vector space (one consisting of something more than just the zero vector). A subset H of V is called a *Hamel basis* for V if H is a linearly independent subset (no element of H is a linear combination of other elements of H) and $\operatorname{sp} H = V$; that is, each $\mathbf{x} \in V$ has a unique representation as a (finite) linear combination of vectors of H.

Theorem 1.4.3. *Every nontrivial linear space V admits a Hamel basis.*

PROOF: Let P be the class of all linearly independent subsets of V, partially ordered by set inclusion. P is nonempty; since V is nontrivial, V contains an element $\mathbf{x} \neq \mathbf{0}$, and the singleton $\{\mathbf{x}\}$ is an element of P. Furthermore, if $\mathcal{C}$ is any chain in P, then the set $M \subset V$ obtained by taking the union of all the sets which comprise the chain, is itself a linearly independent set, and thus $M \in P$. To prove this last statement, recall that a set S is a linearly independent set if and only if every finite subset of S is a linearly independent set. Let $F = \{a_1, \ldots, a_n\}$ be an arbitrary finite subset of M, and note that each $a_i \in F$ belongs to some set in the chain $\mathcal{C}$. Since $\mathcal{C}$ is a chain, ordered by set inclusion, there is a set $A \in \mathcal{C}$ that contains each a_i and hence F. Since A is a linearly independent set, so is F. This completes the proof that $M \in P$.

The set M is an upper bound for the chain $\mathcal{C}$, and since $M \in P$, it follows by Zorn's lemma that P contains a maximal element; call it H. Now $\operatorname{sp} H = V$, for otherwise there would be an $\mathbf{x} \in V \setminus \operatorname{sp} H$, and $H \cup \{\mathbf{x}\}$ would be a linearly independent set in V. Because $H \subset H \cup \{\mathbf{x}\}$, H would not be a maximal element of P; contradiction. Hence H is indeed a Hamel basis for V, and the proof is complete. □

We must point out with some regret that the theorem gives no clue as to how one might set out to find an explicit Hamel basis.

PROBLEMS

1. Prove that for arbitrary sets A, B, C, $A \cap (B \triangle C) = (A \cap B) \triangle (A \cap C)$.

2. Prove Theorem 1.1.9.

3. Let $\{A_n\}_{n=1}^{\infty}$ be a sequence of sets, and put

$$\overline{S} = \bigcap_{n=1}^{\infty} \left(\bigcup_{k=n}^{\infty} A_k \right), \qquad \underline{S} = \bigcup_{n=1}^{\infty} \left(\bigcap_{k=n}^{\infty} A_k \right).$$

Prove that $\overline{S}$ is the set consisting of elements belonging to infinitely many of the sets A_n and that $\underline{S}$ is the set of those elements belonging to all but a finite number of the A_n. $\overline{S}$ is called the *limit superior* of $\{A_n\}$ and $\underline{S}$ the *limit inferior* of $\{A_n\}$.

4. Show that the algebra of quaternions is isomorphic to the algebra of matrices of the form

$$\begin{pmatrix} \alpha & \beta \\ -\bar{\beta} & \bar{\alpha} \end{pmatrix},$$

where α, β are complex numbers and the overbar means complex conjugate. [*Hint*: If $q = a + bi + cj + dk$ is a quaternion, identify it with the matrix whose entries are $\alpha = a + bi$, $\beta = c + di$, $\bar{\alpha} = a - bi$, $-\bar{\beta} = -c + di$, $a, b, c, d \in \mathbb{R}$,

$i^2 = j^2 = k^2 = -1$. See Section 4.4, the very last example, for a very brief discussion of quaternions.]

5. Prove that the field $\mathbb{Q}$ is a lattice, but not a σ-lattice, under the usual order. Show next that $\mathbb{R}$ is a conditionally complete lattice, and find an ordering for $\mathbb{C}$ so that $\mathbb{C}$ will be conditionally complete as a lattice.

6. Prove that if $a > 0, b > 0$, then $\sqrt{2}$ lies between a/b and $(a+2b)(a+b)^{-1}$.

7. Let V be a real linear space and H a Hamel basis for V. Show how to define a positive cone for V that will be consistent with the algebraic structure of V.

8. Suppose φ is a homomorphism from a ring with unity V to a ring with unity W. Prove that if o, e are the primary and secondary identity elements of V, then necessarily $\varphi(o), \varphi(e)$ are the corresponding identity elements in W.

9. Show that the sets $\mathbb{Q}$ and $\mathbb{P}$ are set-isomorphic. ($\mathbb{Q}$ is the set of rationals and $\mathbb{P}$ the set of prime integers.)

10. Prove that the set $\mathbb{A}$ of algebraic numbers is countable.
[*Hint*: How many polynomials of degree n are there with rational coefficients?]

11. Consider the set $\mathcal{F}$ of real-valued functions defined over $[a, b] \subset \mathbb{R}$. Say that $f \sim g$ iff $\int_a^b [f(x) - g(x)]\, dx = 0$. Is $\sim$ an equivalence relation over $\mathcal{F}$? What if $\mathcal{C} \subset \mathcal{F}$ is the subset of continuous functions over $[a, b]$. Is $\sim$ an equivalence relation? Show that the relation $\leftrightarrow$, defined by $f \leftrightarrow g$ iff $\int_a^b |f(x) - g(x)|\, dx = 0$, is, or is not, an equivalence relation over $\mathcal{C}$.

12. Prove that if $\mathcal{S}$ is a class of sets closed with respect to $\cup$ and $\setminus$, then $(\mathcal{S}, \triangle, \cap, =)$ is a commutative ring with unity. If $A \in \mathcal{S}$, what is the "additive inverse" of A?

13. Show that if V is a linear set and $\{\mathbf{x}_i\}_{i=1}^n \subset V$, then the linear span of the $\mathbf{x}_i$ is indeed the smallest linear set containing $\{\mathbf{x}_i\}_{i=1}^n$.

14. If $\{\mathbf{x}_i\}_{i=1}^n$ is a set of points in a real vector space, prove that the convex hull of these n points is the set of all linear combinations of the form $\mathbf{y} = \sum_{i=1}^n \lambda_i \mathbf{x}_i$, $\lambda_i \geqslant 0$, $\sum_{i=1}^n \lambda_i = 1$.

15. Suppose S is a convex set, $S \subset V(\mathbb{R})$, and $\mathbf{e} \in S$ is a point such that for no two distinct points $\mathbf{x}, \mathbf{y} \in S$ is $\mathbf{e} = (1-\lambda)\mathbf{x} + \lambda\mathbf{y}$, where $0 < \lambda < 1$. Such a point $\mathbf{e}$ is called an *extreme point* of S. Prove, or disprove, S is equal to the convex hull of its set of extreme points.
Now suppose S is the convex hull of a finite collection of points $\{\mathbf{x}_i\}_{i=1}^n \subset V(\mathbb{R})$. Prove, or disprove, the above proposition.

16. Show that the set of maps $\mathbb{R} \to \mathbb{R}$ under the ordering described in the text is a conditionally complete lattice.

17. Prove that the set $\mathbb{P}$ of prime integers is countable.

18. Prove that $\mathbb{R}$ and $\mathbb{R}^2$ are set-isomorphic. Can you find a group-isomorphism between the additive group of real numbers $\mathbb{R}$ and the multiplicative group of positive real numbers $\mathbb{R}^+$?

19. In the text we gave an ordering for $\mathbb{C}$ which made a $\mathbb{C}$ a total order. Show that $\mathbb{C}$ with this ordering is non-Archimedean.

20. Construct a subset $S \subset \mathbb{R}$ such that S will contain subsets by which you can illustrate the differences between maximal element, supremum, and upper bound.

21. Exhibit a partial ordering for $\mathbb{C}$ that will make $\mathbb{C}$ a conditionally complete lattice with a positive cone but the order non-Archimedean.

22. Let $\mathbb{F}$ be the algebra of real-valued functions on $\mathbb{R}$, with the compatible order given in the text. Show that this ordering is non-Archimedean.

23. Prove that $\sqrt[3]{2} + \sqrt{3}$ is an algebraic number.

24. Suppose that $f: X \to Y$ and $g: Y \to Z$. If $z \subset Z$, show that $(g \circ f)^{-1}(z) = f^{-1}(g^{-1}(z))$, where $g \circ f$ is the composite map defined by $\forall x \in X$, $g \circ f: x \mapsto g(f(x)) \in Z$.

25. Prove that in a ring $(^-e) \circ (^-e) = e$.

26. Prove that if $\{A_\alpha\}$ is an arbitrary collection of sets and B is a set, then

$$B \cap \left(\bigcup_\alpha A_\alpha \right) = \bigcup_\alpha (B \cap A_\alpha) = \left(\bigcup_\alpha A_\alpha \right) \cap B,$$

$$B \cup \left(\bigcap_\alpha A_\alpha \right) = \bigcap_\alpha (B \cup A_\alpha) = \left(\bigcap_\alpha A_\alpha \right) \cup B.$$

Thus each of $\bigcup$ and $\bigcap$ satisfies a generalized two-sided distributive law with respect to the other.

27. If $\{A_\alpha\}, \{B_\beta\}$ are collections of sets, show that

$$\left(\bigcup_\alpha A_\alpha \right) \cap \left(\bigcup_\beta B_\beta \right) = \bigcup_\alpha \bigcup_\beta (A_\alpha \cap B_\beta) = \bigcup_{\alpha, \beta} (A_\alpha \cap B_\beta),$$

$$\left(\bigcap_\alpha A_\alpha \right) \cup \left(\bigcap_\beta B_\beta \right) = \bigcap_\alpha \bigcap_\beta (A_\alpha \cup B_\beta) = \bigcap_{\alpha, \beta} (A_\alpha \cup B_\beta).$$

28. Complete the proof of Theorem 1.1.5.

29. Prove that $\mathbb{Z}^+$ and $\mathbb{Z}$ are equivalent.

30. Prove that if $n=\frac{1}{2}(i+j-2)(i+j-1)+i=\frac{1}{2}(i'+j'-2)(i'+j'-1)+i'=n'$, then $(i,j)=(i',j')$.

31. Give an example of a partially ordered set $\mathbb{P}$ with a proper subset A such that
(a) A has more than one upper bound in $\mathbb{P}$ which are not comparable.
(b) A has more than one upper bound in $\mathbb{P}$, but no supremum.
(c) A has no upper bound in $\mathbb{P}$.

32. If A is the set of rational numbers whose denominator is 10 and B is the set of all rational numbers whose denominator is 8, what are the sets $A \cup B$ and $A \cap B$?

33. Define a relation $\sim$ on $\mathbb{R}^+ \times \mathbb{R}^+$ as follows:

$$(a,b)\sim(a',b') \qquad \text{iff} \qquad ab'=a'b.$$

Show that if we define addition of pairs by $(a,b)\oplus(c,d)=(ad+bc,bd)$, then $(a,b)\sim(a',b')$, $(c,d)\sim(c',d')$ implies

$$(ad+bc,bd)\sim(a'd'+b'c',b'd').$$

Show that $\sim$ is an equivalence relation over the set of ordered pairs of positive real numbers and that we have in effect defined an addition for these equivalence classes.

34. Show that if $\{A_\alpha\}_{\alpha\in\mathcal{Q}}$ is a family of sets, $\bigcap_\alpha P(A_\alpha)=P(\bigcap_\alpha A_\alpha)$ and $\bigcup_\alpha P(A_\alpha)\subset P(\bigcup_\alpha A_\alpha)$. Show that the indicated inclusion may or may not be proper.

35. If $A_k=\{n\in\mathbb{Z}:n\geqslant k\}$, $k=1,2,3,\ldots$, what is $\bigcap\{A_k:k=1,2,3,\ldots\}$?
If $B_n=[0,1-2^{-n}]$ and $C_n=[0,1-3^{-n}]$, $n\in\mathbb{Z}^+$, show that $\bigcup B_n=[0,1)=\bigcup C_n$.

36. In $\mathbb{Z}^+$, define $m\prec n$ if n divides m. Show that $\mathbb{Z}^+$ so ordered is a partial order and that every chain has an upper bound. What is the set of maximal elements? If the ordering is given by $m\prec n$ if m divides n, is $\mathbb{Z}^+$ partially ordered? What else can you say about this ordering?

37. Let $\mathfrak{F}$ be the set of all real-valued functions of a real variable. Define $\forall f,g\in\mathfrak{F}$

$$f\prec g \quad\Leftrightarrow\quad f=g \quad\text{or}\quad \lim_{x\to\infty}\frac{f(x)}{g(x)}=0.$$

Is $\prec$ a partial order over $\mathfrak{F}$? Demonstrate.

38. Let $\mathbb{P}$ be a partial order, and $S\subset\mathbb{P}$ a subset. S is called *cofinal* if $\forall x\in\mathbb{P}$, $\exists b\in S$ s.t. $x\prec b$. Prove that every totally ordered set has a cofinal well-ordered subset.

39. For arbitrary sets A, B, C prove

$$(A\setminus B)\cap(A\setminus C)=A\setminus(B\cup C),$$
$$(A\setminus B)\setminus(A\setminus C)=A\cap(C\setminus B),$$
$$A\cap(B\triangle C)=(A\cap B)\triangle(A\cap C).$$

40. Let $S=\{x\in\mathbf{Q}: x^2\leqslant 7\}$. What are $\inf S$ and $\sup S$?

41. Let S be points of the form

$$x_n=(-1)^n(3-9/3^n), \qquad n=1,2,3,\ldots.$$

Find $\sup S$ and $\inf S$.

42. Find $\inf S$ and $\sup S$ if $S=\{x_n : x_n=(-1)^n+1/n,\ n\in\mathbf{Z}^+\}$.

43. Prove that if A, B are nonempty sets and $(A\times B)\cup(B\times A)=C\times C$, then $A=B=C$.

44. Show that if A, B, C, D are sets, $(A\times C)\cap(B\times D)=(A\cap B)\times(C\cap D)$ and $(A\times C)\cup(B\times D)\subset(A\cup B)\times(C\cup D)$ and that this last inclusion is generally proper.

45. Suppose C is a totally ordered set and A, B are a pair of nonempty subsets satisfying $A\cup B=C$, $A\cap B=\varnothing$, $a\in A$ and $b\in B\Rightarrow a\prec b$. Then (A, B) is called a *cut* in C. Prove that if (A, B), (A', B') are cuts in C, then either $A\subset A'$ or $A'\subset A$.

46. Show that the set of countable subsets of $\mathbf{R}^2$ is uncountable.

47. Prove that any collection of nondegenerate disjoint intervals of the real line is at most countable.

48. Let $a, b\in\mathbf{Q}$. Prove, or disprove, that the set of all numbers of the form $x=a+b\sqrt{3}$ comprises a field. Do the same for all numbers of the form $y=a+b\pi$.

49. Show that the set of points interior to an ellipsoid is equivalent to the set of points interior to a sphere.

50. Prove, or disprove, that the set of real polynomials in x is totally ordered under the following ordering:

$$p\prec q \qquad \Leftrightarrow \qquad \exists x \text{ s.t. } \forall y>x,\ p(y)\leqslant q(y).$$

51. Prove that if $\mathbb{S}$ is a σ-ring of sets and $\{S_n\}_{n=1}^{\infty}\subset\mathbb{S}$, then

(a) $\bigcap_n S_n\in\mathbb{S}$,

(b) $\underline{S}=\bigcup_{n=1}^{\infty}\bigcap_{k=n}^{\infty}S_k\in\mathbb{S}$,

(c) $\bar{S}=\bigcap_{n=1}^{\infty}\bigcup_{k=n}^{\infty}S_k\in\mathbb{S}$.

52. Let S be the set of all real numbers of the form $2^{-p}+3^{-q}+5^{-r}$, where p, q, and r are arbitrary positive integers. What is $\sup S$? $\inf S$?

53. Let A and B be bounded sets of real numbers. If C is the set $\{c: c=ab, a\in A, b\in B\}$, does $\sup C=(\sup A)\cdot(\sup B)$? If not, give a counterexample.

54. Let S be the set of integers. Define $\forall a, b\in S$, $a\sim b \Leftrightarrow a-b$ is divisible by m, where m is a fixed positive integer. Show whether $\sim$ is an equivalence relation.

55. Explain in some detail how the axiom of choice has been used in the proof of Theorem 1.1.3.

CHAPTER

2

LIMIT AND CONTINUITY IN E^n

2.1 LIMIT OF A FUNCTION

The notion of limit is fundamental to the study of infinitesimal calculus; in fact, the part of mathematics known as point-set topology has been developed and refined primarily to generalize this concept and put it on firmer ground. Chapter 6 deals with topology, but perhaps it would be well to present the limit concept in a concrete form initially.

The word limit often suggests some sort of boundary: "I've reached the limit of my patience!" screams the distraught professor. In mathematics a limit is usually something that is "approached." What that means precisely is generally not clear, but it seems to have something to do with "closeness" or "proximity." We say, "**p** is the limit of **x**," and this suggests that **p** is stationary but **x** is moving, eventually getting as close to **p** as one wishes, and ever after never straying further from **p**. However, this statement, although rather well put, leaves something to be desired. What exactly do we mean by "as close to **p** as one wishes"? One bit of clarification is this: It would be too restrictive to insist that **x** eventually arrive at **p** (closer to **p** than this would be impossible) and remain there. We want the statement "**p** is the limit of **x**" to be synonymous with the statement "**x** approaches **p**," and the most general way to describe the fact that **x** approaches **p** is to say that if one prescribes some arbitrary degree of closeness, and if **x** eventually falls within that degree of closeness to **p** and stays within it, then **x** approaches **p** as a limit. This way of putting things allows for the possibility that x never reaches **p**. But note that in our attempt to define precisely what we mean by "**p** is the limit of **x**" we have had to use a notion of time; we have tacitly assumed that the position of **x** is a function of time and that after a finite length of time (this is what we mean by "eventually"), **x** will always be within a certain prescribed degree of closeness to **p**.

We conclude that it is impossible to define this concept in terms of more primitive concepts; the notion "**x** approaches **p**" must remain for us an undefined concept. But this state of affairs is hardly new; Euclid was unable to define "point" and "line," which have remained in mathematics undefined terms. However, although we cannot define "**x** approaches **p**" in terms of more fundamental concepts we can nevertheless be intuitively certain of what we mean, and we can use this undefined notion to define other concepts.

Suppose we restrict our attention to Euclidean space E^n. As a set, E^n is the same as $\mathbb{R}^n$, n-dimensional Cartesian space, the set of all ordered n-tuples of real numbers, $(x_1,\dots,x_n)$; however, we endow $\mathbb{R}^n$ with a metric ρ, a rule or formula by which we can assign a "distance" between each two points of $\mathbb{R}^n$. The formula we use is this time-honored one: $\forall \mathbf{x}=(x_1,\dots,x_n)$, $\mathbf{y}=(y_1,\dots,y_n)\in\mathbb{R}^n$,

$$\rho(\mathbf{x},\mathbf{y})=\rho((x_1,\dots,x_n),(y_1,\dots,y_n))=\left[\sum_{i=1}^{n}(x_i-y_i)^2\right]^{1/2}.$$

$\mathbb{R}^n$, together with the structure supplied by this metric, is E^n, Euclidean n-space.

Let $\mathbf{p}\in E^n$ be an arbitrary but fixed point. We cannot define in terms of more basic concepts what it means to say "**x** approaches **p**," or $\mathbf{x}\to\mathbf{p}$, but there is no doubt in our own minds what this means. Now suppose that **q** is a point in E^m, Euclidean m-space, $D\subset E^n$ a subset, and $f\colon D\to E^m$ a mapping, or function.

Definition 2.1.1. $\mathbf{q}\in E^m$ is the *limit* of f at $\mathbf{p}\in E^n$ iff it is possible for a variable point **x** to approach **p** and remain in D, and regardless of how $\mathbf{x}\to\mathbf{p}$, while remaining in D, it must necessarily follow that $f(\mathbf{x})$ approach **q**.

This definition says exactly what we mean, but we can rephrase it to make it more precise. Remember that $D=\operatorname{dom} f$; f is not defined outside of D.

Definition 2.1.1′. $\mathbf{q}\in E^m$ is the limit of f at $\mathbf{p}\in E^n$ iff for every $\delta>0$, $\exists\mathbf{x}\in D$ s.t. $0<\rho(\mathbf{x},\mathbf{p})<\delta$, and for any $\varepsilon>0$, $\exists\delta>0$ s.t. $\forall\mathbf{x}\in D$, $0<\rho(\mathbf{x},\mathbf{p})<\delta\Rightarrow\rho(f(\mathbf{x}),\mathbf{q})<\varepsilon$.

The point **p** may or may not belong to D. Also, one should notice that for a given $\varepsilon>0$ and a given point **p**, if **q** is the limit of f at **p**, written

$$\lim_{\mathbf{p}} f=\mathbf{q},$$

we might have to find a very small δ indeed in order that $\rho(f(\mathbf{x}),\mathbf{q})$ be less than ε whenever $0<\rho(\mathbf{x},\mathbf{p})<\delta$ and $\mathbf{x}\in D$. The "sufficiently small" δ evidently depends not only upon the size of ε but also upon the location of $\mathbf{p}$.

Definition 2.1.2. If $f: D\to E^m$ has a limit $\mathbf{q}$ at $\mathbf{p}\in E^n$ and $\mathbf{p}\in D$ with $f(\mathbf{p})=\mathbf{q}$, we say f is *continuous at* $\mathbf{p}$. If $\mathbf{p}$ is an isolated point of D; i.e., there is an $\varepsilon>0$ such that no other point of D satisfies $0<\rho(\mathbf{x},\mathbf{p})<\varepsilon$, we also say f is continuous at $\mathbf{p}$. If f is continuous at each $\mathbf{p}\in D$, we say f is *continuous on* (or *over*) D.

It is easy to see that if f is continuous at $\mathbf{p}$ and $\mathbf{p}$ is not an isolated point of dom f, then f has a limit at $\mathbf{p}$, namely, $f(\mathbf{p})$.

We can rephrase the preceding definition as follows.

Definition 2.1.2′. $f: D\to E^m$ is continuous at $\mathbf{p}$ iff $\mathbf{p}\in D$, and $\forall\varepsilon>0$, $\exists\delta>0$ s.t. $\forall\mathbf{x}\in D$, $\rho(\mathbf{x},\mathbf{p})<\delta\Rightarrow\rho(f(\mathbf{x}),f(\mathbf{p}))<\varepsilon$.

If f is continuous over its domain, it is of course continuous at each point of its domain. This is to say, at each point $\mathbf{p}\in$ dom f, a sufficiently small change in $\mathbf{p}$ will induce a necessarily small change in $f(\mathbf{p})$. Suppose a certain $\varepsilon>0$ has been prescribed and $\mathbf{p}_1,\mathbf{p}_2$ are a pair of points in the domain of f. It could well be that to ensure that $f(\mathbf{p}_1)$ will not change by more than ε if we change $\mathbf{p}_1$ might require that we change $\mathbf{p}_1$ by no more than a gnat's eyelash, whereas we might change $\mathbf{p}_2$ substantially without changing $f(\mathbf{p}_2)$ by more than ε. This is to say, f can be continuous, but the continuity may not be uniform over the whole domain.

On the other hand, if the behavior of f is such that any change in $\mathbf{x}$ within certain prescribed limits will never induce a change in $f(\mathbf{x})$ greater than a certain prescribed amount, regardless of where $\mathbf{x}$ is in dom f, we say that f is uniformly continuous on its domain.

More precisely:

Definition 2.1.3. $f: D\to E^m$ is *uniformly continuous* iff $\forall\varepsilon>0$, $\exists\delta>0$ s.t. $\forall\mathbf{x},\mathbf{y}\in D$, $\rho(\mathbf{x},\mathbf{y})<\delta\Rightarrow\rho(f(\mathbf{x}),f(\mathbf{y}))<\varepsilon$.

Of course the uniform continuity of f implies the continuity of f.

There is an even stronger kind of continuity that a mapping f might enjoy called *absolute continuity*. The uniform continuity of f guarantees that if you change the argument of f just a little bit, there won't be much of a change in the image of the argument. The absolute continuity of f ensures that if you make little changes in a large number of arguments of f, and add up the corresponding changes in the corresponding images, you will still get small total change.

Definition 2.1.4. Let $D \subset E^n$ and $f: D \to E^m$ be a mapping. f is said to be *absolutely continuous* on D iff $\forall \varepsilon > 0$, $\exists \delta > 0$ such that for every finite collection of not necessarily distinct points $\{\mathbf{x}_1, \ldots, \mathbf{x}_{2n}\} \subset D$ for which $\sum_{i=1}^n \rho(\mathbf{x}_{2i}, \mathbf{x}_{2i-1}) < \delta$, we have $\sum_{i=1}^n \rho(f(\mathbf{x}_{2i}), f(\mathbf{x}_{2i-1})) < \varepsilon$.

Certainly absolute continuity on D implies uniform continuity on D, but the converse is not true. In Section 6.9 we give an example of a rather famous function $f: [0,1] \to \mathbb{R}$ which is uniformly continuous on $[0,1]$ but not absolutely continuous on this domain.

Theorem 2.1.5. *Suppose $f: \mathbb{R} \to \mathbb{R}$ is bounded and integrable on each bounded interval of $\mathbb{R}$. Let $[a,b]$ be an arbitrary interval; then $F(x) = \int_a^x f(t)\,dt$, $a \leqslant x \leqslant b$, is absolutely continuous on $[a,b]$.*

PROOF: Let $\varepsilon > 0$ be given. Let $\{x_1, \ldots, x_{2n}\}$ be a finite set of not necessarily distinct points of $[a,b]$ such that $\sum_{i=1}^n |x_{2i} - x_{2i-1}| < \delta$, where δ is a number to be determined shortly. Since f is bounded on $[a,b]$ let $M = \sup\{|f(x)| : x \in [a,b]\}$. (This supremum exists by axiom.) We have

$$\sum_{i=1}^n |F(x_{2i}) - F(x_{2i-1})| = \sum_{i=1}^n \left| \int_{x_{2i-1}}^{x_{2i}} f(t)\,dt \right| \leqslant \sum_{i=1}^n M|x_{2i} - x_{2i-1}| < M\delta.$$

Now assign δ the value ε/M. This completes the proof. □

The classic example of a function which is continuous but not uniformly so on its domain is $f(x) = 1/x$, $0 < x < \infty$. Consider the function

$$f(x) = \begin{cases} x \sin \pi/2x, & 0 < x \leqslant 1, \\ 0, & x = 0. \end{cases}$$

This function is continuous on $[0,1]$ and by the following theorem is uniformly continuous on this interval. Let n be an arbitrary odd positive integer and k a suitably large odd positive integer. Let $\varepsilon > 0$, and consider the set of points $\{x_1, x_2, \ldots, x_k\} \subset [0,1]$, where $x_i = 1/(n+i)$, $1 \leqslant i \leqslant k$. These points determine $k-1$ abutting intervals whose lengths total less than $1/(n+1)$, for every k. However, the sum

$$\begin{aligned} &\sum_{i=1}^{k-1} |f(x_{i+1}) - f(x_i)| \\ &\quad = \sum_{i=1}^{k-1} \left| \frac{1}{n+1+i} \sin \frac{(n+1+i)\pi}{2} - \frac{1}{n+i} \sin \frac{(n+i)\pi}{2} \right| \\ &\quad = 2 \sum_{i=1}^{(k-1)/2} \frac{1}{n+2i}, \end{aligned}$$

and for sufficiently large k this sum is greater than ε. Hence $f(x)$ is not absolutely continuous.

Theorem 2.1.6. *If $f: E^n \to \mathbb{R}$ is continuous and $B \subset E^n$ is a bounded set, then f is uniformly continuous on B.*

We postpone the proof until a later chapter (Theorem 6.7.9).

2.2 SEQUENCES IN E^n

A sequence $\{\mathbf{x}_i\}_{i=1}^{\infty} \subset E^n$ is simply the image of an E^n-valued mapping f whose domain is the set $\mathbb{Z}^+ \subset E^1$. This is to say, for each $i \in \mathbb{Z}^+$, $f(i) = \mathbf{x}_i \in E^n$. Recall that $\mathbb{Z}^+$ is a well-ordered set; this ordering induces an ordering on $f(\mathbb{Z}^+)$ in the obvious way:

$$i \leqslant j \qquad \Leftrightarrow \qquad \mathbf{x}_i \prec \mathbf{x}_j.$$

With this ordering, the sequence $f(\mathbb{Z}^+) = \{\mathbf{x}_i\}_{i=1}^{\infty}$ is well ordered, and one might be tempted to refer to the sequence $\{\mathbf{x}_i\}_{i=1}^{\infty}$ as an "ordered infinity-tuple," or more correctly, an "ordered $\aleph_0$-tuple" since we always understand that a sequence consists of a countable number of terms.

Keep in mind this subtle distinction between sequence and set. As a set, $\{\mathbf{x}_i\} \subset E^n$ consists of distinct points of E^n, whereas a sequence $\{\mathbf{x}_i\}_{i=1}^{\infty}$ might consist of only a finite number of distinct points of E^n. A single point of E^n might occur infinitely often in the sequence; even though the terms $\mathbf{x}_2$ and $\mathbf{x}_4$ of the sequence $\{\mathbf{x}_i\} \subset E^n$ might be one and the same point in E^n, they are distinct terms of the sequence.

A sequence $\{\mathbf{x}_i\} \subset E^n$ is called a *Cauchy* sequence, in honor of the great French mathematician Augustin Louis Cauchy, if, for any $\varepsilon > 0$, there exists an index M such that whenever $\min\{i, j\} \geqslant M$, $\rho(\mathbf{x}_i, \mathbf{x}_j) < \varepsilon$. If there is a point $\mathbf{p} \in E^n$ such that $\forall \varepsilon > 0$, $\exists M$ s.t. $i \geqslant M \Rightarrow \rho(\mathbf{x}_i, \mathbf{p}) < \varepsilon$, we say the sequence *converges* to $\mathbf{p}$ and write

$$\{\mathbf{x}_i\} \to \mathbf{p} \qquad \text{or} \qquad \lim\{\mathbf{x}_i\} = \mathbf{p}.$$

It is easy to prove that every convergent sequence in E^n is a Cauchy sequence and a lot harder to prove the converse. The converse is not true in $\mathbb{Q}^n$ with the Euclidean metric. For example, let $n = 1$, and consider the sequence

$$\{\mathbf{x}_i\} = \{3, 3.1, 3.14, 3.141, 3.1415, \ldots\}.$$

This recognizable sequence is a Cauchy sequence, but it does not converge to any point in $\mathbb{Q}$.

Notice that if we think of the sequence $\{\mathbf{x}_i\}_{i=1}^{\infty} \subset E^n$ as the ordered set of values of a function $f: \mathbb{Z}^+ \to E^n$, we can give the following:

Definition 2.2.1. $\mathbf{p} \in E^n$ is the *limit* of the sequence $\{\mathbf{x}_i\}_{i=1}^{\infty} \subset E^n$ iff $\forall \varepsilon > 0$, $\exists M$ s.t. $\forall i \geqslant M$, $\rho(\mathbf{x}_i, \mathbf{p}) < \varepsilon$ iff $\mathbf{p}$ is the limit of f at ∞.

Of course we must interpret "all numbers $i \in \mathbb{Z}^+$ sufficiently close to ∞" to mean "all numbers $i \in \mathbb{Z}^+$ greater than some sufficiently large M."

Suppose $\mathcal{A}$ is a directed set, countable or uncountable. Consider the indexed set $\{\mathbf{x}_\alpha\}_{\alpha\in\mathcal{A}}$. We can regard this indexed set as a *generalized sequence*, or a *net*, in E^n. We say that the net $\{\mathbf{x}_\alpha\}_{\alpha\in\mathcal{A}}\subset E^n$ converges to a point $\mathbf{p}\in E^n$ iff for any $\varepsilon>0$, there is an $\alpha\in\mathcal{A}$ such that for every $\beta\in\mathcal{A}$ for which $\alpha\prec\beta$ we have $\rho(\mathbf{x}_\beta,\mathbf{p})<\varepsilon$.

We bring this section to a close by introducing several important concepts that have to do with sets and sequences in E^1, or more commonly, $\mathbb{R}$ (we always regard $\mathbb{R}$ as being metrized in the natural way). Recall that $\mathbb{R}$ is a very special structure; it is a field with a compatible chain order, and the order is Archimedean. To completely characterize $\mathbb{R}$ we need one more axiom, the one called the completeness axiom. This axiom states that if $S\subset\mathbb{R}$ is any nonempty bounded set, then $\sup S$ and $\inf S$ exist as unique real numbers. This means that the structure $\mathbb{R}$ can't be augmented without changing it in some way.

The reader should bear in mind that the axioms which characterize the structure $\mathbb{R}$ did not come first. When living things began to think, they noticed the finite cardinal numbers, which had been around since the Big Bang. It took millions of years to augment this system of numbers to obtain the set $\mathbb{R}$, along with its arithmetic and order structure. Only then did mathematicians set about constructing the set of axioms.

Let $B\subset\mathbb{R}$ be a nonempty set. We are going to redefine some concepts we have already defined in the previous chapter, giving more specialized definitions in terms of real numbers.

A real number m is an *upper bound* for B if $m\geqslant x$ for every $x\in B$. A *lower bound* is defined analogously.

A real number μ is called the *supremum*, or *least upper bound*, of B if μ is an upper bound for B, and if for every $\varepsilon>0$ there exists an $x_\varepsilon\in B$ such that $\mu-\varepsilon<x_\varepsilon$. We write $\mu=\sup B=\operatorname{lub} B$. Similarly, $\lambda\in\mathbb{R}$ is called the *infimum*, or the *greatest lower bound*, of B if λ is a lower bound for B, and for every $\varepsilon>0$ there is an $x_\varepsilon\in B$ such that $x_\varepsilon<\lambda+\varepsilon$. We write $\lambda=\inf B=\operatorname{glb} B$.

If $\mu=\sup B\in B$, we call μ the *maximum* of B; if $\lambda=\inf B\in B$, λ is called the *minimum* of B. It is evident that for any finite set $B\neq\varnothing$, $\sup B=\max B$ and $\inf B=\min B$.

A real number l is called an *accumulation point* (*cluster point*, *limit point*) of B if for every $\varepsilon>0$ there is an $x_\varepsilon\in B\setminus\{l\}$ such that $l-\varepsilon<x_\varepsilon<l+\varepsilon$. If $l\in B$ and l is not an accumulation point of B, l is called an *isolated point* of B.

A real number μ is called the *limit superior* of B if for each $\varepsilon>0$, $\mu+\varepsilon$ is greater than all but a finite number of points of B and $\mu-\varepsilon$ is less than infinitely many elements of B. We write $\mu=\limsup B=\overline{\lim}\, B$. A real number λ is called the *limit inferior* of B if for each $\varepsilon>0$, $\lambda-\varepsilon$ is less than all but a finite number of elements of B and $\lambda+\varepsilon$ is greater than infinitely many points of B. We write $\lambda=\liminf B=\underline{\lim}\, B$.

Note that when B is a bounded nonempty set in $\mathbb{R}$, $\overline{\lim}\, B$ is the largest accumulation point of B and $\underline{\lim}\, B$ is the smallest.

To see this, let $\mu=\overline{\lim}\, B$; for each $\varepsilon>0$, the interval $(\mu-\varepsilon, \mu+\varepsilon)$ contains infinitely many points of B. Hence μ is an accumulation point of B. Suppose $\mu'>\mu$ is also an accumulation point of B. Let $\varepsilon=(\mu'-\mu)/2$. The interval $(\mu'-\varepsilon, \mu'+\varepsilon)$ contains at most a finite number of points of B. It is not hard to see that μ' cannot be an accumulation point of B, so $\overline{\lim}\, B$ must be the largest such point. Analogously, $\underline{\lim}\, B$ is the smallest accumulation point of B.

If B is a bounded sequence and if $\underline{\lim}\, B=\overline{\lim}\, B=\lambda$, then λ is the limit of the sequence B. Conversely, if $B=\{x_i\}_{i=1}^{\infty}$ is a sequence converging to λ, then $\lambda=\underline{\lim}\, B=\overline{\lim}\, B$.

In the event that $B\neq\varnothing$ is unbounded above, we agree that $\sup B=+\infty$. If $B\neq\varnothing$ is unbounded below, we agree to write $\inf B=-\infty$. If we write $\overline{\lim}\, B=\infty$ ($\underline{\lim}\, B=-\infty$), we mean that for each real number M, infinitely many elements of B are greater than or equal to M (less than or equal to M).

Suppose $B=\varnothing$. It is true that -10^{10} is greater than any element in B since B contains no elements (one says such a statement is vacuously true), and the same can be said for any real number no matter how small. Hence there is no *least* upper bound for $\varnothing$. We indicate this state of affairs by writing $\sup\varnothing=-\infty$. Analogously, we write $\inf\varnothing=+\infty$.

Recall the notation that we introduced in the last chapter: if B is a bounded subset of a lattice, then

$$\sup B=\bigvee_{x\in B} x, \qquad \inf B=\bigwedge_{x\in B} x.$$

If the lattice is conditionally complete, these two bounds exist in the lattice. Since $\mathbb{R}$ is such a lattice, for any bounded set $B\subset\mathbb{R}$, $\sup B\in\mathbb{R}$ and $\inf B\in\mathbb{R}$.

Theorem 2.2.2. *Suppose $\{x_n\}$ is a bounded sequence of real numbers. Then*

$$\limsup\{x_n\}=\bigwedge_{n=1}^{\infty}\left(\bigvee_{i=n}^{\infty} x_i\right),$$

$$\liminf\{x_n\}=\bigvee_{n=1}^{\infty}\left(\bigwedge_{i=n}^{\infty} x_i\right).$$

PROOF: Since $\{x_n\}$ is bounded, the subsequence $\{x_i\}_{i=n}^{\infty}$ is bounded, so the numbers

$$\mu_n=\bigvee_{i=n}^{\infty} x_i, \qquad \lambda_n=\bigwedge_{i=n}^{\infty} x_i$$

exist for each $n=1,2,3,\ldots$. Consider the sequences of numbers $\{\mu_n\}_{n=1}^{\infty}$, $\{\lambda_n\}_{n=1}^{\infty}$. The first is clearly a decreasing (nonincreasing) sequence and the second an increasing (nondecreasing) sequence. Moreover, $\{\mu_n\}$ is bounded below by λ_1, and $\{\lambda_n\}$ is bounded above by μ_1. Hence

$$\mu = \bigwedge_{n=1}^{\infty} \mu_n, \qquad \lambda = \bigvee_{n=1}^{\infty} \lambda_n$$

exist in $\mathbb{R}$. To show that μ and λ are unique is easy. Assume a bounded set of real numbers has two distinct suprema. The contradiction is immediate. Thus we have proved the existence of the right-hand members of the equations in the statement of the theorem.

We need to show that $\mu = \overline{\lim}\{x_n\}$, $\lambda = \underline{\lim}\{x_n\}$. Let $\varepsilon > 0$ be arbitrary, and consider $\mu + \varepsilon$. Since $\mu = \inf\{\mu_n\}$, there is at least one index n_0 such that $\mu + \varepsilon > \mu_{n_0}$, and since $\{\mu_n\}$ is nonincreasing, $\mu + \varepsilon > \mu_n$ for all $n \geqslant n_0$. Hence $\mu + \varepsilon$ is greater than all but a finite number of μ_n, and since $\mu_{n_0} \geqslant \mu_n \geqslant x_n$ for all $n \geqslant n_0$, we have that $\mu + \varepsilon$ is greater than all but a finite number of the terms of $\{x_n\}$.

Now consider $\mu - \varepsilon$; this number is strictly less than each μ_n. Now $\mu - \varepsilon \leqslant \mu_1 - \varepsilon$, and this number is strictly less than one of the numbers $\{x_1, x_2, \ldots\}$, say x_{n_1}, and we have $\mu - \varepsilon < x_{n_1}$. Similarly, we have $\mu - \varepsilon \leqslant \mu_{n_1+1} - \varepsilon$, and this last is strictly less than one of the numbers $\{x_{n_1+1}, x_{n_1+2}, \ldots\}$, say x_{n_2}. We can continue in this way ad infinitum, building the sequence $\{x_{n_1}, x_{n_2}, x_{n_3}, \ldots\}$, and note that each of these terms is a distinct term of the original sequence and each is strictly greater than $\mu - \varepsilon$. We conclude $\mu = \limsup\{x_n\}$. The proof that $\lambda = \liminf\{x_n\}$ is similar. □

Corollary 2.2.3. *If $\{x_n\} \subset \mathbb{R}$ is a bounded monotone sequence, $\{x_n\}$ converges to $\sup\{x_n\}$ or $\inf\{x_n\}$ according as $\{x_n\}$ is increasing or decreasing.*

PROOF: Assume $\{x_n\}$ is increasing. Since this sequence is a bounded set of real numbers, $\mu = \sup\{x_n\}$ exists. For any $\varepsilon > 0$, $\exists n_\varepsilon$ s.t. $\mu - \varepsilon < x_{n_\varepsilon} \leqslant \mu$. It follows that $\forall n \geqslant n_\varepsilon$, $\mu - \varepsilon < x_n \leqslant \mu$. This means that the sequence converges to μ. Analogously, if $\{x_n\}$ is decreasing, $\{x_n\}$ converges to $\inf\{x_n\}$. □

We shall adopt the convention that an increasing sequence of real numbers $\{x_n\}$ is one for which $i < j \Rightarrow x_i \leqslant x_j$, and we write $\{x_n\} \nearrow$. If $i < j \Rightarrow x_i < x_j$, then $\{x_n\}$ strictly increasing, and we write $\{x_n\} \uparrow$. If $i < j \Rightarrow x_i \geqslant x_j$, then $\{x_n\}$ is a decreasing sequence, and we write $\{x_n\} \searrow$. If $i < j \Rightarrow x_i > x_j$, then $\{x_n\}$ is strictly decreasing, and we write $\{x_n\} \downarrow$.

A *monotone* sequence is either an increasing sequence or a decreasing one; *strictly monotone* refers to the strictly increasing or strictly decreasing sequences.

Theorem 2.2.4 (Bolzano–Weierstrass). *Let S be a bounded infinite set of real numbers. Then S has an accumulation point p.*

PROOF: Since S is bounded, it is contained in some interval $[a, b] \subset \mathbb{R}$. If m is the midpoint of this interval, at least one of the intervals $[a, m], [m, b]$ contains infinitely many points of S. If the right-hand one does, choose it, otherwise choose the left-hand interval. Having chosen the half-interval, denote it by $[a_1, b_1]$, and then divide it in two. At least one of the two halves contains infinitely many points of S. If the right-hand one does, choose it; otherwise, choose the left-hand one, and denote the one chosen by $[a_2, b_2]$. Follow this procedure ad infinitum; this process gives rise to a sequence of intervals $[a_1, b_1], [a_2, b_2], [a_3, b_3], \ldots$, such that the sequence of left endpoints $\{a_1, a_2, a_3, \ldots\}$ is increasing and the sequence of right endpoints $\{b_1, b_2, b_3, \ldots\}$ is decreasing. Each of these monotone sequences is bounded, so $\{a_n\} \nearrow \mu$, $\{b_n\} \searrow \lambda$, and it is not hard to see that $\mu \leq \lambda$. Since it is apparent that $\{(b_n - a_n)\} \to 0$, we must have $\mu = \lambda$. We assert that this number $\mu = \lambda$ is our candidate for p, an accumulation point of S.

Let $\varepsilon > 0$ be arbitrary. Then the interval $(p - \varepsilon, p + \varepsilon)$ must contain for some integer N the interval $[a_N, b_N]$. Hence, for each $\varepsilon > 0$, there are actually infinitely many points of S "within ε" of p, so p is an accumulation point of S. Note that p does not necessarily belong to S. □

Corollary 2.2.5. *Every bounded sequence $\{x_n\}_{n=1}^{\infty}$ of real numbers contains a convergent subsequence $\{x_{n_i}\}_{i=1}^{\infty}$.*

Before we prove this corollary, we make a few remarks about subsequences. If $\{x_n\}_{n=1}^{\infty}$ is a sequence, a subsequence is a subset of the given sequence; we usually denote the subsequence by $\{x_{n_i}\}_{i=1}^{\infty}$. But $\{x_{n_i}\}_{i=1}^{\infty} \subset \{x_n\}_{n=1}^{\infty}$ is more than just a subset; it is itself a sequence in which the relative positions of its elements are the same as they were in the original sequence. This is to say, in $\{x_{n_i}\}_{i=1}^{\infty}$, $x_{n_j} \prec x_{n_k} \Leftrightarrow n_j < n_k \Leftrightarrow j < k$. To prove the corollary, we must show explicitly the existence of a subsequence of $\{x_n\}$ which converges to some real number.

PROOF: By the previous theorem, there exists an accumulation point p of the set of distinct points of $\{x_n\}$ if $\{x_n\}$ contains infinitely many distinct points. (If $\{x_n\}$ contains a point x repeated infinitely often, we can extract a subsequence of the form $\{x_{n_i}\}_{i=1}^{\infty}$, where each x_{n_i} is equal to x. Clearly, this subsequence converges to x.) Suppose $\{x_n\}$ contains infinitely many distinct points and p is an accumulation point for this infinite set. Let $x_{n_1} \in \{x_n\}$ be a point of the sequence such that $|p - x_{n_1}| < 1$. Let x_{n_2} be a point of the subsequence $\{x_k\}_{k=n_1+1}^{\infty} \subset \{x_n\}$ such that $|p - x_{n_2}| < \frac{1}{2}$. In general, let $x_{n_{i+1}}$ be a point of the subsequence $\{x_k\}_{k=n_i+1}^{\infty} \subset \{x_n\}$ such that $|p - x_{n_{i+1}}| < 1/(i+1)$. It is apparent from the way we have chosen $x_{n_1}, x_{n_2}, \ldots$ that these points are all distinct terms of the original sequence $\{x_n\}$, that the order in which they occur is the same as the order in which they occur in the original sequence, that this method of choosing these points is valid since p is an accumulation point of the set of distinct points of $\{x_n\}$, and that $\{x_{n_i}\}_{i=1}^{\infty}$ is indeed a subsequence of $\{x_n\}$ which converges to p. □

2.3 LIMIT SUPERIOR AND LIMIT INFERIOR OF A FUNCTION

Let f be a real valued function on E^n.

Definition 2.3.1. L is the *limit superior* of f at $\mathbf{x}_0 \in E^n$,

$$L = \limsup_{\mathbf{x}\to\mathbf{x}_0} f(\mathbf{x}) = \overline{\lim_{\mathbf{x}_0}} f,$$

iff for each $\varepsilon > 0$ and each $\delta > 0$, there is a point $\mathbf{x}$ satisfying $0 < \rho(\mathbf{x}, \mathbf{x}_0) < \delta$ and $f(\mathbf{x}) > L - \varepsilon$, and furthermore, for any $\varepsilon > 0$, there is a $\delta > 0$ such that for all $\mathbf{x}$ satisfying $0 < \rho(\mathbf{x}, \mathbf{x}_0) < \delta$, $f(\mathbf{x}) < L + \varepsilon$.

l is the *limit inferior* of f at $\mathbf{x}_0 \in E^n$, written

$$l = \liminf_{\mathbf{x}\to\mathbf{x}_0} f(\mathbf{x}) = \underline{\lim_{\mathbf{x}_0}} f,$$

iff $\forall \varepsilon > 0$ and $\forall \delta > 0$, $\exists \mathbf{x}$ s.t.

$$0 < \rho(\mathbf{x}, \mathbf{x}_0) < \delta, \qquad f(\mathbf{x}) < l + \varepsilon,$$

and $\forall \varepsilon > 0$, $\exists \delta > 0$ s.t. $\forall \mathbf{x}$

$$0 < \rho(\mathbf{x}, \mathbf{x}_0) < \delta, \qquad f(\mathbf{x}) > l - \varepsilon.$$

Definition 2.3.2. We say that f is *upper semicontinuous* (u.s.c.) at $\mathbf{x}_0$ iff for each $\varepsilon > 0$, for all $\mathbf{x}$ sufficiently close to $\mathbf{x}_0$ we have $f(\mathbf{x}) < f(\mathbf{x}_0) + \varepsilon$. If for each $\varepsilon > 0$, we have that for all x sufficiently close to $\mathbf{x}_0$, $f(\mathbf{x}) > f(\mathbf{x}_0) - \varepsilon$, we say f is *lower semicontinuous* (l.s.c.) at $\mathbf{x}_0$. If f is semicontinuous at each point of a set S, it is semicontinuous on S, upper or lower as the case may be.

A simple example of a function which is upper, but not lower, semicontinuous at $x = 0$ is

$$f(x) = \begin{cases} \sin x, & x \neq 0, \\ 1, & x = 0. \end{cases}$$

For another example, $f(x) = |x|/x$, $x \neq 0$, is upper semicontinuous at $x = 0$ if $f(0) = 2$, lower semicontinuous if $f(0) = -1$, and neither if $f(0) = 0$.

Theorem 2.3.3. *f is upper semicontinuous at $\mathbf{x} = \mathbf{c}$ iff $\overline{\lim}_{\mathbf{c}} f \leqslant f(\mathbf{c})$. f is lower semicontinuous at $\mathbf{x} = \mathbf{c}$ iff $\underline{\lim}_{\mathbf{c}} f \geqslant f(\mathbf{c})$, provided $\mathbf{c} \in \operatorname{dom} f$ and $\overline{\lim}_{\mathbf{c}} f$, $\underline{\lim}_{\mathbf{c}} f$ exist.*

If $\underline{\lim}_{\mathbf{c}} f = \overline{\lim}_{\mathbf{c}} f = L$, then $\lim_{\mathbf{c}} f = L$, and conversely.

If f is continuous at $\mathbf{x} = \mathbf{c}$, it is both upper and lower semicontinuous at $\mathbf{x} = \mathbf{c}$, and conversely.

PROOF: If f is upper semicontinuous at $\mathbf{x}=\mathbf{c}$, then for each $\varepsilon>0$, $f(\mathbf{x})<f(\mathbf{c})+\varepsilon$ for all $\mathbf{x}$ sufficiently close to $\mathbf{c}$. Now $\overline{\lim}_{\mathbf{c}} f$ exists and equals L iff for each $\varepsilon>0$, $f(\mathbf{x})<L+\varepsilon$ for all $\mathbf{x}\neq\mathbf{c}$ sufficiently close to $\mathbf{c}$, and for each $\varepsilon>0$ and each $\delta>0$, there exists $\mathbf{x}$ satisfying $0<\rho(\mathbf{x},\mathbf{c})<\delta$ and $f(\mathbf{x})>L-\varepsilon$. This is to say, $\overline{\lim}_{\mathbf{c}} f$ is the largest number L such that for each $\varepsilon>0$ and each $\delta>0$, there is an $\mathbf{x}$ with $0<\rho(\mathbf{x},\mathbf{c})<\delta$ and $f(\mathbf{x})>L-\varepsilon$.

Suppose f is u.s.c. at $\mathbf{c}$. This implies that for all $\mathbf{x}$ in the domain of f and sufficiently close to $\mathbf{c}$, $f(\mathbf{x})$ is defined, and, if $\varepsilon_0>0$ is arbitrarily fixed, for all $\mathbf{x}\in\operatorname{dom} f$ sufficiently close to $\mathbf{c}$, $f(\mathbf{x})$ is bounded above by $f(\mathbf{c})+\varepsilon_0$. Let $B_\varepsilon=\{x: x\in\operatorname{dom} f,\ 0<|f(\mathbf{x})-f(\mathbf{c})|<\varepsilon\}$. If $\varepsilon\leqslant\varepsilon_0$, $\lambda=\sup\{f(\mathbf{x}):\mathbf{x}\in B_\varepsilon\}<f(\mathbf{c})+\varepsilon_0$. Note that Definition 2.3.1 implies that $L=\overline{\lim}_{\mathbf{c}} f=\inf_\varepsilon\{\lambda:\varepsilon>0\}$. It follows that $L\leqslant f(\mathbf{c})$.

If $L=\overline{\lim}_{\mathbf{c}} f=\inf_\varepsilon\{\lambda:\varepsilon>0\}$, where $\lambda=\sup\{f(\mathbf{x}):\mathbf{x}\in\operatorname{dom} f,\ 0<|f(\mathbf{x})-f(\mathbf{c})|<\varepsilon\}$, and $L\leqslant f(\mathbf{c})$, then for any $\varepsilon>0$, we have that for all $\mathbf{x}$ sufficiently close to $\mathbf{c}$, $f(\mathbf{x})<f(\mathbf{c})+\varepsilon$, so f is u.s.c. at $\mathbf{x}=\mathbf{c}$.

The proof that f is lower semicontinuous at $\mathbf{c}$ iff $\underline{\lim}_{\mathbf{c}} f\geqslant f(\mathbf{c})$ is similar. $\overline{\lim}_{\mathbf{c}} f=\underline{\lim}_{\mathbf{c}} f=L$ iff $\forall\varepsilon>0$ and $\forall\delta>0$, $\exists\mathbf{x}_1$, with $0<|\mathbf{x}_1-\mathbf{c}|<\delta$, and $\exists\mathbf{x}_2$, with $0<|\mathbf{x}_2-\mathbf{c}|<\delta$, s.t. $f(\mathbf{x}_1)>L-\varepsilon$ and $f(\mathbf{x}_2)<L+\varepsilon$, and $\forall\varepsilon>0$, $\exists\delta$ s.t. $\forall\mathbf{x}\in\operatorname{dom} f$, $0<|\mathbf{x}-\mathbf{c}|<\delta$, $L-\varepsilon<f(\mathbf{x})<L+\varepsilon$, iff $L=\lim_{\mathbf{c}} f$.

Finally, f is continuous at $\mathbf{c}$ iff $\forall\varepsilon>0$, $\exists\delta>0$, s.t. $\forall\mathbf{x}$, $|\mathbf{x}-\mathbf{c}|<\delta$, $f(\mathbf{c})-\varepsilon<f(\mathbf{x})<f(\mathbf{c})+\varepsilon$ iff f is u.s.c. at $\mathbf{c}$ and f is l.s.c. at $\mathbf{c}$. □

Theorem 2.3.4 (Weierstrass). *Let D be a closed and bounded set in E^n and f an upper semicontinuous function on D. Then f attains a maximum value on D.*

PROOF: To say $D\subset E^n$ is closed and bounded is to say that if $\{\mathbf{x}_n\}$ is any convergent sequence in E^n consisting of points of D, then the limit of this sequence is also in D, and D is contained in some ball centered at the origin of E^n. Suppose $\sup_{\mathbf{x}\in D} f(\mathbf{x})=M$. Then for each n, there is an $\mathbf{x}_n\in D$ such that $M-1/n\leqslant f(\mathbf{x}_n)\leqslant M$. This gives rise to a sequence $\{\mathbf{x}_n\}_{n=1}^\infty\subset D$, and $\lim_n f(\mathbf{x}_n)=M$. By Corollary 6.5.9, which extends Corollary 2.2.5 from E^1 to E^n, $\{\mathbf{x}_n\}$ contains a convergent subsequence $\{\mathbf{x}_{n_k}\}_{k=1}^\infty$, so $\{\mathbf{x}_{n_k}\}$ converges to $\mathbf{x}_0\in D$ and $\{f(\mathbf{x}_{n_k})\}_{k=1}^\infty\to M$. It may be that $M=\infty$; in any event we can write

$$M=\lim_{k\to\infty}\{f(\mathbf{x}_{n_k})\}\leqslant\overline{\lim_{k\to\infty}}\{f(\mathbf{x}_{n_k})\}\leqslant\overline{\lim_{\mathbf{x}_0}} f\leqslant f(\mathbf{x}_0)\leqslant M.$$

Hence $f(\mathbf{x}_0)=M$, and thus $M<\infty$. □

We comment that this theorem, known as the Weierstrass maximum theorem, has quite varied applications. Of course, there is the analogous theorem that a lower semicontinuous function attains a minimum on D. Compare this theorem with Theorem 6.7.3.

Definition 2.3.5. Suppose $\{f_n\}_{n=1}^{\infty}$ is a sequence of real-valued functions on E^n. We define four particular functions associated with the sequence as follows:

$\bigvee_{n=1}^{\infty} f_n$ is that function whose value at each $\mathbf{x} \in E^n$ is the supremum of the set of numbers $\{f_n(\mathbf{x})\}_{n=1}^{\infty}$.

$\bigwedge_{n=1}^{\infty} f_n$ is that function whose value at each $\mathbf{x} \in E^n$ is the infimum of the set of numbers $\{f_n(\mathbf{x})\}_{n=1}^{\infty}$.

$\bigvee_{n=1}^{\infty} \bigwedge_{k=n}^{\infty} f_k$ is that function λ (called the limit inferior of the sequence) whose value at each $\mathbf{x} \in E^n$ is the limit inferior of the sequence of numbers $\{f_n(\mathbf{x})\}_{n=1}^{\infty}$; i.e., $\lambda(\mathbf{x}) = \bigvee_{n=1}^{\infty} \bigwedge_{k=n}^{\infty} f_k(\mathbf{x})$.

$\bigwedge_{n=1}^{\infty} \bigvee_{k=n}^{\infty} f_k$ is that function μ (called the limit superior of the sequence) whose value at each $\mathbf{x} \in E^n$ is the limit superior of the sequence of numbers $\{f_n(\mathbf{x})\}_{n=1}^{\infty}$; i.e., $\mu(\mathbf{x}) = \bigwedge_{n=1}^{\infty} \bigvee_{k=n}^{\infty} f_k(\mathbf{x})$.

$\forall \mathbf{x} \in E^n$, $\lambda(\mathbf{x}) \leqslant \mu(\mathbf{x})$. If $\lambda(\mathbf{x}) \equiv \mu(\mathbf{x})$, we say the sequence converges pointwise to a function f, $f(\mathbf{x}) \equiv \lambda(\mathbf{x}) \equiv \mu(\mathbf{x})$.

PROBLEMS

1. Prove that every Cauchy sequence in E^n is bounded.

2. If $f:[a,b] \to \mathbb{R}$ is continuous, then f is bounded on $[a,b]$.

3. If $x \in \mathbb{Q}$, then $x = m/n$, where the fraction m/n is reduced to its lowest terms and $n > 0$, and thus m and n are uniquely determined. Define

$$f(x) = \begin{cases} n & \text{if } x \in \mathbb{Q} \text{ and } x = m/n \text{ in lowest terms}, \quad n > 0, \\ 0 & \text{if } x \text{ is irrational.} \end{cases}$$

Show that $f(x)$, although finite on $\mathbb{R}$, is not bounded on any interval, using the fact that both $\mathbb{Q}$ and $\mathbb{R} \setminus \mathbb{Q}$ are "dense" in $\mathbb{R}$ (i.e., every interval contains both rational and irrational numbers).

4. Suppose f: $E^2 \to \mathbb{R}$ is defined as follows:

$$f(x,y) = \begin{cases} \dfrac{\sin x - \sin y}{\tan x - \tan y}, & \tan x \neq \tan y, \\ \cos^3 x, & \tan x = \tan y. \end{cases}$$

Evaluate, if they exist, the three limits

$$\lim_{x\to 0}\Big(\lim_{y\to 0} f(x,y)\Big), \quad \lim_{y\to 0}\Big(\lim_{x\to 0} f(x,y)\Big), \quad \text{and} \quad \lim_{(x,y)\to(0,0)} f(x,y).$$

The first two are called *iterated* limits.

5. Suppose $f:[a,b]\to\mathbb{R}$ is monotonic and, of course, defined at each point of $[a,b]$. Prove that the points of discontinuity are at most a countable set.

6. Suppose $x_1>0$ and for each n, $x_{n+1}=x_1^{x_n}$. For what values of x_1 will $\lim_{n\to\infty}\{x_n\}$ exist?

7. Let $f:\mathbb{R}\to\mathbb{R}$. There are four basic discontinuities that might plague f.
(1) a removable discontinuity,
(2) a gap discontinuity,
(3) a jump discontinuity, and
(4) an infinite discontinuity.
Characterize each of these and give examples.
What can you say about $f(x)=\sqrt{(x^2-1)/(1-x^2)}$?

8. Suppose $f(x)=(2^x-1)/x$. Prove that $\lim_0 f$ exists. Then, having proved the existence, calculate it.
[*Hint*: Is 2^x everywhere differentiable?]

9. Let $f(x)=x$ if x is rational and $1-x$ if x is irrational. Prove $f:\mathbb{R}\to\mathbb{R}$ is continuous at exactly one point.

10. Suppose $f(x)=\cos^{2m}(n!\pi x)$. Calculate the iterated limit

$$\lim_{n\to\infty}\Big(\lim_{m\to\infty} f(x)\Big),$$

where m,n are positive integers. What about the other iterated limit?

11. Try to prove Theorem 2.1.6 using your ingenuity, instead of referring to Theorem 6.7.9.

12. Can you extend Definition 2.3.1 so as to include the possibility that $L=\overline{\lim}_{x_0} f = +\infty$? $L=-\infty$?

13. What can you say about

$$\lim_{n\to\infty}\frac{1-(1-1/n)^7}{1-(1-1/n)^6}?$$

$$\lim_{n\to\infty}(1/n)^{1/n}?$$

14. Let

$$f(x) = x^2 \text{ if } x \text{ is rational, 0 otherwise.}$$

Prove that f is continuous at $x=0$, differentiable at $x=0$, u.s.c. at each rational x, and l.s.c. at each irrational x.

15. Suppose $f:\mathbb{R}\to\mathbb{R}$ is such that for each real number b, $f^{-1}((-\infty,b))$ is an "open" set in $\mathbb{R}$. Show that f is upper semicontinuous.

16. Let $S_n=(1+1/n)^n$, $n=1,2,3,\ldots$. Show that

$$S_n=2+\frac{1}{2!}\left(1-\frac{1}{n}\right)+\frac{1}{3!}\left(1-\frac{1}{n}\right)\left(1-\frac{2}{n}\right)$$
$$+\cdots+\frac{1}{n!}\left(1-\frac{1}{n}\right)\cdots\left(1-\frac{n-1}{n}\right).$$

Show that $\forall n$, $2\leqslant S_n<3$ and that $\{S_n\}\uparrow$. This means $\{S_n\}$ converges; call the limit e. Show that for any real number α, $n\leqslant\alpha<n+1$,

$$(1+1/(n+1))^n<(1+1/\alpha)^\alpha<(1+1/n)^{n+1},$$

and then conclude that

$$\lim_{\alpha\to\infty}(1+1/\alpha)^\alpha=e.$$

Next, show that

$$\lim_{\alpha\to\infty}(1-1/\alpha)^{-\alpha}=e,$$

so that

$$\lim_{\alpha\to\pm\infty}(1+1/\alpha)^\alpha=e.$$

Finally, show that for any $x\in\mathbb{R}$,

$$\lim_{\alpha\to\infty}(1+x/\alpha)^\alpha=e^x.$$

17. We write $y=\ln x \Leftrightarrow x=e^y$; $x>0$. Then

$$\frac{\Delta y}{\Delta x}=\frac{1}{\Delta x}\ln\left(1+\frac{\Delta x}{x}\right)=\frac{1}{x}\ln\left(1+\frac{\Delta x}{x}\right)^{x/\Delta x}$$

Use Problem 16 to prove $(d/dx)\ln x=1/x$.

18. Prove that if $y=f(x)$ and $f'(x)$ exists, is continuous, and is not zero in an interval (a,b), then on this interval (a,b) we have

$$\frac{dx}{dy}=\left(\frac{dy}{dx}\right)^{-1},$$

where $dx/dy=(f^{-1}(y))'$ and is evaluated at $y=f(x)$.

19. Use Problem 18 to prove that $(d/dx)e^x=e^x$.

20. Let $a_n = n^{1/2n}$, $n = 1,2,3,\ldots$. Define $x_n = a_n - 1$. Show that $x_n \leqslant (\sqrt{n}-1)/n < 1/\sqrt{n}$ for each n. Next, show that $n^{1/n} < 1 + 2/\sqrt{n} + 1/n$ for each n. What then is $\lim_{n\to\infty} \sqrt[n]{n}$?

21. Find $\lim_{x\to\infty}\left\{\sqrt{(x+5)(x+7)} - x\right\}$.

22. Suppose f is continuous on $[a,b]$, differentiable on (a,b), and $f(a) = 0$, $f(x) > 0$ on $(a,b]$. Find a constant M, if one exists, such that for all $x \in (a,b)$

$$0 \leqslant |f'(x)/f(x)| \leqslant M.$$

23. Prove that every real number x has an essentially unique decimal expansion. Next, if $x < y$ are two real numbers, prove the existence of a rational number r, and an irrational number s, with $x < r < s < y$.
[*Hint*: Show that a number represented by an infinite decimal expansion is rational iff the expansion has a repeating cycle of $n \geqslant 1$ digits.]

24. Let S be a set in a partial order P. Prove that if $\sup S$ exists, it is unique.

25. Prove that

$$\lim_{n\to\infty} \frac{x_n - x}{x_n + x} = 0 \quad \Rightarrow \quad \lim_{n\to\infty} x_n = x.$$

26. Calculate

$$\lim_{n\to\infty} \sum_{k=1}^{n} \frac{k}{n^2},$$

$$\lim_{n\to\infty} \sum_{k=1}^{n} \frac{k^2}{n^3},$$

$$\lim_{n\to\infty} \sum_{k=1}^{n} \frac{1}{\sqrt{n^2+k}}.$$

27. If $\lim_p f = A$ and $\lim_p g = B$, prove that $\lim_p (f+g) = A + B$.

28. Find the following limits:

$$\lim_{x\to a} \frac{\sqrt{x-b} - \sqrt{a-b}}{x^2 - a^2}, \qquad (a<b);$$

$$\lim_{x\to\infty} \left(\sqrt{x^2+1} - \sqrt{x^2-1}\right);$$

$$\lim_{x\to\pi} \frac{1-\sin(x/2)}{[\cos(x/2)][\cos(x/4)-\sin(x/4)]};$$

$$\lim_{x\to -1^+} \frac{\sqrt{\pi} - \sqrt{\arccos x}}{\sqrt{x+1}};$$

$$\lim_{x\to 0}\frac{e^{x^2}-\cos x}{x^2};$$

$$\lim_{x\to 0}\left(\sqrt[3]{1+x^2}-\sqrt[4]{1-2x}\right)\Big/(x+x^2).$$

29. Suppose $f:\mathbb{R}\to\mathbb{R}$, $g:\mathbb{R}\to\mathbb{R}$ have limits A and B at $x=p$. Prove that $\lim_p f\cdot g = A\cdot B$ and $\lim_p f/g = A/B$ when $B\neq 0$.

30. Find the limits, if they exist:
(a) $\lim_{(x,y)\to(0,0)}(x^2+y^2)/\left(\sqrt{x^2+y^2+1}-1\right)$,
(b) $\lim_{(x,y)\to(0,0)}[1-\cos(x^2+y^2)]/(x^2+y^2)xy$,
(c) $\lim_{(x,y)\to(0,0)}[\sin(x^3+y^3)]/[x^2+y^2]$,
(d) $\lim_{(x,y)\to(0,0)}[1+x^2y^2]^{-1/(x^2+y^2)}$,
(e) $\lim_{(x,y)\to(0,0)}x^2y^2/(x^4+y^4)$,
(f) $\lim_{(x,y)\to(0,0)}(x^4-y^4)/(x^4+y^4)$.

31. Show that the sequence

$$\left\{\frac{n^n}{n!e^n}\right\}_{n=1}^{\infty}$$

is bounded and decreasing.

32. Show that

$$\lim_{n\to\infty}\frac{1}{n}\left[\frac{2\cdot 4\cdot\cdots\cdot(2n)}{1\cdot 3\cdot\cdots\cdot(2n-1)}\right]^2$$

does or does not exist.

33. Suppose $f(x)$ is strictly positive on the interval $[a,b]$. Can you prove that if f is continuous on $[a,b]$, then $1/f(x)$ is continuous on $[a,b]$?

34. Prove that the product of two continuous functions is continuous and that the product of two discontinuous functions may be continuous. Can you prove the same theorem if "product" is replaced by "sum"?

35. Find the domain of the function

$$f(x)=(3x+5)\left(\sqrt{2x-3}-\sqrt{5x-6}+\sqrt{3x-5}\right)^{-1},$$

and show that f is continuous on this domain.

36. We say f satisfies a Lipschitz condition of order α at x_0 if there exists a constant $M>0$ such that for all x in an interval containing x_0 we have $|f(x)-f(x_0)|\leqslant M|x-x_0|^{\alpha}$. Prove that if f satisfies a Lipschitz condition of order $\alpha>0$ at x_0, f is continuous there, but perhaps not differentiable at x_0, and if f satisfies a Lipschitz condition of order $\alpha>1$ at x_0, f is differentiable at x_0.

37. Find the following limits, if they exist,

$$\lim_{x\to 0}\left[\lim_{y\to 0} f(x,y)\right], \qquad \lim_{y\to 0}\left[\lim_{x\to 0} f(x,y)\right], \qquad \lim_{(x,y)\to(0,0)} f(x,y)$$

for the functions

(a) $f(x,y)=(x^2-y^2)/(x^2+y^2)$, $f(0,0)=0$,
(b) $f(x,y)=(1/x)\sin xy$, $f(0,y)=y$,
(c) $f(x,y)=(x+y)(\sin 1/x)(\sin 1/y)$, $f(0,y)=f(x,0)=0$,
(d) $f(x,y)=(x^2+y^2)/(x^4+y^4)$, $f(0,0)=0$,
(e) $f(x,y)=(1/x^2)(\sin x)(\sin y)$, $f(0,y)=1$,
(f) $f(x,y)=x/(x+y)$, $f(0,0)=\frac{1}{2}$.

38. Find the following limits, if they exist,

$$\lim_{x\to\infty}\left[\lim_{y\to\infty} f(x,y)\right], \qquad \lim_{y\to\infty}\left[\lim_{x\to\infty} f(x,y)\right], \qquad \lim_{(x,y)\to(\infty,\infty)} f(x,y)$$

for the functions

(a) $f(x,y)=(x+y)^{-1}$,
(b) $f(x,y)=x(x+y)^{-1}$,
(c) $f(x,y)=(1/y)\cos x$,
(d) $f(x,y)=\dfrac{x}{y^2}\displaystyle\sum_{n=1}^{[\![y]\!]}\sin\frac{n}{x}$,
(e) $f(x,y)=(-1)^{[\![x]\!]}x/(x+y)$,
(f) $f(x,y)=1/x+1/y$.

(The symbol $[\![x]\!]$ means the greatest integer which is less than or equal to x.)

39. Let $f(x)$be the function on $[0,1]$ such that $f(x)=1$ if x is rational and $\sin 1/x$ otherwise. For what values of x is f u.s.c.? Where is f l.s.c.?

40. Prove that if f is l.s.c. on a closed and bounded set $D\subset E^n$, then f attains a minimum value on D.

41. Exhibit a discontinuous function $g(x)$ which is upper semicontinuous on $[0,1]$, and don't be satisfied with a trivial one.

42. Suppose

$$f(x)=\begin{cases} e^x+1 & \text{if } x \text{ is rational,}\\ \cos x-1 & \text{if } x \text{ is irrational,}\end{cases}$$

for $x\in[0,\pi]$. Where is f semicontinuous, and where is f continuous?

43. If

$$a_n=\begin{cases}(1+4/n)^n, & n \text{ odd,}\\ (1-3/n)^{2n}, & n \text{ even,}\end{cases}$$

find $\overline{\lim}\{a_n\}$ and $\underline{\lim}\{a_n\}$.

44. Suppose $a_n = -\sqrt{n-1} + \sqrt{n+1}$, $n = 1,2,3,\ldots$. For what values of n is a_n rational? Is $\{a_n\}$ monotonic? Does $\lim_n\{a_n\}$ exist? If so, is the limit rational or irrational?

45. Express as a fraction the limit of $2.174123123123\ldots$.

46. Can you find a bounded integrable function $f(x)$ defined on $[0,10]$ such that $\forall x \in (0,10]$, $x\sin(1/x) = \int_0^x f(t)\,dt$? Explain.

47. Prove that every Cauchy sequence in $\mathbb{R}$ converges. Do the same for E^n. Now go back to Problem 32 and try it again.

CHAPTER

3

INEQUALITIES

3.1 SOME BASIC INEQUALITIES

A few classical inequalities are so fundamental that they should be included in every course in analysis. We present them in this chapter with proofs, although on occasion we shall leave a few of the details to the reader. Some of the inequalities have integral analogs; for these we have omitted the proofs. If you don't learn the proofs, which is likely, do learn the names of these inequalities.

Proposition 3.1.1 (Bernoulli's Inequality). $\forall x > -1$, $x \neq 0$, *and for all integers* $n > 1$

$$(1+x)^n > 1+nx.$$

In particular, $e^x > (1+x/n)^n > 1+x$.

PROOF: If $n=2$, $(1+x)^2 > 1+2x$ is certainly true. Assume the inequality holds for some $n \geqslant 2$. Then, since $x > -1$, $1+x > 0$, and we can write

$$(1+x)^{n+1} > (1+nx)(1+x) = 1+(n+1)x+nx^2 > 1+(n+1)x.$$

The proof by induction on n is complete. Note that equality holds if $n=1$ or $x=0$. □

Proposition 3.1.2 (The Geometric–Arithmetic (GAM) Mean Inequality). *Let* $x_1,\ldots,x_n$ *be a set of positive real numbers. Then*

$$\sqrt[n]{x_1x_2\cdots x_n} \leqslant \frac{1}{n}(x_1+x_2+\cdots+x_n).$$

Equality holds if and only if all the numbers are equal.

PROOF: If $x_1 = x_2 = \cdots = x_n$, equality obviously holds. Now suppose the numbers are not all equal and that they have been arranged so that $x_1 \leqslant x_2 \leqslant \cdots \leqslant x_{n-1} \leqslant x_n$. Since the natural logarithm function is strictly increasing the inequality of the theorem is equivalent to the inequality

$$\frac{1}{n}[\ln x_1 + \cdots + \ln x_n] \leqslant \ln\left[\frac{x_1 + \cdots + x_n}{n}\right].$$

The graph of $y = \ln x$ is concave down, so the graph of $y = \ln x$, $x_1 \leqslant x \leqslant x_n$, together with the line segment joining the points $(x_1, \ln x_1)$ and $(x_n, \ln x_n)$, is comprised by the boundary of a convex set in the xy plane; call this set S.

The point $\mathbf{p} = (\Sigma_{i=1}^n x_i/n, \Sigma_{i=1}^n (1/n)\ln x_i)$ is a point of S since $\Sigma_{i=1}^n(1/n) = 1$ and $1/n > 0$ together imply that $\mathbf{p}$ is in the convex hull of the points $\{(x_1, \ln x_1), \ldots, (x_n, \ln x_n)\}$, and S contains this convex hull. But this implies that

$$(\text{ordinate of } \mathbf{p}) = \sum_{i=1}^{n} \frac{1}{n} \ln x_i \leqslant \ln(\text{abscissa of } \mathbf{p}) = \ln \sum_{i=1}^{n} \frac{x_i}{n}.$$

Thus we have

$$\sqrt[n]{x_1 \cdots x_n} \leqslant \frac{1}{n} \sum_{i=1}^{n} x_i.$$

But we actually have strict inequality since we are assuming not all the x_i are equal. Let $\{(x_j, \ln x_j)\}_{j=1}^m$ be the set of distinct points, where $2 \leqslant m \leqslant n$; then $\mathbf{p}$ is the point

$$\mathbf{p} = \left(\sum_{j=1}^{m} \alpha_j x_j, \sum_{j=1}^{m} \alpha_j \ln x_j \right),$$

where $\Sigma_{j=1}^m \alpha_j = 1$ and each $\alpha_j > 0$.

$\mathbf{p}$ is in the convex hull of the m distinct planar points $(x_j, \ln x_j)$; this convex hull is a convex polygon with m vertices. Since each $\alpha_j > 0$, $\mathbf{p}$ lies in the interior of this polygon, a fact we ask the reader to prove. This means $\mathbf{p}$ lies in the interior of the convex set S, and the desired strict inequality follows. □

What follows as a consequence of the GAM inequality you have to see to believe. Suppose indeed that not all the positive numbers $\{x_i\}_{i=1}^n$ of the previous inequality are distinct, so we may write the GAM inequality in the form

$$x_1^{p_1} x_2^{p_2} \cdots x_m^{p_m} \leqslant \left(\frac{p_1 x_1 + \cdots + p_m x_m}{p_1 + \cdots + p_m} \right)^{p_1 + \cdots + p_m}, \qquad (*)$$

where the p_i are positive integers. It is not hard to see that the inequality will hold if the p_i are positive rational numbers as well, and a continuity argument would lead one to believe that the inequality holds if the p_i are positive real numbers, which it does. We call (*) the generalized form of the GAM inequality, where the x_i are distinct positive real numbers and the p_i are positive real numbers.

Now suppose we have m sets of n positive real numbers:

$$\begin{matrix} a_1 & a_2 & \cdots & a_n \\ b_1 & b_2 & \cdots & b_n \\ \vdots & \vdots & & \vdots \\ l_1 & l_2 & \cdots & l_n \end{matrix}$$

and m positive real numbers $\alpha, \beta, \ldots, \lambda$ such that $\alpha+\beta+\cdots+\lambda=1$. We can write

$$\frac{\sum_{i=1}^n a_i^\alpha b_i^\beta \cdots l_i^\lambda}{\left(\sum_{i=1}^n a_i\right)^\alpha \cdots \left(\sum_{i=1}^n l_i\right)^\lambda} = \sum_{i=1}^n \frac{a_i^\alpha}{\left(\sum_{i=1}^n a_i\right)^\alpha} \cdots \frac{l_i^\lambda}{\left(\sum_{i=1}^n l_i\right)^\lambda}.$$

The expression on the right-hand side of this equation, by (*), is less than or equal to

$$\sum_{i=1}^n \left[\frac{\alpha a_i/\sum_{i=1}^n a_i + \cdots + \lambda l_i/\sum_{i=1}^n l_i}{\alpha+\cdots+\lambda}\right]^{\alpha+\cdots+\lambda},$$

which expression reduces to 1 since $\alpha+\beta+\cdots+\lambda=1$. Therefore, we have

Proposition 3.1.3 (The Generalized Hölder Inequality). *If $\{a_i\}_{i=1}^n, \ldots, \{l_i\}_{i=1}^n$ are m sets of positive real numbers and $\{\alpha, \beta, \ldots, \lambda\}$ is a set of m positive real numbers whose sum is* 1, *then*

$$\sum_{i=1}^n a_i^\alpha b_i^\beta \cdots l_i^\lambda \leqslant \left(\sum_{i=1}^n a_i\right)^\alpha \left(\sum_{i=1}^n b_i\right)^\beta \cdots \left(\sum_{i=1}^n l_i\right)^\lambda.$$

equality holds when and only when the $a_i, b_i, \ldots, l_i$ are all proportional. Moreover, if we replace $a_i^\alpha, \ldots, l_i^\lambda$ by $a_i, \ldots, l_i$, respectively, and then replace $\alpha, \ldots, \lambda$ by $1/\alpha, \ldots, 1/\lambda$, respectively, requiring that $1/\alpha+\cdots+1/\lambda=1$ (where now each of $\alpha, \ldots, \lambda$ must be greater than 1), *we get the equivalent inequality*

$$\sum_{i=1}^n a_i b_i \cdots l_i \leqslant \left(\sum_{i=1}^n a_i^\alpha\right)^{1/\alpha} \cdots \left(\sum_{i=1}^n l_i^\lambda\right)^{1/\lambda}.$$

Equality holds if and only if

$$a_i^\alpha : b_i^\beta : \cdots : l_i^\lambda :: a_j^\alpha : b_j^\beta : \cdots : l_j^\lambda$$

for each $i, j=1,2,\ldots,n$.

When $m=2$, the latter form of the inequality is the ordinary Hölder inequality.

Proposition 3.1.4 (Cauchy's Inequality). *If $\{a_i\}_{i=1}^n$, $\{b_i\}_{i=1}^n$ are two sets of n real numbers, then*

$$\left(\sum_{i=1}^n a_i b_i\right)^2 \leqslant \left(\sum_{i=1}^n a_i^2\right)\left(\sum_{i=1}^n b_i^2\right).$$

PROOF: This is just the ordinary Hölder inequality with $\alpha=\beta=2$. □

There is even more to follow. A bit of fancy algebraic footwork occurs, so look sharp. First, we assume $p>1$ is an arbitrary fixed real number, and then we write

$$\sum_{i=1}^n (a_i+b_i)^p = \sum_{i=1}^n a_i(a_i+b_i)^{p-1} + \sum_{i=1}^n b_i(a_i+b_i)^{p-1}.$$

Now apply the Hölder inequality to the right-hand member of this equation to get

$$\sum_{i=1}^n (a_i+b_i)^p \leqslant \left(\sum_{i=1}^n a_i^p\right)^{1/p}\left(\sum_{i=1}^n (a_i+b_i)^p\right)^{(p-1)/p} + \left(\sum_{i=1}^n b_i^p\right)^{1/p}\left(\sum_{i=1}^n (a_i+b_i)^p\right)^{(p-1)/p}.$$

Divide this inequality through by $(\Sigma_{i=1}^n (a_i+b_i)^p)^{(p-1)/p}$ to get

Proposition 3.1.5 (Minkowski's Inequality). *If $\{a_i\}_{i=1}^n$, $\{b_i\}_{i=1}^n$ are two sets of n positive numbers and $p>1$, we have*

$$\left[\sum_{i=1}^n (a_i+b_i)^p\right]^{1/p} \leqslant \left(\sum_{i=1}^n a_i^p\right)^{1/p} + \left(\sum_{i=1}^n b_i^p\right)^{1/p}.$$

If $p=1$, equality obviously holds. We remark that if $0<p<1$, the inequality is reversed.

That so much could come out of the geometric–arithmetic mean inequality is really quite surprising. The author is indebted to H. F. Bohnenblust, who first showed me this development.

Proposition 3.1.6 (Young's Inequality). *Let f be a continuous strictly increasing real-valued function on $[0,\infty)$, with $f(0)=0$ and $\lim_{x\to\infty} f(x)=\infty$. Let $g=f^{-1}$ be the inverse function, and define for all $x\in[0,\infty)$*

$$F(x)=\int_0^x f(u)\,du, \qquad G(x)=\int_0^x g(v)\,dv.$$

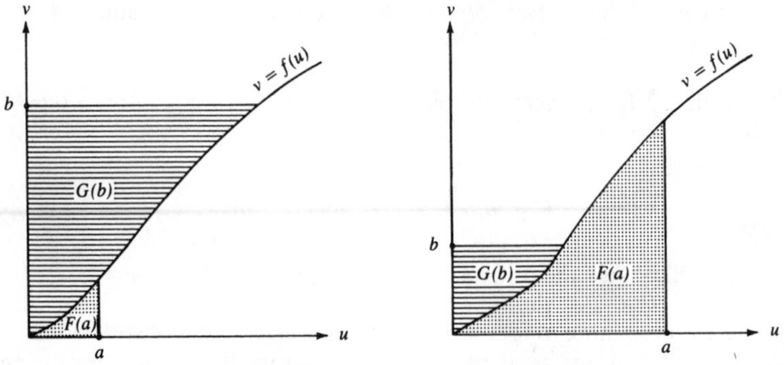

FIGURE 3.1 Young's inequality.

Then for any $a, b \in [0, \infty)$ *we have*

$$ab \leqslant F(a) + G(b).$$

Equality holds if and only if $b = f(a)$.

PROOF: It seems appropriate to prove this inequality with pictures; the contrast should be welcome. If the reader desires a more formal proof, Figure 3.1 will show the way. □

As an example of an application of Young's inequality we give

Proposition 3.1.7. *For* $p > 1$ *and* $q = p/(p-1)$, *so that* $1/p + 1/q = 1$, *if* a *and* b *are nonnegative real numbers, we have*

$$ab \leqslant a^p/p + b^q/q.$$

Equality holds if and only if $a^p = b^q$.

PROOF: Let $f(u) = u^{p-1}$ and $g(v) = v^{1/(p-1)}$. These functions on $[0, \infty)$ fulfill the necessary conditions, so we may apply Young's inequality, with $F(a) = a^p/p$, $G(b) = b^q/q$. Finally, $b = f(a) = a^{p-1}$ iff $b^q = a^p$. □

Proposition 3.1.8 (Jensen's Inequality). *Let* $\{x_i\}_{i=1}^n$ *be a fixed set of* n *positive numbers and* $p > 0$. *The function* $f(p)$ *defined on* $(0, \infty)$ *by*

$$f(p) = \left(\sum_{i=1}^{n} x_i^p \right)^{1/p}$$

is a positive decreasing function, and therefore if $0 < p_1 \leqslant p_2$, *we have the inequality*

$$\left(\sum_{i=1}^{n} x_i^{p_2} \right)^{1/p_2} \leqslant \left(\sum_{i=1}^{n} x_i^{p_1} \right)^{1/p_1}.$$

PROOF: That $f(p)$ is positive is evident. Now put

$$x=\left(\sum_{i=1}^{n} x_i^{p_2}\right)^{1/p_2}>0$$

so that $x^{p_2}=\sum_{i=1}^{n} x_i^{p_2}$. This implies that $x_i^{p_2}\leqslant x^{p_2}$, and thus $x_i \leqslant x$ for each $i=1,2,\ldots,n$. Since $p_2 \geqslant p_1>0$, it follows that

$$\sum_{i=1}^{n}\left(\frac{x_i}{x}\right)^{p_2} \leqslant \sum_{i=1}^{n}\left(\frac{x_i}{x}\right)^{p_1}.$$

The left-hand member of this inequality is equal to 1, so we get that

$$x^{p_1} \leqslant \sum_{i=1}^{n} x_i^{p_1},$$

which immediately lead to

$$\left(\sum_{i=1}^{n} x_i^{p_2}\right)^{1/p_2} \leqslant \left(\sum_{i=1}^{n} x_i^{p_1}\right)^{1/p_1}.$$

This is the Jensen inequality, and it means that the function $f(p)$ is indeed decreasing. Since $f(p)>0$, we can conclude that $f(p)$ approaches a limit as $p \rightarrow \infty$. □

Proposition 3.1.9

$$\lim_{p\rightarrow\infty}\left(\sum x_i^p\right)^{1/p}=\max\{x_i\}.$$

PROOF: Each $x_i>0$ so for any $p>0$, $\sum_{i=1}^{n} x_i^p \leqslant n \max\{x_i^p\}=n(\max\{x_i\})^p$, so $f(p) \leqslant n^{1/p} \max\{x_i\}$. Letting $p \rightarrow \infty$ we get

$$\lim_{p\rightarrow\infty} f(p) \leqslant \max\{x_i\}.$$

But for any $p \geqslant 1$, $\sum_{i=1}^{n} x_i^p \geqslant (\max\{x_i\})^p$, so for all $p \geqslant 1$, $f(p) \geqslant \max\{x_i\}$. We conclude that $\lim_{p\rightarrow\infty} f(p)=\max\{x_i\}$. □

We can use this result to extend the Hölder and Minkowski inequalities slightly by exhibiting the limiting case as $p \rightarrow \infty$. The Hölder inequality is

$$\sum_{i=1}^{n} a_i b_i \leqslant \left(\sum_{i=1}^{n} a_i^p\right)^{1/p}\left(\sum_{i=1}^{n} b_i^q\right)^{1/q},$$

where $p>1$ and $q=p/(p-1)$. Letting $p \rightarrow \infty$ and then $q \rightarrow 1$, we have

$$\sum_{i=1}^{n}(a_i b_i) \leqslant (\max\{a_i\})\left(\sum_{i=1}^{n} b_i\right),$$

or letting $q \to \infty$ and then $p \to 1$ we have

$$\sum_{i=1}^{n} a_i b_i \leqslant \left(\sum_{i=1}^{n} a_i \right) (\max\{b_i\}).$$

For the Minkowski case we have

$$\max\{(a_i + b_i)\} \leqslant \max\{a_i\} + \max\{b_i\}.$$

These inequalities are quite obvious.

Proposition 3.1.10 (Hadamard's Inequality). *Let A be the $n \times n$ matrix with (real or complex) entries (a_{ij}) and A^{t} its transpose. Then*

$$|\det A^{t}|^2 = |\det A|^2 \leqslant \prod_{i=1}^{n} \left(\sum_{j=1}^{n} |a_{ij}|^2 \right),$$

Equality holds iff the rows (columns) are orthogonal.

PROOF: See Theorem 7.8.5. □

We bring this section on inequalities to a close by listing for reference the integral analogs for Hölder's, Cauchy's, Jensen's, and Minkowski's inequalities.

Proposition 3.1.11 (Hölder's Inequality)

$$\int_a^b |f(t)g(t)|\,dt \leqslant \left(\int_a^b |f(t)|^p\,dt \right)^{1/p} \left(\int_a^b |g(t)|^q\,dt \right)^{1/q},$$

$$1 \leqslant p, q \leqslant \infty, \quad \frac{1}{p} + \frac{1}{q} = 1.$$

Proposition 3.1.12 (Cauchy–Schwarz Inequality)

$$\left(\int_a^b |f(t)g(t)|\,dt \right)^2 \leqslant \left(\int_a^b |f(t)|^2\,dt \right) \left(\int_a^b |g(t)|^2\,dt \right).$$

Proposition 3.1.13 (Minkowski's Inequality)

$$\left(\int_a^b |f(t) + g(t)|^p\,dt \right)^{1/p} \leqslant \left(\int_a^b |f(t)|^p\,dt \right)^{1/p} + \left(\int_a^b |g(t)|^p\,dt \right)^{1/p},$$

$$1 \leqslant p \leqslant \infty.$$

Proposition 3.1.14 (Jensen's Inequality). *If $f(x)$ is a function, integrable on $[a, b]$, then*

$$F(p) = \left(\int_a^b |f(t)|^p\,dt \right)^{1/p}$$

is a function of p on $(0, \infty)$ *satisfying* $F(p) \leq F(q)(b-a)^{\frac{1}{p}-\frac{1}{q}}$, $0<p<q<\infty$, *and* $\lim_{p\to\infty} F(p)$ *is the essential supremum of* $|f(t)|$ *on* $[a, b]$.

Just what "essential supremum" means will be made clear later in Section 8.2.

If $0<p<1$ in Inequalities 3.1.11 and 3.1.13, the inequalities are reversed. Finally, we remark that the intervals of integration need not be finite, provided the integrals involved all converge. The same remark may be made regarding the sums from 1 to n; if the sums are convergent series, the inequalities remain valid.

PROBLEMS

1. Prove the Lagrange identity:

$$\left(\sum_{i=1}^{n} a_i b_i\right)^2 = \left(\sum_{i=1}^{n} a_i^2\right)\left(\sum_{i=1}^{n} b_i^2\right) - \sum_{1\leq i<j\leq n} (a_i b_j - a_j b_i)^2$$

for real numbers a_i, b_i.

2. Show that for positive numbers $\{a_i\}_{i=1}^n$ and $\{b_i\}_{i=1}^n$,

$$0<p<1 \Rightarrow \left[\sum_{i=1}^{n} (a_i+b_i)^p\right]^{1/p} \geq \left(\sum_{i=1}^{n} a_i^p\right)^{1/p} + \left(\sum_{i=1}^{n} b_i^p\right)^{1/p}.$$

3. Prove that for all x, $e^x \geq 1+x$.

4. Prove that if $a^2+b^2=1$, $c^2+d^2=1$, then

$$ac+bd\leq 1.$$

5. Show that for $0<x<1$ and every integer $n>0$,

$$x^{n+1}+1/x^{n+1} > x^n + 1/x^n > 2.$$

6. Let r be a real number. Show that

$$\left(\sum_{k=1}^{n} k^r\right)^n > n^n (n!)^r.$$

7. Suppose $\{x_i\}_{i=1}^n$ is a set of positive numbers such that $\sum_{i=1}^n x_i = c$, a constant. Let $\{k_i\}_{i=1}^n$ be a set of fixed positive integers. Find the maximum value of

$$P = \prod_{i=1}^{n} x_i^{k_i}.$$

8. For $0<p<1$, show that Inequality 3.1.11 is reversed. (In this case, $q<0$.)

9. If $0<p<1$ and $1/p+1/q=1$, show that

$$\sum_{i=1}^{n} a_i b_i \geqslant \left(\sum_{i=1}^{n} a_i^p\right)^{1/p}\left(\sum_{i=1}^{n} b_i^q\right)^{1/q},$$

where the as and bs are positive real numbers.

10. Prove that if $0<m<1$ and m is rational, for positive numbers a_i,

$$\frac{1}{n}\sum_{i=1}^{n} a_i^m \leqslant \left[\frac{\sum_{i=1}^{n} a_i}{n}\right]^m.$$

If $m>1$, the inequality is reversed.

11. Show that $2^n(n!)<(n+1)^n<[(n+1)!]^2/(n+1)$.

12. Show that $1\cdot 3\cdot 5\cdot \cdots \cdot(2n-1)<n^n$.

13. Show that $(n!)^3<n^n[(n+1)/2]^{2n}$.

14. Prove that for positive numbers a, b,

$$a^b b^a \leqslant [(a+b)/2]^{a+b}.$$

15. If each $a_i>0$, show that

$$\left[\frac{\sum_{i=1}^{n} a_i}{n}\right]^{\sum_{i=1}^{n} a_i} \leqslant \prod_{i=1}^{n} a_i^{a_i}.$$

16. Prove that if $0<c<b<a$, then

$$\left(\frac{a+c}{a-c}\right)^a < \left(\frac{b+c}{b-c}\right)^b.$$

17. Prove that if $0<n<m$, $(x^m+y^m)^n<(x^n+y^n)^m$, where $x, y>0$.

18. Suppose $n>1$ is an integer. Show that $2^n>1+n2^{(n-1)/2}$.

19. Suppose $\{a_i\}_{i=1}^n$, $\{b_i\}_{i=1}^n$ are real numbers and $B_k=\sum_{j=1}^k b_j$, $k=1,\ldots,n$; show that

$$\sum_{k=1}^{n} a_k b_k = a_n B_n - \sum_{k=1}^{n-1}(a_{k+1}-a_k)B_k.$$

20. Prove Proposition 3.1.7 by using the generalized form of the GAM inequality. [*Hint*: Let $a=\alpha^{1/p}$ and $b=\beta^{1/q}$ and apply $(*)$.]

21. Let $A=(a_{ij})$ be an $n\times n$ matrix. Prove that $|\det A|\leqslant M^n n^{n/2}$, where $\forall i, j$, $M\geqslant |a_{ij}|$. Show by a simple example that equality may hold.

22. Show that

(a) $x(1+x)^{-1} < \ln(1+x) < x : -1 < x,\ x \neq 0.$

(b) $\alpha(x-1) < x^{\alpha} - 1 < \alpha x^{\alpha-1}(x-1) : 1 < x,\ 1 < \alpha.$

(c) $1 + x/2\sqrt{1+x} < \sqrt{1+x} < 1 + x/2 : -1 < x,\ x \neq 0.$

(d) If $a, b, p, q > 0$, $p + q = 1$, then $a^p b^q \leqslant pa + qb$. When does strict inequality hold?

23. Let $x_1 = \sqrt{2}$ and define recursively $x_{n+1} = \sqrt{2 + x_n}$. Prove that $\forall n,\ x_n < x_{n+1}$. Show that if $x_n \geqslant 2$ for $n > 1$, then $x_{n-1} \geqslant 2$. Does $\lim\{x_n\}$ exist? Find it.

Repeat the problem, replacing 2 by an arbitrary $c > 0$.

24. Let $0 < y_1 < x_1$, and define $x_{n+1} = (x_n + y_n)/2$, $y_{n+1} = \sqrt{x_n y_n}$. Show that

(a) $y_n < y_{n+1} < x_1,$

(b) $y_1 < x_{n+1} < x_n,$

(c) $0 < x_{n+1} - y_{n+1} < (x_1 - y_1)/2^n,$

(d) $\lim\{x_n\} = \lim\{y_n\}.$

CHAPTER

4

LINEAR SPACES

4.1 LINEAR AND AFFINE MAPPINGS

There can be little doubt that the reader has had a formal introduction to linear algebra and matrix theory. However, we would be remiss if we did not include for reference some of the basic theory. In Chapter 1 we defined axiomatically a linear space V over a field F, as well as the more general structure, a linear set (module); in this chapter we present a collection of definitions and theorems, leaving many of the proofs to the reader. For the most part, we confine our attention to real linear spaces—spaces whose underlying scalar field is $\mathbb{R}$.

If $V(\mathbb{R})$ is such a space, recall that $S \subset V$ is a linearly independent set of vectors if S contains at least two elements, and no vector (point) of S is a linear combination of other vectors of S. No set S which contains the zero vector can be a linearly independent set.

If $S \subset V$ is a nonempty set, the span of S, $\operatorname{sp} S$, is the set of all linear combinations that can be formed from the finite subsets of S. It is easy to show that $\operatorname{sp} S$ is the smallest linear space that contains S.

If $W \subset V$, then W is a subspace of V iff

$$\forall \mathbf{x}, \mathbf{y} \in W, \quad \forall \alpha, \beta \in \mathbb{R}, \qquad \alpha \mathbf{x} + \beta \mathbf{y} \in W.$$

This is to say, a subspace is a set $W \subset V$ which is closed with respect to linear combinations. Note that a subspace must contain the zero vector. We call $\{\mathbf{0}\}$ and V improper subspaces of V; the others, if any, are proper.

If $\{W_\alpha\}$ is a collection of subspaces of V, then $\bigcap_\alpha W_\alpha$ is also a subspace of V. If A and B are subspaces of V, then the set $\alpha A + \beta B = \{\mathbf{v} \in V : \mathbf{v}$ is of the form $\alpha \mathbf{a} + \beta \mathbf{b}$, $\mathbf{a} \in A$, $\mathbf{b} \in B$, $\alpha, \beta \in \mathbb{R}\}$ is also a subspace of V, since the set $\alpha A + \beta B$ is exactly the same set as $\operatorname{sp}(A, B)$. However, the set $A \cup B$ is generally not a subspace.

Finally, if V, W are linear spaces over $\mathbb{R}$ and $T: V \to W$ is a mapping, T is linear iff $\forall \alpha, \beta \in \mathbb{R}$, $\forall \mathbf{x}, \mathbf{y} \in V$, $T(\alpha\mathbf{x} + \beta\mathbf{y}) = \alpha T\mathbf{x} + \beta T\mathbf{y}$. This is to say, T preserves linear combinations, or the image under T of a linear combination is the same linear combination of the images. Note that it makes no sense to talk about the linearity of a mapping T unless $\operatorname{dom} T$ and $\operatorname{im} T$ have linear structures with a common scalar set.

Theorem 4.1.1. *Suppose V is a linear space and H, H' are two Hamel bases for V. Then* card $H =$ card H'.

PROOF: Whether H, H' are cardinally equivalent or not, certainly one of them is equivalent to a subset of the other. Without loss of generality, assume that there is an injectivity from H into H'.

To complete the proof, we need the aid of several lemmas.

Lemma 4.1.2. *If V is a vector space with a basis, then any linear map on V is completely determined by its action on the basis.*

PROOF OF THE LEMMA: If $\{\mathbf{e}_\alpha\} \subset V$ is the basis and T a mapping, and if for each α, $T\mathbf{e}_\alpha$ is known, then, since any $\mathbf{x} \in V$ is a linear combination of certain vectors from $\{\mathbf{e}_\alpha\}$ and $T\mathbf{x}$ is the same linear combination of the images of these certain basis vectors, $T\mathbf{x}$ is known. □

Lemma 4.1.3. *If T is a linear map on V, T is injective iff* $T\mathbf{x} = \mathbf{0} \Rightarrow \mathbf{x} = \mathbf{0}$.

PROOF OF THE LEMMA: If T is injective, then T can map at most one vector to the zero vector $\mathbf{0}$. Since T must already map $\mathbf{0}$ to $\mathbf{0}$, $T\mathbf{x} = \mathbf{0} \Rightarrow \mathbf{x} = \mathbf{0}$. Conversely, if $T\mathbf{x} = \mathbf{0} \Rightarrow \mathbf{x} = \mathbf{0}$, consider any two vectors $\mathbf{x}, \mathbf{y} \in V$, and suppose $T\mathbf{x} = T\mathbf{y}$. Then $T\mathbf{x} - T\mathbf{y} = T(\mathbf{x} - \mathbf{y}) = \mathbf{0} \Rightarrow \mathbf{x} - \mathbf{y} = \mathbf{0} \Rightarrow \mathbf{x} = \mathbf{y}$. This means T is injective. □

PROOF OF THEOREM 4.1.1. As a consequence of this last lemma, it is clear that if T is an injective linear map on V and $S \subset V$ is a linearly independent set, then TS is a linearly independent set in TV.

Since an injectivity exists from H into H', we define a linear endomorphism on V in the following way. If $H = \{\mathbf{e}_\alpha\}$, let $T\mathbf{e}_\alpha = \mathbf{e}'_\alpha$, where $\mathbf{e}'_\alpha \in H'$ is the image of $\mathbf{e}_\alpha$ under the injectivity. Now $T: V \to TV$ is completely defined, and T is injective, since the only vector of V that gets mapped to $\mathbf{0}$ is $\mathbf{0}$ itself. (Why?) Recall that TV is a subspace of V and that a basis for a linear space is a maximal linearly independent set. Since H is a maximal linearly independent set in V, TH is a linearly independent set in TV. In fact, TH is a basis for TV; this is easy to show. If $\mathbf{y} \in TV$, since $T: V \to TV$ is a linear bijectivity, there is a unique $\mathbf{x} \in V$ with $T\mathbf{x} = \mathbf{y}$. Hence $\mathbf{y}$ has a unique representation in terms of elements of TH.

Now suppose there exists a point $\mathbf{e}' \in H' \setminus TH$, which is to say, suppose TH is a proper subset of $H' = \{\mathbf{e}'_\beta\}$. For some finite subset of H we have

$$\mathbf{e}' = a_1 \mathbf{e}_{\alpha_1} + \cdots + a_k \mathbf{e}_{\alpha_k} \in V,$$

so

$$T\mathbf{e}' = a_1 \mathbf{e}'_{\alpha_1} + \cdots + a_k \mathbf{e}'_{\alpha_k} \in TV.$$

Since H is a basis for V, TH is a basis for TV, and so is $T(TH)$, so there is a finite subset of TTH such that

$$T\mathbf{e}' = b_1 T\mathbf{e}'_{\beta_1} + \cdots + b_m T\mathbf{e}'_{\beta_m}.$$

This implies that $\mathbf{e}' = b_1 \mathbf{e}'_{\beta_1} + \cdots + b_m \mathbf{e}'_{\beta_m}$, i.e., that $\mathbf{e}'$ is a linear combination of points of TH, which is impossible. We conclude $TH = H'$. It follows that $T: V \to V$ is a linear automorphism and that $T: H \to H'$ is a bijectivity. This completes the proof of the theorem. □

Definition 4.1.4. If V is a linear space, $W \subset V$ a subspace, and $\mathbf{x} \in V$, then the set $A = W + \mathbf{x}$ is called an *affine* subspace of V. The space A is said to be parallel to W and is called the *translation* of W by $\mathbf{x}$ if $\mathbf{x} \notin W$ (if $\mathbf{x} \in W$, $A = W + \mathbf{x} = W$).

Theorem 4.1.5. *Let W be a subspace of the linear space $V(\mathbb{R})$. Then*

1. $\forall \mathbf{x} \in V$, $\mathbf{x} \in W + \mathbf{x}$.
2. *If* $\mathbf{y} \in W + \mathbf{x}$, *then* $W + \mathbf{y} = W + \mathbf{x}$.
3. $\forall \mathbf{x}, \mathbf{y} \in V$, $(W + \mathbf{x}) \cap (W + \mathbf{y}) = \varnothing$, *or* $(W + \mathbf{x}) = (W + \mathbf{y})$.
4. $\mathbf{x}$ *and* $\mathbf{y}$ *are in the same translate of* $W \Leftrightarrow \mathbf{x} - \mathbf{y} \in W$.
5. *If* $\{\mathbf{x}_i\}_{i=1}^n$ *is a set of distinct vectors in* V, *and* $A(\mathbf{x}_1, \ldots, \mathbf{x}_n)$ *is the set of all vectors of the form*

$$\mathbf{x} = \sum_{i=1}^n \xi_i \mathbf{x}_i, \qquad \textit{where} \quad \sum_{i=1}^n \xi_i = 1,$$

then $A(\mathbf{x}_1, \ldots, \mathbf{x}_n)$ is an affine subspace of V, the smallest affine subspace which contains all the vectors $\{\mathbf{x}_i\}_{i=1}^n$.

PROOF: We leave the proofs of the first four parts to the reader. For the last, note that $\{\mathbf{x}_i - \mathbf{x}_1\}_{i=2}^n$ is a set of $n-1$ distinct nonzero vectors; let W be the span of these vectors. Evidently, $\{\mathbf{x}_i\}_{i=1}^n \subset W + \mathbf{x}_1$, and W is the set of all vectors of the form $w = \sum_{i=2}^n \xi_i (\mathbf{x}_i - \mathbf{x}_1)$, where the ξ_i are arbitrary real numbers. Therefore, the vectors of $W + \mathbf{x}_1$ are of the form

$$\mathbf{x} = \mathbf{x}_1 + \sum_{i=2}^n \xi_i (\mathbf{x}_i - \mathbf{x}_1) = \left(1 - \sum_{i=2}^n \xi_i\right) \mathbf{x}_1 + \sum_{i=2}^n \xi_i \mathbf{x}_i.$$

Now add the n coefficients of $\mathbf{x}_1, \mathbf{x}_2, \ldots, \mathbf{x}_n$; the sum is 1. Hence each vector of the affine subspace $W+\mathbf{x}_1$, which space contains the points $\{\mathbf{x}_i\}_{i=1}^n$, is a vector of the set $A(\mathbf{x}_1,\ldots,\mathbf{x}_n)$.

On the other hand, if $\mathbf{x}\in A(\mathbf{x}_1,\ldots,\mathbf{x}_n)$, we can write

$$\mathbf{x}-\mathbf{x}_1=\sum_{i=1}^{n}\xi_i(\mathbf{x}_i-\mathbf{x}_1)=\sum_{i=2}^{n}\xi_i(\mathbf{x}_i-\mathbf{x}_1),$$

so $\mathbf{x}-\mathbf{x}_1$ is an element of W, which means $\mathbf{x}\in W+\mathbf{x}_1$. We conclude that $A(\mathbf{x}_1,\ldots,\mathbf{x}_n)=W+\mathbf{x}_1$.

To show that A is the smallest affine space containing the vectors $\{\mathbf{x}_i\}_{i=1}^n$, we first observe from properties 2 and 3 that if $\{\mathbf{x}_i\}_{i=1}^n\subset W+\mathbf{x}_1$, for any subspace W, then $W+\mathbf{x}_1=W+\mathbf{x}_i$ for each $i=1,\ldots,n$. Suppose $\{\mathbf{x}_i\}_{i=1}^n\subset W'+\mathbf{x}_1$, a smaller affine subspace. Then for each $i=1,\ldots,n$, $(\mathbf{x}_i-\mathbf{x}_1)\in W'$. Since $W=\operatorname{sp}\{(\mathbf{x}_i-\mathbf{x}_1)\}_{i=1}^n$, $W\subset W'$, and we have $A(\mathbf{x}_1,\ldots,\mathbf{x}_n)\subset W'+\mathbf{x}_1$. The proof is complete. □

Definition 4.1.6. An *affine mapping* $A: V\to W$ from one linear space to another is one which preserves linear combinations up to a translation constant. This is to say, A is of the form

$$\forall \mathbf{x}\in V: \qquad A(\mathbf{x})=T(\mathbf{x})+\mathbf{c},$$

where $T: V\to W$ is a linear map and $\mathbf{c}\in W$ is a fixed vector.

Theorem 4.1.7. *Suppose $T: V\to W$ is linear. Then*

1. $T(\mathbf{0})=\mathbf{0}$; *T preserves linear combinations.*
2. *T preserves convexity.*
3. *For all subspaces $S\subset V$, TS is a subspace of W.*
4. *For all subspaces $U\subset W$, $T^{-1}(U)$ is a subspace of V.*
5. *For any set $S\subset V$, $T(\operatorname{sp} S)=\operatorname{sp} TS$.*
6. *If T is injective, T^{-1}: $TV\to V$ is linear.*

We omit the proof.

Definition 4.1.8. If $\varphi: X\to Y$ is a mapping from a linear set to another, then the set

$$\varphi^{-1}(\mathbf{0})=\{\mathbf{x}\in X: \varphi(\mathbf{x})=\mathbf{0}\}$$

is called the *kernel* of φ, and we write $\ker\varphi$. If φ is a linear map, we sometimes call $\ker\varphi$ the null space of φ, since in this case $\ker\varphi$ is a subspace of X.

We now present a famous old theorem, which you all will no doubt remember. Recall that a linear space V is of dimension n iff it admits a basis consisting of n vectors; we write, in this case, $\dim V=n$.

Theorem 4.1.9. *Let* $T: V \to W$ *be a linear mapping, with* $\dim V < \infty$. *Then*

$$\dim(\ker T) + \dim(\operatorname{im} T) = \dim V.$$

PROOF: We sketch a proof. Since $\ker T$ is a subspace of V, it admits a basis $\{\mathbf{e}_1, \ldots, \mathbf{e}_j\}$, unless $\ker T = \{\mathbf{0}\}$, in which case T is injective. Let $\dim V = n$. If T is injective, then T: $V \to \operatorname{im} T$ is a bijective map, and it is easy to show that $\dim(\operatorname{im} T) = n$. If the basis for $\ker T$ consists of n vectors, T is the zero mapping, and $\dim(\operatorname{im} T) = 0$.

Extend the basis $\{\mathbf{e}_1, \ldots, \mathbf{e}_j\}$ for $\ker T$ to get a basis for V; $\{\mathbf{e}_1, \ldots, \mathbf{e}_j, \mathbf{e}_{j+1}, \ldots, \mathbf{e}_n\}$. The vectors $T\mathbf{e}_{j+1}, \ldots, T\mathbf{e}_n$ must be linearly independent since none is zero; hence their span is of dimension $n - j$. Now let $\mathbf{y} \in \operatorname{im} T$, $\mathbf{y} \neq \mathbf{0}$, and suppose $\mathbf{x}$ is such that $T\mathbf{x} = \mathbf{y}$. Write $\mathbf{x} = \sum_{i=1}^{n} \xi_i \mathbf{e}_i$; it follows that $\mathbf{y} = \sum_{i=j+1}^{n} \xi_i T\mathbf{e}_i$, and this representation can be shown to be unique. Hence $\{T\mathbf{e}_i\}_{i=j+1}^{n}$ is a basis for $\operatorname{im} T$.

The desired equation follows. □

It is customary to call the dimension of the image of T the *rank* of T, and the dimension of the kernel the *nullity* of T. Hence we could rephrase the statement of the theorem to read: for a linear map from a finite dimensional linear space V into W, rank + nullity = $\dim V$.

Suppose V, W are finite-dimensional real linear spaces, and that we have established bases for these spaces, $\{\mathbf{v}_1, \ldots, \mathbf{v}_m\}$ and $\{\mathbf{w}_1, \ldots, \mathbf{w}_n\}$, respectively. We have already seen that V and W are linearly isomorphic, respectively, to the Cartesian spaces $\mathbb{R}^m$ and $\mathbb{R}^n$, and the natural identification is if $\mathbf{v} = a_1\mathbf{v}_1 + \cdots + a_m\mathbf{v}_m \in V$, then

$$\mathbf{v} \leftrightarrow (a_1, \ldots, a_m) \in \mathbb{R}^m.$$

Furthermore, Lemma 4.1.2 states that if we can specify the action of a mapping $T: V \to W$ at each of the basis vectors of V and then specify that T be linear, we have a linear map over all of V for which all the values are known. This suggests a system for representing a particular linear map T: $V \to W$.

Since we are presuming that we know the m vectors in W $T\mathbf{v}_1, T\mathbf{v}_2, \ldots, T\mathbf{v}_m$, associate each of these m vectors with its corresponding n-tuple from $\mathbb{R}^n$, and write this n-tuple as a column of n numbers. For example, if $T\mathbf{v}_i = a_{1i}\mathbf{w}_1 + \cdots + a_{ni}\mathbf{w}_n$, then write the n-tuple $(a_{1i}, a_{2i}, \ldots, a_{ni})$ as the ith column of a rectangular array of numbers:

$$\begin{bmatrix} a_{11} & a_{12} & \cdots & a_{1i} & \cdots & a_{1m} \\ a_{21} & a_{22} & \cdots & a_{2i} & \cdots & a_{2m} \\ \vdots & \vdots & & \vdots & & \vdots \\ a_{n1} & a_{n2} & \cdots & a_{ni} & \cdots & a_{nm} \end{bmatrix}$$

This rectangular array of numbers indicates clearly what the image is of each of the m basis vectors in V; hence this array, called a *matrix*, seems to be a practical representation for the linear map $T: V \to W$ we had in mind.

Theorem 4.1.10. *The set of linear mappings $L(V, W)$ from a real vector space V of dimension m into a real linear space W of dimension n, is itself a real linear space which is linearly isomorphic to the real linear space $\mathfrak{M}$ of $n \times m$ matrices with real entries. Their common dimension is mn. In the event that $V = W$, these two linear spaces, $L(V)$ and $\mathfrak{M}$, are nonabelian algebras, and as such they are isomorphic.*

The proof of this theorem is a standard part of a course in linear algebra and will be omitted here.

One important class of linear mappings is $L(V, \mathbb{R})$, where $\mathbb{R}$ plays the role of the image space, but is really the underlying scalar set of the vector space V. We call this space of maps the *dual space* of V, and we refer to the elements of $L(V, \mathbb{R})$ as *linear functionals* on V. The space $L(V, V)$, which we denote $L(V)$, is the linear algebra of endomorphisms on V. The secondary operation of the algebra is *composition*. This is to say, if S, $T \in L(V)$, then for each $\mathbf{v} \in V$, $(S \circ T)(\mathbf{v}) = S(T(\mathbf{v}))$. Of course, $(S \oplus T)(\mathbf{v}) = S(\mathbf{v}) + T(\mathbf{v})$ defines the primary operation. If V is n-dimensional, and M_S, M_T are the matrices representing S, T, respectively, then the matrix product $M_S \cdot M_T$ represents the composite map $S \circ T$, and the matrix sum $M_S + M_T$ represents the sum $S \oplus T$. A fascinating problem is the following: if V is a linear space and $T \in L(V)$, what are the invariant subspaces of V with respect to T?

Definition 4.1.11. If $T: V \to V$ is linear, and $W \subset V$ is a subspace, we call W an *invariant subspace* (relative to T) if for each $\mathbf{w} \in W$, $T\mathbf{w} \in W$.

When V is finite dimensional, we can, in theory, find the one-dimensional invariant subspaces by finding the nonzero vectors $\mathbf{v} \in V$ such that $T\mathbf{v}$ is a scalar multiple of $\mathbf{v}$. Such vectors are called *eigenvectors*. The linear spans of one or more of these eigenvectors are invariant subspaces, but there may be other invariant subspaces. When V is infinite dimensional, the problem becomes enormously complex. Suppose V is the space of real-valued functions integrable over the interval $[a, b]$, and $K(x, t)$ is a fixed continuous function over $[a, b] \times [a, b]$. If $v(t) \in V$, we can transform the function v to another function v^* by the integral transformation $v^*(x) = \int_a^b K(x, t) v(t)\, dt$. It is evident that this transformation is linear. Suppose there is a real constant λ for which we can find a nonzero function $v \in V$ such that

$$\lambda v(x) = \int_a^b K(x, t) v(t)\, dt.$$

Then v is an eigenvector, or *eigenfunction* of the space V, relative to the linear map characterized by the "kernel function" $K(x, t)$.

4.2 CONTINUITY OF LINEAR MAPS

In Chapter 2 we took the linear space $\mathbb{R}^n$ and endowed it with the Euclidean metric. As a consequence, each vector of $\mathbb{R}^n$ has a length, namely, its distance from the zero vector. What we would like to do is to abstract this notion in order to assign lengths to vectors which don't seem to bear any resemblance at all to what we think vectors should be. For example, how does one assign a length to a real-valued function?

What we do is to consider what properties are intrinsic to the concept of "length of a vector" and list them as axioms. Certainly the length should be a nonnegative real number, in fact positive for any vector but the zero vector. Moreover, magnifying a vector should magnify its length by the same amount. Finally, following the model of $\mathbb{R}^n$ with the Euclidean length, the length of the sum of two vectors should not be greater than the sum of the lengths of each. We think these are all the properties necessary to characterize the notion of abstract length.

Definition 4.2.1. Let V be a linear space over $\mathbb{R}$ or $\mathbb{C}$. A real-valued function ν on V is a *norm* on V if it satisfies the following axioms.

N1. $\forall \mathbf{v} \in V$, $\nu(\mathbf{v}) \geqslant 0$.
N2. $\forall \mathbf{v} \in V$, $\forall \lambda \in \mathbb{R}(\in \mathbb{C})$, $\nu(\lambda \mathbf{v}) = |\lambda| \nu(\mathbf{v})$.
N3. $\forall \mathbf{v}, \mathbf{w} \in V$, $\nu(\mathbf{v}+\mathbf{w}) \leqslant \nu(\mathbf{v}) + \nu(\mathbf{w})$.
N4. $\nu(\mathbf{v}) = 0 \Leftrightarrow \mathbf{v} = \mathbf{0}$.

A norm is simply an "abstract length" assigner. If ν fails to satisfy axiom N4, but satisfies axioms N1–N3, ν is called a *semi-norm* on V.

A linear space V with a norm function ν is called a *normed linear space*, and our usual way of denoting the norm of a vector $\mathbf{v}$ will be to write $\|\mathbf{v}\|$ instead of $\nu(\mathbf{v})$. Note that if $\mathbb{R}$ and $\mathbb{C}$ are thought of as linear spaces, the ordinary absolute value $|\cdot|$ is a norm on these spaces. In the event that V is a linear algebra, and we have a norm $\|\cdot\|$ on V, for V to be a normed algebra (normed ring) we require in addition that the norm function satisfy

N5. $\forall \mathbf{v}, \mathbf{w} \in V$, $\|\mathbf{v} \cdot \mathbf{w}\| \leqslant \|\mathbf{v}\| \, \|\mathbf{w}\|$.
N6. If V has a unity $\mathbf{e}$, $\|\mathbf{e}\| = 1$.

A norm over an algebra which satisfies, in addition to the above six axioms,

N7. $\forall \mathbf{v}, \mathbf{w} \in V$, $\|\mathbf{v} \cdot \mathbf{w}\| = \|\mathbf{v}\| \, \|\mathbf{w}\|$

is quite special, and is sometimes called an *absolute value*, or a *valuation*, on V.

Note that in any normed space, the norm must satisfy this fundamental inequality:

$$\forall \mathbf{x}, \mathbf{y}: \qquad \|\mathbf{x}\| - \|\mathbf{y}\| \leqslant \big| \|\mathbf{x}\| - \|\mathbf{y}\| \big| \leqslant \|\mathbf{x} \pm \mathbf{y}\| \leqslant \|\mathbf{x}\| + \|\mathbf{y}\|.$$

If V is a normed linear space, we can regard the set

$$B_r = \{\mathbf{v} \in V : \|\mathbf{v}\| \leq r, r > 0\}$$

as a *ball*, centered at the origin of V and having radius r.

Definition 4.2.2. A set $A \subset V$, V an arbitrary normed linear space, is said to be *bounded* if there exists a real number r such that $A \subset B_r$.

A real-valued function f on the vector space E^n is said to be bounded if $\exists M \in \mathbb{R}$ s.t. $\forall \mathbf{x} \in E^n$, $|f(\mathbf{x})| \leq M$. If T is a linear vector-valued function from one normed linear space into another, T can never be bounded in this sense, unless T is the trivial zero mapping, for suppose $\|T\mathbf{x}\| = \varepsilon > 0$ for some $\mathbf{x} \neq \mathbf{0}$. If M is any positive real number, choose an integer n such that $n\varepsilon > M$. Then $T(n\mathbf{x}) = nT\mathbf{x}$ is such that $\|T(n\mathbf{x})\| = n\varepsilon > M$. Since we want something like the concept of a bounded function for linear maps we define a linear map to be bounded if it is bounded on the unit ball B_1.

Definition 4.2.3. Let $T: V \to W$ be a linear map from a normed linear space V into a normed linear space W. T is *bounded* iff $\exists M > 0$ s.t.

$$\forall \mathbf{v} \in V, \qquad \|\mathbf{v}\|_V \leq 1, \quad \|T\mathbf{v}\|_W \leq M$$

iff $\exists M > 0$ s.t.

$$\forall \mathbf{v} \in V, \qquad \|T\mathbf{v}\|_W \leq M \|\mathbf{v}\|_V$$

iff $\exists M > 0$ s.t.

$$\forall \mathbf{v} \in V, \qquad \|\mathbf{v}\|_V = 1, \quad \|T\mathbf{v}\|_W \leq M.$$

Note that if $\mathbf{v}$ is any nonzero vector of a normed space V, then $\mathbf{v}/\|\mathbf{v}\|$ is always a unit vector, i.e., a vector whose norm, or length, is 1.

Definition 4.2.4. Let T be a linear map from V into W, V and W normed. T is continuous at $\mathbf{p} \in V \Leftrightarrow \forall \varepsilon > 0, \exists \delta > 0$ s.t. $\forall \mathbf{x} \in V$, $\|\mathbf{x} - \mathbf{p}\|_V < \delta \Rightarrow \|T\mathbf{x} - T\mathbf{p}\|_W < \varepsilon$. T is uniformly continuous on $V \Leftrightarrow \forall \varepsilon > 0, \exists \delta > 0$ s.t. $\forall \mathbf{x}, \mathbf{y} \in V$, $\|\mathbf{x} - \mathbf{y}\|_V < \delta \Rightarrow \|T\mathbf{x} - T\mathbf{y}\|_W < \varepsilon$.

It should be apparent that the length of the vector $\mathbf{x} - \mathbf{y}$ can be interpreted as an "abstract distance" between the vectors $\mathbf{x}$ and $\mathbf{y}$.

Theorem 4.2.5. *Every linear $T: V \to W$ is bounded, where V and W are normed linear spaces and the dimension of V is finite.*

We must wait a bit for the proof. The impatient reader may refer to Theorem 6.5.13.

Theorem 4.2.6. *Let T be a linear mapping from one normed linear space V into another, W. Then T is continuous iff T is bounded.*

PROOF: Suppose $T: V \to W$ is bounded. $\exists M > 0$ s.t. $\forall \mathbf{x} \in V$, $\|T\mathbf{x}\|_W \leqslant M\|\mathbf{x}\|_V$. (Incidentally, the smallest such M that can exist is the number $\overline{M} = \sup\{\|T\mathbf{x}\|_W : \|\mathbf{x}\|_V = 1\}$) Now for any $\mathbf{x}, \mathbf{y} \in V$, we have $\|T(\mathbf{x}-\mathbf{y})\|_W \leqslant \overline{M}\|\mathbf{x}-\mathbf{y}\|_V$. Suppose $\varepsilon > 0$ is given. Choose $\delta = \varepsilon/(\overline{M}+1)$. Then $\|\mathbf{x}-\mathbf{y}\|_V < \delta \Rightarrow \|T\mathbf{x} - T\mathbf{y}\|_W < \varepsilon$. This is to say, T is uniformly continuous on V.

Now suppose T is not bounded. Then there must exist a sequence of unit vectors $\{\mathbf{u}_n\}$ in V such that for each n, $\|T\mathbf{u}_n\|_W > n$. This means that we have a sequence of vectors $\{\mathbf{u}_n/n\}$ such that for each n, $\|T(\mathbf{u}_n/n)\|_W > 1$. Since for each n, $\|\mathbf{u}_n/n\|_V = 1/n$, it is clear that T is not continuous at $\mathbf{0}$; hence T is not continuous on V. This completes the proof. □

Corollary 4.2.7. *If $T: V \to W$ is continuous at $\mathbf{0}$, T is uniformly continuous on V. Hence T is continuous at $\mathbf{0}$ iff T is bounded on $\mathbf{V}$ iff T is uniformly continuous on V.*

The previous theorem and its corollary allow us to restate Theorem 4.2.5.

Theorem 4.2.8. *If V is a finite-dimensional normed linear space, every linear map $T: V \to W$ is bounded and uniformly continuous on V.*

We have previously defined the dual space of a linear space $V(\mathbb{R})$ to be $L(V, \mathbb{R})$, where $\mathbb{R}$ is here regarded as a linear space over $\mathbb{R}$. In the event that V is normed, we may state

Definition 4.2.9

$$\forall f \in L(V, \mathbb{R}), \qquad \|f\| = \sup\{|f(\mathbf{v})| : \mathbf{v} \in V, \|\mathbf{v}\| = 1\}.$$

The reader can verify that this is indeed a norm on $L(V, \mathbb{R})$, in that the norm axioms are satisfied. However, there is a difficulty; it can be the case that for some $f \in L(V, \mathbb{R})$, $\|f\| = \infty$. Hence, in order to make a normed linear space out of $L(V, \mathbb{R})$, we throw out all the linear functionals f for which $\|f\| = \infty$ and keep the rest. We denote this pruned space by V^*, and call it the *continuous dual*, or *topological dual*, of V. As a set, $V^* \subset L(V, \mathbb{R})$; the containment is proper, in general, but if V is finite dimensional, then $V^* = L(V, \mathbb{R})$. In fact, if $\dim V = n$, then $\dim V^* = n$, as a consequence of Theorem 4.1.10. It is V^* in which we are usually most interested, and if we speak of the dual of a normed space V, it will be V^* that we mean. One refers to $L(V, \mathbb{R})$ as the *algebraic dual* of V; the dual of the dual of a normed linear space V, V^{**}, is what we mean by the *bidual* of V.

If V is a normed real linear space, there is a monomorphism $\varphi: V \to V^{**}$ which is very special in that for each $\mathbf{v} \in V$, $\|\mathbf{v}\|_V = \|\varphi(\mathbf{v})\|_{**}$. Such a mapping is called a *linear isometry*, and we say φ is an *embedding* of V into V^{**}. We don't ask the reader to prove that this is so; we simply state the fact. Note that for an arbitrary $\mathbf{v}\in V$, $f(\mathbf{v})$ is a scalar for each $f\in V^*$. The set of scalars $\{f(\mathbf{v}): f\in V^*\}$ defines a scalar-valued function $F_{\mathbf{v}}$ on V^*; i.e., for each $f\in V^*$, $F_{\mathbf{v}}(f)=f(\mathbf{v})$. It can be shown that $F_{\mathbf{v}}$ is linear, which means that $F_{\mathbf{v}}\in L(V^*,\mathbb{R})$. Define $\|\mathbf{F}_{\mathbf{v}}\|_{**}=\sup\{|F_{\mathbf{v}}(f)|:\|f\|_*=1\}$. $\forall\mathbf{v}\in V$, $\|F_{\mathbf{v}}\|_{**}<\infty$. Now $\varphi(\mathbf{v})=F_{\mathbf{v}}$, for each $\mathbf{v}\in V$, defines a mapping. This mapping is an embedding in that φ is linear and injective, and $\forall\mathbf{v}\in V$, $\|\mathbf{v}\| = \|F_{\mathbf{v}}\|_{**}$.

4.3 DETERMINANTS

We have shown that each linear mapping between finite-dimensional vector spaces with a common scalar ring has a representation in the form of a matrix of scalars, this representation depending upon the particular bases that have been chosen for the spaces. If both spaces have the same dimension, they are algebraically isomorphic, so if one looks at the matrix representing a particular linear mapping, there is no way of telling whether the mapping is an endomorphism, a mapping of one space into another distinct space, or an "identity mapping" of a space onto itself which simply indicates that the coordinates, or components, of each vector are changed in a particular way. This last possibility is usually referred to as a "change of basis" mapping.

Associated with each $n\times n$ square matrix is a scalar called the determinant of the matrix. This scalar is obtained in a rather complicated way; choose exactly one element from each row of the matrix, subject to the restriction that no two elements chosen will belong to the same column, and multiply these n scalars together. Assume that the elements have been chosen from the rows in order, the first from the first row, the second from the second row, and so on. If the elements of the matrix are denoted by a_{ij}, where the first subscript indicates the row number and the second subscript the column number, a typical ordered set of n scalars in one choice-procedure would be $\{a_{1j_1}, a_{2j_2},\ldots,a_{nj_n}\}$. Consider the ordered set of second subscripts $(j_1, j_2,\ldots,j_n)$; these are n distinct numbers from 1 to n. We regard this ordered set as a permutation of the ordered set $(1,2,3,\ldots,n)$. It is called an *even permutation* if one can permute the latter ordered n-tuple to the former by making an even number of interchanges of adjacent numbers; otherwise, it is called an *odd permutation*. We accept as gospel the fact that if one can permute $(1,2,\ldots,n)$ to $(j_1,\ldots,j_n)$ in one way by interchanging adjacent numbers, and then effect the same permutation in any other way by interchanging adjacent numbers, the number of such interchanges will always be even if the first way required an even number of interchanges, or odd if the first way took an odd number of interchanges. For example, one might permute $(1,2,3,4,5)$ to $(3,4,1,2,5)$ by interchanging adjacent numbers in this way: $(1,2,3,4,5)\to(2,1,3,4,5)\to(2,3,1,4,5)\to(3,2,1,4,5)\to(3,2,4,1,5)\to(3,4,2,1,5)\to(3,4,1,2,5)$. This took six adjacent interchanges; do not spend a lot of time trying to effect this permutation with an odd number of such interchanges!

Having chosen the n-tuple of scalars $\{a_{1j_1},\dots,a_{nj_n}\}$ and multiplied these scalars together, we multiply this product by -1 if $(j_1,\dots,j_n)$ is an odd permutation of $(1,\dots,n)$, and if $(j_1,\dots,j_n)$ is an even permutation of $(1,\dots,n)$, we multiply the product by 1. Now there are a total of $n!$ (n factorial) $= n(n-1)\cdots 3\cdot 2\cdot 1$ n-tuples that we can choose from the matrix, so we have $n!$ "adjusted" products. Add them all together; the resulting sum is what we call the *determinant* of the matrix. If A denotes the matrix, we indicate the determinant of A by $\det A$.

This incredible way of defining a scalar-valued mapping $\det: \mathfrak{M}_{n\times n}\to F$ mapping the set of $n\times n$ matrices with entries from a scalar ring F into F boggles the mind. However, the old-timers who thought all this out had some method to their madness. For example, they noticed that if one had a system of n linear equations in n unknowns, which in matrix notation would be written $AX=B$, where A is the matrix of the coefficients of the n unknowns $\{x_1,\dots,x_n\}$, X is the $n\times 1$ matrix whose entries are these unknowns, and B is an $n\times 1$ matrix whose entries are constants, then the solutions all were of the form: $x_i = */\det A$, where $*$ represents some number. The important fact is that all the denominators of the solutions were that one incredibly complicated expression described above. But there is more. The number $*$ turned out to be the determinant of the matrix A with the ith column replaced by the (column) matrix B.

Theorem 4.3.1 (Cramer's Rule). *Suppose we have a system of n linear equations in n unknowns, with coefficients from a field,*

$$\begin{aligned} a_{11}x_1 + \cdots + a_{1n}x_n &= b_1,\\ \vdots \qquad\qquad \vdots\ \ &\ \ \ \vdots\\ a_{n1}x_1 + \cdots + a_{nn}x_n &= b_n. \end{aligned}$$

Then the solutions for this system are given by

$$x_i = \det\left(A^1,\dots,A^{i-1}, B, A^{i+1},\dots A^n\right)/\det A, \qquad i=1,\dots,n,$$

where A is the matrix $(a_{ij})_{i,j=1}^n$ and A^j represents the jth column of the matrix A. If $\det A = 0$, *solutions don't exist, or there are solutions which are not unique*; *if* $\det A \neq 0$, *the solutions are unique.*

From this fundamental definition of the determinant of a matrix, one can deduce all the familiar properties of determinants. Many writers define the determinant axiomatically in terms of these properties. It is not our purpose to develop determinant theory here, but we list these basic properties for reference. Moreover, we are going to take the opportunity to show one delightfully old-fashioned method of evaluating a determinant, but first, an important theorem.

Theorem 4.3.2. *Suppose $T: V \to W$ is linear and V is finite dimensional, with $\dim V = \dim W$. Let A be the matrix which represents T. Then T is bijective iff* $\det A \neq 0$.

PROOF: If $\det A \neq 0$, the previous theorem guarantees that the solution to $AX = B$ is unique. Since $AX = \mathbf{0}$ has $X = \mathbf{0}$ as one solution, this can be the only solution. Hence the associated linear map T sends only $\mathbf{0}$ to $\mathbf{0}$, and thus T is injective. Since TV is a subspace of W and $\dim TV = \dim V = \dim W$, we conclude that $TV = W$; thus T is bijective.

If T is bijective and $\{\mathbf{e}_1, \ldots, \mathbf{e}_n\}$ is a basis for V, then the columns of A, $(A', \ldots, A^n)$ consisting of the coordinates of the images in W of the basis vectors of V under T are linearly independent vectors of $\mathbb{R}^n$. One of the basic properties of determinants to which we have referred is that $\det A \neq 0$ iff the columns of A are linearly independent. □

In the event that $\det A = 0$, we say that the matrix A is singular. We call again attention to the fact that nonsingular linear maps are dimension preserving; singular linear maps on finite-dimensional vector spaces, on the other hand, reduce dimension.

We shall list the basic properties of determinants now as one grand theorem:

Theorem 4.3.3

1. *If $A = (a_{ij})_{i,j=1}^n$ is an $n \times n$ matrix, the transpose of A, A^t, is the matrix whose jth row is the jth column of A and whose ith column is the ith row of A, and* $\det A^t = \det A$.
2. *If two columns of a determinant are interchanged, the value of the determinant changes sign; if two columns are equal, the determinant vanishes.*
3. *Let D_1 and D_2 be two nth-order determinants whose columns are identical except for the jth column, and let D be the nth-order determinant whose columns are identical with those of D_1 and D_2, except for the jth column. Then, if the jth column of D is a linear combination of the jth columns of D_1 and D_2, then the value of D is the same linear combination of the values of D_1 and D_2.*
4. *If each element of a column of a determinant D is multiplied by a constant k, the value of the new determinant is k times the value of the original. In particular, if a column of a determinant consists of all zeros, the value of the determinant is zero.*
5. *The value of a determinant is unchanged if a multiple of one column is added to another column.*
6. *If $A = (a_{ij})$ is an $n \times n$ matrix and we delete the ith row and the jth column, we are left with an $(n-1) \times (n-1)$ matrix M_{ij}, called a*

principal minor of A. For each j

$$\det A = \sum_{i=1}^{n} (-1)^{i+j} a_{ij} \det M_{ij}.$$

7. *As a consequence of properties 4 and 5, the determinant vanishes if the columns are linearly dependent, and conversely.*
8. *In all of the above, "row" may replace "column" and the results remain the same.*

We remind the reader that if A is a matrix, we call the dimension of the image of the linear map T represented by A the rank of T. This number will also be the rank of A. It follows that the rank of a matrix is the order of the largest nonvanishing determinant in A.

We are now going to show you an old way to evaluate the determinant of a matrix, and we sincerely hope that you have never seen this method before. Suppose we have an $n \times n$ matrix:

$$T = \begin{bmatrix} a_1 & a_2 & a_3 & a_4 & \cdots \\ b_1 & b_2 & b_3 & b_4 & \cdots \\ c_1 & c_2 & c_3 & c_4 & \cdots \\ d_1 & d_2 & d_3 & d_4 & \cdots \\ \vdots & \vdots & \vdots & \vdots & \end{bmatrix}.$$

If the pivotal term a_1 is equal to zero, we can interchange the first column and another column to get a nonzero term in this leading position; such an interchange will simply change the sign of the determinant. Now assume that $a_1 \neq 0$. Multiply each column except the first by a_1, so that now we have

$$\det T = \frac{1}{a_1^{n-1}} \det \begin{bmatrix} a_1 & a_1 a_2 & a_1 a_3 & \cdots \\ b_1 & a_1 b_2 & a_1 b_3 & \cdots \\ c_1 & a_1 c_2 & a_1 c_3 & \cdots \\ \vdots & \vdots & \vdots & \end{bmatrix}.$$

Transform this new matrix by subtracting a_2 times the first column from the second column, a_3 times the first column from the third column, and so forth, and this transformed matrix has the same determinant as the former one.

$$\det T = \frac{1}{a_1^{n-1}} \det \begin{bmatrix} a_1 & a_1 a_2 - a_2 a_1 & a_1 a_3 - a_3 a_1 & \cdots \\ b_1 & a_1 b_2 - a_2 b_1 & a_1 b_3 - a_3 b_1 & \cdots \\ \vdots & \vdots & \vdots & \end{bmatrix}.$$

Notice that now the first row of this newest matrix is $(a_1, 0, 0, \dots)$, so

$$\det T = \frac{1}{a_1^{n-2}} \det \begin{bmatrix} a_1b_2 - a_2b_1 & a_1b_3 - a_3b_1 & \cdots \\ a_1c_2 - a_2c_1 & a_1c_3 - a_3c_1 & \cdots \\ \vdots & \vdots & \end{bmatrix}.$$

Since the determinant of a 2×2 matrix

$$\begin{pmatrix} a & b \\ c & d \end{pmatrix} = ad - bc,$$

we can write

$$\det T = \frac{1}{a_1^{n-2}} \det \begin{bmatrix} \det\begin{pmatrix} a_1 & a_2 \\ b_1 & b_2 \end{pmatrix} & \det\begin{pmatrix} a_1 & a_3 \\ b_1 & b_3 \end{pmatrix} & \cdots \\ \det\begin{pmatrix} a_1 & a_2 \\ c_1 & c_2 \end{pmatrix} & \det\begin{pmatrix} a_1 & a_3 \\ c_1 & c_3 \end{pmatrix} & \cdots \\ \vdots & \vdots & \end{bmatrix}.$$

Therefore, to find $\det T$, interchange columns if necessary to get a nonzero term in the leading, or pivotal, position. Then take the reciprocal of the leading term, raised to the $(n-2)$th power, where n is the order of T, and multiply this number into the $(n-1)$th-order determinant whose entries are all the second order subdeterminants in T having the leading term of T as their leading terms. Continue this process until reaching the end of the line, but at each step be sure to factor out common factors from columns or rows in order to keep the numbers smaller. As an example,

$$\det \begin{bmatrix} 2 & 2 & -1 & 3 \\ -2 & 1 & 3 & -2 \\ 2 & -1 & 2 & 1 \\ 3 & -2 & -2 & 1 \end{bmatrix} = \frac{1}{2^2} \det \begin{bmatrix} 6 & 4 & 2 \\ -6 & 6 & -4 \\ -10 & -1 & -7 \end{bmatrix}$$

$$= \det \begin{bmatrix} 3 & 2 & 1 \\ 3 & -3 & 2 \\ 10 & 1 & 7 \end{bmatrix} = \frac{1}{3} \det \begin{bmatrix} -15 & 3 \\ -17 & 11 \end{bmatrix}$$

$$= \det \begin{bmatrix} -5 & 1 \\ -17 & 11 \end{bmatrix}$$

$$= -55 + 17 = -38.$$

We do not claim that this method is any better than other methods; used in combination with other methods, especially when the entries are integers, it seems to work very well. Also, it is a method you might find easy to program for a computer. Incidentally, this process for evaluating a determinant is often called Sylvester's method, after the English mathematician James Sylvester.

4.4 THE GRASSMANN ALGEBRA

In this section we are going to construct an algebra from a given linear set, purely abstractly. The reason for doing this is to make more credible some things we shall do in the next chapter; there we present a "realization" of the abstract model constructed here. Suppose V is an n-dimensional linear set over a ring R; the unity of R we denote by 1, and the zero vector of V by $\mathbf{0}$. Let the ordered set $(\mathbf{e}_1,\ldots,\mathbf{e}_n)$ be a basis for V. In order to get an algebra G over R which will contain $V(R)$, we need to define a secondary operation, the symbol for which will be the wedge $\wedge$. It is enough to define $\wedge$ over the set of basis vectors; we can then extend $\wedge$ to all of $V(R)$ by stipulating that $\wedge$ be distributive from both sides with respect to vector addition. We prescribe the following set of axioms.

G1. $\forall i, j=1,\ldots,n$, $\mathbf{e}_i\wedge\mathbf{e}_j$ is a unique element of G. Moreover, $(\mathbf{e}_i\wedge\mathbf{e}_j)+(\mathbf{e}_j\wedge\mathbf{e}_i)=\mathbf{0}$, and this implies that $\forall i=1,\ldots,n$, $\mathbf{e}_i\wedge\mathbf{e}_i=\mathbf{0}$.

G2. The "wedge product" is associative; we extend the definition of $\wedge$ to products of p, $2<p\leqslant n$, basis vectors in this way: for $i_1<i_2<\cdots<i_p$, $\mathbf{e}_{i_1}\wedge\mathbf{e}_{i_2}\wedge\cdots\wedge\mathbf{e}_{i_p}$ is a unique element of $G(R)$, and

$$\left(\mathbf{e}_{j_1}\wedge\mathbf{e}_{j_2}\wedge\cdots\wedge\mathbf{e}_{j_p}\right)=(-1)^{\sigma}\left(\mathbf{e}_{i_1}\wedge\mathbf{e}_{i_2}\wedge\cdots\wedge\mathbf{e}_{i_p}\right),$$

where $\sigma=0$ or 1, according as $(j_1,\ldots,j_p)$ is an even or odd permutation of $(i_1,\ldots,i_p)$.

G3. We adjoin to $G(R)$ the element $\mathbf{1}$ and define $\mathbf{1}\wedge\mathbf{1}=\mathbf{1}$ and for each $i=1,\ldots,n$, $\mathbf{1}\wedge\mathbf{e}_i=\mathbf{e}_i\wedge\mathbf{1}=\mathbf{e}_i$. $\mathbf{1}$ is thus the unity of $G(R)$.

G4. If λ, μ are any ring elements, then $\wedge$ obeys, $\forall i, j, k=1,\ldots,n$,

$$(\lambda\mathbf{e}_i)\wedge(\mu\mathbf{e}_j)=\lambda\mu(\mathbf{e}_i\wedge\mathbf{e}_j),$$

$$\mathbf{e}_i\wedge(\lambda\mathbf{e}_j+\mu\mathbf{e}_k)=\lambda(\mathbf{e}_i\wedge\mathbf{e}_j)+\mu(\mathbf{e}_i\wedge\mathbf{e}_k),$$

$$(\lambda\mathbf{e}_i+\mu\mathbf{e}_j)\wedge\mathbf{e}_k=\lambda(\mathbf{e}_i\wedge\mathbf{e}_k)+\mu(\mathbf{e}_j\wedge\mathbf{e}_k).$$

G5. The following ordered set of 2^n elements comprise a basis for $G(R)$:

$$\mathbf{1},\quad \{\mathbf{e}_i\}_{i=1}^{n},\quad \{\mathbf{e}_i\wedge\mathbf{e}_j\}_{1\leqslant i<j\leqslant n},\quad \ldots,$$

$$\left\{\mathbf{e}_{i_1}\wedge\cdots\wedge\mathbf{e}_{i_p}\right\}_{1\leqslant i_1<\ldots<i_p\leqslant n},\quad \ldots,$$

$$\mathbf{e}_1\wedge\cdots\wedge\mathbf{e}_n.$$

This linear algebra $G(R)$, of dimension 2^n, constructed from the linear set $V(R)$ of dimension n, is called the Grassmann algebra for $V(R)$. Sometimes it is simply referred to as an exterior algebra, or the wedge algebra, of V. A typical element of $G(R)$ is of the form

$$\omega = \lambda \cdot \mathbf{1} + \sum_{i=1}^{n} \lambda_i \mathbf{e}_i + \sum_{1 \leqslant i \leqslant j \leqslant n}^{n} \lambda_{ij}(\mathbf{e}_i \wedge \mathbf{e}_j) + \cdots$$

$$+ \sum_{1 \leqslant i_1 < \cdots < i_p \leqslant n}^{n} \lambda_{i_1 \cdots i_p}(\mathbf{e}_{i_1} \wedge \cdots \wedge \mathbf{e}_{i_p}) + \cdots + \lambda'(\mathbf{e}_1 \wedge \cdots \wedge \mathbf{e}_n).$$

where all λ belong to R.

As a linear set, $G(R)$ has $n+1$ mutually disjoint (except for the zero vector) linear subsets, namely $\mathrm{sp}\{\mathbf{1}\}$, $\mathrm{sp}\{\mathbf{e}_i\}_{i=1}^n, \ldots, \mathrm{sp}\{\mathbf{e}_1 \wedge \cdots \wedge \mathbf{e}_n\}$. We write: $G(R) = \mathrm{sp}\{\mathbf{1}\} \oplus \mathrm{sp}\{\mathbf{e}_i\}_{i=1}^n \oplus \cdots \oplus \mathrm{sp}\{\mathbf{e}_1 \wedge \cdots \wedge \mathbf{e}_n\}$, and say that $G(R)$ is the direct sum of these $n+1$ linear sets, meaning specifically that these linear sets are disjoint, except for $\mathbf{0}$, and that each element of $G(R)$ has a unique representation as a sum of $n+1$ terms, exactly one term from each of these $n+1$ component linear sets.

Let us take a few minutes to construct an exterior algebra from the very familiar vector space over $\mathbb{R}$, $\mathbb{R}^2$. Denote the two usual basis vectors $(1,0)$ and $(0,1)$ by $\mathbf{i}$ and $\mathbf{j}$, respectively, and define $\mathbf{i} \wedge \mathbf{j}$ to be a unique element of the algebra which we shall denote by $\mathbf{k}$. Of course, we adjoin $\mathbf{1}$ to the algebra for a unity. We now have a linear algebra, having as basis $(\mathbf{1}, \mathbf{i}, \mathbf{j}, \mathbf{k})$.

Let (x_1, x_2), (y_1, y_2) be two arbitrary vectors of $\mathbb{R}^2$. In our algebra, $(x_1, x_2) \wedge (y_1, y_2) = (x_1\mathbf{i} + x_2\mathbf{j}) \wedge (y_1\mathbf{i} + y_2\mathbf{j}) = x_1 y_2 \mathbf{k} - x_2 y_1 \mathbf{k}$, a multiple of $\mathbf{k}$. Note that the coefficient of $\mathbf{k}$ is equal to the area of the parallelogram determined by the two vectors (x_1, x_2), (y_1, y_2) in E^2. In the Grassmann algebra, $\mathbf{i} \wedge \mathbf{k} = \mathbf{j} \wedge \mathbf{k} = \mathbf{0}$, which is not very interesting. But instead of prescribing that the Grassmann axioms hold strictly, we shall weaken the axiom G1 a bit by not insisting that $\mathbf{i} \wedge \mathbf{i} = \mathbf{j} \wedge \mathbf{j} = \mathbf{0}$. In place of this axiom we stipulate that $\mathbf{i} \wedge \mathbf{j} + \mathbf{j} \wedge \mathbf{i} = \mathbf{0}$ and that $\mathbf{i} \wedge \mathbf{i} = \mathbf{j} \wedge \mathbf{j} = -\mathbf{1}$. Now it must follow that $\mathbf{j} \wedge \mathbf{k} = \mathbf{i} = -(\mathbf{k} \wedge \mathbf{j})$, $\mathbf{k} \wedge \mathbf{i} = \mathbf{j} = -(\mathbf{i} \wedge \mathbf{k})$, and $\mathbf{k} \wedge \mathbf{k} = -\mathbf{1}$. This exterior algebra constructed from $\mathbb{R}^2$ is the famous algebra of quaternions. If q is the quaternion: $x_0 + x_1\mathbf{i} + x_2\mathbf{j} + x_3\mathbf{k}$, then the conjugate of q, denoted $\bar{q}$, is the quaternion

$$\bar{q} = x_0 - x_1\mathbf{i} - x_2\mathbf{j} - x_3\mathbf{k}.$$

If you carry out the wedge multiplication, you will find that $q \wedge \bar{q} = \bar{q} \wedge q = \sum_{\nu=0}^{3} x_\nu^2$, a real number, and hence it follows that for any nonzero quaternion q, the "wedge" inverse is given by $q^{-1} = \bar{q}/q \wedge \bar{q}$. Such an algebra in which each nonzero element has a "multiplicative" inverse is called a *division algebra*. This nonabelian algebra of quaternions contains $\mathbb{C}$ as a (commutative) subalgebra.

PROBLEMS

1. Prove the first four parts of Theorem 4.1.5.

2. Prove Theorem 4.1.7.

3. The matrix

$$T=\begin{bmatrix} 1 & 2 & 0 \\ 4 & 1 & -3 \\ -3 & -1 & -1 \end{bmatrix}$$

maps the unit cube $K=[0,1]^3$ to a parallelopiped in E^3. Show that the volume of $TK=|\det T|$.

4. Try to construct a (nonassociative) algebra from the three unit vectors $\mathbf{i},\mathbf{j},\mathbf{k}$ of E^3 which will contain the quaternions as a noncommutative subalgebra. [*Hint*: Consider Problem 4 of Chapter 1.]

5. Suppose the three vectors $\{\mathbf{e}_1,\mathbf{e}_2,\mathbf{e}_3\}\subset E^3$ comprise a basis. Write a matrix that will give the new coordinates of the vector $\mathbf{x}=3\mathbf{e}_1-2\mathbf{e}_2-5\mathbf{e}_3$ in terms of a basis $\{\mathbf{b}_1,\mathbf{b}_2,\mathbf{b}_3\}$, where $\mathbf{b}_1=\mathbf{e}_1+\mathbf{e}_2+\mathbf{e}_3$, $\mathbf{b}_2=4\mathbf{e}_1-2\mathbf{e}_2-\mathbf{e}_3$, and $\mathbf{b}_3=5\mathbf{e}_1+3\mathbf{e}_2+\mathbf{e}_3$.

6. If

$$A=\begin{bmatrix} 1 & 4 & 1 & -2 \\ 1 & 3 & 5 & -1 \\ 2 & 1 & 4 & -1 \\ 4 & 3 & 2 & 1 \end{bmatrix},$$

what is A^{-1}? Evaluate $\det A$.

7. Let

$$T=\begin{bmatrix} 2 & 3 & 7 \\ 0 & -1 & -2 \\ 0 & 0 & 2 \end{bmatrix}$$

be an automorphism in E^3. Find the invariant subspaces of E^3 relative to T.

8. Prove that if $\|\cdot\|$ is a norm on a vector space V, then

$$\forall \mathbf{x},\mathbf{y}\in V, \qquad \big|\|\mathbf{x}\|-\|\mathbf{y}\|\big| \leqslant \|\mathbf{x}+\mathbf{y}\| \leqslant \|\mathbf{x}\|+\|\mathbf{y}\|.$$

9. If $(V,\|\cdot\|)$ is a normed linear space, show that the mapping $V^*\to\mathbb{R}$ given by $\forall f\in V^*$, $\|f\|=\sup\{|f(\mathbf{v})|:\mathbf{v}\in V,\ \|\mathbf{v}\|=1\}$ is indeed a norm for V^*.

10. Let $T: V\to W$ be a linear mapping from one normed space into another; suppose that for each $\mathbf{v}\in V$, $\|T\mathbf{v}\|_W=\|\mathbf{v}\|_V$. Prove that such a mapping, called a *linear isometry*, is injective. Is it continuous?

11. If $(i_1,i_2,\dots,i_n)$ is a permutation of $(1,2,\dots,n)$, show that this permutation is even or odd, i.e., that to effect the permutation always requires an even number of interchanges of adjacent numbers or an odd number of such interchanges.

12. Let A be an $n \times n$ matrix. Prove that $\det A = 0$ iff the columns of A are linearly dependent n-tuples in $\mathbb{R}^n$.

13. Suppose the equations

$$x^2 + yx + z^3 + 5t^4 = 6, \qquad x^3 + x^2yz + tx = 7$$

implicitly define z and t in terms of x and y. Use Cramer's rule to solve for $\partial t/\partial x$.

14. Prove that the linear image of a parallelopiped is a (perhaps degenerate) parallelopiped.

15. Let S be a set of vectors from a linear space V. Prove that $\operatorname{sp} S$ is the smallest linear space which contains S and is contained in V.

16. In E^4, describe the smallest affine flat that contains the vectors $(-1,0,5,10)$, $(1,2,3,4)$, $(0,-1,0,1)$, and $(1,1,1,1)$. Does this affine subspace contain the vector $(0,0,2,4)$?

17. Let A be a 3×3 matrix such that for every unit vector $\mathbf{u} \in E^3$, $\|A\mathbf{u}\| < 1$. Use Hadamard's inequality to show that $|\det A| < 1$. Show that $|\det A| < 1$ does not imply that A is norm reducing.

18. Define for all $x > 0$, $\varphi^*(x) = \int_0^\infty x^3 e^{-xt} \varphi(t)\, dt$. This integral equation defines a linear mapping which sends $\varphi \mapsto \varphi^*$. Show that $\varphi(t) \equiv t$ is an eigenfunction for this mapping. Can you find any others?

19. Prove Theorem 4.3.1.

20. If A, B are $n \times n$ matrices, prove that $\det AB = (\det A)(\det B)$.

21. Express

$$\det\begin{bmatrix} ax + by & az + bw \\ cx + dy & cz + dw \end{bmatrix}$$

as a sum of four determinants, and evaluate to get $(ad - bc)(xw - yz)$.

22. Show that if A, B are real linear spaces with a common zero vector, then $A \cap B$, $\operatorname{sp}(A \cup B)$ are, respectively, the largest linear space contained in both A and B and the smallest linear space containing both A and B.

23. If V is a linear space with a Hamel basis H, we call the cardinal number of H the *algebraic dimension* of V, and write $\dim V$. Show that for arbitrary subspaces A, B of a linear space V, we always have $\dim A + \dim B = \dim \operatorname{sp}(A, B) + \dim(A \cap B)$.

Suppose A and B are linear spaces with a common scalar field and zero vector. If C is another such linear space such that $A \cap B \subset C \subset \operatorname{sp}(A, B)$, we call C an *intermediate space*, relative to A and B. If A and B are subspaces of $\mathbb{R}^5$ spanned respectively by $\{(1,1,1,1,1),\ (2,-1,3,1,4),\ (-3,1,-1,1,1)\}$ and

$\{(1,0,0,1,2),\ (0,1,0,0,0),\ (2,2,1,2,3)\}$, and C is the span of $\{(-2,3,0,2,2),$ $(1,2,1,1,1), (-3,2,-1,1,1)\}$, show that $\mathrm{sp}(A,B)=\mathbb{R}^5, A\cap B=\mathrm{sp}\{(1,1,1,1,1)\}$ and that C is an intermediate space relative to A and B which is distinct from either A or B.

24. Find the vector of minimum length in E^5 terminating in the affine space determined by the three points (1,3,4,0,0), (−1,3,0,1,0) and (4,0,1,1,1). What is its length?

25. The numbers 277,695; 166,345; 131,852; 191,352; 385,101; and 945,387 are all divisible by 17. Show that

$$\det\begin{bmatrix} 2 & 7 & 7 & 6 & 9 & 5 \\ 1 & 6 & 6 & 3 & 4 & 5 \\ 1 & 3 & 1 & 8 & 5 & 2 \\ 1 & 9 & 1 & 3 & 5 & 2 \\ 3 & 8 & 5 & 1 & 0 & 1 \\ 9 & 4 & 5 & 3 & 8 & 7 \end{bmatrix}$$

is also divisible by 17. You may do this by evaluating the determinant, of course, but use Theorem 4.3.3 instead.

26. Evaluate

$$\det\begin{bmatrix} 2 & 1 & 1 & 1 & 1 \\ 1 & 3 & 1 & 1 & 1 \\ 1 & 1 & 4 & 1 & 1 \\ 1 & 1 & 1 & 5 & 1 \\ 1 & 1 & 1 & 1 & 6 \end{bmatrix}$$

27. Solve by inspection the quartic

$$\det\begin{bmatrix} 1 & 1 & 2 & 3 \\ 1 & 2-X^2 & 2 & 3 \\ 2 & 3 & 1 & 5 \\ 2 & 3 & 1 & 9-X^2 \end{bmatrix}=0.$$

28. The following determinant is called the Vandermonde determinant. Show that

$$V(X_1,\ldots,X_n)=\det\begin{bmatrix} 1 & 1 & \cdots & 1 \\ X_1 & X_2 & \cdots & X_n \\ X_1^2 & X_2^2 & \cdots & X_n^2 \\ X_1^3 & X_2^3 & \cdots & X_n^3 \\ \vdots & \vdots & & \vdots \\ X_1^{n-1} & X_2^{n-1} & \cdots & X_n^{n-1} \end{bmatrix}=\prod_{1\leqslant i<j\leqslant n}(X_j-X_i).$$

29. The class P of polynomials with rational coefficients is a commutative ring with unity. If p, q are two such polynomials, can you show that if $p\cdot q$ is the zero polynomial, then necessarily one of p, q must have been the zero polynomial?

If a, b are two relatively prime integers (no common factor other than 1), then there exist integers m, n, such that $am-bn=1$. Find two polynomials p and q, of minimal degree, such that $p(X^2+3X+2)-q(X^3-1)\equiv 1$.

(The algebraic structure of P is that of an *integral domain*. The mathematician Wedderburn used to call such a structure a *domain of integrity*, a term which this writer feels has a much nicer ring to it!)

30. Find a singular matrix C that will map the vector $(1,2,3)$ to the vector $(4,-1,4)$ and $(-4,0,0)$ to $(8,1,8)$.

31. A scalar matrix is an $n\times n$ matrix of the form (a_{ij}), where $a_{ij}=0$, $i\neq j$, and $(a_{ii})=k$, $i=1,\ldots,n$.

Prove that an $n\times n$ matrix A commutes with every $n\times n$ matrix B iff A is a scalar matrix.

32. If V is a linear space and A, B are nonempty subsets of V, show that $\operatorname{sp} A\subset \operatorname{sp} B \Leftrightarrow A\subset B$ is false. Show that $\operatorname{sp}(A\cap B)\subset(\operatorname{sp} A)\cap(\operatorname{sp} B)$, and give an example to show that the inclusion may be proper.

33. Suppose V, W are subspaces of a linear space X. Prove that for vectors $\mathbf{x},\mathbf{y}\in X$, $(\mathbf{x}+V)\cap(\mathbf{y}+W)$ is empty or is an affine space parallel to $V\cap W$.

34. If $T: E^3\to E^1$ is given by $T(X_1,X_2,X_3)=2X_1-3X_2+X_3$, show that T is linear. What is $\ker T$? What is $T^{-1}(1)$?

35. If $T: E^3\to E^3$ is given by $T(X_1,X_2,X_3)=(X_1,0,X_2)$, show that T is linear, describe $\ker T$ and $\operatorname{im} T$, and if $V\subset E^3$ is the set of vectors having 0 as their third component, describe $T(V)$ and $T^{-1}(V)$.

36. Let $T: V\to W$ be linear and $\mathbf{w}\in \operatorname{im} T$. Show that $T^{-1}(\mathbf{w})$ is an affine subspace parallel to $\ker T$. If $A\subset W$ is an affine subspace, show that $T^{-1}(A)$ is empty or an affine subspace of V.

37. Suppose $T: V\to W$ and $\operatorname{im} T$ is finite dimensional. Show that there exists a subspace $U\subset V$ such that $T(U)$ and $T(V)$ are linearly isomorphic. Show that each parallel of U intersects each parallel of $\ker T$ in exactly one vector.

38. Let P be a linear endomorphism on a vector space V. P is called a *projection*, or a *projector*, if $PP=P$ on V. Show that if P is a projector, $(\ker P)\cap(\operatorname{im} P)=\mathbf{0}$, and that each $\mathbf{v}\in V$ has a unique representation of the form

$$\mathbf{v}=\mathbf{x}+\mathbf{y},\qquad \mathbf{x}\in \operatorname{im} P,\quad \mathbf{y}\in\ker P.$$

If I is the identity map on V, show that $Q=I-P$ is also a projector with $\ker P=\operatorname{im} Q$ and $\operatorname{im} P=\ker Q$, and that $\forall \mathbf{v}\in V$, $PQ(\mathbf{v})=QP(\mathbf{v})=\mathbf{0}$.

If $J: V\to V$ is such that $JJ=I$, J is called an *involution* on V. Show that $J=I-2P$ is an involution for any projector P on V.

CHAPTER

5

FORMS IN E^n

To many of you this chapter will seem strange indeed; the ideas will be new, and you will be uncomfortable with them, mainly because what we are purporting to be very basic calculus is material with which you have had little or no experience. But be of good cheer; after the initial shock it really is quite enjoyable.

There are two main parts to this chapter. The first introduces the notion of orientation of parallelotopes, and the second addresses itself to the notion of k-forms in E^n. These k-forms are the things we are going to integrate, initially over oriented parallelotopes, and then later over more smooth and undulating sets called manifolds. By a *parallelotope* we mean a higher-dimension parallelopiped. In E^3 three independent vectors emanating from the origin determine a parallelopiped; analogously, in E^n k independent vectors emanating from the origin determine a k-parallelotope. In general, by a k-parallelotope in E^n we mean a translate of such an object. We define a point to be a 0-parallelotope, a line segment to be a 1-parallelotope, and so forth.

It should be clear that a k-parallelotope P in E^n is a nonsingular affine image of the unit cube K in E^k, and that this figure is a bounded convex set. By the unit cube $K \subset E^k$ we mean the set $[0,1]^k = [0,1] \times \cdots \times [0,1]$ (with k factors in the Cartesian product). If P is translated by a constant vector so that one of its vertices is at the origin, then P', the translated P, is determined by the k vectors emanating from the origin of E^n, each of these vectors being the linear image of one of the unit basic vectors of E^k. We can write $P' = TK$, or $P = AK = TK + C$, where A represents the affine map, and C is the translation vector.

5.1 ORIENTATION OF PARALLELOTOPES

Having defined what we mean by a k-parallelotope in E^n, we now have to come up with some reasonable way of defining an orientation for it. Why do we want to orient these things in the first place? Very loosely speaking, the manifolds in E^n over which we are going to do integration need to be oriented; it does make a difference, as you know, whether you go from 1 to 2 or from 2 to 1 when integrating the exponential function over the manifold in E^1 bounded by the points $x=1$ and $x=2$. A manifold in E^n is a kind of limiting configuration of parallelotopes delicately pasted together. If we can orient parallelotopes and paste them together so that the orientations of the parts determine an orientation of the whole, we are very close indeed to obtaining an orientation for the manifold being approximated.

We begin by stating right off that the "exceptional" parallelotope of dimension zero, the point, cannot be assigned an orientation. We remark in passing that a point has no "boundary," a geometric point-set of one less dimension which "limits its extent." We next consider the 1-cube in E^1. This is simply the interval $[0,1]$; we orient it by decreeing that the origin, 0, will be its initial point and 1 will be its terminal point. We have thus made a directed line segment out of a line segment and indicated "positive direction" or "positive orientation" by stipulating which is the initial point and which is the terminal point.

In E^n any line segment can be given a positive orientation provided we can tell which of its endpoints is the "initial" point. (Such is not the case in general for other k-parallelotopes in E^n, $2 \leqslant k < n$, as we shall see.) Now, the boundary of a line segment is a doublet, a set of two distinct points in E^n. If this set, say $\{P, Q\}$, is regarded as an ordered pair (P, Q), then we have oriented the boundary of the line segment PQ. By stipulating that the first point of the ordered pair is to be considered the initial point of the corresponding line segment, we have induced an orientation of the line segment by orienting its boundary. Conversely, if the line segment is already oriented, that orientation induces one on the doublet which is its boundary.

The next step is to orient the 2-cube in E^2. Let K be the unit square in E^2, $K=[0,1]\times[0,1]$. We say that K is positively oriented if the "face" of K which lies in the coordinate subspace $x_2=0$ is positively oriented, the "face" which lies in the coordinate subspace $x_1=0$ is negatively oriented, and the remaining "faces" are oriented so that their orientations are

FIGURE 5.1 The positive orientation for $K \subset E^2$.

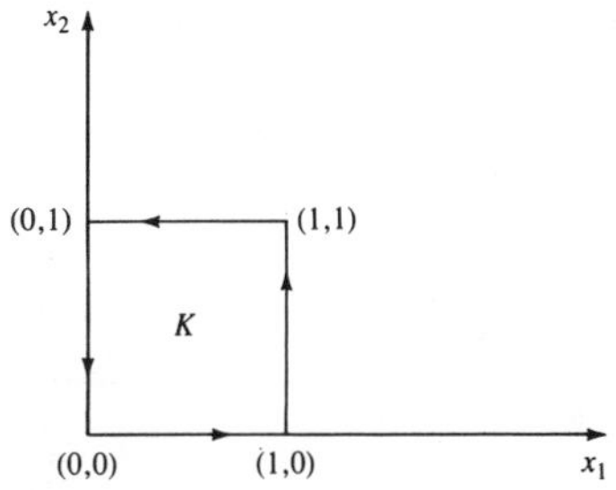

opposite to those of the faces they oppose. Figure 5.1 will suffice to make this clear.

Thus the square K is oriented if you can start at one corner and walk completely around the boundary, always going with, or against, the orientation of the boundary. Note that K is positively oriented if as you follow the orientation of the boundary, the interior of K is on your left. It is clear that if K has a positive orientation, the orientation of K becomes negative if the orientation of the boundary is reversed.

We have oriented $K=[0,1]\times[0,1]\subset E^2$ by orienting its boundary, but it is apparent that if K is already oriented, that orientation induces an orientation on the boundary (Figure 5.2). One should also take note of the fact that the boundary of K is a closed configuration which itself has no boundary. Hence the orientation of the boundary of K is not induced by an orientation of its own boundary.

Since we have effectively oriented the 1-cube and the 2-cube by orienting their boundaries, we shall proceed to orient the n-cube in E^n by orienting its boundary. By the unit n-cube in E^n we mean the set

$$K=[0,1]^n=[0,1]\times\cdots\times[0,1],$$

and we shall designate pairs of opposite faces of K in the following way:

$$F_{k,0}=\{\mathbf{x}\in E^n:\mathbf{x}=(x_1,\ldots,x_{k-1},0,x_{k+1},\ldots,x_n),\,0\leqslant x_i\leqslant 1\},$$

$$F_{k,1}=\{\mathbf{x}\in E^n:\mathbf{x}=(x_1,\ldots,x_{k-1},1,x_{k+1},\ldots,x_n),\,0\leqslant x_i\leqslant 1\}.$$

$F_{k,0}$ is the face lying in the coordinate hyperplane of E^n whose equation is $x_k=0$, and $F_{k,1}$ is the indicated translation of $F_{k,0}$, lying in an affine flat of E^n, parallel to $F_{k,0}$. Thus K has $2n$ faces, and each face is an $(n-1)$-cube. Assume that we have a well-given definition for the positive orientation of the unit $(n-1)$-cube. We now postulate that a translation of an oriented cube does not affect the orientation.

FIGURE 5.2 The orientation for $K\subset E^2$ induces an orientation for the boundary.

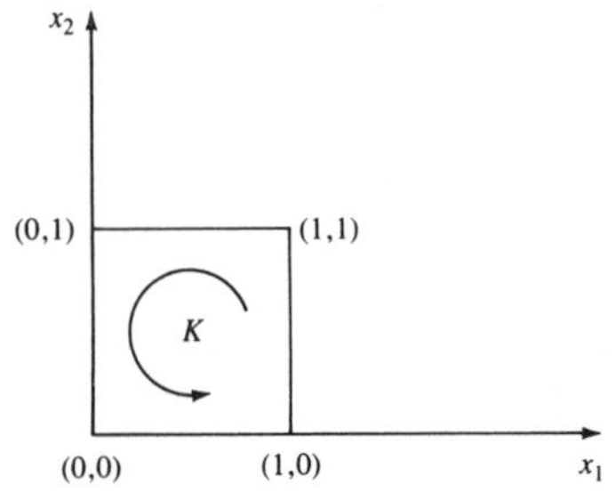

Let K be an arbitrary unit n-cube in E^n, $n \geqslant 2$. Consider arbitrary pairs of opposite faces $F_{k,0}$, $F_{k,1}$. If k is an odd integer, we place a negative orientation on $F_{k,0}$, and a positive orientation on its translate $F_{k,1}$. If k is an even integer, we give a positive orientation to $F_{k,0}$, and a negative orientation to its translate $F_{k,1}$. Now we have an orientation on the boundary of K; the orientation on K induced by this orientation of the boundary is the defined positive orientation for K. If we reverse the orientation of each face, the cube K will be regarded as negatively oriented. If K is already positively oriented, it is to be understood that the orientation on each face induced by the orientation of K is consistent with the face orientations as we have described them.

Do we really have a consistent orientation for the boundary of K? Our induction hypothesis is that on each individual face which lies in a coordinate subspace we have a bona fide orientation, and we have postulated that any translate of an orientable cube can be oriented, so all the faces of K are certainly oriented. The only problem that might arise is this: When we paste the faces together to get the boundary of K, do the orientations on the face boundaries being pasted together cancel out? To form the boundary of the 2-cube, we pasted together four oriented line segments, joining an initial point of one with the terminal point of another, "canceling" the orientations of the boundaries of these line segments. (Although we do not have a way to orient a point, we may regard the initial point of an oriented line segment as negatively oriented and the terminal point as positively oriented.) Let's see what happens on the common boundary of two intersecting faces of our n-cube.

It is clear that, for each k, $F_{k,0} \cap F_{k,1} = \varnothing$, so we consider a fairly general pair $F_{i,0}$ and $F_{j,1}$, where $1 \leqslant i < j \leqslant n$, and let

$$\begin{aligned} E &= F_{i,0} \cap F_{j,1} \\ &= \{\mathbf{x} \in E^n : \mathbf{x} = (x_1, \ldots, x_{i-1}, 0, x_{i+1}, \ldots, x_{j-1}, 1, x_{j+1}, \ldots, x_n), 0 \leqslant x_\nu \leqslant 1\}. \end{aligned}$$

E is an $(n-2)$-cube in the $(n-2)$-dimensional affine subspace of E^n given by the equations $x_i = 0$, $x_j = 1$. Hence E can be oriented, but E is part of the boundary of $F_{i,0}$, so E already has an orientation induced on it by the orientation of $F_{i,0}$. Suppose i is odd and j is even. As a face of $F_{i,0}$, E is the set

$$\{\mathbf{x} \in F_{i,0} : \mathbf{x} = (x_1, \ldots, x_{i-1}, x_{i+1}, \ldots, x_{j-1}, 1, x_{j+1}, \ldots, x_n), 0 \leqslant x_\nu \leqslant 1\}.$$

Here the constant coordinate 1 is in the $(j-1)$th position, and $j-1$ is odd, so the orientation of E should be positive, if the cube $F_{i,0}$ is positively oriented. But $F_{i,0}$ is negatively oriented, so the orientation on E, inherited from $F_{i,0}$, must be negative.

The face $F_{j,1}$ is negatively oriented. As a face of $F_{j,1}$, E is the set

$$\{\mathbf{x}\in F_{j,1}:\mathbf{x}=(x_1,\ldots,x_{i-1},0,x_{i+1},\ldots,x_{j-1},x_{j+1},\ldots,x_n),\ 0\leqslant x_\nu\leqslant 1\}.$$

Here the constant coordinate 0 is in the ith position, and i is odd, so the orientation of E should be negative, but since the orientation of $F_{j,1}$ is negative, the face E must inherit a positive orientation from $F_{j,1}$. Thus the orientations on the common intersection of the two face induced by the orientations of the two faces do indeed cancel out. Needless to say, there are a number of other cases to be checked out, which must, unfortunately, be left to the reader.

We shall make a momentary digression. Take a rectangular strip of paper and draw the rectangles on both sides of the paper as indicated by Figure 5.3. Then paste the edges AB and $A'B'$ together, pasting A to A' and B to B'. Now orient all the rectangles, if you can, so that the orientations along the common boundaries cancel out. You will probably conclude that this surface, called a Möbius band, cannot be oriented. On the other hand, consider a torus, the surface of a conventional doughnut. You can put a rectangular grid on this surface and orient all the rectangles so that the orientations along common edges do cancel out. A torus is thus orientable. What about such a grid put on a Klein bottle?

Now back to the job at hand. We have set down a scheme for assigning a positive orientation to each unit n-cube in E^n, and we have implied that associated with an orientation is an opposite orientation, the negative of the former. Orientation is like an electric charge; if we superpose an oriented figure onto a congruent one with opposite orientation, the orientations cancel out. We are ready to describe how we shall orient n-parallelotopes in E^n. If P is such a figure, it is a nonsingular affine image of K, the unit n-cube in E^n. Let T be the linear part of an affine map that sends K to P; T is nonsingular and may be represented by a nonsingular matrix A. We say that P is positively (negatively) oriented, relative to the particular affine mapping, if $\det A>0$ ($\det A<0$). Unless P is the unit n-cube itself, or a translate of K, we don't speak of P as being positively or negatively oriented

FIGURE 5.3 A Möbius band, cut along AB.

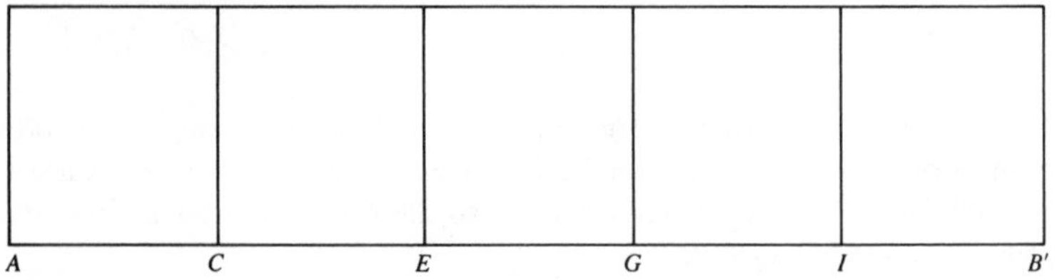

in its own right; P is only oriented, positively or negatively, with respect to the linear part of a particular affine map sending K to P.

Suppose now that P is a k-dimensional parallelotope, $1 \leq k < n$, in E^n. Because it simplifies matters, we assume that one vertex of P is at the origin; now P is just the linear image under a map $T: E^k \to E^n$ which maps the unit k-cube onto P. The matrix for T is an $n \times k$ matrix having rank k. Since $k < n$, this matrix has no determinant; therefore we cannot assign a positive or negative orientation to P, relative to T. However, we can assign such orientations to the projections of P onto the $\binom{n}{k}$ coordinate subspaces of dimension k, and here is how we do it. Suppose the matrix for T is

$$A = \begin{bmatrix} a_{11} & a_{12} & \cdots & a_{1k} \\ a_{21} & a_{22} & \cdots & a_{2k} \\ \vdots & \vdots & & \vdots \\ a_{n1} & a_{n2} & \cdots & a_{nk} \end{bmatrix}.$$

The columns of A are the n-tuples representing the k determining vectors of P in order. Now the projection of P onto the k-dimensional subspace of E^n spanned by the unit vectors $\{\mathbf{e}_{i_1}, \mathbf{e}_{i_2}, \ldots, \mathbf{e}_{i_k}\}$ is the image of the unit k-cube under the linear map whose matrix is

$$A'_{i_1 \cdots i_k} = \begin{bmatrix} a_{i_1 1} & \cdots & a_{i_1 k} \\ \vdots & & \vdots \\ a_{i_k 1} & \cdots & a_{i_k k} \end{bmatrix},$$

where all the rows are zero except the k rows $i_1, i_2, \ldots, i_k$. Let $A'_{i_1 \ldots i_k}$ be the obvious $k \times k$ submatrix. If $\det A'_{i_1 \ldots i_k}$ is positive, the projection $P_{i_1 \ldots i_k}$ of P onto $\mathrm{sp}\{\mathbf{e}_{i_1}, \ldots, \mathbf{e}_{i_k}\}$ will be regarded as positively oriented with respect to T. If this determinant is negative, the orientation of the projection is negative, relative to T, and if the determinant is zero, the projection is degenerate, i.e., not k-dimensional, and we disregard the question of its orientation. We shall prove in Chapter 7 that the value of $\det A'_{i_1 \ldots i_k}$ is actually equal to the signed measure of $P_{i_1 \ldots i_k}$.

If $P \subset E^n$ is a 2-parallelotope and $P_{i_1 i_2}$ is the projection of P onto the $x_{i_1} x_{i_2}$-coordinate subspace, $1 \leq i_1 < i_2 \leq n$, our method for determining whether $P_{i_1 i_2}$ is positively or negatively oriented is equivalent to the following: imagine that E^n has been embedded in E^{n+1}, so that E^n lies in the span of the first n basis vectors of E^{n+1}, and that a vertex of P is at the origin of E^{n+1}. Let P be oriented, and assume $P_{i_1 i_2}$ is not degenerate. Now the boundary of $P_{i_1 i_2}$ in $\mathrm{sp}\{\mathbf{e}_{i_1} \mathbf{e}_{i_2}\}$ is oriented; if, as the fingers of the right hand point in the direction of the orientation of the boundary while the interior of $P_{i_1 i_2}$ is on the palm side of the right hand, the (extended) right thumb is pointing in the direction of the basic unit vector $\mathbf{e}_{n+1}$, the orientation of $P_{i_1 i_2}$ is positive.

5.2 1-FORMS IN E^n

The space E^n is an n-dimensional real vector space, and we have noted that the dual space is likewise an n-dimensional space. A very special subset of the dual space is the set of coordinate mappings $\{\Pi_i\}_{i=1}^n$, defined as follows: $\forall \mathbf{x} \in E^n$, $\forall i=1,\ldots,n$, $\Pi_i(\mathbf{x})=x_i$. It is easy to see that the Π_i are indeed linear maps from E^n to $\mathbb{R}$. That they comprise a linearly independent set of linear functionals may be shown this way: Suppose $\Sigma_{i=1}^n \alpha_i \Pi_i$ is the zero-functional. Apply it to each of the basis vectors in turn; for $j=1,\ldots,n$, $0=(\Sigma_{i=1}^n \alpha_i \Pi_i)\mathbf{e}_j=\alpha_j$. This means that the only vanishing linear combination of the Π_i is the trivial one (all the α_i are zero). Since there are n of these linearly independent functionals from $(E^n)^*$, this set can be taken for a basis for the dual space.

Incidentally, the word "functional" seems to have been coined originally to distinguish from ordinary functions those whose domains are sets of functions and whose ranges lie in function spaces. In more modern times mathematicians think of a functional as a mapping from a function space to a space of scalars, and in fact the domain need not actually be a space of functions; it is frequently a space of vectors. In this latter case, the functional may be linear or nonlinear. For example, if $x(t)$ is an arbitrary integrable function, the following are functionals on the space of integrable functions:

$$|x(t_0)|, \quad \int_a^b \alpha(t)x(t)\,dt, \quad \int_a^b \alpha(y,t)x(t)\,dt+\beta(y)x(y), \quad [x(t_0)]^2.$$

If $x=(x_1,\ldots,x_n)\in E^n$, one usually thinks of x as a vector emanating from the origin, if one is thinking of E^n as a vector space in the first place. However, if E^n is thought of as $\mathbb{R}^n$ with the Euclidean metric, then E^n is a point space. At each point $\mathbf{P}$ in E^n one might have a vector, or a directed line segment, initiating at $\mathbf{P}$ and terminating at some point $\mathbf{Q}$. Such a vector we call a free vector, and a collection of free vectors, exactly one at each point of E^n, we call a vector field on E^n. If $\mathbf{P}=(P_1,\ldots,P_n)$, $\mathbf{Q}=(Q_1,\ldots,Q_n)$ are two points of E^n, then $\overrightarrow{\mathbf{PQ}}$ is a free vector, and for each $i=1,\ldots,n$, $\Pi_i(\overrightarrow{\mathbf{PQ}})=Q_i-P_i$. Mathematically, the free vector $\overrightarrow{\mathbf{PQ}}$ is equivalent to the mathematical vector $\mathbf{Q}-\mathbf{P}$, where here $\mathbf{Q}$ and $\mathbf{P}$ are thought of as vectors, and not points, of E^n. Thus we can think of Π_i as a (linear) operator that assigns to each free vector $\overrightarrow{\mathbf{PQ}}$ in E^n the "signed measure" of the oriented projection of $\overrightarrow{\mathbf{PQ}}$ on the ith coordinate axis.

Early in the study of calculus the differential of a real-valued function of a real variable was defined. Great pains were taken to make the point that dx, the differential of x, could not be defined; it, like the symbol x itself, was simply an indeterminate, a variable if you like, which ranged over $\mathbb{R}$. But if one had a differentiable function f, then the differential of f, df, could be defined in terms of the indeterminate dx and the derivative f'. It was, if you recall, defined by the formula $df=f'dx$, and its value at any x was the formal product $f'(x)dx$. Thus df was really a function of two variables, x and dx, linear in dx. If f was a differentiable real-valued function on E^n, then the analog of the indeterminate dx was the undefined vector $(dx_1,\ldots,dx_n)$, and we defined $df = (\mathbf{grad}\, f)\cdot d\mathbf{x} = \Sigma_{i=1}^n \partial f/\partial x_i dx_i = \Sigma_{i=1}^n D_i f dx_i$. We are now going to give a meaning to these undefined symbols dx_i.

Definition 5.2.1. By the symbol dx_i we mean the functional $\Pi_i: E^n \to \mathbb{R}$. Hence dx_i assigns to each directed line segment in E^n the signed measure of its projection onto the ith coordinate axis. We call the linearly independent set $\{dx_i\}_{i=1}^n$ the set of basic 1-forms on E^n.

Definition 5.2.2. A 1-form on E^n is an expression of the form:

$$\omega = \sum_{i=1}^{n} A_i(\mathbf{x})\, dx_i,$$

where each A_i is a real-valued function on E^n. (The symbol ω is pronounced omega.) The 1-form ω is said to be constant, integrable, continuous, and differentiable if the same can be said for each of the A_i.

Suppose ω is a constant 1-form on E^n, with $A_i(\mathbf{x}) \equiv a_i$ for each i. Let $\overrightarrow{\mathbf{PQ}}$ be an arbitrary directed line segment, or free vector, in E^n. Then

$$\omega(\overrightarrow{\mathbf{PQ}}) = \sum_{i=1}^{n} a_i dx_i(\overrightarrow{\mathbf{PQ}}) = \sum_{i=1}^{n} a_i(Q_i - P_i).$$

What physical situation could this equation describe? Suppose $\mathbf{F} = (F_1,\ldots,F_n)$ is a constant force field on E^n, and $\overrightarrow{\mathbf{PQ}}$ an oriented line segment. If $\omega = \Sigma_{i=1}^n F_i\, dx_i$, then $\omega(\overrightarrow{\mathbf{PQ}})$ represents the amount of work done by the force field in moving a unit point mass from $\mathbf{P}$ to $\mathbf{Q}$.

Suppose now that ω is a continuous 1-form on $\overrightarrow{\mathbf{PQ}}$. Each A_i is actually uniformly continuous on the line segment $\overrightarrow{\mathbf{PQ}}$. Let λ be the length of $\overrightarrow{\mathbf{PQ}}$. Partition the segment $\mathbf{P} = \mathbf{P}_0, \mathbf{P}_1, \ldots, \mathbf{P}_m = \mathbf{Q}$, so that if $\varepsilon > 0$ has been given, the difference on each subinterval $\overrightarrow{\mathbf{P}_{j-1}\mathbf{P}_j}$ between max $A_i(\mathbf{x})$ and min $A_i(\mathbf{x})$, for each $i = 1,\ldots,n$, is less than $\varepsilon/\lambda n$. This can be done because of the uniform continuity of each A_i on $\overrightarrow{\mathbf{PQ}}$. (In the next chapter we shall prove that continuous functions on such sets are uniformly continuous.) If λ_j is the length of the subsegment $\overrightarrow{\mathbf{P}_{j-1}\mathbf{P}_j}$, then the value of $\omega(\overrightarrow{\mathbf{P}_{j-1}\mathbf{P}_j})$ lies between $\underline{\omega}(\overrightarrow{\mathbf{P}_{j-1}\mathbf{P}_j}) = \Sigma_{i=1}^n \min A_i(\mathbf{x})\, dx_i(\overrightarrow{\mathbf{P}_{j-1}\mathbf{P}_j})$ and $\overline{\omega}(\overrightarrow{\mathbf{P}_{j-1}\mathbf{P}_j}) = \Sigma_{i=1}^n \max A_i(\mathbf{x})\, dx_i(\overrightarrow{\mathbf{P}_{j-1}\mathbf{P}_j})$. The difference of these two values is less than $(\varepsilon/\lambda n)(n\lambda_j) = \varepsilon\lambda_j/\lambda$. It follows that the difference between a maximal value for ω on $\overrightarrow{\mathbf{PQ}}$ and a minimal value, relative to that partition, is less than ε since $\Sigma_{j=1}^m \lambda_j = \lambda$. Hence in this case we can write

$$\omega(\overrightarrow{\mathbf{PQ}}) = \lim_{\max \lambda_j \to 0} \sum_{j=1}^{m} \sum_{i=1}^{n} A_i(\mathbf{x})\, dx_i\big(\overrightarrow{\mathbf{P}_{j-1}\mathbf{P}_j}\big) = \int_{\overrightarrow{\mathbf{PQ}}} \sum_{i=1}^{n} A_i(\mathbf{x})\, dx_i = \int_{\overrightarrow{\mathbf{PQ}}} \omega.$$

Physically, this might represent the work done by a continuous force field **F** on E^n in moving a unit point mass from **P** to **Q** along $\overrightarrow{\mathbf{PQ}}$. We are naturally led to the following:

Definition 5.2.3. If $\omega = \Sigma_{i=1}^n A_i(\mathbf{x})\,dx_i$ is a 1-form on E^n and $\overrightarrow{\mathbf{PQ}}$ is an oriented line segment, then

$$\int_{\overrightarrow{\mathbf{PQ}}} \omega = \lim_{\|\pi\| \to 0} \sum_{j=1}^{m} \left(\sum_{i=1}^{n} A_i(\mathbf{x})\, dx_i\left(\overrightarrow{\mathbf{P}_{j-1}\mathbf{P}_j}\right) \right) = \omega(\overrightarrow{\mathbf{PQ}}),$$

where π is the partition $\mathbf{P} = \mathbf{P}_0, \mathbf{P}_1, \mathbf{P}_2, \ldots, \mathbf{P}_m = \mathbf{Q}$, $\|\pi\|$ is the maximum of the lengths of the segments $\overrightarrow{\mathbf{P}_{j-1}\mathbf{P}_j}$, and $\mathbf{x} \in \overrightarrow{\mathbf{P}_{j-1}\mathbf{P}_j}$, if this limit exists. If ω is a constant 1-form, then

$$\int_{\overrightarrow{\mathbf{PQ}}} \omega = \sum_{i=1}^{n} A_i(Q_i - P_i) = \omega(\overrightarrow{\mathbf{PQ}}).$$

Consider the following special example. Let **v** be the oriented line segment from the origin to the point $\mathbf{v} = (v_1, \ldots, v_n) \in E^n$. Let $\mathbf{A}(\mathbf{x})$ be defined by the formula $\forall \mathbf{x} \in E^n \setminus \{\mathbf{0}\}$, $\mathbf{A}(\mathbf{x}) = \mathbf{x}/\|\mathbf{x}\|$, $\mathbf{A}(\mathbf{0}) = \mathbf{0}$. Then the vector-valued function $\mathbf{A}(\mathbf{x}) = (1/\|\mathbf{x}\|)(x_1, \ldots, x_n)$, and integration of the 1-form $\omega = \Sigma_{i=1}^n A_i(\mathbf{x})\,dx_i$ over **v** gives the value

$$\int_{\mathbf{v}} \omega = \sum_{i=1}^{n} A_i(\mathbf{x}) v_i = \sum_{i=1}^{n} \frac{v_i}{\|\mathbf{v}\|} v_i = \|\mathbf{v}\|.$$

This is not a surprising result since $\mathbf{A}(\mathbf{x})$ is a unit force field whose direction at any point of the line segment **v** is the direction of **v**. Hence the total work done by this unit force field in moving a unit point mass from **0** to **v** is just the product of the magnitude of the force, unity, and the length of the line segment through which the force acts, $\|\mathbf{v}\|$.

If γ is an arc in E^n, we say γ is rectifiable if we can approximate γ by a polygonal line. More precisely stated, suppose **P** and **Q** are the endpoints of γ, and we partition γ by $\pi = \{\mathbf{P} = \mathbf{P}_0, \mathbf{P}_1, \ldots, \mathbf{P}_m = \mathbf{Q}\}$ and consider the length L of the polygonal line determined by π. If, as $\|\pi\|$, the maximum of the lengths of the segments $\overrightarrow{\mathbf{P}_{j-1}\mathbf{P}_j}$, approaches zero, L approaches a limit, we say γ is *rectifiable*, with length equal to the limit of L. It should be proved by the reader that if γ is rectifiable, and Γ is an approximating polygonal line, then γ and Γ are uniformly close together.

Now suppose ω is a continuous 1-form on E^n and hence uniformly continuous on bounded sets. Let γ be a rectifiable arc; γ is a bounded set in E^n and is of course contained in a bounded set, so ω is uniformly continuous over a neighborhood of γ. If Γ is an approximating polygonal line, we have already seen that $\int_\Gamma \omega$ exists. Since the length of γ is defined to be the limit of the lengths of approximating polygonal lines, it is reasonable to expect that $\lim_{\Gamma \to \gamma} \int_\Gamma \omega$ exists and that we should define that limit to be the value of $\int_\gamma \omega$. Let's see if we can show that the limit really does exist.

Since $\omega=\sum_{i=1}^{n}A_i(\mathbf{x})\,dx_i$ is uniformly continuous on bounded sets, so is each $A_i(\mathbf{x})$. Let λ be the length of γ, and let $\varepsilon>0$ be arbitrarily small. Let π_1, π_2 be two arbitrary partitions of γ with $\boldsymbol{\Gamma}_1$, $\boldsymbol{\Gamma}_2$ the associated polygonal lines, but each sufficiently fine so that a ball of radius δ, centered at any partition point of either partition contains at least two partition points of each partition as well as the subarcs of γ joining these points, where δ is a number sufficiently small such that for all $\mathbf{x}$, $\mathbf{y}$ in a ball containing γ we have for each $i=1,2,\ldots,n$, $\|\mathbf{x}-\mathbf{y}\|<\delta\Rightarrow|A_i(\mathbf{x})-A_i(\mathbf{y})|<\varepsilon/2n\lambda$. Now we consider the partial integrals $\int_{\boldsymbol{\Gamma}_1}A_i(\mathbf{x})\,dx_i$ and $\int_{\boldsymbol{\Gamma}_2}A_i(\mathbf{x})\,dx_i$. It is not hard to see that $dx_i(\boldsymbol{\Gamma}_1)=dx_i(\boldsymbol{\Gamma}_2)$, i.e., that the lengths of the projections of the two polygonal lines on the x_i axis are equal since these projections have identical endpoints. Let λ_i be the length of the projection; then the difference of these two partial integrals is less than $\varepsilon\lambda_i/2n\lambda$ in absolute value. Since $\max_i\lambda_i\leqslant\lambda$, we can conclude that the difference $|\int_{\boldsymbol{\Gamma}_1}\omega-\int_{\boldsymbol{\Gamma}_2}\omega|<\varepsilon/2$. Since $\varepsilon>0$ was arbitrary, we conclude that a sequence of partitions of $\{\pi_k\}$ whose norms approach zero gives rise to a Cauchy sequence of numbers $\{\int_{\boldsymbol{\Gamma}_k}\omega\}$, which sequence must converge in $\mathbb{R}$.

Definition 5.2.4. If γ is a rectifiable arc in E^n, $\{\boldsymbol{\Gamma}_k\}$ is a sequence of approximating polygonal lines whose norms approach zero, and ω is a 1-form on E^n, then

$$\int_\gamma\omega=\lim_{\boldsymbol{\Gamma}_k\to\gamma}\int_{\boldsymbol{\Gamma}_k}\omega,$$

provided the integrals on the right-hand side exist and the limit exists.

We summarize our results in the following theorem.

Theorem 5.2.5. *Let γ be an oriented rectifiable arc, or curve if the initial and terminal points coincide, in E^n. If ω is a continuous 1-form on E^n, the integral $\int_\gamma\omega$ exists. If we reverse the orientation of γ and indicate this by writing $-\gamma$, then $\int_{-\gamma}\omega=-\int_\gamma\omega$.*

The last statement of the theorem is clear when one considers that for a direct line segment $\overrightarrow{\mathbf{PQ}}$, it is true that $\int_{\overrightarrow{\mathbf{PQ}}}\omega=-\int_{\overrightarrow{\mathbf{QP}}}\omega$.

5.3 SOME APPLICATIONS OF 1-FORMS

The first application we are going to consider is one in the plane. Suppose $\mathbf{v}=(v_x,v_y)$ is a constant flow vector in the plane, i.e., $\mathbf{v}$ gives the direction and the amount of flow across a unit normal per unit of time of a thin sheet of incompressible fluid moving in the plane. Let $\mathbf{u}$ be a fixed oriented line segment in the plane. We are interested in finding the amount of flow across $\mathbf{u}$, but first we must agree as to what is positive and what is negative flow across. The ground rules will be if, as we stand at the initial point of **u** and look toward the terminal point, the flow is from left to right, the flow across will be positive; otherwise it will be negative. Of course, if the flow is in the direction of **u** or $-$**u**, the flow across will be zero.

Consider Figure 5.4. The free vector **u** has x and y components u_x and u_y, and **v** has components v_x and v_y. The total flow across both vectors $\mathbf{u}_x$ and $\mathbf{u}_y$ is

$$-v_y \times (\text{signed length of } \mathbf{u}_x) + v_x \times (\text{signed length of } \mathbf{u}_y),$$

which is the same thing as $-v_y\,dx(\mathbf{u}) + v_x\,dy(\mathbf{u})$. This quantity represents the total net flow "into" the triangle across its legs; since the fluid is incompressible, the same amount must flow "out" across the hypotenuse. We conclude that if $\mathbf{v} = (v_x, v_y)$ is a constant flow vector in the plane and **u** is a free vector, the flow per unit of time across **u** is given by $-v_y\,dx(\mathbf{u}) + v_x\,dy(\mathbf{u})$. If $\mathbf{v} = (v_x(x, y), v_y(x, y))$ is nonconstant, but continuous, we can partition **u**, pass to the limit, and obtain the result that the net total flow across **u** per unit of time is equal to $\int_{\mathbf{u}} -v_y(x, y)\,dx + v_x(x, y)\,dy$. Finally, if γ is an oriented rectifiable arc in E^2 and **v** is a continuous flow vector, then the net total flow across γ per unit of time is given by

$$\int_\gamma -v_y(x, y)\,dx + v_x(x, y)\,dy.$$

A second application is one in E^n. If **F** is a constant force field and $\overrightarrow{\mathbf{PQ}}$ a free vector in E^n, we have already mentioned the integral $\int_{\overrightarrow{\mathbf{PQ}}} F_1\,dx_1 + \cdots + F_n\,dx_n$ represents the amount of work done by the field in moving a

FIGURE 5.4 Flow across in E^2.

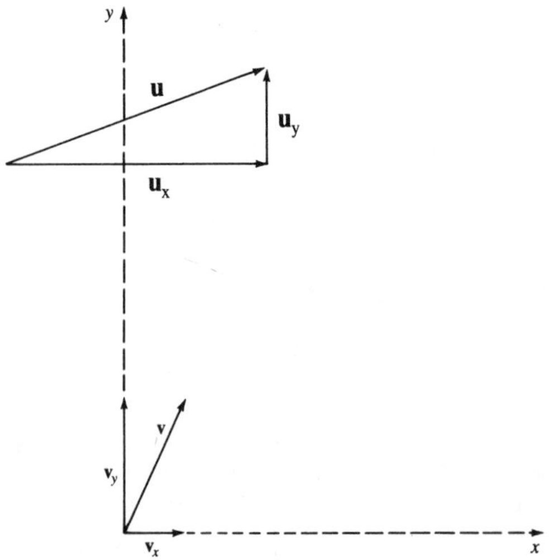

unit point mass from **P** to **Q** along $\overrightarrow{\mathbf{PQ}}$. Now suppose that the field **F** is continuous and γ is an oriented rectifiable arc, or curve if the endpoints coincide. Then the integral

$$\int_\gamma F_1(\mathbf{x})\,dx_1 + \cdots + F_n(\mathbf{x})\,dx_n$$

exists and represents the work done in moving a unit point mass along γ from the initial point to the terminal point.

Suppose now that **F** is a unit vector whose direction at each point $\mathbf{x} \in \gamma$ is in the direction of the tangent vector to γ at **x** (the direction of the tangent vector follows the orientation of γ, and we are tacitly assuming here that γ is sufficiently smooth so as to have a nonzero tangent vector at each point). Then numerically the above integral is the same as the length of γ. This is just what you would expect. Work is the product of a force and a distance over which it acts; in this case the magnitude of the force is always unity, and its direction of application is always in the direction of γ. Hence the total work done is simply the length of γ.

To find the length of the arc γ, we need to find in some way or other the unit tangent vector to γ, $\mathbf{T}(\mathbf{x}) = (T_1(\mathbf{x}), \ldots, T_n(\mathbf{x}))$, $\mathbf{x} \in \gamma$. This vector may then play the role of the unit force field. Now γ will almost invariably be described in terms of a parameter. This is to say, we shall be given a set of equations, called parametric equations, of the form

$$x_1 = f_1(t), \quad \ldots, \quad x_n = f_n(t), \qquad a \leqslant t \leqslant b.$$

These equations define a mapping $\boldsymbol{\Gamma}: [a,b] \to E^n$, where specifically, $\boldsymbol{\Gamma}([a,b]) = \gamma \subset E^n$, with $\boldsymbol{\Gamma}(a), \boldsymbol{\Gamma}(b)$, respectively, the initial and terminal points of γ. It turns out that this mapping $\boldsymbol{\Gamma}$, defined by these n equations, maps $[a,b]$ onto the arc γ such that there is a nonzero tangent vector at each point of the arc as long as all the f_i are differentiable and the n derivatives never vanish simultaneously. Assuming this to be the case, we can write down the unit tangent vector to γ, but in terms of the parameter t:

$$\mathbf{T}^*(t) = \left(f_1'(t), \ldots, f_n'(t) \Big/ \sqrt{\sum_{i=1}^{n} [f_i'(t)]^2} \right).$$

The vector $\mathbf{T}^*(t)$ is what we call the pullback of $\mathbf{T}(\mathbf{x})$. You see, the mapping $\boldsymbol{\Gamma}$ which describes the arc γ sends t into x_is. We take $\mathbf{T}(\mathbf{x})$ and "pull back" the x_is to t, so that $\mathbf{T}(\mathbf{x})$ becomes a vector-valued function of t, namely, $\mathbf{T}^*(t)$. To pull back the 1-form ω to ω^*, we have to know how to pull back each dx_i. As you might guess, dx_i really does pull back to $f'(t)\,dt$, and we can write

$$\int_\gamma \omega = \int_\gamma \sum_{i=1}^{n} T_i(\mathbf{x})\,dx_i = \int_{\overrightarrow{ab}} \omega^* = \int_a^b \sqrt{\sum_{i=1}^{n} [f_i'(t)]^2}\,dt.$$

Needless to say, what we have done here will be justified later in the course.

As an example of a planar flow problem we present the following idealized one. Suppose that the sun is at the center rather than at a focus of our orbit and a flow of strange particles whose flux density obeys an inverse cube law emanates from the sun and crosses our orbit. What is the flow across per unit of time if the flow vector is

$$\mathbf{v}(x,y)=\left(\frac{kx}{(x^2+y^2)^2},\frac{ky}{(x^2+y^2)^2}\right)$$

for a certain constant k? Let the orbit be given by the parametric equations

$$x=a\cos t,\quad y=b\sin t,\qquad t\in[0,2\pi),$$

where a and b are certain positive constants.

We let γ be the quarter-orbit, $0\leqslant t\leqslant\pi/2$, and compute the integral

$$4\int_\gamma\left(\frac{-ky}{(x^2+y^2)^2}\,dx+\frac{kx}{(x^2+y^2)^2}\,dy\right).$$

To carry out the computation, we must make substitutions, which is to say, we must find all the pullbacks and then compute the transformed integral. The above integral pulls back to

$$4\int_0^{\pi/2}\left[\frac{-kb\sin t}{(a^2\cos^2t+b^2\sin^2t)^2}(-a\sin t)+\frac{ka\cos t}{(a^2\cos^2t+b^2\sin^2t)^2}(b\cos t)\right]dt$$

$$=4abk\int_0^{\pi/2}\frac{dt}{(a^2\cos^2t+b^2\sin^2t)^2}=\frac{k\pi(a^2+b^2)}{a^2b^2}.$$

This is the total flow across the orbit per unit of time. (We remark that the last integral is a standard one that appears in most integral tables. In Problem 8.20 we shall consider methods of evaluating such integrals.)

5.4 0-FORMS IN E^n

By a 0-form on E^n we simply mean a real-valued function on E^n. The function identically equal to 1 will be our basic 0-form, just as the set $\{dx_1,\ldots,dx_n\}$ comprised the basic 1-forms on E^n. If ω is a 0-form, its value at a point $\mathbf{v}\in E^n$ is $\omega(\mathbf{v})$, simply the function value at $\mathbf{v}$. Hence 0-forms are in general nonlinear functionals on the set of points (0-parallelotopes) of E^n.

At this time it is convenient to introduce the notion of a zero-dimensional oriented set. We have stated that a single point is not orientable; on the other hand, if $\{\mathbf{P}_i\}$ is a finite or countable set of points containing more than one point, we consider this set to be (positively) oriented once the set has been sequentially ordered and the point $\mathbf{P}_1$ has been designated the initial point.

Having defined the integral of a 1-form over an oriented 1-parallelotope, we proceed to define the integral of a 0-form over an oriented zero-dimensional set.

Definition 5.4.1. Let $(\mathbf{P}_1,\mathbf{P}_2)$ be an oriented pair of points and ω a 0-form on E^n. Then $\omega((\mathbf{P}_1,\mathbf{P}_2))=\int_{(\mathbf{P}_1,\mathbf{P}_2)}\omega=\omega(\mathbf{P}_2)-\omega(\mathbf{P}_1)$. More generally, if $S=\{\mathbf{P}_i\}_{i=1}^n$ is an oriented 0-dimensional set, $\omega(S)=\int_S\omega=\sum_{i=1}^{n-1}(\omega(\mathbf{P}_{i+1})-\omega(\mathbf{P}_i))=\omega(\mathbf{P}_n)-\omega(\mathbf{P}_1)$. Finally, if $S=\{\mathbf{P}_i\}_{i=1}^\infty$ is a sequence of distinct points, $\omega(S)=\int_S\omega=\lim_{n\to\infty}\omega(\mathbf{P}_n)-\omega(\mathbf{P}_1)$ if this limit exists.

Since it follows logically that we should do this, we make a slight extension of this definition: if $(\mathbf{P}_1,\mathbf{P}_2)$ is an ordered pair of points and it happens that $\mathbf{P}_1=\mathbf{P}_2$, we define the integral $\int_{(\mathbf{P}_1,\mathbf{P}_2)}\omega$ to be zero.

Everyone knows that the integral of a real-valued function over a finite set of points is zero, so if ω is a real-valued function, how can we have $\int_{(\mathbf{P}_1,\mathbf{P}_2)}\omega=\omega(\mathbf{P}_2)-\omega(\mathbf{P}_1)$? Think again! In elementary calculus one does not integrate 0-forms. What we are doing here is something new. These new integrals aren't of any apparent use at the moment, but they fit very nicely into the grand plan of things later on.

Suppose ω is a 0-form on E^n. Let's assume for the present that ω is what we call a C^1 function; i.e., ω and all its first partial derivatives are continuous. The differential of ω can only be, if we want to stay in step with the rest of mathematics,

$$d\omega=\sum_{i=1}^{n} A_i(x)\,dx_i,$$

where for each $i=1,\ldots,n$, $A_i(x)=(\partial\omega/\partial x_i)(x)$. Thus $d\omega$ is a continuous 1-form, and we can compute the integral

$$\int_{\overrightarrow{\mathbf{PQ}}} d\omega.$$

Let $\mathbf{\Gamma}:[0,1]\to\overrightarrow{\mathbf{PQ}}\subset E^n$ be the mapping described by the parametric equations

$$x_1=x_1(t),\quad \ldots,\quad x_n=x_n(t),\qquad 0\leqslant t\leqslant 1,$$

where $\mathbf{\Gamma}(0)=\mathbf{P}$, $\mathbf{\Gamma}(1)=\mathbf{Q}$. The function $\omega(x)$ pulls back under $\mathbf{\Gamma}$ to a function of t, $g(t)$, and by the chain rule for differentiation,

$$g'(t)=\left[\frac{\partial\omega}{\partial x_1}x_1'(t)+\cdots+\frac{\partial\omega}{\partial x_n}x_n'(t)\right],$$

so

$$dg(t)=\sum_{i=1}^{n}\frac{\partial\omega}{\partial x_i}dx_i=d\omega.$$

This is to say, the pullback of $d\omega$ is dg; thus

$$\omega^* = g, \qquad d\omega^* = dg, \qquad (\overrightarrow{\mathbf{PQ}})^* = [0,1].$$

It is a fact that

$$\int_{\overrightarrow{\mathbf{PQ}}} d\omega = \int_{(\overrightarrow{\mathbf{PQ}})^*} d\omega^*,$$

so

$$\int_{\overrightarrow{\mathbf{PQ}}} d\omega = \int_{[0,1]} g'(t)\,dt = g(1) - g(0) = \omega(\mathbf{Q}) - \omega(\mathbf{P}) = \int_{(\mathbf{P},\mathbf{Q})} \omega.$$

If we introduce the notation ∂ to indicate "the boundary of," we can write for a C^1 0-form ω and a free vector $\overrightarrow{\mathbf{PQ}}$,

$$\int_{\overrightarrow{\mathbf{PQ}}} d\omega = \int_{\partial(\overrightarrow{\mathbf{PQ}})} \omega.$$

This equation extends immediately, in light of what we have already done, to the following.

Theorem 5.4.2 (Stokes's Theorem). *Let ω be a C^1 0-form on E^n, γ be an oriented rectifiable arc in E^n, and $\partial(\gamma)$ be the boundary of γ with its inherited orientation. Then*

$$\int_{\partial(\gamma)} \omega = \int_\gamma d\omega.$$

Do you recognize this famous theorem from elementary calculus? It looked something like this:

$$f(b) - f(a) = \int_a^b f'(x)\,dx,$$

and it was called the fundamental theorem of integral calculus. It is fitting that we refer to the generalized Stokes's theorem as the fundamental theorem of this course; this theorem states that the integral of the differential of a C^1 $(k-1)$-form over a k-dimensional oriented manifold with boundary is equal to the integral of the form over the boundary of the manifold.

It is well to remark that if γ is an oriented rectifiable arc with coincident endpoints, Stokes's theorem guarantees that the integral of a 1-form ω over γ will be equal to zero whenever there is a C^1 0-form whose differential is ω.

Definition 5.4.3. A 1-form on E^n, $\omega = \sum_{i=1}^n A_i(\mathbf{x})\,dx_i$, is said to be *exact* if there exists a 0-form φ on E^n such that $d\varphi = \omega$ on E^n. The form ω is called *closed* if $d\omega = 0$.

Another important observation to be made is if ω is an exact continuous 1-form, then the integral of ω over γ is independent of γ except for the endpoints. Hence, if we have to evaluate $\int_\gamma \omega$, the obvious first step is to see if ω is exact. If it is, look at the endpoints **P** and **Q** of γ; the value of the integral is independent of the path from **P** to **Q**, so choose one that makes the evaluation easier.

If a 1-form ω is exact, it should follow that $d\omega \equiv 0$, for consider: if φ is a 0-form with $d\varphi = \omega$, then Stokes's theorem implies that

$$\int_\gamma \omega = \int_R d\omega = \int_{\partial\gamma} \varphi,$$

where R is some two-dimensional manifold for which γ is the boundary. But the boundary of a boundary is at most a pair of coincident points in this case, so $\int_R d\omega$ is zero for every possible 2-manifold having γ as a boundary. In later sections we shall see that what we suspect is indeed true, so we shall conclude that if ω is a differentiable 1-form and $d\omega \not\equiv 0$, then ω is not exact. But what in the world is $d\omega$? This is a natural lead-in to the next section.

5.5 2-FORMS IN E^n

By differentiating 0-forms on E^n we obtained the exact 1-forms on E^n, and it seems reasonable to believe that there are many nonexact 1-forms (indeed there are!). To construct 2-forms we should start with the differentiable 1-forms ω and formally obtain the differentials $d\omega$; these we shall call the exact 2-forms. A basic 1-form is a "constant" element of the linear set of 1-forms over the ring of real-valued functions on E^n, so we would naturally define the differential of such a thing to be zero. On the other hand, an element such as $A_i(\mathbf{x})\,dx_i$ is not "constant" in the same sense. To differentiate this element, we formally differentiate it as an ordinary product:

$$\begin{aligned} d(A_i(\mathbf{x})\cdot dx_i) &= \left[\frac{\partial A_i(\mathbf{x})}{\partial x_1}dx_1 + \cdots + \frac{\partial A_i(\mathbf{x})}{\partial x_n}dx_n\right]dx_i + 0 \\ &= D_1A_i(\mathbf{x})\,dx_1\,dx_i + \cdots + D_nA_i(\mathbf{x})\,dx_n\,dx_i. \end{aligned}$$

If $\omega = \Sigma_{i=1}^n A_i(\mathbf{x})\,dx_i$, it follows that

$$d\omega = \sum_{i=1}^n \sum_{j=1}^n D_jA_i(\mathbf{x})\,dx_j\,dx_i.$$

For us the symbol dx_i is no longer an undefined term; it is a functional on the set of all free vectors, or oriented line segments, in E^n. We shall define the symbol $dx_i\,dx_j$ to stand for the functional that assigns to each oriented parallelogram in E^n the signed measure of its projection on the ijth coordinate subspace of E^n. Recall that in general we can't speak of a parallelogram in E^n as being positively or negatively oriented, but when we can so orient the coordinate subspace projections, it is customary to refer to the parallelogram as an oriented one. Our definition for $dx_i\,dx_j$ does not seem so unreasonable: if P is a parallelogram with a vertex at the origin, determined by the two vectors **u**, **v**, then dx_i, operating on a line segment, gives a measure in the x_i direction, and dx_j, operating on a line segment, gives a measure in the x_j direction. The product of two such measures is an area in the x_ix_j-plane. But as we shall see, $dx_i\,dx_j$ is a very special kind of product.

The functionals dx_i are linear when we consider them as mappings of the vector space $\mathbb{R}^n \to \mathbb{R}$; strictly speaking, they are not linear functionals on the set of directed line segments because this set is not really a linear space. Analogously, the functionals $dx_i\,dx_j$ are not linear; we don't have a way to form linear combinations of parallelograms.

Suppose P is the oriented parallelogram in E^5 whose "primary" side is the vector $(3,1,7,-1,2)$ and whose other determining side is the vector $(-1,-2,3,4,5)$. P then is the oriented parallelogram which is the image of the positively oriented 2-cube under the mapping T whose matrix is

$$A=\begin{bmatrix} 3 & -1 \\ 1 & -2 \\ 7 & 3 \\ -1 & 4 \\ 2 & 5 \end{bmatrix}.$$

$dx_2\,dx_4(P)$ is the signed measure of the projection of P onto $\operatorname{sp}(\mathbf{e}_2,\mathbf{e}_4)$. The mapping sending the 2-cube to the projection has for its matrix

$$A_{24}=\begin{bmatrix} 0 & 0 \\ 1 & -2 \\ 0 & 0 \\ -1 & 4 \\ 0 & 0 \end{bmatrix}.$$

The projection of P onto $\operatorname{sp}(\mathbf{e}_2,\mathbf{e}_4)$ is positively oriented, because

$$\det\begin{pmatrix} 1 & -2 \\ -1 & 4 \end{pmatrix}=2>0.$$

We shall see in Chapter 7 that the Euclidean measure of this projection of P is 2, the value of the subdeterminant; since it is positively oriented, we conclude

$$dx_2\,dx_4(P)=2.$$

What if we had wanted to calculate $dx_4\,dx_2(P)$? The projection of course is the same, but we are suggesting that the "natural" orientation of $\operatorname{sp}(\mathbf{e}_2,\mathbf{e}_4)$ be reversed. Hence we should interpret $dx_4\,dx_2(P)$ to be the negative of $dx_2\,dx_4(P)$. Indeed, this is what we do.

To calculate all the $dx_i dx_j(P)$, we note that first $dx_i dx_j(P) = -dx_j dx_i(P)$ and that $dx_i dx_i(P) = 0$, since the two-dimensional measure of the projection of P onto the x_i axis is zero. Hence we need only to calculate the values for which $1 \leqslant i < j \leqslant 5$.

$$dx_1 dx_2(P) = -5, \qquad dx_1 dx_3(P) = 16,$$
$$dx_1 dx_4(P) = 11, \qquad dx_1 dx_5(P) = 17,$$
$$dx_2 dx_3(P) = 17, \qquad dx_2 dx_4(P) = 2, \qquad dx_2 dx_5(P) = 9,$$
$$dx_3 dx_4(P) = 31, \qquad dx_3 dx_5(P) = 29,$$
$$dx_4 dx_5(P) = -13.$$

We have simply put down the values of the determinants of the 2×2 submatrices consisting of the ith and jth rows.

Having calculated these ten measures of the projections of P onto the coordinate subspaces, we can apply the Pythagorean theorem to get the area of P; which is to say, we calculate

$$\left[(-5)^2 + (16)^2 + \cdots + (29)^2 + (-13)^2\right]^{1/2} = 2\sqrt{759}\,.$$

(Do you believe it? If not, use a linear algebra formula for the area of P and see what you get.)

Definition 5.5.1. A *basic* 2-form on E^n, denoted by the symbol $dx_i dx_j$, $1 \leqslant i < j \leqslant n$, is a functional on the class of oriented 2-parallelotopes in E^n, assigning to each such figure the number which is the signed measure of the parallelotope's projection on the coordinate subspace $\mathrm{sp}(\mathbf{e}_i, \mathbf{e}_j)$. For arbitrary i, $j = 1, 2, \ldots, n$, $dx_i dx_j = -dx_j dx_i$ and $dx_i dx_i = 0$.

A *general* 2-form on E^n is an expression of the form

$$\omega = \sum_{1 \leqslant i < j \leqslant n} A_{ij}(\mathbf{x})\, dx_i dx_j,$$

where the $A_{ij}(\mathbf{x})$ are real-valued functions on E^n.

Note that the set of 0-forms on E^n constitute a one-dimensional linear set over the ring of real-valued functions on E^n, the set of 1-forms an n-dimensional linear set, and the set of 2-forms a $\binom{n}{2}$-dimensional linear set over this ring. Recall the Grassmann algebra.

When writing general 2-forms, we naturally write $dx_i dx_j$ in lexicographical order, i.e., $i < j$, since our basis for the linear set of 2-forms consists of these symbols. However, in ordinary Euclidean 3-space, where the axes are usually denoted by x, y, and z and (important for orientation) the right-hand rule is observed, we often see a general 2-form written in this order:

$$\omega = A_1(x, y, z)\, dy\, dz + A_2(x, y, z)\, dz\, dx + A_3(x, y, z)\, dx\, dy.$$

If ω is a constant 2-form, we "evaluate" on parallelograms as we did in our example. In the event that ω is continuous, we define $\omega(P)$ to be the limit, if the limit exists, of the sums of the evaluations we get by partitioning P into a finite number of subparallelograms, and regarding ω as essentially constant on each subparallelogram. This is to say,

$$\omega(P)=\int_P \omega$$

and, even more generally, if S is an oriented surface that can be approximated suitably by pasting together oriented parallelograms, then $\omega(S)=\int_S \omega$. If $-S$ is S with the orientation reversed, then we have $\omega(-S)=-\omega(S)$. We hasten to add that the evaluation of a 2-form over a simple parallelogram is not a simple task.

Not all 2-forms are exact, but suppose ω is an exact 2-form. Then there is a differentiable 1-form $\boldsymbol{\sigma}$, such that $d\boldsymbol{\sigma}=\omega$. If

$$\boldsymbol{\sigma}=\sum_{i=1}^{n} A_i(\mathbf{x})\,dx_i,$$

then

$$\begin{aligned} d\boldsymbol{\sigma} &= \sum_{i=1}^{n}\sum_{j=1}^{n} D_j A_i(\mathbf{x})\,dx_j\,dx_i \\ &= \sum_{1\leqslant i<j\leqslant n} B_{ij}(\mathbf{x})\,dx_i\,dx_j, \end{aligned}$$

where

$$B_{ij}(\mathbf{x})=\left[D_i A_j(\mathbf{x})-D_j A_i(\mathbf{x})\right], \qquad i<j,$$

since $dx_i\,dx_j=-dx_j\,dx_i$ and $dx_i\,dx_i=0$. Thus, when ω is an exact 2-form, the "components," or "coordinates" of ω are those particular differences of the partial derivatives of components of the 1-form from which ω is derived. Now most of you know that if φ is a C^2-function (φ and its derivatives through the second are continuous), then the "mixed partial derivatives" of φ are equal. Hence, if $\boldsymbol{\sigma}$ had been an exact 1-form derived from φ, the functions $A_i(\mathbf{x})$ would have been the first partials of φ, and $D_i A_j(\mathbf{x})$, $D_j A_i(\mathbf{x})$ would be two ij-mixed partials and hence identically equal. Thus we have presented more conclusive evidence that the differential of an exact 1-form is zero.

5.6 AN APPLICATION IN E^3

Let $\mathbf{F}=(F_1, F_2, F_3)$ be a constant flow vector in ordinary 3-space; this is to say, at each point in space there is a flow whose direction is that of $\mathbf{F}$, and the "absolute" amount of flow through a unit cross-sectional area per unit of time is $\|\mathbf{F}\|$. Let P be any oriented parallelogram in E^3. We agree to regard the flow through P as positive if the direction of flow through P is

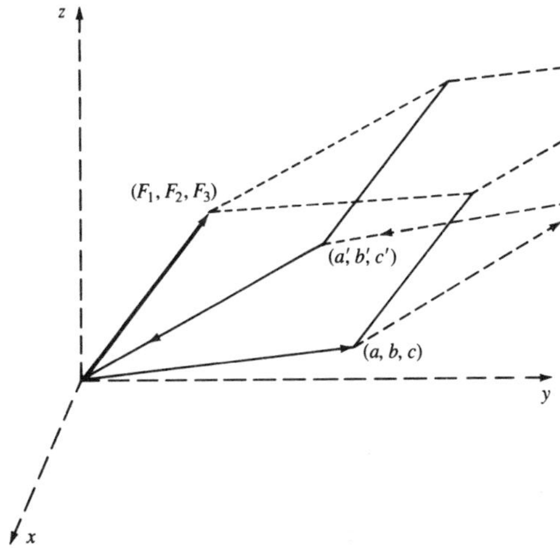

FIGURE 5.5 Flow-through in E^3.

the same as the general direction in which the (extended) thumb of your right hand is pointing when you hold your hand in such a way that the fingers point in the direction of the orientation of P and the interior of P is on the palm side of your hand. We assert that the total flow through P per unit of time is given by the 2-form

$$\omega(P) = F_1 dy\, dz(P) - F_2\, dx\, dz(P) + F_3\, dx\, dy\,(P).$$

To see that this formula holds, consider that the total flow through P per unit of time from the flow $\mathbf{F}$ is the sum of the flows contributed by the components F_1, F_2, F_3 through the projections of P on the yz, xz, and xy planes, respectively. Figure 5.5 may help to clarify matters.

Suppose P is the parallelogram determined by (a, b, c) and (a', b', c'), oriented as indicated in the figure, and (F_1, F_2, F_3) is the flow vector. (In this case, the flow-through is positive, by our convention). Now the total flow-through per unit of time is just the volume of the parallelopiped determined by the three vectors since it must be the same as the flow by $\mathbf{F}$ through the projection of P on the plane normal to (F_1, F_2, F_3), or the flow through P with respect to a flow vector which is the projection of (F_1, F_2, F_3) onto the normal to P. The volume of this parallelopiped is, as we shall prove later, equal to the absolute value of the determinant of the determining vectors. Hence the total flow-through, positive in this case, is

$$\left| \det \begin{bmatrix} F_1 & F_2 & F_3 \\ a & b & c \\ a' & b' & c' \end{bmatrix} \right|.$$

But the determinant is equal to

$$F_1(bc'-b'c)-F_2(ac'-a'c)+F_3(ab'-a'b).$$

The matrix of the map that sends the oriented square to P, with the orientation indicated, is

$$\begin{bmatrix} a & a' \\ b & b' \\ c & c' \end{bmatrix}.$$

By our convention,

$$dx\,dy(P)=(ab'-a'b), \qquad dx\,dz(P)=(ac'-a'c),$$
$$dy\,dz(P)=(bc'-b'c).$$

Thus the total flow-through per unit of time can be written

$$F_1 dy\,dz(P)-F_2 dx\,dz(P)+F_3 dx\,dy(P).$$

Note that if $\mathbf{F}$ is a constant unit vector, normal to P, the absolute value of the flow-through is the ordinary area of P. In the event that $\mathbf{F}$ is a continuous flow vector, we can partition P, pass to the limit, and obtain

$$\omega(P)=\int_P F_1 dy\,dz - F_2 dx\,dz + F_3 dx\,dy$$

as the value of the total flow-through per unit of time. Finally, if S is a suitable oriented two-dimensional region, we can write

$$\omega(S)=\int_S \omega = \int_S F_1\,dy\,dz - F_2\,dx\,dz + F_3\,dx\,dy$$
$$= \text{total flow through } S \text{ per unit of time.}$$

This then is our model for flow-through problems in E^3. It can be extended to higher dimensions; if $\mathbf{F}=(F_1,\ldots,F_n)$ is a flow vector and S is a suitable $(n-1)$-dimensional manifold in E^n, and

$$\omega=\sum_{i=1}^{n}(-1)^{i-1}F_i\,dx_1\cdots d\check{x}_i\cdots dx_n$$

is the $(n-1)$-form analogous to the 2-form we have been considering, then $\int_S\omega$ will give the total flow through the hypersurface S per unit of time. Incidentally, the symbol $dx_1\cdots d\check{x}_i\cdots dx_n$ means $dx_1\cdots dx_{i-1}dx_{i+1}\cdots dx_n$.

If you have an insane desire to calculate the three-dimensional measure of the 3-sphere S, i.e., the boundary of the 4-ball B, given by the equation $x^2+y^2+z^2+w^2=R^2$, you need only evaluate

$$\frac{1}{R}\int_S x\,dy\,dz\,dw - y\,dx\,dz\,dw + z\,dx\,dy\,dw - w\,dx\,dy\,dz$$

since $\mathbf{F}=(x/R, y/R, z/R, w/R)$ is a vector field normal to S and of unit length at each point of S. (You should obtain $2\pi^2R^3$.)

We need to clarify one small point at this time. Suppose $A(\mathbf{x})$ is a continuous real-valued function on E^2 and ω is the 2-form $A(\mathbf{x})\,dx_1\,dx_2$. Suppose S is a suitable region, and we consider $\int_S \omega = \int_S A(\mathbf{x})\,dx_1\,dx_2$. In this case, $-\omega = A(\mathbf{x})\,dx_2\,dx_1$, and it must follow that

$$\int_S A(\mathbf{x})\,dx_1\,dx_2 = -\int_S A(\mathbf{x})\,dx_2\,dx_1.$$

In ordinary "multiple integration" it makes no difference if we change the order of integration. Clearly, we must keep in mind that there is always a distinction between the integration of a 2-form over an oriented manifold and the integration of a real-valued function over a region.

In cases where confusion might arise, we shall use a single integral sign for the integration of forms, and multiple integral signs for the ordinary integration. Thus, if we are in E^2 and $A(\mathbf{x})$ is an integrable function and S an orientable region, $\int_S A(\mathbf{x})\,dx_1\,dx_2 = -\int_S A(\mathbf{x})\,dx_2\,dx_1$, but $\int\int_S A(\mathbf{x})\,dx_1\,dx_2 = \int\int_S A(\mathbf{x})\,dx_2\,dx_1$; as long as S is positively oriented,

$$\int_S A(\mathbf{x})\,dx_1\,dx_2 = \int\int_S A(\mathbf{x})\,dx_1\,dx_2.$$

5.7 A SUBSTANTIAL EXAMPLE

Let $P(\mathbf{A},\mathbf{B},\mathbf{C},\mathbf{D})$ be an oriented parallelogram in E^4, where

$$\mathbf{A} = (1,2,0,-1), \qquad \mathbf{C} = (-1,-12,6,-5),$$
$$\mathbf{B} = (3,1,0,0), \qquad \mathbf{D} = (-3,-11,6,-6)$$

are the vertices in order. Let ω be the 1-form on E^4

$$\omega = -(x_2+x_3+x_4)\,dx_1 - (x_3+x_4)\,dx_2 - x_4\,dx_3.$$

Let $\varphi: E^2 \to E^4$ be the affine map that sends the positively oriented square $K = [0,1]\times[0,1]$ onto P, with $\varphi(0,0) = \mathbf{A}$, $\varphi(1,0) = \mathbf{B}$, $\varphi(0,1) = \mathbf{D}$.

First, we want to exhibit φ in matrix form, then find $d\omega$, and then compute the pullbacks ω^* and $(d\omega)^*$ under φ. Having done that, we shall evaluate the integrals $\int_P d\omega$, $\int_K (d\omega)^*$, $\int_{\partial P}\omega$, $\int_{\partial K}\omega^*$. We shall also compare $(d\omega)^*$ with $d(\omega^*)$.

If P were not a parallelogram, we would be temporarily stymied, but it is. We first translate P by the vector $-\mathbf{A}$ in order to get a parallelogram with a vertex at the origin; the determining vectors of this translate are $(\mathbf{B}-\mathbf{A}) = (2,-1,0,1)$ and $(\mathbf{D}-\mathbf{A}) = (-4,-13,6,-5)$. Hence the required matrix representation for $\varphi: u \mapsto x$ is given by the expression

$$\begin{bmatrix} x_1 \\ x_2 \\ x_3 \\ x_4 \end{bmatrix} = \begin{bmatrix} 2 & -4 \\ -1 & -13 \\ 0 & 6 \\ 1 & -5 \end{bmatrix} \begin{bmatrix} u_1 \\ u_2 \end{bmatrix} + \begin{bmatrix} 1 \\ 2 \\ 0 \\ -1 \end{bmatrix}. \tag{5.7.1}$$

This matrix equation is really four simple linear equations expressing x_i in terms of u_1 and u_2, $i=1,2,3,4$. To find the dx_i expressions in terms of the differentials du_1, du_2 is most simply done, and we can again use a matrix equation to display the results.

$$\begin{bmatrix} dx_1 \\ dx_2 \\ dx_3 \\ dx_4 \end{bmatrix} = \begin{bmatrix} 2 & -4 \\ -1 & -13 \\ 0 & 6 \\ 1 & -5 \end{bmatrix} \begin{bmatrix} du_1 \\ du_2 \end{bmatrix}. \tag{5.7.2}$$

The matrix representing $d\varphi$ in this case is the same as the matrix of the linear part of the affine map φ.

To find all the values of $dx_i dx_j P$, $1 \leqslant i < j \leqslant 4$, is easy; we just evaluate all the subdeterminants in $d\varphi$.

$$dx_1 dx_2 P = -30, \qquad dx_1 dx_3 P = 12, \qquad dx_1 dx_4 P = -6,$$
$$dx_2 dx_3 P = -6, \qquad dx_2 dx_4 P = 18, \qquad dx_3 dx_4 P = -6.$$

Since $K = \varphi^{-1}(P)$ is clearly such that $du_1 du_2(K) = 1$, we can express the six equations immediately above by the single matrix equation

$$\begin{bmatrix} dx_1 dx_2 \\ dx_1 dx_3 \\ dx_1 dx_4 \\ dx_2 dx_3 \\ dx_2 dx_4 \\ dx_3 dx_4 \end{bmatrix} = \begin{bmatrix} -30 \\ 12 \\ -6 \\ -6 \\ 18 \\ -6 \end{bmatrix} du_1 du_2. \tag{5.7.3}$$

We have

$$\omega = -(x_2 + x_3 + x_4)\, dx_1 - (x_3 + x_4)\, dx_2 - x_4\, dx_3. \tag{5.7.4}$$

To find $d\omega$, we do straightforward differentiation as defined earlier.

$$\begin{aligned} d\omega &= -dx_2 dx_1 - dx_3 dx_1 - dx_4 dx_1 - dx_3 dx_2 - dx_4 dx_2 - dx_4 dx_3 \\ &= \sum_{1 \leqslant i < j \leqslant 4} dx_i dx_j. \end{aligned} \tag{5.7.5}$$

We are now ready to calculate the pullbacks relative to φ. First, we remark that

$$\begin{gathered} P^* = K, \\ (\overrightarrow{\mathbf{AB}})^* = \overrightarrow{(0,0)(1,0)}, \qquad (\overrightarrow{\mathbf{BC}})^* = \overrightarrow{(1,0)(1,1)}, \\ (\overrightarrow{\mathbf{CD}})^* = \overrightarrow{(1,1)(0,1)}, \qquad (\overrightarrow{\mathbf{DA}})^* = \overrightarrow{(0,1)(0,0)}. \end{gathered} \tag{5.7.6}$$

To compute ω^* is tiresome; we substitute in (5.7.4) the pullbacks for the x_i

and dx_i that we can read off from (5.7.1) and (5.7.2) to get

$$\begin{aligned}\omega^* &= -(-u_1-13u_2+2+6u_2+u_1-5u_2-1)(2du_1-4du_2)\\ &\quad -(6u_2+u_1-5u_2-1)(-du_1-13du_2)-(u_1-5u_2-1)(6du_2)\\ &=(u_1+25u_2-3)\,du_1+(7u_1-5u_2-3)\,du_2.\end{aligned} \tag{5.7.7}$$

To find $(d\omega)^*$, we substitute in (5.7.5) the pullbacks for the $dx_i\,dx_j$ from (5.7.3) to get

$$(d\omega)^* = -18\,du_1\,du_2. \tag{5.7.8}$$

We call attention to the fact that if one substitutes in (5.7.5) the pullbacks for the dx_i from (5.7.2) and does the expanding, remembering that $du_1\,du_2 = -du_2\,du_1$ and that $du_1\,du_1 = du_2\,du_2 = 0$, the result is the same: $(d\omega)^* = -18du_1\,du_2$.

If we formally differentiate ω^*, given by (5.7.7), we get

$$\begin{aligned}d\omega^* &= du_1\,du_1+25\,du_2\,du_1+7du_1\,du_2-5\,du_2\,du_2\\ &= -18\,du_1\,du_2\end{aligned}$$

so we have verified that $(d\omega)^* = d(\omega^*)$.

We have some integration left to do. Since $d\omega$ is a constant 2-form, the value of the integral of $d\omega$ over P is just the value of $d\omega$ at P. This is to say,

$$\int_P d\omega = d\omega(P) = \sum_{1\leqslant i<j\leqslant 4} dx_i\,dx_j(P) = -18. \tag{5.7.9}$$

To evaluate $\int_K (d\omega)^*$ is also easy. Since K is positively oriented, we need only compute the ordinary double integral.

$$\int_K (d\omega)^* = \int_0^1\int_0^1 -18\,du_1\,du_2 = -18. \tag{5.7.10}$$

We next evaluate $\int_{\partial K}\omega^*$. This is indeed the simplest kind of line integral to evaluate, and we get

$$\begin{aligned}\int_{\partial K}\omega^* &= \int_{\partial K}(u_1+25u_2-3)\,du_1+(7u_1-5u_2-3)\,du_2\\ &=\left(u_1^2/2-3u_1\right)\Big|_0^1+\left(7u_2-\tfrac{5}{2}u_2^2-3u_2\right)\Big|_0^1\\ &\quad+\left(u_1^2/2+25u_1-3u_1\right)\Big|_1^0+\left(-\tfrac{5}{2}u_2^2-3u_2\right)\Big|_1^0=-18.\end{aligned} \tag{5.7.11}$$

Finally, we are ready to tackle $\int_{\partial P}\omega$. Now we really can't evaluate this directly, because ω is not a constant 1-form, so we parametrize the four edges of P, pull ω back, relative to our parametrization, and integrate. This is just what we did when we found $\int_{\partial K}\omega^*$, so it hardly seems worth the effort to do it again. However, we shall sketch the idea, using different parametrizations. Let ∂_1 be the path $(1,2,0,-1)$ to $(3,1,0,0)$, and parametrize ∂_1, by the set of equations

$$\begin{aligned} x_1 &= 1+t, & x_2 &= 2-t/2, \\ x_3 &= 0, & x_4 &= -1+t/2, \end{aligned} \qquad 0 \leqslant t \leqslant 2.$$

Then

$$\int_{\partial_1} \omega = \int_0^2 [-1-(-1+t/2)(-\tfrac{1}{2})-(-1+t/2)(0)]\, dt = -\tfrac{5}{2}.$$

Note that as t goes from 0 to 2, the point (x_1, x_2, x_3, x_4) moves from **A** to **B**, so we have the correct orientation.

Now pick three other arbitrary parametrizations for the remaining three edges, make sure that the orientations are correct, evaluate the integrals, add everything together, and, *mirabile dictu*, we get

$$\int_{\partial K} \omega = -18. \tag{5.7.12}$$

It is certainly worth remarking that we have checked out a number of theoretical results, at least as far as we can with one example. We summarize

$$\int_{P^*} (d\omega)^* = \int_P d\omega = \int_{P^*} d(\omega^*) = \int_{\partial P} \omega = \int_{\partial(P^*)} \omega^* = \int_{(\partial P)^*} \omega^*.$$

$$\omega(P) = \omega^*(P^*), \qquad \text{for constant } \omega.$$

One last comment; refer back to (5.7.3). We assert that the 6×1 matrix on the right-hand side of this equation is a vector whose norm, or length, is the Euclidean area of P. In this example, the length of that vector is $g = 6\sqrt{41}$. Now P in fact is a rectangle, so you can calculate rather easily that $6\sqrt{41}$ is the area of P. What we are saying is that the measure of P is the square root of the sum of the squares of the measures of the projections of P on the coordinate subspaces. Little did Pythagoras know how clever he really was! If f is a real-valued function on E^4 and we wanted to integrate f over P (for example, to calculate the mass of P if f is a density function), a good conjecture would be that we would evaluate

$$\int_K f^* g \, du_1 \, du_2,$$

where f^* is the pullback of f under φ and $g = 6\sqrt{41}$.

5.8 k-FORMS IN E^n

A basic k-form in E^n is symbolized by an expression of the form

$$dx_{i_1} dx_{i_2} \cdots dx_{i_k},$$

where $1 \leqslant i_1 < i_2 < \cdots < i_k \leqslant n$. We define the k-form to be a functional on the class of oriented k-parallelotopes in E^n; its value at such a figure P is the signed measure of the projection of P onto the $x_{i_1}x_{i_2}\cdots x_{i_k}$-coordinate subspace of E^n, and this value is invariant under translation. Moreover, since this signed measure of the projection of P, $P_{i_1\cdots i_k}$, is the value of a particular determinant, as we shall prove later, it is natural to define the symbol $dx_{j_1}dx_{j_2}\cdots dx_{j_k}$ by the equation

$$dx_{j_1}dx_{j_2}\cdots dx_{j_k}=(-1)^{\sigma}dx_{i_1}dx_{i_2}\cdots dx_{i_k},$$

where $(j_1, j_2,\dots,j_k)$ is a permutation of the lexicographically ordered k-tuple $(i_1,i_2,\dots,i_k)$ and σ is 0 or 1, according as the permutation is even or odd.

Definition 5.8.1. A k-form in E^n is an expression of the form

$$\omega=\sum_{1\leq i_1<i_2<\cdots<i_k\leq n} A_{i_1\dots i_k}(\mathbf{x})\,dx_{i_1}\cdots dx_{i_k}.$$

If all the $A_{i_1\dots i_k}$ are constant (continuous, differentiable, C^1), ω is called a *constant* (*continuous*, *differentiable*, C^1) k-form.

If ω is a constant k-form, we can evaluate ω at P to get

$$\omega(P)=\sum_{1\leq i_1<\cdots<i_k\leq n} A_{i_1\dots i_k}dx_{i_1}\cdots dx_{i_k}(P),$$

and if the orientation of P is reversed and we indicate this by writing $-P$, then $\omega(-P)=-\omega(P)$.

If ω is continuous, we can partition P, refine the partition, pass to a limit, and obtain

$$\omega(P)=\int_P\sum_{1\leq i_1<\cdots<i_k\leq n} A_{i_1\dots i_k}(\mathbf{x})\,dx_{i_1}\cdots dx_{i_k}=\int_P\omega.$$

If, moreover, S is a suitable orientable k-dimensional surface which can be approximated in a certain special way by oriented k-parallelotopes pasted together in such a way as to remain consistently oriented, then

$$\omega(S)=\int_S\omega.$$

There are $\binom{n}{k}$ basic k-forms in E^n; these generate an $\binom{n}{k}$-dimensional linear set over the ring of real-valued functions on E^n. It should by now be clear that if we throw all the basic k-forms on E^n, $k=0,1,2,\dots,n$, into a bowl, stir in the ring of real-valued functions on E^n, and define a wedge product $\wedge$ by

$$\forall i,j=1,\dots,n,\quad i\neq j,\qquad dx_i\wedge dx_j=dx_i dx_j=-dx_j dx_i=-(dx_j\wedge dx_i),$$

$$\forall i=1,\dots,n,\qquad dx_i\wedge dx_i=dx_i dx_i=0,$$

$$\forall i=1,\dots,n,\qquad 1\wedge dx_i=dx_i\wedge 1=dx_i,$$

we have indeed concocted a Grassmann algebra over the ring of real-valued functions on E^n, its basis being all the 2^n k-forms. In fact, this exterior algebra is the direct sum of the $n+1$ linear sets

$$
\begin{aligned}
V^{(0)} &= \mathrm{sp}\{1\},\\
&\vdots\\
V^{(k)} &= \mathrm{sp}\{dx_{i_1}\wedge\cdots\wedge dx_{i_k}\}_{1\le i_1<\cdots<i_k\le n},\\
&\vdots\\
V^{(n)} &= \mathrm{sp}\{dx_1\wedge\cdots\wedge dx_n\}.
\end{aligned}
$$

But it is even something more. Let us formally define a "differential operator" d which will map linearly each differential k-form of $V^{(k)}$, $k=0,1,\dots,n-1$, into a $(k+1)$-form in $V^{(k+1)}$ in the following way:

Definition 5.8.2. If

$$\omega = \sum_{1\le i_1<\cdots<i_k\le n} A_{i_1\dots i_k}(\mathbf{x})\,dx_{i_1}\cdots dx_{i_k}$$

is a differentiable k-form in E^n, i.e., each $A(\mathbf{x})$ is a differentiable function, where $0\le k\le n$, then

$$d\omega = \sum_{1\le i_1<\cdots<i_k\le n}\left(\sum_{j=1}^{n} D_jA_{i_1\dots i_k}(\mathbf{x})\,dx_j\right)dx_{i_1}\cdots dx_{i_k}.$$

It is to be understood that after the expansion, each term involving $dx_j dx_{i_1}\cdots dx_{i_k}$ is zero if j is equal to one of the other k indices, and if j is not equal to any of the other k indices, then the factors are permuted into lexicographic order, and a sign change is introduced if that permutation is odd. Evidently, d maps $V^{(n)}$ to the zero element, or more properly said, d maps the differential forms of $V^{(n)}$ to zero. A k-form ω is called *exact* if there is a $(k-1)$-form σ such that $d\sigma=\omega$ on E^n, and ω is called *closed* if $d\omega\equiv 0$. Evidently, an exact form is closed.

We may now refer to our algebra of forms as a graded differential exterior algebra.

5.9 ANOTHER EXAMPLE

Let P be the oriented 4-parallelotope in E^6 determined by the four linearly independent vectors, in order, $(1,0,0,0,1,0)$, $(1,1,0,0,0,1)$, $(1,1,1,0,0,0)$, $(1,1,1,1,1,1)$. P is the linear image of the positively oriented cube K in E^4, and the matrix equation of the map is

$$\begin{bmatrix} x_1 \\ x_2 \\ x_3 \\ x_4 \\ x_5 \\ x_6 \end{bmatrix} = \begin{bmatrix} 1 & 1 & 1 & 1 \\ 0 & 1 & 1 & 1 \\ 0 & 0 & 1 & 1 \\ 0 & 0 & 0 & 1 \\ 1 & 0 & 0 & 1 \\ 0 & 1 & 0 & 1 \end{bmatrix} \begin{bmatrix} u_1 \\ u_2 \\ u_3 \\ u_4 \end{bmatrix}.$$

From this equation we have immediately the six pullbacks of the dx_i. To calculate, for example, $(dx_2\, dx_5)^*$, we simply find the exterior product of dx_2^* and dx_5^*, which is

$$(du_2 + du_3 + du_4) \wedge (du_1 + du_4) = -du_1 du_2 - du_1 du_3 - du_1 du_4 + du_2 du_4 + du_3 du_4.$$

A matrix method for obtaining this result is this: Put down the second and the fifth rows of the above matrix and compute all the 2×2 determinants: these determinants are the coefficients of $du_i du_j$, where i and j are the column numbers of this 2×5 matrix, and $i<j$.

$$\begin{bmatrix} 0 & 1 & 1 & 1 \\ 1 & 0 & 0 & 1 \end{bmatrix} \to \overset{ij=\ \ 12\quad 13\quad 14\quad 23\quad 24\quad 34}{(\ -1,\ -1,\ -1,\ 0,\ 1,\ 1\ )}.$$

$dx_i dx_j dx_k$, $1 \leq i < j < k \leq 6$ is calculated analogously. One can multiply out the expressions, using the exterior multiplication, for example,

$$\begin{aligned}(dx_2\, dx_4\, dx_5)^* &= dx_2^*\, dx_4^*\, dx_5^* = (du_2 + du_3 + du_4) \wedge (du_4) \wedge (du_1 + du_4) \\ &= du_1 du_2 du_4 + du_1 du_3 du_4.\end{aligned}$$

Alternatively, one can write the submatrix consisting of rows 2, 4, and 5 of the transformation matrix and evaluate the subdeterminant of columns i, j, and k to get the coefficients of $du_i du_j du_k$

$$\begin{bmatrix} 0 & 1 & 1 & 1 \\ 0 & 0 & 0 & 1 \\ 1 & 0 & 0 & 1 \end{bmatrix} \to \overset{ijk=\ 123\quad 124\quad 134\quad 234}{(\ \ 0,\ \ 1,\ \ 1,\ \ 0\ )}.$$

To find the pullback $(dx_i dx_j dx_k dx_l)^*$, the coefficient of $du_1 du_2 du_3 du_4$ is just the value of the determinant of the submatrix consisting of rows i, j, k, and l, as we already know.

Suppose $\boldsymbol{\sigma}$ is the 3-form:

$$\boldsymbol{\sigma} = x_1 \sum_{2\leq j<k<l\leq 6} dx_j dx_k dx_l + x_2 \sum_{3\leq j<k<l\leq 6} dx_j dx_k dx_l + x_3\, dx_4\, dx_5\, dx_6.$$

How do we compute $\int_{\partial P} \boldsymbol{\sigma}$?

First, we compute $\boldsymbol{\sigma}^*$, the pullback. This means we pull back x_1 to $u_1 + u_2 + u_3 + u_4$, x_2 to $u_2 + u_3 + u_4$, and x_3 to $u_3 + u_4$. Then we must pull back all the $dx_j dx_k dx_l$, a simple but tiring task. We've already pulled back

$dx_2\, dx_4\, dx_5$, so assume we've done the rest to get

$$\begin{aligned}\sigma^* &= u_1[3\, du_1\, du_2\, du_3 + 4\, du_1\, du_2\, du_4 + 7\, du_2\, du_3\, du_4] \\ &\quad + u_2[4\, du_1\, du_2\, du_3 + 6\, du_1\, du_2\, du_4 + 9\, du_2\, du_3\, du_4] \\ &\quad + (u_3 + u_4)[4\, du_1\, du_2\, du_3 + 7\, du_1\, du_2\, du_4 + 9\, du_2\, du_3\, du_4].\end{aligned}$$

There are eight faces in ∂K over which we integrate σ^*, half of them with positive orientation, half with negative. On $F_{1,0}$ the face we start with, the orientation is negative, $u_1 = 0$ and $du_1 = 0$, so

$$\int_{F_{1,0}} \sigma^* = -9\int_0^1\int_0^1\int_0^1 (u_2 + u_3 + u_4)\, du_2\, du_3\, du_4 = -\tfrac{27}{2}.$$

Next, on $F_{1,1}$, the orientation is positive, $u_1 = 1$ and $du_1 = 0$, so

$$\int_{F_{1,1}} \sigma^* = 7\int_{F_{1,1}} du_2\, du_3\, du_4 + 9\int_{F_{1,1}} (u_2 + u_3 + u_4)\, du_2\, du_3\, du_4 = 7 + \tfrac{27}{2} = \tfrac{41}{2}.$$

Continuing in this way, on $F_{2,0}$ and $F_{2,1}$ the integral is zero since $du_2 = 0$.

$$\int_{F_{3,0}} \sigma^* = -\tfrac{17}{2}, \qquad \int_{F_{3,1}} \sigma^* = \tfrac{31}{2},$$

$$\int_{F_{4,0}} \sigma^* = \tfrac{11}{2}, \qquad \int_{F_{4,1}} \sigma^* = -\tfrac{19}{2}.$$

Adding all these together, we get

$$\int_{\partial K} \sigma^* = 10 = \int_{\partial P} \sigma.$$

That's a lot of work. Notice that

$$d\sigma = \sum_{1 \leq i < j < k < l \leq 6} dx_i\, dx_j\, dx_k\, dx_l,$$

a constant 4-form. Stokes's theorem says

$$\int_{\partial P} \sigma = \int_P d\sigma.$$

To evaluate the right-hand integral is easy, in this case. Each $dx_i\, dx_j\, dx_k\, dx_l(P)$ can be computed. Add them all up, and you get 10. Of course, the pullback of $d\sigma$ is

$$(d\sigma)^* = 10\, du_1\, du_2\, du_3\, du_4,$$

and this is the same as $d(\sigma^*)$.

Our final observation is that the measure of P is the g factor, and here $g = 4$; if the density at each point of P is given by $\delta(\mathbf{x}) = x_1 x_3 x_4^2$, the mass of P would be computed by evaluating

$$\int_{P^*} \delta^* g \, du_1 \, du_2 \, du_3 \, du_4$$

$$= \int_0^1 \int_0^1 \int_0^1 \int_0^1 (u_1 + u_2 + u_3 + u_4)(u_3 + u_4) u_4^2 \cdot g \, du_1 \, du_2 \, du_3 \, du_4.$$

PROBLEMS

1. Let $\mathbf{F} = (x^2, y^2, z^2)$ be a force field in E^3 and $\mathbf{V}$ the free vector $\overrightarrow{\mathbf{PQ}}$, where $\mathbf{P} = (1,3,2)$, $\mathbf{Q} = (4,-1,1)$. Find the work done by $\mathbf{F}$ in moving a point mass from $\mathbf{P}$ to $\mathbf{Q}$.

2. Same problem as 1, except that the path from $\mathbf{P}$ to $\mathbf{Q}$ is along the arc given parametrically by

$$x = 3t^2 + 1, \quad y = -4t + 3, \quad z = -t^2 + 2, \qquad 0 \leq t \leq 1.$$

Do this by using Theorem 5.4.2, and also by evaluating $\int_0^1 \omega^*$, the pullback of $\int_\gamma \omega$.

3. Find the three-dimensional measure of the surface of the ball in E^4 having radius R. Integrate this result to get the four-dimensional measure of the ball.

4. Let

$$P = \begin{bmatrix} 1 & 2 & 3 \\ -3 & -1 & 4 \\ 4 & 4 & -6 \\ 2 & 1 & 1 \\ 1 & -1 & 4 \\ 6 & -2 & 0 \end{bmatrix}$$

be a parallelotope in E^6. Find its three-dimensional volume.

5. If you already know about dot products in E^n and can recall a little trigonometry, derive a formula for the area of a parallelogram determined by the vectors $\mathbf{v}, \mathbf{w} \in E^n$. Use your formula to see if the area of the parallelogram determined by $(3,1,7,-1,2)$ and $(-1,-2,3,4,5)$ is $2\sqrt{759}$.

6. Find an example of a closed form that is not exact.
[*Hint*: If ω is exact, then $\int_{\partial R} \omega = \int_R d\omega = 0$. Suppose R is the square in E^2 given by $\{(x,y) : |x| \leq 1, |y| \leq 1\}$ and

$$\omega = \frac{x}{x^2 + y^2} dy - \frac{y}{x^2 + y^2} dx.$$

Show that $d\omega = 0$, so ω is closed. Suppose $\boldsymbol{\sigma}$ is a zero form on E^2 such that $d\boldsymbol{\sigma} = \omega$. Then $\int_{\partial(\partial R)} \boldsymbol{\sigma} = \int_{\partial R} d\boldsymbol{\sigma} = \int_{\partial R} \omega$. But $\partial(\partial R)$ is empty, so $\int_{\partial R} \omega$ must equal zero. Now compute $\int_{\partial R} \omega$; you will get 2π. Contradiction! Hence no such $\boldsymbol{\sigma}$ can exist. Ergo ω is not exact in this case, even though it is closed.]

7. Let $\sigma = \arctan(y/x)$. Find $d\sigma$, and compare it to the ω of the preceding problem. Why is ω not an exact 1-form on E^2?

Let R be a box in E^2 which does not contain (0,0) in its interior or boundary. (Pick an easy one, like $[\pi/3, \pi/2]\times[\pi/3, \pi/2]$.) Compute $\int_{\partial R} d\sigma$.

These two problems point out that if ω is a form in E^n, to apply the Stokes theorem you had better be sure that both ω and $d\omega$ exist (and are integrable) throughout the region in question and its boundary.

8. Sketch the positively oriented unit cube K in E^3, indicating the orientations of each face. Let K' be the image of K under the mapping T whose matrix is

$$\begin{bmatrix} 0 & 1 & 0 \\ 1 & 0 & 0 \\ 0 & 0 & 1 \end{bmatrix}.$$

Sketch K', and indicate the orientations of the faces.

9. Prove that if γ is a rectifiable arc and Γ is an approximating polygonal line, then γ and Γ are "uniformly close" together. This is to say, if an arbitrarily small fixed $\varepsilon > 0$ has been given, we can find an approximating polygonal line Γ such that if at each vertex of Γ we center a disk of radius ε, both Γ and γ will be contained in the union of these disks. Having shown this, can you find some reasonable expression in terms of ε that will show how close the approximation of the length of γ is?

10. Suppose A $(2,1,-1,-3)$, B $(4,2,0,-2)$, C $(5,0,0,-8)$, D $(3,3,0,-5)$ are four points in E^4. These points determine a skew quadrilateral ABCD. Find its two-dimensional measure.

11. Let γ be a plane arc described parametrically by $x = t^2+1, y = t^3+1, 0 \leqslant t \leqslant 2$, and $\mathbf{v} = (\sqrt{x}, y)$ be a flow vector in the plane. Find the net total flow across γ per unit of time.

12. If γ is a plane curve given by $x = 8\cos^3 t$, $y = 8\sin^3 t$, $0 \leqslant t < \pi/2$, and $\mathbf{F} = (-\sqrt[3]{x}, \sqrt[3]{y})$ is a force field, calculate the work done by the force $\mathbf{F}$ in moving a unit point mass from (8,0) to (0,8) along γ. Then calculate the work done by $\mathbf{F}$ in moving the point mass from (8,0) to (0,8) along the straight-line path.

13. Suppose γ is the arc of the parabola $y^2 = 6x$ cut off by the parabola $x^2 = 6y$. Find the length of γ. Now suppose $\delta(x, y)$ is a density function for γ. Find the mass of γ if $\delta(x, y) = y$. (Recall that mass equals density times measure.)

14. If ω is the 0-form $(x\cos y - y\cos x)/(1+\sin x+\sin y)$, express $d\omega$. If σ is the 3-form

$$\sigma = \ln(x^2+y^2+z^2+w^2+1)\,dx\,dy\,dz + \tan^{-1}xw\,dx\,dy\,dw + \sec^2 xy\,dx\,dz\,dw$$
$$+ x^2y^2z^2w^2\,dy\,dz\,dw,$$

what is $d\sigma$?

15. Suppose P is the oriented parallelopiped in E^4 determined by the vectors in the order given: **A** $(3,1,2,1)$, **B** $(-1,1,-2,3)$, **C** $(2,1,0,-1)$. Let $\boldsymbol{\sigma}$ be the 2-form $w\,dx\,dy + 2xy\,dx\,dz + xyz\,dx\,dw + x^2\,dy\,dz + y^2\,dy\,dw + z^2\,dz\,dw$. Evaluate (by pullbacks) $\int_{\partial P}\boldsymbol{\sigma}$ and $\int_P d\boldsymbol{\sigma}$.

16. The points $(0,0,0,0)$, $(1,1,-1,4)$, $(4,3,0,4)$, $(3,2,1,0)$ determine a parallelogram P in E^4. Verify this. Find its area. If $f(x,y,z,w) = xyzw$ is a charge function, find the total charge on P by finding the g factor, the pullback f^*, and integrating f^*g over P^*.

17. Suppose $P \subset E^4$ is a parallelogram, the affine image of the unit square $K \subset E^2$ under the map

$$\begin{bmatrix} u \\ v \end{bmatrix} \mapsto \begin{bmatrix} 1 & 2 \\ 0 & 1 \\ -1 & 1 \\ 3 & -1 \end{bmatrix} \begin{bmatrix} u \\ v \end{bmatrix} + \begin{bmatrix} 1 \\ 1 \\ 1 \\ 1 \end{bmatrix} = \begin{bmatrix} x_1 \\ x_2 \\ x_3 \\ x_4 \end{bmatrix}.$$

We define the first moment of P about the set $S \subset E^4$ to be the product of the average distance of P from S and the measure of P. This is tantamount to saying that the first moment of P about S is the integral of the distance function over P. Suppose S is the $x_1x_2x_3$-coordinate subspace; then the distance of any point $(x_1,x_2,x_3,x_4) \in P$ from S is simply x_4. Calculate, for each $i = 1,2,3,4$,

$$\bar{x}_i = \int_P x_i\,d\mu/(\text{measure of } P)$$

by integrating the pullback of x_i, multiplied by the g factor, over K.

Geometrically, the point $(\bar{x}_1,\bar{x}_2,\bar{x}_3,\bar{x}_4)$ is the *centroid* of P. Note that the preimage of this centroid is the centroid of K.

The second moment of P about a set S is the product of the measure of P and the mean square of the distance of P from S. This second moment is often called the *moment of inertia* of P about S, when measure equals mass. Find the moment of inertia of P about the origin, assuming that the mass of P and the measure of P are the same (i.e., the density of P is 1 at each point).

18. Evaluate

$$\int_\gamma \left[\frac{(1+y^2)}{x^3}\,dx - \frac{(y+x^2y)}{x^2}\,dy \right],$$

where γ is an oriented arc with endpoints $(1,0)$ and $(5,2)$. γ may not meet the y axis.

19. Evaluate

$$\int_{\mathbf{C}} \left[\frac{dy}{\sqrt{y^2-x^2}} - \frac{x\,dx}{y\sqrt{y^2-x^2}+y^2-x^2} \right]$$

where **C** is a suitably restricted path from $(3,5)$ to $(5,13)$.

20. Let $\omega = \ln(x^2 + y^2)$. Find $d\omega$. Let γ be the unit circle centered at the origin. Compute $\int_\gamma d\omega$. Is $d\omega$ closed and exact on E^2? Is $d\omega$ exact on $E^2 \setminus \{(0,0)\}$? Let $\mathbf{C}$ be the closed path starting at $(0,-1)$ and going horizontally to $(\sqrt{3}/2, -1)$, then vertically to $(\sqrt{3}/2, \frac{1}{2})$, and then around the unit circle counterclockwise back to $(0,-1)$. Evaluate $\int_{\mathbf{C}} d\omega$.

21. Let $\omega = \ln(x^2 + y^2)$. Let γ be the line segment from $(1,0)$ to $(0,1)$. Evaluate $\int_\gamma d\omega$ and $\int_{\partial\gamma} \omega$.

Now let γ be the circle centered at $(\frac{1}{2}, 0)$ having radius 1. Evaluate $\int_\gamma d\omega$ by parametrizing γ and pulling back $d\omega$.

22. Again let $\omega = \ln(x^2 + y^2)$. Let (a,b), (c,d) be any two points except the origin and γ, a C^1 arc not passing through the origin connecting these two points. Show that

$$\int_\gamma d\omega = \ln\left[\frac{c^2 + d^2}{a^2 + b^2}\right] = \int_{\partial\gamma} \omega$$

by parametrizing γ as follows:

$$x = f(t), \quad 0 \leqslant t \leqslant 1, \qquad f(0) = a, \quad f(1) = c,$$

$$y = g(t), \quad 0 \leqslant t \leqslant 1, \qquad g(0) = b, \quad g(1) = d.$$

Find the pullback of $d\omega$ and integrate.

23. Let $\omega = \arctan(y/x)$ and (a,b), (c,d) be two points in E^2, $a \neq 0$, $c \neq 0$. Suppose γ is a C^1 arc initiating at (a,b) and terminating at (c,d), parametrized in the same way as the arc γ in the previous problem. Find the pullback of $d\omega$ and integrate it. If you manipulate your result a bit, you should get

$$\int_\gamma d\omega = \tan^{-1}\left(\frac{ad - bc}{ab + cd}\right) + k\pi, \qquad k \in \mathbf{Z}.$$

We call your attention to the fact that the logarithm function in the previous problem is single-valued, whereas the arctangent function is not. If γ never meets the y axis, can you use Stokes's theorem?

24. Suppose ω is a C^1 k-form on an open region $D \subset E^n (k \geqslant 1)$, and ω is closed. A lemma by Poincaré assures that at each point $\mathbf{p} \in D$, there is an open set G containing $\mathbf{p}$ and contained in D on which ω is exact. In fact, we give for reference the following definition and theorem.

Definition 5.P.1. A subset S of a linear space V is said to be *star-shaped* relative to a point $\mathbf{p} \in S$ if for every $\mathbf{x} \in S$, the point $\mathbf{p} + t(\mathbf{x} - \mathbf{p}) \in S$, $0 \leqslant t \leqslant 1$. We say S is *starlike* if it is star-shaped relative to some point of S.

It is not hard to see that any convex set is starlike, and that if G is an open set in E^n containing a point $\mathbf{p}$, there is an open set D, with $\mathbf{p} \in D \subset G$, and D is starlike.

Theorem 5.P.2. *If D is an open starlike set in E^n, and ω is a closed k-form on D, $k \geqslant 1$, then ω is exact on D.*

Don't try to prove this theorem; simply consider the previous four problems in the light of this theorem.

CHAPTER

6

TOPOLOGY

In Chapter 1 we talked at some length about various structures we could put on sets, algebraic structures, and order structures. In this chapter we introduce the idea of topological structure for a set. The motivation behind what we are about to do is this: If we can successfully generalize—abstract—those fundamental notions of calculus, limit and continuity, we may gain many insights into calculus that would otherwise be completely beyond our powers to grasp. Historically, topology is a branch of geometry, but it has grown to such an extent as to be regarded as a principal part of mathematics, taking its place along side of algebra, geometry, and analysis. The three main divisions of the discipline are combinatorial topology, algebraic topology, and point-set topology. What we are going to study falls into the last category.

Since we are going to be concerned with topological properties of mappings, it might be well to recall to the reader just what a mapping is. First, we need two nonempty sets V, W. Next, we need a correspondence f from V into W which assigns to each $\mathbf{v} \in V$ correspondents in W. Finally, we need an axiom to limit in some respect the variety of choices of correspondences.

Axiom. To each $\mathbf{v} \in V$, f assigns exactly one correspondent $\mathbf{w} \in W$.

The ordered triple (V, W, f) is a mapping if f obeys the axiom. Of course, it is f that we call the mapping. V and W are called the domain and range of f, respectively. If $U \subset V$ is a subset, the triple (U, W, f), or more commonly, $f_{|U}$, is called the *restriction* of f to U. If $S \supset V$, it could be that there is a mapping $(S, W, \bar{f})$ such that $\bar{f}_{|V} \equiv f$. In this case we call $\bar{f}$ an *extension* of f to S.

Generally $f(V)$ is a proper subset of W; if $f(V)=W$, we call f a *surjective*, or onto, mapping. We sometimes denote the set $f(V)$ by im f, meaning the image of f, and the domain of f by dom f. Now, it generally happens that there exist $\mathbf{v}_1, \mathbf{v}_2 \in V$ with $\mathbf{v}_1 \neq \mathbf{v}_2$, such that $f(\mathbf{v}_1)=f(\mathbf{v}_2)$. However, if $\forall \mathbf{v}_1, \mathbf{v}_2 \in V$, $\mathbf{v}_1 \neq \mathbf{v}_2 \Rightarrow f(\mathbf{v}_1) \neq f(\mathbf{v}_2)$, we say that f is *injective*, or one-to-one. A mapping f which is both injective and surjective is called *bijective*, or one-to-one onto.

If $B \subset W$, the set $\{\mathbf{v} \in V: f(\mathbf{v}) \in B\}$ is called the *preimage* of B under f and is denoted $f^{-1}(B)$. Note that in general f^{-1} is not a mapping because it is not single-valued, unless f is bijective. We can write

$$f^{-1}: W \rightarrow V \quad \text{is a mapping} \quad \Leftrightarrow \quad f: V \rightarrow W \quad \text{is bijective;}$$

$$f^{-1}: \operatorname{im} f \rightarrow V \quad \text{is a mapping} \quad \Leftrightarrow \quad f: V \rightarrow W \quad \text{is injective.}$$

There are occasions when mathematicians speak of bivalued or many-valued mappings, but these will not appear in this course.

Because we are going to do calculus, which is "infinitesimal mathematics" as opposed to "finite mathematics," we need the notions of continuity and limit for mappings. As we pointed out in Chapter 2, these notions depend upon a notion of "nearness." We now set about developing these notions quite abstractly.

6.1 THE OPEN-SET TOPOLOGY

Let $\varphi: V \rightarrow W$ be a mapping and $\mathbf{x} \in V$ an arbitrary point. For φ to be continuous at $\mathbf{x}$ means that if $\mathbf{y}$ is sufficiently close to $\mathbf{x}$, then necessarily $\varphi(\mathbf{y})$ is close to $\varphi(\mathbf{x})$. But obviously we don't have any idea what "close" means in V, or in W. It should be enough to say for all $\mathbf{y}$ in a sufficiently small neighborhood of $\mathbf{x}$, $\varphi(\mathbf{y})$ is necessarily in a certain restricted neighborhood of $\varphi(\mathbf{x})$. But then, what is a neighborhood? You see, V and W are just sets, or perhaps algebraic order structures, but (unless a metric structure has been introduced as we did for the structure $\mathbb{R}^n$) there is no structure that defines a notion of closeness.

We are going to introduce a topological structure for a nonempty set S that will alleviate our problem. From the power set $P(S)$ we single out a distinguished collection of subsets $\mathfrak{T}$, and we stipulate that the class $\mathfrak{T}$ satisfy the following set of three axioms:

T1. $\varnothing \in \mathfrak{T}$, $S \in \mathfrak{T}$.
T2. If $\{T_\alpha\} \subset \mathfrak{T}$ is an arbitrary collection of sets of $\mathfrak{T}$, then $\bigcup_\alpha T_\alpha \in \mathfrak{T}$.
T3. $A, B \in \mathfrak{T} \Rightarrow A \cap B \in \mathfrak{T}$.

If the collection $\mathfrak{T}$ of subsets of S satisfies these axioms, we say $\mathfrak{T}$ is a topology for S, and S together with $\mathfrak{T}$ is a topological space.

If you recall something that was said in Chapter 1, axiom T1 is perhaps redundant. If $\{T_\alpha\}$ is an empty collection of elements of $\mathfrak{T}$, then $\bigcup_\alpha T_\alpha = \varnothing$, so axiom T2 ensures that $\varnothing \in \mathfrak{T}$.

If we restated axiom T3 to read "$\mathfrak{T}$ is closed with respect to finite intersections," then the intersection of the sets of an empty collection of sets of $\mathfrak{T}$ would be all of S. In any event, remember these axioms. A topology for S is a collection of subsets of S closed with respect to arbitrary unions and finite intersections; $\varnothing$ and S do belong to the collection.

Note that if S contains more than one point, there are a number of collections $\mathfrak{T}$ that will serve as a topology for S. One such collection, $\mathfrak{T} = \{\varnothing, S\}$, is called the trivial topology for S. Another, called the discrete topology, is the power set itself, $P(S)$. This last topology is obviously the richest in the number of sets it contains. If we are going to topologize a set S, we usually try to hit a happy medium; this is to say, we look for a topology that will be rich enough in sets to do what we want it to do, but not much more. This will make more sense shortly, as you will see.

Having a collection $\mathfrak{T}$ in hand, we call these sets the "open" sets of S, and any subset of S whose complement is in $\mathfrak{T}$ we call a "closed" set in S. The two sets $\varnothing$ and S, being simultaneously open and closed, are sometimes called clopen by whimsical people, a slight abuse of the language. Notice that there is no particularly apparent quality that a set enjoys to be open; it is open simply because it happens to be in the collection $\mathfrak{T}$. The collection of intervals of the form $(a,b) = \{x : x \in E^1,\ a < x < b\}$, together with all the sets that are unions of such intervals, are sets that we are used to calling open sets in E^1; hence such sets probably "look" open to you. But consider the set $\mathbb{Z}$ of integers, and define $A \subset \mathbb{Z}$ to be *open* iff $A = \varnothing$ or $(\mathbb{Z} \setminus A)$ is a finite set. The collection of all such sets A is a topology for $\mathbb{Z}$, called the cofinite topology. (Verify this.) Now the sets of this topology do not "look" open, they just behave that way!

We are now ready to define what we mean by a neighborhood of a point of a topological space.

Definition 6.1.1. Let $(V, \mathfrak{T})$ be a topological space and $\mathbf{v} \in V$ an arbitrary point. A set $N \subset V$ is called a *neighborhood* of $\mathbf{v}$ provided that $\exists T \in \mathfrak{T}$ s.t. $\mathbf{v} \in T \subset N$.

Evidently, a nonempty open set is automatically a neighborhood of each of its points. Conversely, if a set N is a neighborhood of each of its points, then N is open. This last statement is easy to prove; for each $\mathbf{x} \in N$, there exists an open set $T_\mathbf{x} \in \mathfrak{T}$ such that $\mathbf{x} \in T_\mathbf{x} \subset N$. Hence $\bigcup_{\mathbf{x} \in N} T_\mathbf{x} \subset N$ since $T_\mathbf{x} \subset N$ for each $\mathbf{x}$. Obviously, the inclusion goes the other way as well since $N = \bigcup_{\mathbf{x} \in N} \{\mathbf{x}\} \subset \bigcup_{\mathbf{x} \in N} T_\mathbf{x}$. Since $N = \bigcup_{\mathbf{x} \in N} T_\mathbf{x}$, N is open, being the union of open sets. We state this result as a theorem characterizing open sets in terms of neighborhoods.

Theorem 6.1.2. *In a topological space $(V, \mathfrak{T})$, a set $N \neq \varnothing$ is open if and only if N is a neighborhood of each of its points.*

Consider again the space $\mathbb{Z}$ endowed with this topology: $A \subset \mathbb{Z}$ is open if and only if $A = \varnothing$ or A is of the form

$$A_n = \{i \in \mathbb{Z} : i \geqslant n, n \in \mathbb{Z}\}.$$

There is no way we can find a neighborhood of 10 that doesn't contain 20. Thus it seems that 20 is about as close to 10 as it can get. However, there are a number of neighborhoods of 20 that don't include 10, so 10 is not really very close to 20. A bad situation, this. Or suppose we define a topology for $\mathbb{R}$ by the following rule: A subset $T \subset \mathbb{R}$ is open if $T = \mathbb{R}$, or if T is an arbitrary union of sets of the form $(n, n+1)$, where $n \in \mathbb{Z}$. It seems that the collection $\mathfrak{T}$ that we obtain in this way is indeed a topology for $\mathbb{R}$. Evidently, the sequence $\{1/n\}_{n=1}^{\infty}$ converges to 10^{10} since every neighborhood of this number contains all but a finite number of terms of the sequence. Just as clearly, the sequence converges to -2. The sequence $\{n/(2n+1)\}_{n=1}^{\infty}$ apparently converges to each real number in the set $(0, 1)$. This topology is not very useful if we are thinking about things like limits of sequences and convergence.

Definition 6.1.3. A topological space $(V, \mathfrak{T})$ rich enough in open sets such that for each pair of distinct points $\mathbf{x}, \mathbf{y} \in V$ there exist disjoint open sets $A_{\mathbf{x}}, B_{\mathbf{y}} \in \mathfrak{T}$ such that $\mathbf{x} \in A_{\mathbf{x}}$, $\mathbf{y} \in B_{\mathbf{y}}$ is called a *Hausdorff space*.

From what we have just hinted at, it seems that if V and W are domain and range, respectively, of a mapping, or a whole family of mappings, then we had better have Hausdorff topologies on these sets if we want to consider questions of limit and continuity. How do we construct such topologies? It is not a trivial task.

Definition 6.1.4. We say that a family of sets $\mathfrak{B}$ generates a family of sets $\mathfrak{T}$ if every $T \in \mathfrak{T}$ is a union of an arbitrary number of sets of $\mathfrak{B}$. In particular, if $\mathfrak{T}$ is a topology for a space V and $\mathfrak{B}$ generates $\mathfrak{T}$, we say $\mathfrak{B}$ is a *base* for that topology for V.

Note that if $\mathfrak{B}$ is a base for a topology for V, $\mathfrak{B}$ need not contain the empty set, but if $\mathbf{x} \in V$ is an arbitrary point, there must be at least one set $B_{\mathbf{x}} \in \mathfrak{B}$, with $\mathbf{x} \in B_{\mathbf{x}}$, since V itself must be a union of sets of $\mathfrak{B}$.

Definition 6.1.5. If a nonempty family $\mathfrak{S}$ of subsets of V is such that the collection $\mathfrak{B}$ of all sets which can be formed by taking intersections of finitely many sets of $\mathfrak{S}$ is a base for a topology for V, we call $\mathfrak{S}$ a *sub-base* for that topology.

Note that if $\mathfrak{S}$ is a sub-base for a topology for V, the base $\mathfrak{B}$ which $\mathfrak{S}$ generates by intersections does contain V since V is the intersection of the sets of an empty collection of sets of $\mathfrak{S}$. The sets of $\mathfrak{S}$ are called sub-basic open sets, and those of the collection $\mathfrak{B}$ which generates the topology are called the basic open sets.

And now the reason for introducing these last two concepts. If we have a set V, and we can pick out certain subsets of V, other than $\varnothing$ and V itself, which we know we want to be open, we can let this collection be a sub-base $\mathfrak{S}$ for a topology $\mathfrak{T}$. In this way we can topologize V in the most efficient way; all the sets we want to be open will be open, together with enough additional sets which will make the whole collection a topology. For example, in $\mathbb{R}$ we know that we want sets of the form (a, b), $a, b \in \mathbb{R}$, $a < b$, to be "open" sets, so we let $\mathfrak{S}$ be the collection of all such intervals. From $\mathfrak{S}$ we generate by finite intersections a collection $\mathfrak{B}$, and having $\mathfrak{B}$, we generate a collection $\mathfrak{T}$. $\mathfrak{T}$ is the standard topology for $\mathbb{R}$. But first, there is something to prove. We need the following theorem.

Theorem 6.1.6. *Let V be a nonempty set and $\mathfrak{S}$ a nonempty collection of subsets of V. If $\mathfrak{B}$ is the collection of subsets of V which we can obtain by taking intersections of finite numbers of sets of $\mathfrak{S}$, $\mathfrak{B}$ is a base for a topology $\mathfrak{T}$ for V, and we have $\mathfrak{S} \subset \mathfrak{B} \subset \mathfrak{T}$. Moreover, $\mathfrak{T}$ is unique.*

PROOF: That $\mathfrak{S} \subset \mathfrak{B} \subset \mathfrak{T}$ is obvious; any set $S \in \mathfrak{S}$ is the intersection of sets of a finite collection of sets of $\mathfrak{S}$ consisting of the one single set S alone; thus $S \in \mathfrak{B}$. Any $B \in \mathfrak{B}$ is the union of sets in the singleton collection $\{B\}$, so $B \in \mathfrak{T}$. The rest of the proof we leave to the reader. We've already shown that $V \in \mathfrak{T}$ and that $\varnothing \in \mathfrak{T}$, so the $\mathfrak{T}$ obtained from $\mathfrak{S}$ does satisfy axiom T1. That $\mathfrak{T}$ satisfies axiom T2 is easy to show, but that $\mathfrak{T}$ satisfies T3 is somewhat harder to show. Uniqueness will be easier to show a bit later. $\square$

This theorem having been proved, we now have an efficient method for at least describing the topological structure with which we wish to endow a set. For example, let V be a normed linear space. The sets

$$B_r(\mathbf{p}) = \{\mathbf{x} \in V : \|\mathbf{x} - \mathbf{p}\| < r\},$$

which we have described as the "open balls of radius r centered at $\mathbf{p}$," are naturally open because they behave like sets which are neighborhoods of each of their points. This is to say, if $\mathbf{y} \in B_r(\mathbf{p})$, then there exists a real number $\varepsilon > 0$ such that $B_\varepsilon(\mathbf{y}) \subset B_r(\mathbf{p})$. Hence we take as $\mathfrak{S}$ the collection of all such balls, from $\mathfrak{S}$ we obtain a collection $\mathfrak{B}$, and then we generate from $\mathfrak{B}$ a collection $\mathfrak{T}$. $\mathfrak{T}$ is a topology, the theorem says, and we call this $\mathfrak{T}$ the norm topology for V. Evidently, $(V, \mathfrak{T})$ is a Hausdorff space.

Often a collection $\mathfrak{S}$ which we take to be a sub-base for a topology $\mathfrak{T}$ is already a base for $\mathfrak{T}$, in that the sets of the collection $\mathfrak{B}$ we obtain from $\mathfrak{S}$ could have been obtained by taking arbitrary unions of sets of $\mathfrak{S}$. When this is so is not always apparent, so the following theorem is of some importance.

Theorem 6.1.7. *Suppose* $\mathfrak{B} \subset \mathfrak{T}$ *is a collection of open sets of a topological space* $(V, \mathfrak{T})$. *Then* $\mathfrak{B}$ *is a base for* $\mathfrak{T}$ *iff for each* $T \in \mathfrak{T}$ *and each* $\mathbf{x} \in T$, *there exists a* $B \in \mathfrak{B}$ *such that* $\mathbf{x} \in B \subset T$.

PROOF: Assume $\mathfrak{B}$ is a base for $\mathfrak{T}$. Let $\mathbf{x}, T$, with $\mathbf{x} \in T$ and T open, be arbitrary. Then there is a collection $\{B_\alpha\} \subset \mathfrak{B}$ such that $T = \bigcup_\alpha B_\alpha$. For at least one α we have $\mathbf{x} \in B_\alpha$. On the other hand, if T is an arbitrary open set, $\mathbf{x}$ is an arbitrary point of T, and there is a $B_\mathbf{x} \in \mathfrak{B}$ such that $\mathbf{x} \in B_\mathbf{x} \subset T$, it is easy to see that $\bigcup_{\mathbf{x} \in T} B_\mathbf{x} = T$. Thus $\mathfrak{B}$ must be a base for $\mathfrak{T}$, and the proof is complete. □

Definition 6.1.8. Let V be a topological space, $\mathbf{x} \in V$ a point, and $S \subset V$ a subset. We say $\mathbf{x}$ is an *adherent* (*contact*) point of S iff for every neighborhood $N_\mathbf{x}$ of $\mathbf{x}$, $N_\mathbf{x} \cap S \neq \varnothing$. $\mathbf{x}$ is an *isolated* point of S if there is at least one neighborhood $N_\mathbf{x}$ of $\mathbf{x}$ such that $N_\mathbf{x} \cap S = \{\mathbf{x}\}$. $\mathbf{x}$ is a *cluster point* (*limit point*, *accumulation point*) of S iff $\mathbf{x}$ is an adherent point but not an isolated point of S.

Equivalently, $\mathbf{x}$ is a cluster point of S iff for each neighborhood $N_\mathbf{x}$ of $\mathbf{x}$, $(N_\mathbf{x} \setminus \{\mathbf{x}\}) \cap S \neq \varnothing$. The set of all cluster points of S is called the *derived set* of S; we denote it by S'. $\mathbf{x}$ is a *condensation point* of S iff $\forall N_\mathbf{x}$, $N_\mathbf{x} \cap S$ is an uncountable set. (Some people with a strong sense of order specialize "accumulation point": they say that $\mathbf{x}$ is an accumulation point of a set S if $\forall N_\mathbf{x}$, $N_\mathbf{x} \cap S$ contains a countable number of points of S. This distinction between cluster point and accumulation point hardly seems worth the effort!)

We have defined a closed set in a topological space to be one whose complement is open. A useful characterization of closed set is the following.

Theorem 6.1.9. *A set* S *is closed in a topological space* V iff S *contains all its limit points.*

PROOF: Suppose S is closed. Then $V \setminus S$ is open and thus a neighborhood of each of its points. Hence no $\mathbf{x} \in V \setminus S$ can be a limit point of S, so S must contain any limit points it might have. On the other hand, if S contains all its limit points, it must be true that for each $\mathbf{x} \in V \setminus S$, there is an open neighborhood $N_\mathbf{x}$ of $\mathbf{x}$ such that $N_\mathbf{x} \subset V \setminus S$. Hence $V \setminus S$ can be expressed as the union of a collection of open sets; specifically, $V \setminus S = \bigcup_{\mathbf{x} \in V \setminus S} N_\mathbf{x}$. This means $V \setminus S$ is open, and S is closed. The proof is complete. □

It might be well to emphasize the fact that if a set S has no limit points, then it certainly contains all of them, so it is a closed set.

Definition 6.1.10. Let $\mathbf{x}$ and S be, respectively, a point and a set in a topological space V. Then $\mathbf{x}$ is an *interior point* of S if S is a neighborhood of $\mathbf{x}$. $\mathbf{x}$ is an *exterior point* of S if it is an interior point of $V \setminus S$. $\mathbf{x}$ is a *frontier point* of S if it is adherent to both S and $V \setminus S$.

We denote the set of interior points of S by $\operatorname{int} S$, or $\mathring{S}$, the exterior of S by $\operatorname{ext} S$, the set of frontier points of S by $\operatorname{Fr} S$, or ∂S, and the set of adherent points of S by $\bar{S}$. $\bar{S}$ is usually called the *closure*, or the *adherence*, of S. It is sometimes denoted $\operatorname{Cl} S$.

Theorem 6.1.11. $\operatorname{int} S$ *is the largest open set contained in* S, *and* $\bar{S}$ *is the smallest closed set containing* S. ∂S *is closed.* S' *is closed if* S *is closed.*

PROOF: Let $\mathbf{x} \in \operatorname{int} S$ be arbitrary. There exists an open neighborhood $N_{\mathbf{x}}$, with $\mathbf{x} \in N_{\mathbf{x}} \subset S$. Every point of $N_{\mathbf{x}}$ is thus in $\mathring{S}$, and it follows that $\mathring{S} \subset \bigcup_{\mathbf{x} \in \operatorname{int} S} N_{\mathbf{x}} \subset \mathring{S}$, so $\operatorname{int} S$ is a union of open sets and hence open. If B is any open set in S, then S is a neighborhood of each point of B, so $B \subset \mathring{S}$. This means $\mathring{S}$ is the largest open set in S.

That $\bar{S} \supset S$ is obvious. Let $\mathbf{x} \in V \setminus \bar{S}$ be arbitrary. If $\mathbf{x}$ is an adherent point of $\bar{S}$, then for each open neighborhood $N_{\mathbf{x}}$ of $\mathbf{x}$, $N_{\mathbf{x}} \cap \bar{S} \neq \varnothing$. Since $\mathbf{x}$ itself is not in $\bar{S}$ (nor in S), this intersection must contain some point $\mathbf{y} \in \bar{S}$, and since $N_{\mathbf{x}}$ is open, it is a neighborhood of $\mathbf{y}$ and thus contains a point of S. Since this is the case for each such $N_{\mathbf{x}}$, $\mathbf{x}$ itself must be an adherent point of S. Contradiction. Therefore $\bar{S}$ must contain all its limit points, and so is closed. If $A \supset S$ is closed, A contains S, all the limit points of S, and all the limit points of A. Therefore A certainly contains $\bar{S}$; hence $\bar{S}$ is the smallest closed set containing S.

By definition, $\partial S = \bar{S} \cap \overline{(V \setminus S)}$, so ∂S is a closed set. (An application of the De Morgan theorem proves that the intersection of the sets of an arbitrary collection of closed sets is closed, and the union of the sets of a finite collection of closed sets is closed.) We leave the proof of the last statement to the reader. □

We remark that a closed set is the disjoint union of its set of cluster points and its set of isolated points. If a closed set S has no isolated points, it is called a *perfect* set; i.e., S is *perfect* iff $S = S'$.

6.2 CONTINUITY AND LIMIT

We are now ready to define continuity of a mapping $\varphi: V \to W$ at a point $\mathbf{v} \in V$. We assume now that V, W are topological spaces, each with its own topology.

Definition 6.2.1. $\varphi: V \to W$ is continuous at $\mathbf{v} \in V$ if for each neighborhood B of $\varphi(\mathbf{v})$ there is a neighborhood A of $\mathbf{v}$ such that $\varphi(A) \subset B$. (There is no loss of generality if we assume the neighborhoods to be open). If φ is continuous at each $\mathbf{v}\in V$, we say φ is continuous on V.

Definition 6.2.2. Let U, W be topological spaces, with $V\subset U$. $\varphi: V\to W$ has a limit $\lambda\in W$ at $\mathbf{v}\in U$ if $\mathbf{v}$ is a cluster point of $\operatorname{dom}\varphi$ and for each neighborhood B of λ there exists a *punctured* (*deleted*) neighborhood $\dot{A}$ of $\mathbf{v}$ (i.e., A is a neighborhood of $\mathbf{v}$ and $\dot{A}=A\backslash\{\mathbf{v}\}$) such that $\varphi(V\cap\dot{A})\subset B$. Again, there is no loss of generality if we take open neighborhoods. We write

$$\lim_{\mathbf{v}}\varphi=\lambda.$$

Both these definitions presume that V is indeed the whole domain of definition of φ. However, it often happens that we have this situation: two topological spaces V, W are already fixed, and a whole set of mappings is under consideration, mappings which map subsets of V into subsets of W. Suppose $\varphi: X\to W$ is such a mapping, where $X\subset V$. What we need to do is make a topological space out of X in a way that will be consistent with the fact that $X\subset V$. We shall endow X with a topology known as the *relative topology*, a topology that X inherits from V.

Definition 6.2.3. Let $(V, \mathfrak{T})$ be a topological space, and $X\subset V$ a subset. We make a space out of X by endowing it with the relative topology; to wit,

$$\mathfrak{S}=\{S: S=X\cap T,\ T\in\mathfrak{T}\}.$$

$\mathfrak{S}$ is a topology for X, the topology inherited from $(V, \mathfrak{T})$.

This way of making a consistent topological space out of a subset of a topological space allows us to extend our definitions for continuity and limit to a wider class of mappings.

We should remark that in light of this last definition a single isolated point of the domain of a mapping can be an open neighborhood of itself, so a mapping is continuous at each isolated point of its domain. However, a mapping does not have a limit at an isolated point of its domain.

The following theorem is an important one.

Theorem 6.2.4. *Let $\varphi: V\to W$ be a mapping from one topological space into another. The following statements are equivalent.*

a. *φ is continuous on V.*
b. *For all open $B\subset W$, $\varphi^{-1}(B)$ is open in V.*
c. *For all closed $D\subset W$, $\varphi^{-1}(D)$ is closed in V.*
d. *For each set $A\subset V$, $\varphi(\overline{A})\subset\overline{\varphi(A)}$.*

PROOF: a $\Rightarrow$ b: Let $B\subset W$ be open and $\mathbf{v}\in\varphi^{-1}(B)$. Then $\varphi(\mathbf{v})\in B$, and there exists an open neighborhood N, with $\varphi(\mathbf{v})\in N\subset B$. Since φ is continuous at

$\mathbf{v}$, there is an open neighborhood M, with $\mathbf{v} \in M$, such that $\varphi(M) \subset N$. This means $M \subset \varphi^{-1}(B)$, and $\varphi^{-1}(B)$ is thus a neighborhood of $\mathbf{v}$. Since $\mathbf{v} \in \varphi^{-1}(B)$ was arbitrary, $\varphi^{-1}(B)$ is a neighborhood of each of its points, and hence is open.

b $\Rightarrow$ a: Let $\mathbf{v} \in V$, and let N be any neighborhood of $\varphi(\mathbf{v})$. Let $A = \varphi^{-1}(\mathring{N})$. A is open, and $\mathbf{v} \in A$. Hence $\varphi(A) \subset N$, and φ is thus continuous at $\mathbf{v}$. Since $\mathbf{v}$ was arbitrary, φ is continuous on V.

b $\Rightarrow$ c: Let $D \subset W$ be closed. Then $W \setminus D$ is open, and $\varphi^{-1}(W \setminus D)$ is open. But $\varphi^{-1}(W \setminus D) = V \setminus \varphi^{-1}(D)$, so $\varphi^{-1}(D)$ must be closed.

c $\Rightarrow$ b: Let $B \subset W$ be open, so $W \setminus B$ is closed, and $\varphi^{-1}(W \setminus B) = V \setminus \varphi^{-1}(B)$ is closed. Hence $\varphi^{-1}(B)$ is open.

c $\Rightarrow$ d: Let $A \subset V$ be any set. $\overline{\varphi(A)}$ is closed in W, so $\varphi^{-1}(\overline{\varphi(A)})$ is closed in V. But $A \subset \varphi^{-1}(\overline{\varphi(A)})$, and it follows that $\bar{A} \subset \varphi^{-1}(\overline{\varphi(A)})$ since $\bar{A}$ is the smallest closed set containing A. Therefore $\varphi(\bar{A}) \subset \overline{\varphi(A)}$.

d $\Rightarrow$ c: Let $D \subset W$ be closed and put $C = \varphi^{-1}(D)$. Then $\varphi(\bar{C}) \subset \overline{\varphi(C)} = \overline{\varphi(\varphi^{-1}(D))} \subset \bar{D} = D$ implies that $\varphi^{-1}\varphi(\bar{C}) \subset \varphi^{-1}(D) = C$. Therefore C must be closed since $\bar{C} \subset \varphi^{-1}\varphi(\bar{C}) \subset C$. □

Here is a useful theorem; it allows us to determine continuity by considering just the sub-basic open sets instead of all the open sets.

Theorem 6.2.5. *$\varphi: V \to W$ is continuous if and only if the preimage of every sub-basic open set in W is open in V.*

PROOF: We refer the reader to Theorem 1.1.10 and wish you luck. (It really is an easy proof.)

Let us now consider how one might compare two topologies on the same ground set. Recall that we said at the beginning of the chapter that we frequently have a choice in deciding just how to topologize a particular set. For example, suppose we have a collection of mappings $\{\varphi_\alpha\}_{\alpha \in \mathscr{A}}$ having a common domain and range, and suppose further that there is an obvious topology that we must put on the range space. Assuming that the concept of continuity is important to our consideration of the class of mappings $\{\varphi_\alpha\}$, it would be reasonable to topologize the domain so as to make each of the φ_α continuous. This is to say, if $\mathfrak{T}$ is the topology on the range, then for each $T \in \mathfrak{T}$ and each α, there is a set $\varphi_\alpha^{-1}(T)$. Let the totality of all such sets be a sub-base for a topology on the domain. The topology so generated will be such that all the φ_α will be continuous.

Suppose now that $\mathfrak{T}_1, \mathfrak{T}_2$ are two topologies on a nonempty set V. We say $\mathfrak{T}_1$ is *finer* than $\mathfrak{T}_2$ if every set T in $\mathfrak{T}_2$ is also a set of $\mathfrak{T}_1$. The collection of all topologies on V is a lattice, with respect to the partial order relation "is finer than." The discrete topology is the finest possible topology on V, the trivial topology being the coarsest. There is a unique topology $\mathfrak{T}_3$ that is the supremum of $\mathfrak{T}_1$ and $\mathfrak{T}_2$; we write $\mathfrak{T}_3 = \mathfrak{T}_1 \vee \mathfrak{T}_2$. $\mathfrak{T}_3$ is the coarsest topology which is simultaneously finer than $\mathfrak{T}_1$ and $\mathfrak{T}_2$. To obtain $\mathfrak{T}_3$, we simply generate it from the union of $\mathfrak{T}_1$ and $\mathfrak{T}_2$. It is a fact that the collection of topologies on V is a complete lattice.

If $\mathfrak{T}_1$ is finer than $\mathfrak{T}_2$ and $\mathfrak{T}_2$ is finer than $\mathfrak{T}_1$, then $\mathfrak{T}_1 = \mathfrak{T}_2$. Often it is hard to discover whether two topologies are equal; we need some sort of test that will tell us when they may be regarded as one and the same. The next theorem is a big help.

Definition 6.2.6. Suppose $\mathfrak{B}_1, \mathfrak{B}_2$ are bases for the topologies $\mathfrak{T}_1, \mathfrak{T}_2$, respectively, on a set V. We say the two bases are *equivalent* if for each set T in either one of the topologies and each point $\mathbf{x} \in T$, there is a B in the base which generates the other topology such that $\mathbf{x} \in B \subset T$.

Theorem 6.2.7. *Equivalent bases on a set V generate the same topology.*

PROOF: The proof is easy and is left to the reader.

In a general topological space $(V, \mathfrak{T})$ we often think of a sequence $\{\mathbf{x}_n\}_{n=1}^{\infty} \subset V$ as the image of a mapping $\sigma: \mathbb{Z}^+ \to V$; for each $n \in \mathbb{Z}^+$, $\sigma(n) = \mathbf{x}_n$. Strictly speaking, $\{\mathbf{x}_n\}_{n=1}^{\infty}$ is not a subset of V because some of the terms may be one and the same point of V (σ may not be injective), but if it is clearly understood that $\{\mathbf{x}_n\}$ is a sequence, we shall write $\{\mathbf{x}_n\} \subset V$. Now $\mathbb{Z}^+$ is a well-ordered set; we could consider "generalized sequences" in V, i.e., a partially ordered collection of points of V, $\{\mathbf{x}_\alpha\}$, where the index set $\mathfrak{A}$, the domain of the mapping σ, is an upward directed partial order, and $\alpha \prec \beta \Leftrightarrow \sigma(\alpha) = \mathbf{x}_\alpha \prec \mathbf{x}_\beta = \sigma(\beta)$ in the generalized sequence, and card $\mathfrak{A}$ is arbitrary. We won't, except to call attention to the fact that an ordinary vector-valued function $f: \mathbb{R} \to E^n$ is a "generalized sequence." $\mathbb{R}$ is the partially ordered index set (actually chain ordered, of course), card $\mathbb{R} = \aleph_1$, and $f(x) \prec f(y)$ in the "sequence" iff $x \leqslant y$ in $\mathbb{R}$. You may imagine, if $n=1$, that the sequence is just the whole set of function values strung out in a continuous line, an ordered "$\aleph_1$-tuple" of numbers.

As a mapping, $\sigma: \mathbb{Z}^+ \to V$ is always continuous, since the relative topology on $\mathbb{Z}$ is the discrete topology. Remember that if $f: X \to Y$, you increase the probability that f will be continuous if you refine the topology on X or coarsen the topology on Y. Since σ is continuous, the only interesting limit possibility is the limit of σ at ∞; this is tantamount to the limit of $\{\mathbf{x}_n\}$ as $n \to \infty$.

Definition 6.2.8. Let $\sigma = \{\mathbf{x}_n\}_{n=1}^{\infty} \subset W$ be a sequence. $\lim_\infty \sigma = \mathbf{x} \Leftrightarrow \forall N_{\mathbf{x}}$, $\exists N_\infty$ s.t. $\sigma(N_\infty) \subset N_{\mathbf{x}}$, where by $N_{\mathbf{x}}$ we mean a W-neighborhood of $\mathbf{x}$ and by N_∞ we mean the set of all positive integers larger than some positive integer n_0. Equivalently,

$$\lim_{n\to\infty} \{\mathbf{x}_n\}_{n=1}^{\infty} = \mathbf{x} \quad \Leftrightarrow \quad \forall N_{\mathbf{x}}, \exists n_0 \text{ s.t. } \forall n > n_0, \quad \mathbf{x}_n \in N_{\mathbf{x}}.$$

It is evident that if a sequence in a Hausdorff space has a limit, that limit is unique. We remind the reader that if such a sequence has a limit, that limit may or may not be a cluster point of the set consisting of the distinct terms of the sequence. If it is a cluster point, it is the only one.

We bring this section to a close by defining a few more terms.

Definition 6.2.9. A mapping $\varphi: V \to W$ is called *open* if for all open $A \subset V$, $\varphi(A)$ is open in W.

A mapping $\varphi: V \to W$ is called *closed* if for all closed $A \subset V$, $\varphi(A)$ is closed in W. If φ is a linear mapping and V, W are linear topological spaces, we further require that for any sequence $\{\mathbf{x}_n\} \subset A$ converging to $\mathbf{x}_0 \in A$, the sequence $\{\varphi(\mathbf{x}_n)\} \subset \varphi(A)$ converge to $\varphi(\mathbf{x}_0) \in \varphi(A)$, if it converges.

If $\varphi: V \to W$ is a continuous injectivity and $\varphi^{-1}: \varphi(V) \to V$ is continuous, we say φ is *bicontinuous*. If $\varphi: V \to W$ is a bicontinuous bijectivity, we call φ a *homeomorphism*, or a *topological isomorphism.*

6.3 METRICS AND NORMS

In a general topological space we have no absolute notion of distance or measure; it may nevertheless be possible, given a set V, or even a topological space $(V, \mathfrak{T})$, to assign numerical values to point pairs in such a way that the numbers so assigned seem to be "distances" between the points of the pairs.

Definition 6.3.1. Suppose we have a real-valued function ρ defined on the set $V \times V$ and we stipulate that ρ obey the following axioms $\forall \mathbf{x}, \mathbf{y}, \mathbf{z} \in V$:

M1. $\rho(\mathbf{x}, \mathbf{y}) \geqslant 0$.
M2. $\rho(\mathbf{x}, \mathbf{y}) = \rho(\mathbf{y}, \mathbf{x})$.
M3. $\rho(\mathbf{x}, \mathbf{y}) + \rho(\mathbf{y}, \mathbf{z}) \geqslant \rho(\mathbf{x}, \mathbf{z})$.
M4. $\rho(\mathbf{x}, \mathbf{y}) = 0 \Leftrightarrow \mathbf{x} = \mathbf{y}$.

Then ρ is called a *metric* for V.

ρ does indeed seem to have the intrinsic properties that any "distance function" ought to have. A set V, endowed with a metric ρ, is called a metric space. If $\mathbf{p} \in V$ is an arbitrary point and the distance function ρ has been established, then for each real number $r > 0$, and each point $\mathbf{p} \in V$, we can form the set

$$B_{\mathbf{p}}(r) = \{\mathbf{v} \in V: \rho(\mathbf{v}, \mathbf{p}) < r\}.$$

To generate a topology for V we take as a sub-base the collection of all such sets. The topology so generated is called the metric space topology (relative to the metric ρ); if V was already a topological space, the topology induced by ρ may not agree with the already existing topology. For example, the metric topology is always a Hausdorff topology.

A trivial example of a metric space is the following. Let S be a nonempty set. Define a metric ρ for S by the condition

$$\rho(\mathbf{x},\mathbf{y})=1 \quad \Leftrightarrow \quad \mathbf{x}\neq\mathbf{y}.$$

It follows that the metric topology for this space is the discrete topology. Can you see why?

If V is a linear space, we have already shown how we can introduce a norm function on V (Definition 4.2.1). A normed linear space can be made into a topological space just as a metric space can since it is a metric space once we observe that we can define a metric ρ on a normed linear space by setting

$$\forall \mathbf{x},\mathbf{y}\in V, \qquad \rho(\mathbf{x},\mathbf{y})=\|\mathbf{x}-\mathbf{y}\|.$$

Although every normed linear space is a metric space, the converse is not true. The trivial example of a metric space given earlier (the distance between two distinct points is always 1) is one which can not be made into a normed space, even if the underlying structure is a linear space. The minute this space is normed, the metric topology induced by the norm will be different from the metric topology induced by the trivial metric (which was, you recall, the discrete topology).

Suppose $\{\mathbf{x}_n\}$ is a sequence in a metric space (V,ρ). Then the following definition holds.

Definition 6.3.2. A sequence $\{\mathbf{x}_n\}$ in a metric space is called a *Cauchy sequence* if for each $\varepsilon>0$ $\exists n_0$ s.t. $\forall m,n\geqslant n_0$, $\rho(\mathbf{x}_m,\mathbf{x}_n)<\varepsilon$.

Cauchy sequences by their very nature "look" convergent, but in some metric spaces, the "limit" of a Cauchy sequence might not exist in the space. For example, in the metric space $\mathbf{Q}$, the Cauchy sequence

$$\{\mathbf{e}_n\}_{n=1}^{\infty}=\{(1+1/n)^n\}_{n=1}^{\infty}$$

does not converge to a number in $\mathbf{Q}$. This brings us to another definition.

Definition 6.3.3. A metric space V is said to be *complete* if every Cauchy sequence in the space converges to a limit which is also in V.

Theorem 6.3.4. *Suppose $\{\mathbf{x}_n\}$ is a sequence in a metric space V and $\mathbf{x}_0\in V$ is a cluster point of the sequence. Then $\{\mathbf{x}_n\}$ contains a subsequence $\{\mathbf{x}_{n_k}\}$ which converges to $\mathbf{x}_0$.*

PROOF: When we say that $\mathbf{x}_0 \in V$ is a cluster point of the sequence $\{\mathbf{x}_n\} \subset V$ it may not be clear that we mean that $\mathbf{x}_0$ is a cluster point of the set of distinct points of the sequence. If the point $\mathbf{x}_0$ occurs infinitely often as a point of the sequence, then it is easy to see that there is a subsequence $\{\mathbf{x}_{n_i}\}_{i=1}^{\infty} \subset \{\mathbf{x}_n\}$, where for each $i=1,2,3,\ldots$, $\mathbf{x}_{n_i} = \mathbf{x}_0$. If $\mathbf{x}_0$ occurs only finitely often (perhaps not at all) in the sequence, the hypothesis of the theorem is that every deleted neighborhood of $\mathbf{x}_0$ contains a point of the sequence (other than $\mathbf{x}_0$).

Let $\{\dot{B}_k\}_{k=1}^{\infty}$ be a decreasing sequence of open balls of radius $1/k$ centered at $\mathbf{x}_0$, from each of which $\mathbf{x}_0$ has been deleted. Let $\mathbf{x}_{n_1} \in \{\mathbf{x}_n\}_{n=1}^{\infty}$ be a term of the sequence which is in $\dot{B}_1$. Now $\dot{B}_2$ must contain a point $\mathbf{x}_{n_2} \in \{\mathbf{x}_n\}_{n=n_1+1}^{\infty}$ since each $\dot{B}_k$ necessarily contains infinitely many points of $\{\mathbf{x}_n\}$. Similarly, there is a point $\mathbf{x}_{n_3} \in \{\mathbf{x}_n\}_{n=n_2+1}^{\infty} \subset \{\mathbf{x}_n\}$ which is in $\dot{B}_3$. Continuing in this way, we have a set $\{\mathbf{x}_{n_k}\}$ of distinct terms of the original sequence, and the order in which the terms $\mathbf{x}_{n_k}$ appear in $\{\mathbf{x}_{n_k}\}$ is the same as they appear in the original sequence. Hence $\{\mathbf{x}_{n_k}\}$ is a subsequence of $\{\mathbf{x}_n\}$. That $\mathbf{x}_0$ is the limit of $\{\mathbf{x}_{n_k}\}$ is clear. □

Theorem 6.3.5. *Let V, W be metric spaces and $A \subset V$ a set. Suppose $a \in A'$ and $\varphi: A \to W$ is a mapping. Then $\lim_a \varphi = b$ iff for each sequence $\{\mathbf{x}_n\} \subset A \setminus \{a\}$ which converges to a, we have $\{\varphi(\mathbf{x}_n)\}$ converging to b in W.*

PROOF: $\Rightarrow$: The proof in this direction is clear; the implication is a direct consequence of the definitions. Now suppose $\lim_a \varphi \neq b$. Let $\{B_n\}_{n=1}^{\infty}$ be a sequence of open balls in W, centered at b, with radius $1/n$. This sequence gives rise to a decreasing sequence of sets $\{\varphi^{-1}(B_n)\} \subset V$. Let $\{\dot{A}_m\}_{m=1}^{\infty}$ be a sequence of punctured open balls centered at a, where the radius of $\dot{A}_m$ is $1/m$. Now if $\lim_a \varphi$ were equal to b, for each n, there would be an m_n such that $A \cap \dot{A}_{m_n} \subset \varphi^{-1}(B_n)$. But b is not the limit; indeed φ may have no limit at a. Therefore there must be a set B_k such that for no m_k is $A \cap \dot{A}_{m_k} \subset \varphi^{-1}(B_k)$. We now construct a sequence converging to a as follows.

Choose $\mathbf{x}_1$ from the set $A \cap (\dot{A}_1 \setminus A_2) \setminus \varphi^{-1}(B_k)$, and in general, choose $\mathbf{x}_n$ from the set $A \cap (\dot{A}_n \setminus A_{n+1}) \setminus \varphi^{-1}(B_k)$. It is apparent that the sequence $\{\mathbf{x}_n\} \to a$, and $\varphi(\{\mathbf{x}_n\}) \subset W \setminus B_k$. The proof is complete. □

We have defined what we mean by a complete metric space. The reader should keep in mind that normed linear spaces are also metric spaces, so they too may be complete or incomplete. It is a fact that any metric space V which might happen to be incomplete may be "completed" by adjoining sufficiently many points that in the augmented space $\hat{V}$ every Cauchy sequence does have a limit. Every student of mathematics at some point must go through the trial of learning the proof of the following theorem.

Theorem 6.3.6. *For every metric space V there is a unique space $\hat{V}$, called the completion of V, into which V may be (isometrically) embedded. One writes*

$$\forall V, \quad V \text{ a metric space}, \qquad \exists | \hat{V} \ \text{s.t.} \ V \subsetneq \hat{V}.$$

The completion of $\mathbb{Q}$ is $\mathbb{R}$; this is the simplest example we can think of. We should add that the completion of a space V is a minimal complete space which contains V, or a copy of V.

We shall have occasion to use the notion of diameter of a set later in this chapter, so we give the definition at this time. It is after all a metric space concept.

Definition 6.3.7. If $S \neq \varnothing$ is a set in a metric space (V, ρ), by the diameter of S we mean the nonnegative number:

$$\operatorname{diam} S = \sup\{\rho(\mathbf{x},\mathbf{y}) : \mathbf{x},\mathbf{y} \in S\}.$$

We close this section by repeating that if V is a metric space, with ρ as its distance function, then the topology on V is that topology induced by the metric ρ. If $(V, \mathfrak{T})$ is a topological space, and we find a metric ρ for it, we don't then call V a metric space, unless the topology induced by ρ is $\mathfrak{T}$, or we agree to retopologize V with the new metric topology. A set V is always metrizable (see our trivial example), but a topological space is said to be metrizable only if a metric can be found which induces the already existing topology.

Every linear space can be normed; for example, if H is a Hamel basis, let each element of the Hamel base have norm 1. If

$$\mathbf{x} = \sum_{i=1}^{n} \xi_i h_i, \qquad h_i \in H,$$

is the unique representation for a vector $\mathbf{x}$, relative to H, let $\|\mathbf{x}\| = \max_i |\xi_i|$.

This is indeed a well-defined norm. But we call only those topological linear spaces normable for which a norm function can be found which induces the original topology.

6.4 PRODUCT TOPOLOGIES

Suppose we have a family of sets $\{A_\alpha\}$; the set $\times_\alpha A_\alpha$ has been defined in Chapter 1, and this Cartesian product is particularly nice when the index set is finite. If we choose, we can topologize it. But what if the A_α are already topological spaces? We most probably would want the topology of the product space to be such that the topologies of each A_α would be consistent with the relative topologies these A_α would inherit from the topology of $\times_\alpha A_\alpha$ as "subsets" of the product space. (Strictly speaking, A_{α_0} is not a subset of $\times_\alpha A_\alpha$, but we can identify A_{α_0} with the subset $\times_\alpha S_\alpha$, where for each α, $S_\alpha \subset A_\alpha$, and in particular, $S_\alpha = \{a_\alpha\}$, $a_\alpha \in A_\alpha$, $\alpha \neq \alpha_0$, and $S_{\alpha_0} = A_{\alpha_0}$.) There is a standard way of doing this. If we take as a sub-base sets in $\times_\alpha A_\alpha$, which are of the form $S = \pi_\alpha^{-1}(B)$, where B is any sub-basic set of A_α, and π_α is the projection mapping $\times_\alpha A_\alpha \to A_\alpha$, we generate a topology for the product that has the desired properties. This product topology is the "natural" topology for products. For example, the basic open sets of E^1 are the open intervals; in E^3 they are the open boxes, the Cartesian products of three intervals. Note that with the product topology on $\times_\alpha A_\alpha$, each projection mapping π_α: $\times_\alpha A_\alpha \to A_\alpha$ is continuous and open.

Now here is where the disturbing part of the discussion manifests itself. Suppose the index set is infinite; what does an open set "look like" in the product topology? In the next chapter you will meet a space which as a set looks like $\times_{i=1}^{\infty} E_i^1$, the Cartesian product of $\aleph_0$ copies of E^1. Now the preimage of an interval $(a, b) \subset E_1^1$ under the projection map π_1 is the subset of the product space consisting of all the ordered $\aleph_0$-tuples whose first entries are real numbers between a and b and whose remaining entries are arbitrary real numbers. This is to say, geometrically, the preimage of an interval is an infinite-dimensional slab, infinite in extent in every dimension except the first. If this is a sub-basic set, the intersections of any finite number of such things will still be geometric objects in the product space which are infinite in extent in all but a finite number of dimensions. Every basic open set of the product will be this way, and hence the same will be true for unions of such sets. Thus the product topology of $\times_{i=1}^{\infty} E_i^1$ is such that no open set can be "bounded" in the ordinary sense.

In general, an open set in the space $\times_\alpha A_\alpha$ is a subset of the form $\times_\alpha S_\alpha$, where $S_\alpha = A_\alpha$ for all but a finite number of the αs (in mathematical language, "all but a finite number" translates to "almost all") and each S_α is open in A_α. It is this fact that makes the topologies of infinite-dimensional normed linear spaces intrinsically different from the product topologies that one might put on them.

6.5 COMPACTNESS

The need to construct some abstraction to take the place of "closeness" in a set where there is no notion of distance whatsoever led to the idea of starting with a collection of "undefined" subsets, called open, satisfying certain axioms, and by means of them to build "neighborhoods." Now we can say, as a consequence of the topology we have constructed, a point $\mathbf{x}$ is as close to the set S as it can possibly be if $\mathbf{x} \in \bar{S}$, but such is not the case if $\mathbf{x} \in \text{ext}\, S$. We really can't do much better than this, but then it would be fruitless to try to compare degrees of closeness when we have absolutely no notion of comparative distances. Topologically speaking, the surface of a doughnut and the surface of a coffee cup are indistinguishable since each surface can be smoothly deformed into the other, and if you can't tell the difference between a centimeter and a kilometer, how could you tell which surface you were looking at? (Get some coffee: if you can drink from it, it's the cup, if you can dunk it, it's the doughnut.) For a simpler example, compare a soccer ball and an American football. Topologically, they will be indistinguishable since the lengths of all their axes will appear to be the same positive number.

Having had a certain measure of success abstracting "closeness," let's try to improve upon what we have by abstracting "boundedness." That is, let's look for some way to characterize a set S in an abstract topological space $(V, \mathfrak{T})$ that will be analogous to the characterization of a bounded set in E^n. If we can do this, we shall at least be able to say that if S is unbounded and $\mathbf{x} \in S$, there are points of S as far away from $\mathbf{x}$ as they can possibly be, whereas if S is bounded, every point of S is not as far away from $\mathbf{x}$ as possible. An obvious idea is this: a nonempty set $S \subset V$ is bounded if there exists a neighborhood of a point $\mathbf{x} \in S$ which contains S. But this is too trivial; V itself is such a neighborhood. An improvement would be this definition: $S \subset V$ is bounded if and only if there exists an open set $A \neq V$ with $\bar{S} \subset A$. The problem that first presents itself now is that V itself must be unbounded, and some topological spaces, by their very nature, are clearly bounded (for example, $[0, 1]$ with its inherited topology from E^1). However, this definition does allow for the implication that if $N_{\mathbf{x}}$ is a bounded neighborhood of $\mathbf{x}$, then there are points of V "not as close to $\mathbf{x}$" as the neighborhood points are.

Perhaps we have done the best we can in this direction, but our accomplishment has not been very fruitful. We shall try another approach. A bounded set S in E^n is contained in a ball; hence it would be reasonable to say that S is "finite in extent" in all possible "directions." We shall try to abstract this concept. Suppose such a set $S \subset E^n$ is being guarded by an arbitrarily large number of policemen (infinitely many, perhaps) stationed at fixed points of S, many if not all of whom are quite near-sighted but not completely blind. Regardless of how near-sighted they are or where their fixed stations are, it would seem that if S were "finite in extent," we could dispense with the services of all but a finite number of them and S would still be under satisfactory surveillance. On the other hand, if S were of infinite extent, such would not be the case.

Now we shall describe this situation in abstract mathematical terms. We replace the near-sighted policemen, whose vision at their limits is fuzzy, with open sets and make the following attempt at a definition for "abstract boundedness." A set S in a topological space $(V, \mathfrak{T})$ is "finite in extent" if S is such that whenever $\mathcal{C}$ is an arbitrary collection of open sets whose union covers S, then all but a finite number of the sets of the collection $\mathcal{C}$ may be discarded, and the union of the remaining sets will still cover S. (Such a collection $\mathcal{C}$ is called an open covering of S.) Have we abstracted boundedness satisfactorily? No; let $S \subset \mathbb{R}$ be the set $\{1, \frac{1}{2}, \frac{1}{3}, \dots\}$, and let the topology for $\mathbb{R}$ be the usual metric topology. This set is certainly bounded, and "finite in extent," but not finite in extent by our criterion. Let $\mathcal{C}$ be the open cover for S consisting of intervals of the form

$$\left(\frac{1}{n} - \frac{1}{(n+1)^4}, \frac{1}{n} + \frac{1}{(n+1)^4}\right), \qquad n = 1, 2, 3, \dots.$$

$\mathcal{C}$ certainly covers S, but each interval is disjoint from the others, so we cannot discard even one of the sets of $\mathcal{C}$ and still have a cover for S.

One cannot help but notice that had the set S contained the point 0, a limit point of S, then S would have been "finite in extent" by our definition, so we make one more modification. We shall say that "finite in extent" is an abstraction for "closed-and-boundedness." Again we have failed. Consider the set H of square-summable real sequences, a typical point being $\mathbf{x} = \{x_n\}_{n=1}^{\infty}$, with $\sum_{n=1}^{\infty} x_n^2 < \infty$. H is a normed linear space if we define

$$\alpha\mathbf{x} + \beta\mathbf{y} = \{\alpha x_n + \beta y_n\}_{n=1}^{\infty}$$

for all $\mathbf{x} = \{x_n\}$, $\mathbf{y} = \{y_n\} \in H$, and $\alpha, \beta \in \mathbb{R}$, and define

$$\|\mathbf{x}\| = \left(\sum_{n=1}^{\infty} x_n^2\right)^{1/2}.$$

If $S \subset H$ is the set $\{\mathbf{e}_1, \mathbf{e}_2, \mathbf{e}_3, \dots\}$ of unit vectors, where for each i, $\mathbf{e}_i$ is the sequence with 1 in the ith place and zeros everywhere else, the set S is in the closed unit ball of H. S is closed because S has no limit points; thus S is closed and bounded. Since the distance between $\mathbf{e}_i$ and $\mathbf{e}_j$, $i \neq j$, is $\sqrt{2}$, we can cover S with a collection of open balls of radius $\sqrt{2}/4$, each centered at a point of S. This covering cannot be reduced to a finite subcovering.

Instead of giving up, we admit that the idea of abstracting boundedness, or closed-and-boundedness was not so good, but the abstract characterization of finite in extent seems too good to throw away. We shall keep the characterization, but discard the descriptive term "finite in extent" because it no longer seems appropriate.

Definition 6.5.1. Let S be a set in a topological space $(V, \mathfrak{T})$. S is said to be *compact* if every open covering $\mathcal{C}$ of S can be reduced to a finite subcovering. If S is such that $\bar{S}$ is compact, we call S *relatively*, or *conditionally*, *compact*.

This criterion for compactness is known as the *Heine–Borel criterion*. Compactness is an intrinsic topological property, and how it is related to closedness and boundedness we shall see immediately.

Theorem 6.5.2. *A compact set K in a Hausdorff space V is closed.*

PROOF: If $K=V$, we are done. Otherwise, let $\mathbf{b}\in V\setminus K$. For each $\mathbf{x}\in K$, there are disjoint pairs of open neighborhoods $A_{\mathbf{x}}$ and $B_{\mathbf{x}}$ of $\mathbf{x}$ and $\mathbf{b}$, respectively, so the collection $\{A_{\mathbf{x}}\}_{\mathbf{x}\in K}$ is an open covering of K and $\mathbf{b}$ is not in the union of these sets. A finite number of these sets, say $\{A_{\mathbf{x}_1}, A_{\mathbf{x}_2},\dots,A_{\mathbf{x}_n}\}$, is a collection whose union contains K but not $\mathbf{b}$. Consider the set

$$B=B_{\mathbf{x}_1}\cap\cdots\cap B_{\mathbf{x}_n}.$$

B is open, contains $\mathbf{b}$, and $B\cap K=\varnothing$. This means $\mathbf{b}\in\operatorname{ext}K$. Since $\mathbf{b}\in V\setminus K$ was arbitrary we conclude K is closed. □

Theorem 6.5.3. *If K is compact and S closed, then $K\cap S$ is compact. In particular, a closed subset of a compact set is compact.*

PROOF: Let $\mathcal{C}$ be an arbitrary open cover of $K\cap S$. Then $\mathcal{C}\cup\{V\setminus S\}$ is an open covering of K and may be reduced to a finite subcovering $\mathcal{C}'$. If the set $V\setminus S$ is in $\mathcal{C}'$, throw it out; the remaining sets will cover $K\cap S$. Hence $K\cap S$ is compact. □

Theorem 6.5.4. *A compact set K in a metric space V is closed and bounded.*

PROOF: Since a metric space is a Hausdorff space, K is closed. Let K be covered by a collection of open balls of radius 1, and reduce this cover to a finite subcover of n such balls. It is clear that K is contained in a ball whose radius is $2n$, the sum of the diameters of the covering balls. Thus K is bounded. □

Theorem 6.5.5. *If $S\subset E^n$ is closed and bounded, then S is compact.*

PROOF: Since S is bounded, it is contained in a closed cube $B=\{(x_1,\dots,x_n): |x_i|\leq R,\ \forall i\}$ for some sufficiently large real number R; we will show that the closed and bounded set B is compact. There are 2^n orthants in E^n ("orthant" in E^n is the analog of a quadrant in E^2, or an octant in E^3), so the cube B can be thought of as being comprised of 2^n closed subcubes, one in each orthant. If B_1 is one of these sub-boxes, note that $\operatorname{diam}B_1=R\sqrt{n}$. Furthermore, B_1 can be further partitioned into 2^n congruent closed sub-boxes, each having diameter $\frac{1}{2}R\sqrt{n}$. Before we proceed with the proof, we present a lemma.

Lemma 6.5.6. *If $\{F_k\}_{k=1}^{\infty}$ is a decreasing sequence of closed sets in E^n, i.e., $\forall k$, $F_k\supset F_{k+1}$, and $\operatorname{diam}F_k\to 0$, then $\bigcap_{k=1}^{\infty}F_k$ is a singleton.*

PROOF: This proof depends upon the fact that the space E^1, or $\mathbb{R}$, with its natural topology is complete. Every Cauchy sequence of real numbers converges to a real number. Let $\mathbf{x}_k$ be an arbitrary point of F_k. Let $\varepsilon>0$, and consider the sequence $\{\mathbf{x}_k\}$. It is a Cauchy sequence since for all m, k greater than some particular k_0, we have $\rho(\mathbf{x}_k, \mathbf{x}_m) < \varepsilon$, where ρ is the Euclidean metric. We can also say that for all $k, m \geqslant k_0$, the absolute values of the differences of the respective coordinates of $\mathbf{x}_k, \mathbf{x}_m$ are all less than ε. Hence the coordinates of the points of the sequence $\{\mathbf{x}_k\}$ comprise n Cauchy sequences of real numbers. These sequences converge to an n-tuple which we denote by $\mathbf{x}_0 \in E^n$. This point $\mathbf{x}_0$ is the limit of the sequence $\{\mathbf{x}_k\}$ since for all $k \geqslant k_0$, $\rho(\mathbf{x}_k, \mathbf{x}_0) < n\varepsilon$. We conclude that since $\varepsilon > 0$ was arbitrarily small, $\{\mathbf{x}_k\} \to \mathbf{x}_0$.

Since, for any $N > 0$, F_N is closed and F_N contains all the points of the sequence beyond the Nth term, $\mathbf{x}_0$, the limit of the sequence, is in F_N since $\mathbf{x}_0$ must be a cluster point of F_N. The fact that $\mathbf{x}_0 \in F_N$ for each $N > 0$ means $\mathbf{x}_0 \in \bigcap_{k=1}^{\infty} F_k$. Hence we have $\bigcap_{k=1}^{\infty} F_k \neq \varnothing$.

That the intersection cannot contain two points is clear; suppose $\mathbf{y} \neq \mathbf{x}_0$ and $\mathbf{y} \in \bigcap F_k$. If $\operatorname{diam}\{\mathbf{x}_0, \mathbf{y}\} = \rho(\mathbf{x}_0, \mathbf{y}) = \delta$, then for some k_0, we have $\operatorname{diam} F_{k_0} < \delta$, so $\mathbf{y} \notin F_{k_0}$. The proof of the lemma is complete. □

We return to the proof of the theorem. If B is not compact, there is an open covering of B, call it $\mathcal{C}$, which cannot be reduced to a finite subcovering. If this is the case, then for at least one of the 2^n closed subcubes into which we may partition B, the covering $\mathcal{C}$ cannot be reduced to a finite subcovering. Choose one such subcube and call it B_1. Partition B_1 in a similar fashion, into 2^n closed congruent subcubes; for at least one of these closed subcubes the open cover $\mathcal{C}$ cannot be reduced to a finite subcover. Choose one such cube, call it B_2, and then make a similar partition of B_2. Proceeding in this way we obtain a decreasing sequence of closed sets $\{B, B_1, B_2, \ldots\}$, and for each k, $\operatorname{diam} B_{k+1} = \frac{1}{2}\operatorname{diam} B_k$, so $\operatorname{diam} B_k \to 0$. By the lemma, there is a single point $\mathbf{p}$ contained in all the B_k, including B itself. Since $\mathcal{C}$ covers B, there must be one open set $A \in \mathcal{C}$ such that $\mathbf{p} \in A$. Since A is open, there is an $\varepsilon > 0$ such that the ball $B_{\mathbf{p}}(\varepsilon) = \{\mathbf{x} \in E^n: \rho(\mathbf{x}, \mathbf{p}) < \varepsilon\}$ is contained in A. But for some k_0, $\operatorname{diam} B_{k_0} < \varepsilon/3$, and $\mathbf{p} \in B_{k_0}$, so we have $\mathbf{p} \in B_{k_0} \subset B_{\mathbf{p}}(\varepsilon) \subset A$. Contradiction. B_{k_0} was a cube with the property that no finite subcover of $\mathcal{C}$ would cover it. We conclude that B is compact.

Since $S \subset B$ is closed, Theorem 6.5.3 guarantees that S is compact, and the proof is complete. □

Definition 6.5.7. A set S in a topological space V is said to have the *Bolzano–Weierstrass property* if every infinite subset of S has an accumulation point in S.

Theorem 6.5.8. *A compact set K in a metric space V has the Bolzano–Weierstrass property.*

PROOF: Suppose $T \subset K$ is an infinite subset and $T' \cap K = \varnothing$. Then for each $\mathbf{x} \in K$, there is an open ball $B_{\mathbf{x}}$ such that $B_{\mathbf{x}} \cap (T \setminus \{\mathbf{x}\}) = \varnothing$. The collection $\{B_{\mathbf{x}}\}_{\mathbf{x} \in K}$ is an open cover for K and thus can be reduced to a finite subcover. But each $B_{\mathbf{x}}$ contains at most one point of T, a contradiction—T is an infinite set. □

Corollary 6.5.9. *Let K be a compact set in a metric space V and $\{\mathbf{x}_n\}$ a sequence of points of K. Then $\{\mathbf{x}_n\}$ contains a convergent subsequence.*

PROOF: If $\{\mathbf{x}_n\}$ contains the same point of K infinitely often, we need look no further for a convergent subsequence. If not, then the distinct points of $\{\mathbf{x}_n\}$ comprise an infinite set $T \subset K$. Since K has the Bolzano–Weierstrass property, T has an accumulation point in K; call it $\mathbf{x}_0$. Let $\{B_n(\mathbf{x}_0)\}_{n=1}^{\infty}$ be a sequence of open balls of radius $1/n$ centered at $\mathbf{x}_0$.

From the sequence $\{\mathbf{x}_n\}_{n=1}^{\infty}$ choose the first point, $\mathbf{x}_{n_1}$, which is contained in $B_1(\mathbf{x}_0)$. Such a first point must exist, since $\mathbf{x}_0$ is an accumulation point of T. From the subsequence $\{\mathbf{x}_n\}_{n=n_1+1}^{\infty}$, take the first point, $\mathbf{x}_{n_2}$, which is contained in $B_2(\mathbf{x}_0)$. Although $\mathbf{x}_{n_1}$ and $\mathbf{x}_{n_2}$ may be one and the same point of K, they are distinct points of the sequence since $n_1 < n_2$. Continue in this way by choosing from the subsequence $\{\mathbf{x}_n\}_{n=n_2+1}^{\infty}$ the first point, $\mathbf{x}_{n_3}$, which is contained in $B_3(\mathbf{x}_0)$.

This process gives rise to a sequence $\{\mathbf{x}_{n_k}\}_{k=1}^{\infty} \subset \{\mathbf{x}_n\}$ having the properties that $i < j \Rightarrow n_i < n_j$, so that $\{\mathbf{x}_{n_k}\}_{k=1}^{\infty}$ is indeed a subsequence. For any $\varepsilon > 0$, there exists an N such that $1/N < \varepsilon$; hence every term $\mathbf{x}_{n_k}$ of the sequence for which $k \geqslant N$ is within ε of $\mathbf{x}_0$. Thus $\{\mathbf{x}_{n_k}\}_{k=1}^{\infty} \to \mathbf{x}_0 \in K$. □

Definition 6.5.10. A set S in a metric space V is said to be *totally bounded* if for each $\varepsilon > 0$, there exists a finite set of points $\{\mathbf{x}_i\}_{i=1}^{n} \subset S$ such that the collection $\{B_\varepsilon(\mathbf{x}_i)\}_{i=1}^{n}$ of open balls of radius ε centered at the n points $\mathbf{x}_i$ is an open cover for S.

Theorem 6.5.11. *Suppose K is a set in a complete metric space. Then K is compact iff K is closed and totally bounded.*

PROOF: Suppose K is compact. Since V is a Hausdorff space, K is closed. For any $\varepsilon > 0$, an open cover consisting of ε balls centered at the points of K can be reduced to a finite subcover, so K is totally bounded.

Now suppose K is closed and totally bounded. Let $\mathcal{C}$ be an arbitrary open cover for K. Suppose that K is not compact and $\mathcal{C}$ is an open cover which cannot be reduced to a finite subcover. There exists a finite subset of K such that the collection of open balls of radius $\frac{1}{2}$ centered at these points is an open cover for K. For at least one of the points of the finite subset of K, say $\mathbf{x}_1$, it must be true that no finite subcover from $\mathcal{C}$ will cover the set $K_1 = K \cap B_{1/2}(\mathbf{x}_1)$

The set K_1 is closed and totally bounded but not compact. There is a finite subset of K_1 such that the open balls of radius $\frac{1}{4}$ centered at these points will cover K_1, and it must be true that for at least one of these points, say $\mathbf{x}_2$, no finite subcover of $\mathcal{C}$ will cover the set $K_2 = K_1 \cap \overline{B_{1/4}(\mathbf{x}_2)}$. Thus K_2 is not compact, but it is closed and totally bounded.

Continue in this way indefinitely. We have a decreasing sequence of sets $K \supset K_1 \supset K_2 \supset \cdots$, each closed and totally bounded, and diam $K_n \leqslant 2^{-(n-1)}$. The sequence of centers $\{\mathbf{x}_n\}_{n=1}^{\infty}$ is a Cauchy sequence, and since V is complete and K is closed, the sequence converges to $\mathbf{x}_0 \in K$. Moreover, $\mathbf{x}_0 \in K_n$ for each n. Now $\mathbf{x}_0$ belongs to at least one open set $G \in \mathcal{C}$. For a sufficiently large n, we have $\mathbf{x}_0 \in K_n \subset G$, a contradiction—$K_n$ cannot be covered by a finite subcollection of $\mathcal{C}$. We conclude that K must be compact. □

We shall use this theorem to prove a lemma that we need for the proof of a later theorem. Before we present the lemma, we shall tell you what we mean by the normed linear space l_1^n. We start with the linear space $\mathbb{R}^n$ and define a norm by the formula

$$\forall \mathbf{x} = (\xi_1, \ldots, \xi_n) \in \mathbb{R}^n, \qquad \|\mathbf{x}\|_1 = \sum_{i=1}^{n} |\xi_i|.$$

The space l_1^n is the real linear space $\mathbb{R}^n$ normed by $\| \ \|_1$.

Lemma 6.5.12. *The space l_1^n is complete and the closed unit ball in l_1^n is compact.*

PROOF: Let $B \subset l_1^n$ be the closed unit ball. We need first of all to show that l_1^n is complete. It is not hard to see that if $\{\mathbf{x}_k\}_{k=1}^{\infty}$ is a Cauchy sequence in l_1^n it necessarily follows that for each $i = 1, 2, \ldots, n$, the sequence of ith coordinates, $\{\xi_{ik}\}_{k=1}^{\infty}$, is a Cauchy sequence of real numbers and hence converges to a number ξ_i. Hence $\{\mathbf{x}_k\}$ must converge to the point $\mathbf{x}_0 = (\xi_1, \ldots, \xi_n\} \in l_1^n$. Thus the space is complete.

Let $Q = \{\mathbf{x} \in l_1^n : \mathbf{x} = (\xi_1, \ldots, \xi_n),\ 0 \leqslant \xi_i \leqslant 1,\ \forall i = 1, \ldots, n\}$. If we can show that Q is totally bounded, this will suffice to prove that B is totally bounded.

Let $\varepsilon > 0$. Let $L \subset Q$ be the set of lattice points in Q of the form

$$(k_1/m, \ldots, k_i/m, \ldots, k_n/m),$$

where for each i, k_i is an integer satisfying $0 \leqslant k_i \leqslant m$ and m is a sufficiently large fixed integer so that $2n/m < \varepsilon$. It is clear that L is a finite set of points and that the set of open balls centered at the points of L, each with radius ε, make up an open cover for Q. Thus Q, and hence B, is totally bounded. Since B is closed and l_1^n is complete, B is compact. □

Theorem 6.5.13. *Let V be an n-dimensional normed linear space and W an arbitrary normed linear space. Then every linear map $\varphi: V \to W$ is continuous.*

PROOF: Assume V has a normalized basis $\{\mathbf{e}_1,\ldots,\mathbf{e}_n\}$. Let S be the particular linear map $l_1^n \to V$ defined as follows:

$$\forall \mathbf{x} \in l_1^n, \quad \mathbf{x} = (\xi_1,\ldots,\xi_n), \qquad S\mathbf{x} = \sum_{i=1}^{n} \xi_i \mathbf{e}_i \in V.$$

Evidently, S is a linear bijectivity; moreover,

$$\|S\mathbf{x}\|_V \leqslant \sum_{i=1}^{n} |\xi_i|\,\|\mathbf{e}_i\| = \sum_{i=1}^{n} |\xi_i| = \|\mathbf{x}\|_1,$$

so S is norm reducing and hence continuous.

Suppose W is an arbitrary normed linear space and $R: l_1^n \to W$. Let $\{\mathbf{b}_i\}_{i=1}^n$ be the standard basis vectors in l_1^n, so that if $\mathbf{x} \in l_1^n$ is an arbitrary unit vector, we can write $\mathbf{x} = \sum_{i=1}^n \xi_i \mathbf{b}_i$ instead of $\mathbf{x} = (\xi_1,\ldots,\xi_n)$. Apply R to each $\mathbf{b}_i$ and put $K = \max_i\{\|R\mathbf{b}_i\|_w\}$. Then $\|R\mathbf{x}\|_w \leqslant K\,\|\mathbf{x}\|_1 = K$. This is to say, R is bounded and hence continuous.

Denote $S^{-1}: V \to l_1^n$ by T, and suppose that T is not continuous at $\mathbf{0} \in V$. Then, for some $\varepsilon > 0$, we can find a sequence $\{\mathbf{x}_k\} \subset V$ which converges to $\mathbf{0}$ but such that, for each k, $\|T\mathbf{x}_k\|_1 \geqslant \varepsilon$. This gives rise to another sequence

$$\{\mathbf{z}_k\} = \left\{ \frac{\mathbf{x}_k}{\|T\mathbf{x}_k\|_1} \right\} \subset V,$$

which still converges to $\mathbf{0}$, but the image under T of this sequence is a sequence on the unit sphere of l_1^n. This unit sphere is compact and hence has the Bolzano–Weierstrass property. We now have this situation: in V, the sequence $\{\mathbf{z}_k\} \to \mathbf{0}$, and in l_1^n, the sequence $\{T\mathbf{z}_k\}$, being an infinite set of distinct points since T is bijective, contains a subsequence $\{T\mathbf{z}_{k_i}\}$ which converges to a point $\mathbf{y}$ on the unit sphere. Apply S to this convergent subsequence; we get the sequence $\{\mathbf{z}_{k_i}\}$ which converges to $\mathbf{0}$. We conclude that $S(\mathbf{y}) = \mathbf{0}$. Contradiction; $\mathbf{y} \neq \mathbf{0}$, and S is bijective. Hence $T: V \to l_1^n$ must be continuous.

Any map $\varphi: V \to W$ is the composite of some $R: l_1^n \to W$ and $T: V \to l_1^n$, so $\varphi = R \circ T$, being the composite of two continuous functions, is continuous (Theorem 6.7.6), and the proof is complete. □

Corollary 6.5.14. *Let V be an n-dimensional linear space and $\|\ \|_1$, $\|\ \|_2$ two distinct norms on V. Then the norm topologies $\mathfrak{T}_1$ and $\mathfrak{T}_2$ generated by these norms are equal.*

PROOF: In the proof of the theorem we showed that the normed space V, regardless of the norm, was homeomorphic to l_1^n. Hence $(V,\mathfrak{T}_1)$ and $(V,\mathfrak{T}_2)$ are homeomorphic with each other, which is to say, one is topologically indistinguishable from the other. Let $(\mathbf{e}_1,\ldots,\mathbf{e}_n)$ be a basis for V, and let S: $l_1^n \to V$ be the bijectivity described in the proof of the theorem. S is a homeomorphism, regardless of whether V has $\mathfrak{T}_1$ or $\mathfrak{T}_2$ for its topology. We shall show that $\mathfrak{T}_1=\mathfrak{T}_2$ by showing that the bases $\mathfrak{B}_1$ and $\mathfrak{B}_2$ generated by the two norms are equivalent.

Let $B_1\in\mathfrak{B}_1$ be arbitrary and $\mathbf{x}\in B_1$ an arbitrary point. Then $S^{-1}(B_1)$ is open in l_1^n and contains $\boldsymbol{\alpha}=S^{-1}(\mathbf{x})$. Let $C_2\in\mathfrak{B}_2$ be a ball containing $\mathbf{x}$; $S^{-1}(C_2)$ is open in l_1^n and contains $\boldsymbol{\alpha}$. The set $G=S^{-1}(B_1)\cap S^{-1}(C_2)\subset l_1^n$ is open and contains $\boldsymbol{\alpha}$, so $S(G)$ is open in the $\mathfrak{T}_2$ topology, contains $\mathbf{x}$, and is contained in $B_1\cap C_2$. Therefore there exists $B_2\in\mathfrak{B}_2$, with $\mathbf{x}\in B_2\subset S(G)\subset B_1$. A repetition of this argument may be used to show that for any $B_2\in\mathfrak{B}_2$ and any $\mathbf{x}\in B_2$, there exists $B_1\in\mathfrak{B}_1$, with $\mathbf{x}\in B_1\subset B_2$. This proves the two bases are equivalent, so their respective generated topologies are equal. □

Corollary 6.5.15. *Every finite-dimensional normed linear space over* $\mathbb{R}$ *is complete.*

PROOF: Let V be an arbitrary real normed linear space of n dimensions. The space l_1^n is complete, and every map φ: $l_1^n\to V$ is continuous. Let S: $l_1^n\to V$ be the map defined at the beginning of the proof of Theorem 6.5.13. This S is a continuous linear bijectivity, and so is the inverse $T=S^{-1}: V\to l_1^n$.

Let $\{\mathbf{v}_n\}_{n=1}^{\infty}\subset V$ be a Cauchy sequence, and consider $\{T\mathbf{v}_n\}\subset l_1^n$. This sequence is a Cauchy sequence in l_1^n. To see this, let $\varepsilon>0$ be given, and since T is bounded, let $M=\sup\{\|T\mathbf{v}\|_1:\|\mathbf{v}\|=1\}$. Hence $\|T(\mathbf{v}_m-\mathbf{v}_n)\|_1\leqslant M\|\mathbf{v}_m-\mathbf{v}_n\|$. For all m, n sufficiently large, $\|\mathbf{v}_m-\mathbf{v}_n\|<\varepsilon/M$; hence, for such m, n, we have $\|T\mathbf{v}_m-T\mathbf{v}_n\|<\varepsilon$, which is to say $\{T\mathbf{v}_n\}$ is a Cauchy sequence in l_1^n.

Put $\mathbf{z}_n=T\mathbf{v}_n$, and let $\mathbf{z}_0\in l_1^n$ be the limit of $\{\mathbf{z}_n\}$. $S(\mathbf{z}_0)=\mathbf{v}_0\in V$. Since S is norm reducing, it is easy to see that for all n sufficiently large, $S(T\mathbf{v}_n)=\mathbf{v}_n$ is necessarily close to $\mathbf{v}_0$. This means $\{\mathbf{v}_n\}$ converges to $\mathbf{v}_0\in V$. We conclude that V is complete. □

Theorem 6.5.16. *Let B be the closed unit ball in the real normed linear space V. Then B is compact iff V is finite dimensional.*

PROOF: Suppose V is n-dimensional. Let S: $l_1^n\to V$ be the homeomorphism described previously. Then $S^{-1}(B)$ is a bounded neighborhood of $\mathbf{0}$ in l_1^n. If K is the closed unit ball in l_1^n, K is compact, and for some $\lambda>0$, $\lambda K\supset S^{-1}(B)$. It is not hard to see that since λK is totally bounded and $S^{-1}(B)$ is a closed convex region contained in λK, $S^{-1}(B)$ is also closed and totally bounded; i.e., $S^{-1}(B)$ is compact. By Theorem 6.7.1, B is compact.

The converse is a bit more difficult to prove. We need to establish two preliminary results.

Lemma 6.5.17. *If V is a normed linear space and $M \subset V$ is a finite-dimensional subspace, then M is a closed subset of V.*

PROOF: Corollary 6.5.15 establishes the fact that M is complete; hence, if $\mathbf{x}_0$ is a limit point of M, there is a sequence of points of M converging to $\mathbf{x}_0$, so $\mathbf{x}_0$ must belong to M. Thus M is closed in V. □

Lemma 6.5.18 (Riesz). *If V is a normed linear space and M a proper finite-dimensional subspace of V, then for each t, $0<t<1$, there exists an $\mathbf{x}_t \in V$, with $\|\mathbf{x}_t\|=1$, such that $\|\mathbf{x}-\mathbf{x}_t\| \geqslant t$, $\forall \mathbf{x} \in M$.*

PROOF: Choose $\mathbf{x}_1 \in V \setminus M$, and let $d=\inf\{\|\mathbf{x}-\mathbf{x}_1\| : \mathbf{x} \in M\}$. Since M is closed, $d>0$. There exists an $\mathbf{x}_0 \in M$ such that $d<\|\mathbf{x}_0-\mathbf{x}_1\| \leqslant d/t$ since $0<t<1$. Let $\mathbf{x}_t=(\mathbf{x}_0-\mathbf{x}_1)/\|\mathbf{x}_0-\mathbf{x}_1\|$, so that $\|\mathbf{x}_t\|=1$. Evidently $\mathbf{x}_t \notin M$. (Can you see why?)

If $\mathbf{x} \in M$, so also does $\|\mathbf{x}_0-\mathbf{x}_1\|\mathbf{x}+\mathbf{x}_0$, and thus

$$\|\mathbf{x}-\mathbf{x}_t\| = \left\| \mathbf{x} - \frac{\mathbf{x}_1-\mathbf{x}_0}{\|\mathbf{x}_0-\mathbf{x}_1\|} \right\| = \frac{\|(\|\mathbf{x}_0-\mathbf{x}_1\|\mathbf{x}+\mathbf{x}_0)-\mathbf{x}_1\|}{\|\mathbf{x}_0-\mathbf{x}_1\|} \geqslant \frac{d}{\|\mathbf{x}_0-\mathbf{x}_1\|} \geqslant t.$$

□

To complete the proof of the theorem, assume that B, the closed unit ball in V, is compact. Let S be the unit sphere in V, i.e., the closed subset of B consisting of all the unit vectors of V, so that S is also compact. The family of open balls of radius $\frac{1}{2}$ and centers on S is an open cover for S and can be reduced to a finite subcover. Let $\{\mathbf{x}_1,\ldots,\mathbf{x}_n\}$ be the centers of the balls of the subcover and M the span of these n unit vectors. M is a closed subspace of V. But M must be all of V, for if not, by the previous lemma, there would be a point $\mathbf{x}_0 \in S$ whose distance from M is greater than $\frac{1}{2}$, so that $\mathbf{x}_0$ would belong to none of the balls of the finite subcover covering S. Since $\dim M \leqslant n$ and $M=V$, V is finite dimensional. □

This theorem provides an extremely useful characterization of finite-dimensional normed linear spaces.

We close this section with a few remarks. When we discussed the product topology we mentioned the projection mappings π_α which map the product space $\times_\alpha A_\alpha$ onto A_α. If this was not clear to the reader at the time, we repeat that the product topology was constructed in just such a way as to make each of these projection mappings continuous. We are going to prove shortly that the continuous image of a compact set is compact, so if K is a compact set in $\times_\alpha A_\alpha$, then for each α, $\pi_\alpha K$ is compact in A_α. The converse of this, known as Tychonoff's theorem, states that if we have a product space $\times_\alpha A_\alpha$ and $K_\alpha \subset A_\alpha$ is compact for each α, then the set $\times_\alpha K_\alpha$ is compact in $\times_\alpha A_\alpha$. We do not prove this theorem, but we suggest that the reader prove that if $A \subset E^m$ and $B \subset E^n$ are compact sets, then $A \times B \subset E^m \times E^n = E^{m+n}$ is compact. (You might start by showing that $A \times B$ is closed and bounded.)

We began this discussion on compactness by trying to abstract the notion of boundedness, and we were not very successful. However, in attempting to characterize the rather loose notion of "finite in extent" we ended up with a concept better described by the word compactness. In Hausdorff spaces, compactness is a stronger notion than either closedness or boundedness. We remark that in finite-dimensional normed vector spaces, such as E^n, "compact" and "closed and bounded" are equivalent terms. In more general complete metric spaces, conditionally compact and totally bounded are equivalent notions.

6.6 DENSE SETS, CONNECTED SETS, SEPARABILITY, AND CATEGORY

We have met open sets, closed sets, compact sets, discrete sets (each point of a discrete set is an isolated point), bounded sets, and of course sets which are none of the above. One special type of set in a topological space is the dense set.

Definition 6.6.1. If $(V, \mathfrak{T})$ is a topological space, a set $S \subset V$ is called *dense* if $\bar{S} = V$. If $B \subset V$ is a set, a set $S \subset V$ is said to be *dense in B* if $B \subset \bar{S}$.

Definition 6.6.2. If a set S is dense in no nonempty open set of V, we say S is *nowhere dense*. This is equivalent to saying that S is nowhere dense if $\mathrm{int}(\bar{S}) = \varnothing$.

A set that is nowhere dense cannot be a neighborhood. Discrete sets are nowhere dense, but the converse is not necessarily true. If a set is not nowhere dense, then it is dense in some basic open set. (One is allowed to say "not nowhere dense," as opposed to "somewhere dense.")

The set $S = \{1, \frac{1}{2}, \frac{1}{3}, \frac{1}{4}, \dots\}$ is nowhere dense in $\mathbb{R}$. $\mathbb{Q}$ is dense in $\mathbb{R}$, and every set A is dense in its closure.

Intuitively, the idea of a connected set seems simple enough, but to give a topological definition is harder than one would expect. We go about it backwards.

Definition 6.6.3. If S is a set in a topological space V, S is *disconnected* iff S can be decomposed into two nonempty parts A and B such that

$$S = A \cup B, \qquad \varnothing = (\bar{A} \cap B) \cup (A \cap \bar{B}).$$

S is *connected* iff for any two open (closed) sets A_1 and A_2, with $S \cap A_1 \neq \varnothing$, $S \cap A_2 \neq \varnothing$ and $S \subset (A_1 \cup A_2)$, we have $A_1 \cap S \cap A_2 \neq \varnothing$.

If S is already a topological space, then S is disconnected iff there exists a proper subset of S which is clopen. This follows from the definition above; if A is a proper clopen subset of S, then so is its complement $B=S\setminus A$. Therefore, to determine whether or not a subset $S\subset V$ is disconnected, it is sometimes better to put the relative topology on S, and then look for a pair of complementary, nonempty, clopen sets in S.

If S is a nonempty set, the nonempty parts of S which are themselves connected and maximal in this respect are called *components* of S. A pair (A, B) such as appears in Definition 6.6.3 is called a *disconnection* for S.

We shall prove in Section 6.7 that the continuous image of a connected set is connected. Hence, if S is connected in a product space $\times_\alpha A_\alpha$, then the projections $\pi_\alpha S=S_\alpha\subset A_\alpha$ are connected sets. On the other hand, connected sets are like compact sets in this sense: if for each α, $S_\alpha\subset A_\alpha$ is connected, then $\times_\alpha S_\alpha$ is connected in $\times_\alpha A_\alpha$. We do not intend to prove this.

Theorem 6.6.4. $\mathbb{R}$, *with its usual topology, is connected, as is every open interval* (a,b). *Conversely, every nonempty connected set in* $\mathbb{R}$ *is an interval, or a singleton.*

PROOF: Let (a,b) be an arbitrary interval in $\mathbb{R}$. We will allow the possibility that $a=-\infty$ or $b=+\infty$. Assume (a,b) is disconnected and that (A,B) is a disconnection, and let $b'\in B$.

Case I. There exists a point of A which is less than b'.

Let $A_1=\{\mathbf{x}\in A:\mathbf{x}<b'\}$, and let $\lambda=\sup A_1$. Then $a<\lambda\leq b'<b$ and $\lambda\in\bar{A}_1\subset\bar{A}$. If $\lambda=b'$, we have $\lambda\in\bar{A}\cap B$, so (A,B) would not be a disconnection. If $\lambda<b'$, the interval $(\lambda,b']\subset B$, so $\lambda\in\bar{B}$. Since $a<\lambda<b$, λ must belong to one of A or B, and we would have $\lambda\in(\bar{A}\cap B)\cup(A\cap\bar{B})$. Hence, if case I is the situation at hand, (A,B) is not a disconnection of (a,b).

Case II. $\forall\mathbf{x}\in A$, $b'<\mathbf{x}$. Let $\mu=\inf A$, and we have $a<b'\leq\mu<b$. If $\mu=b'$, then $\mu\in B$. Since $\mu\in\bar{A}$, we have $\bar{A}\cap B\neq\varnothing$. If $\mu>b'$, then $[b',\mu)\subset B$, and $\mu\in\bar{B}\cap\bar{A}$. Since $\mu\in A$ or $\mu\in B$, we have $(\bar{A}\cap B)\cup(A\cap\bar{B})\neq\varnothing$. Thus, in case II, (A,B) is not a disconnection. Hence $\mathbb{R}$, or any open interval $(a,b)\subset\mathbb{R}$, is connected.

Now let $S\subset\mathbb{R}$ be an arbitrary nonempty set. If S is not an interval nor a singleton, and there exists a real number $t\notin S$ such that for some $a\in S$, $a<t$ and, for some $b\in S$, $b>t$. Define $A=\{\mathbf{x}\in S:\mathbf{x}<t\}$, $B=\{\mathbf{x}\in S:\mathbf{x}>t\}$. Then $\lambda=\sup A\leq t\leq\mu=\inf B$. Certainly A and B are nonempty disjoint sets whose union is S, and since $t\notin S$, it is clear that $\bar{A}\cap B=\varnothing=A\cap\bar{B}$. Hence S is disconnected. This completes the proof. □

Theorem 6.6.5. *If C is a convex set in E^n, C is connected. In particular, balls and boxes are connected.*

PROOF: If C is disconnected in E^n, let (A, B) be a disconnection. If $\mathbf{a}\in A$ and $\mathbf{b}\in B$, the line segment $\overline{ab}\subset C$. Since each point of this segment is in A or in B, we can easily formulate a proof analogous to the proof of the previous theorem. Let α be given by

$$\alpha=\sup\{\rho(\mathbf{x},\mathbf{a}):\mathbf{x}\in A\cap\overline{ab}\}.$$

Let $p\in\overline{ab}$ be the unique point of the line whose distance from a is α. Then $p\in\bar{A}$, and $(\bar{A}\cap B)\cup(A\cap\bar{B})\neq\varnothing$. Contradiction. □

Theorem 6.6.6. *If S is a connected set, so is $\bar{S}$, as well as any set T, with $S\subset T\subset\bar{S}$.*

PROOF: Assume $\bar{S}$ is disconnected, and let (A, B) be a disconnection. Then $S=(S\cap A)\cup(S\cap B)$, and $\overline{(S\cap A)}\cap(S\cap B)\subset\bar{S}\cap\bar{A}\cap S\cap B=S\cap\bar{A}\cap B=\varnothing$, $(S\cap A)\cap\overline{(S\cap B)}=S\cap A\cap\bar{B}=\varnothing$, so S is disconnected. Contradiction. Hence $\bar{S}$ is connected, and it is clear that any intermediate set T is connected. □

There is another type of connectedness that we must mention: arcwise connectedness. We are concerned with this kind of connectedness in Euclidean spaces.

Definition 6.6.7. An arc in E^n is the continuous image of an interval in E^1. A set $S\subset E^n$ is said to be *arcwise connected* if for every two distinct points $\mathbf{x}\in S$, $\mathbf{y}\in S$, there is an arc in S initiating at $\mathbf{x}$ and terminating at $\mathbf{y}$. As a consequence of Theorem 6.7.2, an arcwise connected set is necessarily connected. The proof of this is easy, and we leave the details to the reader.

Let $S\subset E^2$ be the set $A\cup B$, where A is the interval $[-1,1]$ on the y axis and B is the Cartesian graph of the equation $y=\sin(1/x)$, $0<x\leqslant 1$. S is connected, but not arcwise connected. The reader may see why with no difficulty.

Let S_1 be the set of irrational points in $[0,1]$. Then S_1 is as disconnected as it can be. In the relative topology for S_1, $S_1\cap[0,\frac{1}{2}]$ and $S_1\cap[\frac{1}{2},1]$ are two complementary closed sets. Now consider the set S_2 of irrational points in $[0,1]\times[0,1]\subset E^2$, a point (x, y) being irrational if one of x or y is irrational. Not only is S_2 connected, it is arcwise connected.

Definition 6.6.8. A topological space V is said to be *separable* if it contains an at most countable dense subset.

Definition 6.6.9. A topological space $(V,\mathcal{T})$ is called *second countable* if there is an at most countable base for $\mathcal{T}$.

Theorem 6.6.10. *A separable metric space is second countable.*

PROOF: Let T be an open set in a separable metric space and $\mathbf{x} \in T$ an arbitrary point. Let $\{p_i\}_{i=1}^{\infty}$ be a countable dense subset; there is some p_i and some rational number r such that the ball centered at p_i with radius r contains x and is contained in T. Thus the set of all open balls having rational radii centered at the points of $\{p_i\}$ is a base for $\mathfrak{T}$ and is countable. □

Corollary 6.6.11. *E^n is a separable, second countable, metric space. Moreover, an equivalent countable base for the topology is the set of open boxes having edges of rational length, centers at rational points of E^n, and edges parallel to the coordinate axes.*

We leave the proof as an exercise.

Definition 6.6.12. A *region* is a set that is the union of an open connected set and none, some, or all of its boundary points. In E^n, a region is arcwise connected. A region R in E^n is said to be *simply connected* if for any $S \subset R$, where S is a homeomorphic image of the circle, S may be continuously shrunk to a single point of R without ever leaving the region R. A region in E^n which is not simply connected is said to be *multiply connected*.

Examples of multiply connected regions are the ball in E^2 with center removed and a solid torus in E^3. Any convex region in E^n is simply connected. The unit sphere in E^3, $S = \{(x, y, z): x^2 + y^2 + z^2 = 1\}$, is simply connected.

Definition 6.6.13. Let $(S, \mathfrak{T})$ be a topological space and $T \subset S$ a subset. T is said to be of *first category* if T can be expressed as a union of an at most countable number of nowhere dense sets. Otherwise, T is said to be of *second category*.

Theorem 6.6.14 (Baire). *A complete metric space S is of second category.*

PROOF: Assume the contrary. $S = \bigcup_{n=1}^{\infty} A_n$, where each A_n is nowhere dense. Since $\bar{A}_1 \neq S$, let $b_1 \in S \setminus \bar{A}_1$, an open set. For some $d_1 < 1$, there is an open ball $B(b_1, d_1)$ centered at b_1 with radius d_1 such that $B(b_1, d_1) \subset S \setminus \bar{A}_1$. Let $B_1 = B(b_1, d_1/2)$ and $\bar{B}_1 \cap \bar{A}_1 = \varnothing$.

Since $B_1 \setminus \bar{A}_2$ is open and nonempty, choose $b_2 \in B_1 \setminus \bar{A}_2$, and for some open ball $B(b_2, d_2)$, with $d_2 < \frac{1}{2}$, we have $B(b_2, d_2) \subset B_1 \setminus \bar{A}_2$. Put $B_2 = B(b_2, d_2/2)$ and $\bar{B}_2 \cap \bar{A}_2 = \varnothing$. Furthermore, $B_1 \supset B_2$. Continue in this way to get a decreasing sequence of open sets $\{B_n\}$ such that for each n, $\bar{B}_n \cap \bar{A}_n = \varnothing$, and diam $B_n < 1/n$. Now $\bigcap_{n=1}^{\infty} \bar{B}_n$ is nonempty. We need now the following lemma.

Lemma 6.6.15 (The Cantor Intersection Theorem). *If S is a complete metric space and $\{F_n\}_{n=1}^{\infty}$ is a decreasing sequence of nonempty closed sets with* diam $F_n \to 0$, *then $\bigcap_{n=1}^{\infty} F_n$ is a singleton.*

PROOF: To say that S is complete is to say that every Cauchy sequence in S converges to a limit in S. We can assume without loss of generality that diam $F_1 < \infty$. For each n, choose $\mathbf{x}_n \in F_n$. Then $\{\mathbf{x}_n\}$ is clearly a Cauchy sequence and hence converges to $\mathbf{x}_0 \in S$. Since each F_n is closed and $\mathbf{x}_k \in F_n$ for $k \geqslant n$, we have $\mathbf{x}_0 \in F_n$ for every n. Hence $\mathbf{x}_0 \in \bigcap_{n=1}^{\infty} F_n$.

If $\mathbf{y} \in S$ and $\mathbf{y} \neq \mathbf{x}_0$, then $\rho(\mathbf{x}_0, \mathbf{y}) = \varepsilon > 0$. For some n_0, diam $F_{n_0} < \varepsilon$, so $\mathbf{y} \notin F_{n_0}$. Therefore $\mathbf{y} \notin \bigcap_{n=1}^{\infty} F_n$, and we have $\bigcap_{n=1}^{\infty} F_n = \{\mathbf{x}_0\}$. This proves the lemma. □

To return to the proof of the theorem, let $\mathbf{x} \in \bigcap_{n=1}^{\infty} \overline{B}_n$. Then for some n_0, $\mathbf{x} \in A_{n_0}$ since $S = \bigcup_{n=1}^{\infty} A_n$. But this means $\mathbf{x} \in \overline{B}_{n_0} \cap \overline{A}_{n_0}$; contradiction. For each n, we had $\overline{A}_n \cap \overline{B}_n = \varnothing$. This completes the proof of the famous "bearcat" theorem. □

The reader should keep in mind that Lemma 6.6.15, the Cantor intersection theorem, is really one of the fundamental theorems that lie at the heart of real analysis. We have had several occasions to use the result, for example, in the proof of Lemma 6.5.6.

6.7 SOME PROPERTIES OF CONTINUOUS MAPS

Theorem 6.7.1. *Let $\varphi: V \to W$ be a continuous map and $K \subset V$ compact. Then $\varphi(K)$ is compact in W.*

PROOF: Let $\mathcal{C}$ be an open covering for $\varphi(K)$. If $\mathcal{C} = \{B_\alpha\}$, then for each α, $\varphi^{-1}(B_\alpha) = A_\alpha$ is open, and $\{A_\alpha\}$ is an open cover for K. Reduce this cover to a finite subcover $\{A_{\alpha_1}, \ldots, A_{\alpha_m}\}$. Then $\{B_{\alpha_i}\}_{i=1}^{m}$ is the desired finite subcover from $\mathcal{C}$, $\varphi(A_{\alpha_i}) = B_{\alpha_i}$ for each $i = 1, \ldots, m$, and $\varphi(K) \subset \bigcup_{i=1}^{m} B_{\alpha_i}$. □

Theorem 6.7.2. *Let $\varphi: V \to W$ be continuous and $K \subset V$ connected. Then $\varphi(K)$ is connected in W.*

PROOF: To simplify the proof, we restrict φ to the set K. Suppose $\varphi(K)$ is disconnected in W. Let (B_1, B_2) be a disconnection of $\varphi(K)$.

$$\varphi(K) = B_1 \cup B_2 \quad \Rightarrow \quad \varphi^{-1}(\varphi(K)) = K = \varphi^{-1}(B_1) \cup \varphi^{-1}(B_2).$$

Let $A_1 = \varphi^{-1}(B_1) \subset K$ and $A_2 = \varphi^{-1}(B_2) \subset K$. Since $\overline{B}_1 \cap B_2 = \varnothing$ and $\varphi(\overline{A}_1) \subseteq \overline{\varphi(A_1)} = \overline{B}_1$, we have $\varphi(\overline{A}_1) \cap \varphi(A_2) = \varnothing$. But then $\varphi^{-1}(\varphi(\overline{A}_1) \cap \varphi(A_2)) = \overline{A}_1 \cap A_2 = \varnothing$. Analogously, we get $A_1 \cap \overline{A}_2 = \varnothing$. This means K is disconnected, a contradiction. Therefore $\varphi(K)$ is connected. □

These two theorems show that continuous mappings preserve compactness and connectedness. Hence we have one of the fundamental theorems of elementary calculus, one which is almost always taken on faith but which is essential even to begin calculus.

Theorem 6.7.3. *If f is a continuous real-valued function on a finite closed interval (box) I in E^n, then f attains a minimum value, a maximum value, and all intermediate values between these two extreme values.*

In general, a continuous mapping is not open, nor is it closed. Constant maps are continuous but not open. On the other hand, the continuous function $f(x)=x^2(1+x^2)^{-1}$ maps the closed set $[0,\infty)$ onto the set $[0,1)$, which is neither open or closed. However, if f is a homeomorphism, then f preserves all topological properties.

Theorem 6.7.4. *Let V, W be Hausdorff spaces, $D\subset V$ a dense set, and φ: $D\to W$ a continuous mapping. Suppose there exists a mapping $\bar{\varphi}: V\to W$ which is an extension of φ [i.e., for each $\mathbf{x}\in D$, $\bar{\varphi}(\mathbf{x})=\varphi(\mathbf{x})$] and $\bar{\varphi}$ is also continuous. Then $\bar{\varphi}$ is unique in that it is the only continuous extension of φ to V.*

PROOF: When we say φ is continuous on D, a subset of V, of course we mean that D is to be considered as a topological space with the relative topology inherited from V. Let $\mathbf{x}\in V\setminus D$; $\mathbf{x}$ is a cluster point of D since D is dense; $\varphi(\mathbf{x})$ is not defined but $\bar{\varphi}(\mathbf{x})$ is. Since $\lim_{\mathbf{x}}\bar{\varphi}=\bar{\varphi}(\mathbf{x})$, it must follow that $\lim_{\mathbf{x}}\varphi=\bar{\varphi}(\mathbf{x})$. Since W is a Hausdorff space, if $b=\lim_{\mathbf{x}}\varphi$, then $b=\bar{\varphi}(\mathbf{x})$, so for each $\mathbf{x}\in V\setminus D$, $\lim_{\mathbf{x}}\varphi$ is uniquely determined. Hence there is no other possible extension of φ which is continuous. □

The point of this little theorem is simply to show that if $\varphi: V\to W$ is a continuous mapping from one Hausdorff space into another, then φ is actually determined by its values on a dense subset of V.

Theorem 6.7.5. *Let f be a uniformly continuous mapping of a dense set D of a complete metric space V into a complete metric space W. Then there exists a uniquely determined mapping F which is uniformly continuous on V and is an extension of f.*

PROOF: Let $\{\mathbf{x}_n\}\subset D$ be an arbitrary Cauchy sequence, converging to $\mathbf{x}\in V$. Then $\{f(\mathbf{x}_n)\}$ is a Cauchy sequence in W and hence converges to a limit in W. Now define $F(\mathbf{x})=\lim_n\{f(\mathbf{x}_n)\}$. Of course, if $\mathbf{x}\in D$, $F(\mathbf{x})=f(\mathbf{x})$.

If $\{\mathbf{x}'_n\} \subset D$ is another Cauchy sequence converging to $\mathbf{x} \in V$, the uniform continuity of f guarantees that $\lim_n\{f(\mathbf{x}_n)\} = \lim_n\{f(\mathbf{x}'_n)\}$, so $F(\mathbf{x})$ is well defined. Now F is defined over all of V, since each point of $V \setminus D$ is a limit point of D in V.

We want to show now that F is uniformly continuous on V. Let $\varepsilon > 0$, and suppose $\mathbf{x}, \mathbf{y} \in V$ are such that $\rho(\mathbf{x}, \mathbf{y}) < \delta$, δ a real number to be determined later. Since $F(\mathbf{x}) = \lim_{\mathbf{u} \to \mathbf{x}} f(\mathbf{u})$ and $F(\mathbf{y}) = \lim_{\mathbf{v} \to \mathbf{y}} f(\mathbf{v})$, where $\mathbf{u}, \mathbf{v} \in D$, we can write

$$\rho(F(\mathbf{x}), F(\mathbf{y})) = \rho\Big(\lim_{\mathbf{u} \to \mathbf{x}} f(\mathbf{u}), \lim_{\mathbf{v} \to \mathbf{y}} f(\mathbf{v})\Big)$$

$$\leqslant \rho(\lim f(\mathbf{u}), f(\mathbf{u}')) + \rho(f(\mathbf{u}'), f(\mathbf{v}')) + \rho(f(\mathbf{v}'), \lim f(\mathbf{v})),$$

where $\mathbf{u}', \mathbf{v}' \in D$ and these points are chosen such that $\rho(\mathbf{x}, \mathbf{u}') < \delta$, $\rho(\mathbf{y}, \mathbf{v}') < \delta$, and $\rho(\mathbf{u}', \mathbf{v}') < 3\delta$. Now choose δ small enough so that each term of the right-hand side of the above inequality is less than $\varepsilon/3$. This can be done because f is uniformly continuous on D. Thus $\rho(F(\mathbf{x}), F(\mathbf{y})) < \varepsilon$, and F is uniformly continuous.

That F is unique is a consequence of Theorem 6.7.4, and we are done. □

Theorem 6.7.6. *If V, W, Z are topological spaces, and the mappings $\varphi: V \to W$, $\psi: W \to Z$ are continuous, then the composite map $\theta = \psi\varphi: V \to Z$ is also continuous.*

PROOF: Let A be an arbitrary set in V. Then $\varphi(\bar{A}) \subset \overline{\varphi(A)} \Rightarrow \psi(\varphi(\bar{A})) \subset \psi(\overline{\varphi(A)}) \subset \overline{\psi(\varphi(A))}$ since φ and ψ are continuous (Theorem 6.2.4). But this says $\theta(\bar{A}) \subset \overline{\theta(A)}$, so θ is continuous. □

We are about to prove the important theorem that a continuous mapping from a compact metric space into a metric space is uniformly continuous. We need an additional notion:

Definition 6.7.7. Let M be a metric space, or possibly a subset of a metric space, and suppose $\mathcal{C}$ is an open cover for M. A real number $\delta > 0$ is called a *Lebesgue number* for $\mathcal{C}$ if each subset of M whose diameter is less than δ is contained in at least one of the open sets of $\mathcal{C}$.

Theorem 6.7.8. *If K is a compact set in a metric space V, every open covering $\mathcal{C}$ for K has a Lebesgue number.*

PROOF: Let $\mathcal{C} = \{G_\alpha\}$ be an arbitrary open cover for K. Since each G_α is open and $K \subset \bigcup_\alpha G_\alpha$, it is true that for each $\mathbf{x} \in K$, there is a $d_x > 0$ such that the ball $B(\mathbf{x}, d_x) = \{\mathbf{y} \in V: \rho(\mathbf{y}, \mathbf{x}) < d_x\} \subset G_\alpha$ for at least one α. Thus $\{B(\mathbf{x}, d_x)\}_{\mathbf{x} \in K}$ is an open cover for K. But then so is the collection $\{B(\mathbf{x}, \frac{1}{2}d_x\}_{\mathbf{x} \in K}$ an open cover for K, from which we can extract a finite subcover

$$\{B(x_1, \tfrac{1}{2}d_{x_1}), \ldots, B(x_n, \tfrac{1}{2}d_{x_n})\}.$$

Let $\varepsilon = \min\{\frac{1}{2}d_{x_1}, \ldots, \frac{1}{2}d_{x_n}\}$. Now suppose that $\mathbf{x} \in K$ is arbitrary. Then $\mathbf{x} \in B(\mathbf{x}_j, \frac{1}{2}d_{x_j})$ for some j, $1 \leqslant j \leqslant n$. If $\mathbf{y} \in B(\mathbf{x}, \varepsilon)$ is arbitrary, then

$$\rho(\mathbf{y}, \mathbf{x}_j) \leqslant \rho(\mathbf{y}, \mathbf{x}) + \rho(\mathbf{x}, \mathbf{x}_j) < \varepsilon + \tfrac{1}{2}d_{x_j} \leqslant d_{x_j},$$

which is to say, $\mathbf{y} \in B(\mathbf{x}_j, d_{x_j})$. Hence $B(\mathbf{x}, \varepsilon) \subset B(\mathbf{x}_j, d_{x_j}) \subset G_\alpha$ for some α.

Now let $\delta = \varepsilon/3$. We assert that every subset of K whose diameter is less than or equal to δ is contained in at least one of the G_α. This is clear; let S be such a set. If $\mathbf{x} \in S$, every $\mathbf{y} \in S$ is within δ of $\mathbf{x}$, so $S \subset B(\mathbf{x}, \varepsilon) \subset G_\alpha$ for some α. □

Now to the important theorem that we promised.

Theorem 6.7.9. *Every continuous mapping f from a compact set K contained in a metric space V into a metric space W is uniformly continuous on K.*

PROOF: Suppose ρ_1, ρ_2 are the metrics on V, W, respectively. Let $\varepsilon > 0$. For each $\mathbf{y} \in f(K)$, let $B(\mathbf{y}, \varepsilon/3)$ be the open ball centered at $\mathbf{y}$ with radius $\varepsilon/3$. The preimages of these open sets comprise an open cover $\mathcal{C}$ for K. Let δ be the Lebesgue number for $\mathcal{C}$. If $\mathbf{x}, \mathbf{x}' \in K$ with $\rho_1(\mathbf{x}, \mathbf{x}') < \delta$, then $\operatorname{diam}\{\mathbf{x}, \mathbf{x}'\} < \delta$, so this doublet is contained in one of the open sets of $\mathcal{C}$. It follows that $\rho_2(f(\mathbf{x}), f(\mathbf{x}')) < \varepsilon$. Since $\mathbf{x}, \mathbf{x}'$ were arbitrary, we have that f is uniformly continuous on K. □

6.8 NORMAL SPACES AND THE TIETZE EXTENSION THEOREM

In a general topological space $(V, \mathfrak{T})$, one can describe fairly well to what extent points can be "separated," or sets can be "separated," by the following set of five "separation characterizing" axioms. This scheme was first introduced by P. S. Alexandrov and H. Hopf in their treatise *Topologie* I (Berlin, 1935).

T0. If $\mathbf{x}, \mathbf{y}$ are any two distinct points of V, at least one of them is contained in an open set $T \in \mathfrak{T}$ which does not contain the other.

T1. If $\mathbf{x}, \mathbf{y}$ are any two distinct points of V, there are open neighborhoods $T_\mathbf{x}, T_\mathbf{y}$ of $\mathbf{x}, \mathbf{y}$, respectively, such that $\mathbf{x} \notin T_\mathbf{y}$ and $\mathbf{y} \notin T_\mathbf{x}$.

T2. If $\mathbf{x}, \mathbf{y}$ are distinct points of V, there exist open neighborhoods $T_\mathbf{x}, T_\mathbf{y}$ of $\mathbf{x}, \mathbf{y}$, respectively, such that $T_\mathbf{x} \cap T_\mathbf{y} = \varnothing$.

T3. If F is any nonempty closed set in V and $\mathbf{x} \notin F$ is a point in V, there exist open sets A, B, with $F \subset A$, $\mathbf{x} \in B$, and $A \cap B = \varnothing$.

T4. If F, G are disjoint nonempty closed sets in V, there exist open sets A, B with $F \subset A$, $G \subset B$, and $A \cap B = \varnothing$.

If $(V, \mathfrak{T})$ satisfies T0, V is called a Kolmogoroff space, and, as you must certainly recognize, if V satisfies T2, V is a Hausdorff space.

Theorem 6.8.1. *A topological space satisfies axiom* T1 *iff each singleton is a closed set.*

The proof, which is quite easy, we leave as an exercise.

Definition 6.8.2. If $(V, \mathfrak{T})$ satisfies T1 and T3, V is called a *regular space*. Note that T3 does not imply T1 unless we already know that each point is a closed set.

Definition 6.8.3. A Hausdorff space S is called *normal* if it satisfies T4; i.e., if for any two disjoint closed nonempty subsets F_1, F_2, there exist disjoint open sets G_1, G_2 such that $F_1 \subset G_1$, $F_2 \subset G_2$. This is equivalent to saying that for any closed set A and any open set $\tilde{B}$ with $A \subset \tilde{B}$ there is an open set U with $A \subset U \subset \bar{U} \subset \tilde{B}$.

Theorem 6.8.4. *A metric space M is normal.*

PROOF: M is a Hausdorff space. Now let A, B be two disjoint nonempty closed subsets of M. Since no point of A is an adherent point of B, we can do the following. For each $\mathbf{a} \in A$, there is a neighborhood $N_{\mathbf{a}}$ which is disjoint from B; hence we can find an $\varepsilon_{\mathbf{a}}$ such that the closed ball $\overline{B_{\mathbf{a}}(\varepsilon_{\mathbf{a}})}$ centered at $\mathbf{a}$ with radius $\varepsilon_{\mathbf{a}}$ is disjoint from B. This means that the distance of $\mathbf{a}$ from any point of B is more than $\varepsilon_{\mathbf{a}}$. Let $B_{\mathbf{a}}(\frac{1}{3}\varepsilon_{\mathbf{a}})$ be the open ball centered at $\mathbf{a}$ with radius $\frac{1}{3}\varepsilon_{\mathbf{a}}$, and let $U = \bigcup_{\mathbf{a} \in A} B_{\mathbf{a}}(\frac{1}{3}\varepsilon_{\mathbf{a}})$. Evidently, $U \cap B = \varnothing$, and U is an open set containing A.

Let V be an open set containing B constructed in a similar fashion; $V \cap A = \varnothing$. We have $U \cap V = \varnothing$, for if $\mathbf{x} \in U \cap V$, then for some $\mathbf{a} \in A$, $\mathbf{b} \in B$, $\mathbf{x} \in B_{\mathbf{a}}(\frac{1}{3}\varepsilon_{\mathbf{a}}) \cap B_{\mathbf{b}}(\frac{1}{3}\varepsilon_{\mathbf{b}})$. This is to say, the distance between $\mathbf{a}$ and $\mathbf{b}$ is less than $\frac{2}{3}$ of the larger of $\varepsilon_{\mathbf{a}}$, $\varepsilon_{\mathbf{b}}$. This is impossible; the distance between $\mathbf{a}$ and $\mathbf{b}$ is greater than the larger of $\varepsilon_{\mathbf{a}}$, $\varepsilon_{\mathbf{b}}$. We conclude that the metric space M is normal. □

We are going to present a famous theorem, called the Tietze extension theorem, which makes our earlier extension theorems pale by comparison.

Theorem 6.8.5 (Tietze). *Let F be a nonempty closed set in a normal space V, and f a real-valued continuous mapping on F such that $f(F) \subset [a, b]$ for some $a, b \in \mathbb{R}$. Then f has a continuous extension $\bar{f}$ defined and continuous over all of V, with $\bar{f}(V) \subset [a, b]$.*

PROOF: If $a = b$, f is constant, and trivially it can be continuously extended, so we assume $a < b$, and we also assume without loss of generality that

$f^{-1}(a) \subset F$, $f^{-1}(b) \subset F$. Put $A = f^{-1}(a)$, $B = f^{-1}(b)$; these are closed subsets of F, and since F is closed in V, A and B are also closed in V. We denote the complement of B by $\tilde{B}$. A and B are disjoint, so (refer to Definition 6.8.3) there exists an open set $V_{1/2}$ such that

$$A \subset V_{1/2} \subset \overline{V}_{1/2} \subset \tilde{B} \qquad \text{and} \qquad f^{-1}([a,(a+b)/2)) \subset V_{1/2}.$$

The use of the strange subscript will become more clear in a moment.

Since $\overline{V}_{1/2}$ and B are disjoint closed sets, as are A and $\tilde{V}_{1/2}$, there exist open sets $V_{3/4}$ and $V_{1/4}$ such that $A \subset V_{1/4} \subset \overline{V}_{1/4} \subset V_{1/2} \subset \overline{V}_{1/2} \subset V_{3/4} \subset \overline{V}_{3/4} \subset \tilde{B}$ and $f^{-1}([a,(3a+b)/4)) \subset V_{1/4}$, $f^{-1}([a,(a+3b)/4)) \subset V_{3/4}$. Continuing in this way, we get a sequence of sets

$$A \subset V_{1/2^k} \subset \overline{V}_{1/2^k} \subset \cdots \subset V_{m/2^k} \subset \overline{V}_{m/2^k} \subset \cdots \subset \overline{V}_{(2^k-1)/2^k} \subset \tilde{B} = V_1,$$

where for each $m = 1, \ldots, 2^k - 1$,

$$f^{-1}\left(\left[a, \frac{(2^k-m)a+mb}{2^k}\right)\right) \subset V_{m/2^k}.$$

Define a function g_k on V as follows:

$$g_k(\mathbf{x}) = \begin{cases} a & \text{if } \mathbf{x} \in A = V_0, \\ b & \text{if } \mathbf{x} \in B, \\ \dfrac{(2^k-m)a+mb}{2^k} & \text{if } \mathbf{x} \in V_{m/2^k} \backslash V_{(m-1)/2^k}, \quad 1 \leq m \leq 2^k. \end{cases}$$

For each k, g_k is a well-defined function on V. Now let $k \to \infty$, and define $\bar{f}$ by $\bar{f}(\mathbf{x}) = \lim_{k\to\infty} g_k(\mathbf{x})$ for each $\mathbf{x} \in V$. The function $\bar{f}$ is well defined. For any fixed $\mathbf{x} \in V$, the values $g_k(\mathbf{x}) \searrow$ as $k \to \infty$, but these values are bounded below, so they must converge. Hence the functions g_k converge, as we say, pointwise to $\bar{f}$.

We need to show that $\bar{f}$ is continuous. The sub-basic sets of $[a, b]$ are sets of the form $[a, t)$ or $(t, b]$. It is not hard to see that the preimage under $\bar{f}$ of $[a, t)$ is the open set $V_{(t-a)/(b-a)}$. The reader can verify that $\bar{f}^{-1}((t, b])$ is also open. That $\bar{f} \equiv f$ on F follows from the way we defined the open sets $V_{m/2^k}$. This completes our proof. □

Corollary 6.8.6 (Urysohn's Lemma). *If A, B are disjoint closed nonempty subsets of a normal space V, and $[\alpha, \beta]$ is a compact interval in $\mathbb{R}$, there is a continuous function $g: V \to [\alpha, \beta]$ such that*

$$g(A) = \alpha, \qquad g(B) = \beta.$$

PROOF: Regard $A \cup B$ as a closed set in V, and let f be the function defined by $f(\mathbf{x}) = \alpha$ if $\mathbf{x} \in A$ and $f(\mathbf{x}) = \beta$ if $\mathbf{x} \in B$. Then f is trivially continuous on $A \cup B$. By Tietze's theorem it can be continuously extended to all of V. □

It goes without saying that E^n is a normal space, so these two important results apply in all Euclidean spaces. If A and B are any two nonempty disjoint closed sets in E^n, there is a continuous function f mapping E^n into $[0,1]$ such that $f(A)=0$ and $f(B)=1$. It should also be noted that the following corollary is an immediate consequence of the Tietze theorem.

Corollary 6.8.7. *Let F be a nonempty closed set in a normal space V, and f: $F \to E^n$ a bounded continuous map. Then there is a bounded continuous extension $\bar{f}$ of f mapping V into E^n.*

PROOF: The product topology on E^n is the same as the Euclidean topology; thus the projection mappings, or coordinate mappings, on E^n are continuous and open. For each $\mathbf{x} \in F$, $\mathbf{f}(\mathbf{x})$ is a point $\mathbf{y} \in E^n$, $\mathbf{y}=(y_1,\dots,y_n)$. Let p_i be the ith coordinate mapping; then $p_i \circ \mathbf{f}$ is a bounded continuous map from F into $\mathbb{R}$. By the Tietze theorem, $p_i \circ \mathbf{f}$ can be continuously extended to all of V, and the image of the extension will be the same as the image of $p_i \circ \mathbf{f}$. Define $\bar{\mathbf{f}}=(\overline{p_1 \circ \mathbf{f}},\dots,\overline{p_n \circ \mathbf{f}})$. $\bar{\mathbf{f}}: V \to E^n$ is the desired extension. We prove later (Theorem 9.1.4) that a mapping into E^n is continuous iff each component mapping is continuous. □

We remark that Theorem 6.8.5 holds true for continuous mappings f: $F \to \mathbb{R}$ which are not necessarily bounded, where F is a closed subset of a normal space V, so we can drop the restriction of boundedness in the above corollary. For a proof of this fact we refer the reader to Schurle (1979).

6.9 THE CANTOR TERNARY SET

It seems appropriate that we bring this rather lengthy chapter on topology to a close by presenting an absolutely marvelous set, certainly one of the most famous in all topology. From a practical point of view, this set is very useful because it is unusual enough to provide beautiful counterexamples for conjectures that mathematicians often make. We can best describe this set, the Cantor ternary set, by showing its actual construction.

Let $C_1=[0,1]$, the closed unit interval in E^1, and let $C_2=C_1 \setminus (\frac{1}{3},\frac{2}{3})= [0,\frac{1}{3}] \cup [\frac{2}{3},1]$. That is, we get C_2 by deleting the "open middle third" of C_1, and C_2 is evidently the union of two disjoint component intervals. Let C_3 be the set obtained by removing the "open middle thirds" from each of the components of C_2,

$$C_3=[0,\tfrac{1}{9}] \cup [\tfrac{2}{9},\tfrac{3}{9}] \cup [\tfrac{6}{9},\tfrac{7}{9}] \cup [\tfrac{8}{9},1].$$

Similarly, C_4 is what is left after removing the open middle thirds from each of the four closed component intervals of C_3; C_4 is the union of eight disjoint closed component intervals:

$$[0,\tfrac{1}{27}] \cup [\tfrac{2}{27},\tfrac{3}{27}] \cup [\tfrac{6}{27},\tfrac{7}{27}] \cup [\tfrac{8}{27},\tfrac{9}{27}] \cup [\tfrac{18}{27},\tfrac{19}{27}] \cup [\tfrac{20}{27},\tfrac{21}{27}] \cup [\tfrac{24}{27},\tfrac{25}{27}] \cup [\tfrac{26}{27},1].$$

Continue in this way indefinitely, generating a sequence $\{C_n\}_{n=1}^{\infty}$ of closed sets. Now define

$$C = \bigcap_{n=1}^{\infty} C_n.$$

It is not hard to see that each C_n is closed, being the union of 2^{n-1} closed intervals each of length $(\frac{1}{3})^{n-1}$. Hence C is closed and has length equal to $\lim_{n\to\infty}(\frac{2}{3})^{n-1}$, which is zero. Since C cannot possibly contain an interval, it has an empty interior, so C is nowhere dense. On the other hand, C contains no isolated points, so C is a perfect set. Since C contains no interval, no two points of C belong to the same component; hence, the components of C are points. Such a set is said to be *totally disconnected*. Finally, C is an uncountable set; this may be hard for you to prove, so we sketch the reasoning.

Refer back to the proof of Theorem 1.1.6; from a remark made there it will be clear that each $x \in (0,1)$ has an essentially unique ternary representation. If a point x is in the first middle third removed, i.e., if $x \in (\frac{1}{3}, \frac{2}{3})$, the ternary expansion for x will have the digit 1 in the first place to the right of the ternary point. if $x \in (\frac{1}{9}, \frac{2}{9}) \cup (\frac{7}{9}, \frac{8}{9})$, then there is a digit 1 in the second place, and so on. Hence, if $x \in C$, its ternary expansion will contain the digit 1 only if it is a terminating expansion with a 1 in the last position. Hence there is a natural set-isomorphism between the set of numbers in $(0,1)$ with nonterminating binary expansions and the set of nonterminating ternary expansions of numbers in C. This means C contains at least as many points as there are irrational numbers in $[0,1]$. It follows that $\operatorname{card} C = \aleph_1$.

Theorem 6.9.1. *The Cantor ternary set is an uncountable, perfect, nowhere dense, totally disconnected, compact set having measure zero.*

Now we are going to define a real-valued function on the set $[0,1]$; the set C will play a key role in the definition of this function. Recall that we constructed C by removing, in order, the open sets $(1/3,2/3)$; $(1/9,2/9)$, $(7/9,8/9)$; $(1/3^3,2/3^3)$, $(7/3^3,8/3^3)$, $(19/3^3,20/3^3)$, $(25/3^3,26/3^3)$; $(1/3^4,2/3^4)$, and so on. Label these intervals $I(\frac{1}{2})$; $I(\frac{1}{4})$, $I(\frac{3}{4})$; $I(\frac{1}{8})$, $I(\frac{3}{8})$, $I(\frac{5}{8})$, $I(\frac{7}{8})$;..., respectively. This strange way of labeling these intervals has the advantage that if $r/2^s < m/2^n$, then the interval $I(r/2^s)$ is to the left of the interval $I(m/2^n)$. Thus all the removed intervals are ordered by this labeling. It is not obvious where the index $m/2^n$ comes from, so we shall tell you. Note that the left endpoint of a removed interval of the nth level is of the form

$$x = \frac{a_1}{3} + \frac{a_2}{3^2} + \cdots + \frac{a_{n-1}}{3^{n-1}} + \frac{1}{3^n},$$

where each a_i, $i = 1,\ldots,n-1$, is the digit 0 or 2. The index $m/2^n$ associated

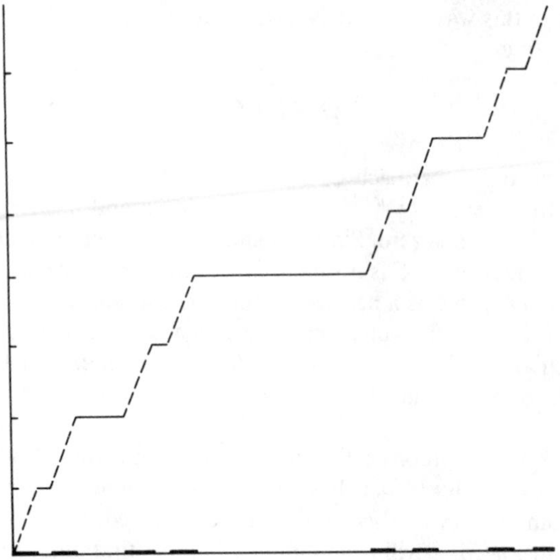

FIGURE 6.1 The third step in the construction of the Cantor function.

with that interval is gotten by the formula

$$\left(a_1^{n-1}+a_2^{n-2}+\cdots+a_{n-1}+1\right)/2^n=m/2^n.$$

It should not be too hard to show that the mapping φ that sends the left endpoints x to the corresponding numbers $m/2^n$ is a strictly increasing function of the left endpoints.

Define f on $[0,1]\backslash C$ by $\forall I(m/2^n)$, $f(x)=m/2^n$ for each $x\in I(m/2^n)$. Now extend f to the whole interval by defining

$$f(0)=0,$$
$$f(x)=\sup\{f(\xi):\xi\in[0,1]\backslash C,\xi<x\}.$$

It is not hard to show that $f\nearrow$ on $[0,1]$ and that f is continuous, and hence uniformly continuous, on $[0,1]$. But, given any $\delta>0$, there is a finite collection of disjoint (perhaps abutting) closed subintervals in $[0,1]$, $\{[a_i,b_i]\}_{i=1}^m$, whose combined length is less than δ, for which we have

$$\sum_{i=1}^{m}|f(b_i)-f(a_i)|=1.$$

This means that f is not absolutely continuous on $[0,1]$.

Note that f is constant on the removed subintervals of $[0,1]$. This means $f'(x)=0$ "almost everywhere" on $[0,1]$; that is, f is differentiable on $[0,1]$, except for a set of measure zero, namely, the Cantor set, and since $f'(x)\equiv 0$

on the complement of the Cantor set, f' is continuous there. A theorem from elementary calculus states that $f'(x)$ must therefore be integrable on all of $[0,1]$; it is, and $\int_0^1 f'(x)\,dx=0$. But $f(1)-f(0)=1$. (What is wrong?) We refer to this function as the Cantor function. Figure 6.1 suggests what the graph of $y=f(x)$ looks like.

PROBLEMS

1. Suppose A is a set in a topological space S. Show that one can get at most 14 distinct sets from A by using complementation and closure.
[*Hint*: Show $\overline{\text{int}\,A}=\text{int}(\overline{\text{int}\,A})$ and $\text{int}(\overline{A})=\text{int}(\overline{\text{int}\,\overline{A}})$.]

2. Let $S=\{a,b,c,d,e,f,g,h\}$. Let $\mathfrak{S}=\{\varnothing, S, \{a\}, \{a,b\}, \{c,d,e\}, \{f,g,h\}\}$. Is $\mathfrak{S}$ a topology for S? If not, generate a topology from $\mathfrak{S}$. Having done this, find the closures of the sets $\{a,c\}$, $\{g,h\}$, $\{a\}$, $\{c,d,e,f,g,h\}$, $\{b,c\}$, $\{e\}$.

3. Prove that if A is open, then $A\cap\overline{B}\subset\overline{A\cap B}$.

4. Suppose $f\colon E^1\to E^1$ is differentiable on $[a,b]$ but not necessarily C^1. Suppose at some interior point $\mathbf{x}_0\in(a,b)$, $f'(\mathbf{x}_0)=\alpha$ and at some $\mathbf{x}_1\in(\mathbf{x}_0,b)$, $f'(\mathbf{x}_1)=\beta$, with $\alpha<\beta$. Then f' assumes all values between α and β on $(\mathbf{x}_0,\mathbf{x}_1)$.

5. Prove that if S is a compact Hausdorff space, S is normal.

6. Prove the following. Suppose X is a *Baire space*; i.e., the intersection of any finite or countable collection of open dense subsets of X is a dense subset of X. Suppose further that $f\colon X\to Y$ is continuous and open and $f(X)$ is dense in Y. Then Y is a Baire space. (Hence "Baire-ness" is a topological property.)

7. Prove that in a T-1 space (see Problem 16) if F is closed, then $F=\bigcap\{G: G$ is open, $F\subset G\}$, and if G is open, then $G=\bigcup\{F\colon F$ is closed, $F\subset G\}$.

8. Let $S=\{a,b,c,d\}$, $\mathfrak{T}_1=\{\varnothing, S, \{a,b\}, \{a,b,c\}\}$, and $\mathfrak{T}_2=\mathfrak{T}_1\cup\{a,b,d\}$. Find the adherence points, or the closures of the sets $\{c\}$, $\{c,d\}$, $\{a,d\}$, $\{b,c\}$, $\{b\}$, $\{a\}$, $\{a,b\}$, $\{a,b,d\}$ in $\mathfrak{T}_1$ and in $\mathfrak{T}_2$.

9. If each F_α is closed, prove that $\bigcap_\alpha F_\alpha$ is closed.

10. Let $S=\{a,b,c\}$ and $\mathfrak{T}=\{\varnothing, S, \{a\}, \{a,b\}, \{a,c\}\}$. For each $A\subset S$, find $\text{int}\,A$, ∂A, $\overline{A}$, and A' (the set of limit points of A).

11. Let $f\in\mathcal{C}[0,1]$, the space of continuous real-valued functions on $[0,1]$. Define $\|f\|_1=\int_0^1|f(x)|\,dx$. Now let $\{f_n\}\subset\mathcal{C}[0,1]$, where for each n,

$$f_n(x)=\begin{cases}1, & 0\leqslant x\leqslant\frac{1}{2},\\ 1-2^n\left(x-\frac{1}{2}\right), & \frac{1}{2}<x\leqslant\frac{1}{2}+1/2^n,\\ 0, & \frac{1}{2}+1/2^n<x\leqslant 1.\end{cases}$$

Show that each $f_n \in \mathcal{C}[0,1]$, that $\lim_{m,n\to\infty} \| f_m - f_n \|_1 = 0$, but that the function f, defined by $\forall x \in [0,1]$, $f(x) = \lim_{n\to\infty} f_n(x)$, does not belong to $\mathcal{C}[0,1]$. Now suppose $\mathcal{C}[0,1]$ is normed by $\| f \|_u = \sup_{x\in[0,1]} |f(x)|$. Show that under this new norm, $\{f_n\}$ is not a Cauchy sequence; i.e., that as $m, n \to \infty$, $\| f_n - f_m \|_u \nrightarrow 0$.

12. Prove or disprove that if S is arcwise connected, S is connected and that if S is open and connected, S is arcwise connected.

13. The Cantor set described in Section 6.9 is not the only Cantor set. See if you can construct a Cantor set, following the same general idea of the construction, which has a positive measure.

14. Prove that if V is a topological space and S is a closed subset, then S' is closed. Can you find a restriction to put on the topology of V that will make S' closed for every set $S \in P(V)$? If V is endowed with the trivial topology (and V consists of more than one point), show that the set $\{\mathbf{x}\}'$ for each $\mathbf{x} \in V$ is not closed. What if the topology is the discrete topology? Next, consider the case where V is a metric space. Finally, what if V is just a Hausdorff space?

15. Prove that the "cofinite topology" for $\mathbb{Z}$ is indeed a topology.

16. A topological space $(V, \mathfrak{T})$ enjoying that property that for each pair of distinct points $\mathbf{x}, \mathbf{y} \in V$, there exists a set $T_{\mathbf{x}} \in \mathfrak{T}$ such that $\mathbf{y} \notin T_{\mathbf{x}}$, and there exists a set $T_{\mathbf{y}} \in \mathfrak{T}$ such that $\mathbf{x} \notin T_{\mathbf{y}}$ is called a T-1 space. Prove that if V is a T-1 space, then singletons are closed sets (Theorem 6.8.1).

17. Complete the proof of Theorem 6.1.6.

18. Let V be a metric space and $\mathfrak{S}$ the collection of open balls of V. $\mathfrak{S}$ is a sub-base for the metric space topology. Prove that $\mathfrak{S}$ is in fact a base for this topology.

19. Prove Theorem 6.2.5.

20. Prove the theorem that if $\mathbf{x}$ is a cluster point of a metric space V, then every neighborhood of $\mathbf{x}$ contains infinitely many points of V.

21. If $\mathfrak{T}_1, \mathfrak{T}_2$ are two topologies for a space V, there is a unique topology $\mathfrak{T}_3$ which is the infimum of $\mathfrak{T}_1$ and $\mathfrak{T}_2$; $\mathfrak{T}_3 = \mathfrak{T}_1 \wedge \mathfrak{T}_2$. Show that $\mathfrak{T}_3 = \mathfrak{T}_1 \cap \mathfrak{T}_2$ and that $\mathfrak{T}_3$ is the finest topology which is simultaneously more coarse than $\mathfrak{T}_1$ or $\mathfrak{T}_2$.

22. Prove that if a sequence in a Hausdorff space converges to a limit in that space, the limit is unique.

23. Prove that if X and Y are topological spaces and $A \subset X$, $B \subset Y$ are compact, then $A \times B$ is compact in $X \times Y$.

24. Prove that if X, Y are topological spaces and $A \subset X$, $B \subset Y$ are connected, $A \times B$ is connected in $X \times Y$.

25. Prove Corollary 6.6.11.

26. Let X be a compact space, and $\mathfrak{T}$ a family of continuous real-valued functions on X such that
(a) $f, g \in \mathfrak{T} \Rightarrow f \cdot g \in \mathfrak{T}$;
(b) for all $\mathbf{x} \in X$, there exists a neighborhood $N_{\mathbf{x}}$, and there exists $f \in \mathfrak{T}$ such that f vanishes identically on $N_{\mathbf{x}}$.
Prove that $\mathfrak{T}$ contains the zero function.

27. Suppose X is a compact connected set and $A \subset X$ is closed. Prove that there exists a minimal set B such that B is closed and connected and $A \subset B \subset X$.

28. Let $f: X \to Y$ be continuous and $\{K_n\}_{n=1}^{\infty}$ a decreasing sequence of compact sets in X. Prove that $f(\bigcap K_n) = \bigcap f(K_n)$. (Compare Theorem 1.1.10c.)

29. Let T be a linear map from one normed space V to another normed space W. We say T is norm reducing if $\forall \mathbf{x} \in V$, $\|T\mathbf{x}\|_W \leqslant \|\mathbf{x}\|_V$. Show that T is continuous if it is norm reducing. Next, show that if there is a $k > 0$ such that for all $x \in V$, $\|T\mathbf{x}\|_W \leqslant k\|\mathbf{x}\|_V$, then T is continuous.

30. Prove that if $\{F_i\}_{i=1}^{n}$ is a finite collection of closed sets, $\bigcup_{i=1}^{n} F_i$ is closed.

31. Prove that the collection of all topologies on a set V is a complete lattice.

32. Prove Theorem 6.2.7.

33. Prove or disprove that a continuous mapping preserves convergence. Prove or disprove that a continuous mapping preserves Cauchy-ness.

34. Prove that the set S described just after Definition 6.6.7 is connected but not arcwise connected. Next, show that the set of irrational points in E^2 is arcwise connected. ($(x, y) \in E^2$ is irrational if at least one of x and y is irrational.)

35. Let $\mathcal{C}[0,1]$ be the normed linear space of real-valued functions continuous on $[0,1]$, the norm being given by

$$\|f\| = \sup\{|f(x)| : x \in [0,1]\}.$$

Certainly $\mathcal{C}[0,1]$ contains the set P of polynomials with rational coefficients. Show that P is countable. Next, try to prove that P is dense in $\mathcal{C}[0,1]$. This may be virtually impossible, so try this. Let Π be the class of polygonal functions on $[0,1]$; these are continuous functions whose graphs consist of a finite number of straight line segments. Show that for any $\varepsilon > 0$, $f \in \mathcal{C}[0,1]$, there is a $\pi \in \Pi[0,1]$ such that $\|f - \pi\| < \varepsilon$.

Suppose this function $\pi(x)$ has vertices at $0 = x_0, x_1, \ldots, x_n = 1$. Can you show that for the right coefficients a_i

$$\pi(x) = a_0|x - x_0| + \cdots + a_{n-1}|x - x_{n-1}| + a_n?$$

If you have progressed this far, the next step would be to see if you can find a polynomial $p(t)$ that will approximate $|t|$ on $[-1,1]$. Then

$p(x-x_i)$ will approximate $|x-x_i|$ on $[0,1]$. If you could do this, then $h(x) = \sum_{i=0}^{n-1} a_i p(x-x_i) + a_n$ would approximate $\Pi(x)$.

Finally, a polynomial $g(x)$ with rational coefficients exists which will approximate $h(x)$. This $g(x) \in P$ is an approximator for $f(x)$, and you can conclude P is dense in $\mathcal{C}[0,1]$, so $\mathcal{C}[0,1]$ is separable and hence second countable.

This problem is really too hard. See Theorem 11.4.4 for a proof of the Weierstrass approximation theorem. This theorem assures that there is a polynomial which approximates f uniformly on $[0,1]$, which in turn can be approximated by one with rational coefficients.

36. Give an example of a T-3 space which is not regular.

37. If $f(x)$ is the Cantor function, with $f'(x)$ continuous on $[0,1]$ except for a set of measure zero, why is not $\int_0^1 f'(x)\,dx = f(1) - f(0)$?

38. Let $f\colon X \to Y$ be a mapping and $A \subset X$, $B \subset X$ sets. Suppose f_{1_A}, f_{1_B} are continuous. Show that if A and B are both open, or both closed, then $f_{1_{A\cup B}}$ is continuous.

39. For topological spaces, prove that $T2 \Rightarrow T1 \Rightarrow T0$. If V is a T-1 space, prove that $T4 \Rightarrow T3 \Rightarrow T2 \Rightarrow T1 \Rightarrow T0$.

40. Try to find an example of a regular space which is not normal. [*Hint*: Consider $\mathbb{R}^2$ with a topology generated by sets of the form $[a,b) \times [c,d)$.]

41. Let $(X, \mathfrak{T})$ be a topological space and R an equivalence relation on X. Perhaps R is simply determined by some generalized partition of X. We call X modulo R, denoted by X/R, the *quotient space* of X, relative to R. This space is sometimes called an identification space, because we identify each element of X with the other elements of X which are in the same equivalence class; the "points" of X/R are the equivalence classes in X determined by (or that determine) R. The mapping $\varphi\colon X \to X/R$ maps each x to its equivalence class. Prove that this "identification map" φ is continuous if we topologize X/R as follows: $G \subset X/R$ is open iff $\varphi^{-1}(G)$ is open in X. This is the standard "quotient topology" for X/R.

Next, suppose $g\colon (X, \mathfrak{T}) \to Y$ is surjective. Topologize Y as follows: $U \subset Y$ is open iff $g^{-1}(U)$ is open in X. Now g is a continuous surjection. Next, let R be a binary relation on X defined by $\forall \mathbf{x}, \mathbf{y} \in X$, $\mathbf{x} R \mathbf{y} \Leftrightarrow g(\mathbf{x}) = g(\mathbf{y})$. Prove R is an equivalence relation on X. Now let φ be the identification map $X \to X/R$, where X/R is endowed with the quotient topology. Prove that $\bar{g}\colon X/R \to Y$ is a homeomorphism, where $\bar{g}$ is defined by $\forall \mathbf{x} \in X$, $\bar{g}(\varphi(\mathbf{x})) = g(\mathbf{x})$.

42. A topological space is said to be locally compact if for all $\mathbf{x}$ and for every neighborhood $N_{\mathbf{x}}$ of $\mathbf{x}$, there is a compact set K such that $\mathbf{x} \in \mathring{K} \subset K \subset N_{\mathbf{x}}$. Prove that if $(X, \mathfrak{T})$ is a locally compact Hausdorff space and $Y \subset X$ is a subspace with the induced topology, then Y is locally compact iff Y is the intersection of an open set and a closed set in X.

43. Let X be a compact metric space. Prove that X is separable.
[*Hint*: Suppose $S_p = \{\mathbf{x} \in X : \forall \mathbf{x}, \mathbf{x}' \in S_p, \rho(\mathbf{x}, \mathbf{x}') \geqslant p\}$. If S_p is an infinite set, does it have an accumulation point in X? What does this imply about the distance between certain points of S_p? If $\mathbf{x} \in X$, is $B(\mathbf{x}, p) \cap S_p = \varnothing$? Consider the union of the sets $\{S_{1/n}\}_{n=1}^{\infty}$.]

44. Let X, Y be complete metric spaces and $f: X \to Y$ be continuous. Prove that if $A \subset X$ is conditionally compact, so also is $f(A)$.

45. Prove that the continuous image of a separable space is separable.

46. Prove that if S is any set of real numbers, the derived set S' is closed.

47. Prove or disprove that the collection of "half-open" boxes in E^n of the form $B = \times_{i=1}^{n} \{[a_i, b_i) : \forall i,\ a_i, b_i \in \mathbb{Q},\ a_i < b_i\}$ is a countable base for a topology, and if G is an open set in E^n, with the ordinary topology, then G is a countable union of "half-open" boxes.

48. If $\{\mathbf{x}_n\}$ is a Cauchy sequence and f is a uniformly continuous map, then $\{f(\mathbf{x}_n)\}$ is a Cauchy sequence.

49. A set S is said to be *locally connected* if for each $\mathbf{x} \in S$ and each neighborhood $N_{\mathbf{x}}$, there is an open set G such that $\mathbf{x} \in G \cap S \subset N_{\mathbf{x}}$ and $G \cap S$ is connected. Let S be the set described immediately after Definition 6.6.7. S is connected. Prove this. Show that S is not locally connected by considering any sufficiently small open disk centered at $(0,0)$.

50. (a) Let (a_1, b_1) be an arbitrary point in E^2. Remove all points of E^2 which are at an irrational distance from (a_1, b_1). Move to a point (a_2, b_2) which remains, and remove all of the remaining points which are at an irrational distance from (a_2, b_2). Can you wipe out the plane this way? If so, how many steps will it take?
(b) Same as (a) with "irrational" replaced by "rational."

51. Let $\mathbb{S}$ be the topology for $\mathbb{R}$ generated by the half-open intervals of the form $[a, b)$, $a < b$. $\mathbb{S}$ is sometimes called the *Sorgenfrey topology* for $\mathbb{R}$. Do the following sequences converge in $(\mathbb{R}, \mathbb{S})$?
(a) $\{1/n\}$,
(b) $\{(-1)^n + 1/n\}$,
(c) $\{(-1)^n/n\}$,
(d) $\{a^n : 0 \leqslant a \leqslant 1\}$,
(e) $\{(1/n)\sin n\sqrt{2}\}$,
(f) $\{3 + 2/n\}$,
(g) $\{3 - 2/n\}$.

52. Prove that the intersection of two connected compact sets in E^2 may be disconnected. What about in E^1?

53. A set S is said to be *Lindelöf*, or have the *Lindelöf property*, if every open cover of S can be reduced to a countable subcover. Prove that E^n is a Lindelöf space. Then prove that any second countable space is Lindelöf.

54. Suppose $\mathbb{R}^2$ is equipped with the Sorgenfrey topology, i.e., the topology generated by the half-open rectangles $[x_1, x_2)\times[y_1, y_2)$, and $D=\{(x,-x): x\in\mathbb{R}\}$, the line $y=-x$ in the plane, is endowed with the relative topology. Show that this topology on D is the discrete topology. Is D a Lindelöf space?

55. Prove that if $f: E^n \to (Y, \rho)$ is continuous, where (Y, ρ) is a metric space, then f is uniformly continuous on each bounded set $B\subset E^n$.

56. Prove that open subsets of locally compact spaces are, as subspaces, locally compact.

57. Suppose X is a second countable locally compact space. Then X is the union of an increasing sequence of compact sets.

58. Is the continuous image of a Hausdorff space a Hausdorff space?

59. Prove that a continuous bijectivity from a compact space X onto a Hausdorff space Y is a homeomorphism.

60. A subspace A of a space X is called a *retract* if there exists a continuous surjection $f: X\to A$ such that f is the identity on A. Prove that a retract of a Hausdorff space is closed. Next, prove that if $A\subset X$ is a retract and $g: A\to Y$ is continuous, then g may be extended to a continuous function $\bar{g}: X\to Y$.

61. Find a retraction from $\mathbb{R}$ to $[-1,1]$.

62. Prove that $(\mathbb{R}, \mathbb{S})$, the Sorgenfrey line, is not metrizable, but that it is a T-3 and Lindelöf space.

63. Let $(X, \mathfrak{T})$ be a topological space, and denote by the strange symbol ∞ a point not in X. Let $\hat{X}=X\cup\{\infty\}$. Let $\hat{\mathfrak{T}}=\{T\in\hat{X}: T\in\mathfrak{T}$ or $\hat{X}\setminus T$ is compact in $X\}$. Show that $(\hat{X}, \hat{\mathfrak{T}})$ is a compact space and that X is dense in $\hat{X}$ iff X is not compact. The space $\hat{X}$ is called the *Alexandroff one-point compactification* of X when X is not already compact.

64. Prove that X is compact iff every family $\mathcal{C}$ of closed sets in X with the finite intersection property has a nonempty intersection. (A family of sets $\mathcal{C}$ has the *finite intersection property* if the intersection of every finite subfamily of $\mathcal{C}$ is nonempty).

65. Suppose (X, d), (Y, ρ), are two metric spaces, and $f: X\to Y$ is an isometry from X onto Y (i.e., $\forall \mathbf{x},\mathbf{z}\in X$, $\rho(f(\mathbf{x}), f(\mathbf{z}))=d(\mathbf{x},\mathbf{z})$). Prove that f is a homeomorphism.

66. Prove or disprove that if a set D in a topological linear space is convex, then $\bar{D}$ is convex, and $\mathring{D}$ is convex.

67. Let $\varphi=(f_1,\ldots,f_n)$ be a mapping from E^m to E^n. Prove φ is continuous iff for each i, $f_i: E^m\to\mathbb{R}$ is continuous. Do not use the proof given in Theorem 9.1.4.

CHAPTER

7

INNER-PRODUCT SPACES

7.1 REAL INNER PRODUCTS

Suppose we have a linear space V over a field and the field specifically is $\mathbb{R}$. There may exist a "bilinear" functional μ that maps $V \times V$ into $\mathbb{R}$ and satisfies the prescribed set of axioms listed below $\forall \mathbf{x}, \mathbf{y}, \mathbf{z} \in V(\mathbb{R})$, $\forall \alpha \in \mathbb{R}$; if such is the case, we say that μ is an *inner product* for $V(\mathbb{R})$.

I1. $\mu(\mathbf{x}+\mathbf{y}, \mathbf{z}) = \mu(\mathbf{x}, \mathbf{z}) + \mu(\mathbf{y}, \mathbf{z})$.
I2. $\mu(\mathbf{x}, \mathbf{y}) = \mu(\mathbf{y}, \mathbf{x})$.
I3. $\mu(\alpha\mathbf{x}, \mathbf{y}) = \alpha\mu(\mathbf{x}, \mathbf{y})$.
I4. $\mu(\mathbf{x}, \mathbf{x}) \geqslant 0$; $\mu(\mathbf{x}, \mathbf{x}) = 0 \Leftrightarrow \mathbf{x} = \mathbf{0}$.

We call μ bilinear because it is linear in both its variables. Together axioms I1–I3 ensure that $\forall a, b, a', b' \in \mathbb{R}$, $\forall \mathbf{x}, \mathbf{y}, \mathbf{x}', \mathbf{y}' \in V$,

$$\mu(a\mathbf{x}+b\mathbf{y}, a'\mathbf{x}'+b'\mathbf{y}') = aa'\mu(\mathbf{x}, \mathbf{x}') + ab'\mu(\mathbf{x}, \mathbf{y}') + a'b\mu(\mathbf{y}, \mathbf{x}') + bb'\mu(\mathbf{y}, \mathbf{y}').$$

A linear space $V(\mathbb{R})$, together with an inner product defined over it, is called a real inner-product space. The inner product satisfies a famous inequality known as the *Schwarz inequality* [compare it with the Cauchy inequality (Section 3.1.4)]:

Theorem 7.1.1. *Let* $\mathbf{x}, \mathbf{y}$ *be arbitrary vectors in a real inner-product space, where we denote the inner product of* $\mathbf{x}$ *and* $\mathbf{y}$ *by* $(\mathbf{x}, \mathbf{y})$. *Then*

$$|(\mathbf{x}, \mathbf{y})| \leqslant \sqrt{(\mathbf{x}, \mathbf{x})(\mathbf{y}, \mathbf{y})}\,.$$

Equality holds if and only if one vector is a scalar multiple of the other.

PROOF: If $(\mathbf{x}, \mathbf{y}) = 0$, the statement of the theorem is true, so assume $(\mathbf{x}, \mathbf{y}) \neq 0$ and consider the following real quadratic expression in a real variable t:

$$\left(t\mathbf{x}+\frac{(\mathbf{x},\mathbf{y})}{|(\mathbf{x},\mathbf{y})|}\mathbf{y},\, t\mathbf{x}+\frac{(\mathbf{x},\mathbf{y})}{|(\mathbf{x},\mathbf{y})|}\mathbf{y} \right) = t^2(\mathbf{x},\mathbf{x})+2t|(\mathbf{x},\mathbf{y})|+(\mathbf{y},\mathbf{y}).$$

The expression on the left-hand side is zero iff

$$t\mathbf{x} = -\frac{(\mathbf{x},\mathbf{y})}{|(\mathbf{x},\mathbf{y})|}\mathbf{y},$$

i.e., $\mathbf{y}$ is a scalar multiple of $\mathbf{x}$. If it is, then the quadratic formula, together with the positivity of the inner product, ensures that the discriminant of the expression on the right-hand side is zero. This is to say,

$$|(\mathbf{x},\mathbf{y})|^2 = (\mathbf{x},\mathbf{x})(\mathbf{y},\mathbf{y}).$$

On the other hand, if $\mathbf{y}$ is not a scalar multiple of $\mathbf{x}$, then the discriminant must be negative:

$$|(\mathbf{x},\mathbf{y})|^2 < (\mathbf{x},\mathbf{x})(\mathbf{y},\mathbf{y}).$$

This completes the proof. □

If V is a linear space with an inner product, then V can be normed, and hence metrized, in the following way. For each $\mathbf{x} \in V$ define

$$\|\mathbf{x}\| = \sqrt{(\mathbf{x},\mathbf{x})}\,.$$

The axioms that a norm must satisfy (Definition 4.2.1) are certainly satisfied in this case.

$$\|\mathbf{x}\| = \sqrt{(\mathbf{x},\mathbf{x})} \geqslant 0,$$

$$\|\mathbf{x}\| = 0 \qquad \Leftrightarrow \qquad \mathbf{x} = \mathbf{0},$$

$$\|a\mathbf{x}\| = \sqrt{(a\mathbf{x},a\mathbf{x})} = |a|\sqrt{(\mathbf{x},\mathbf{x})} = |a|\,\|\mathbf{x}\|,$$

$$\begin{aligned}\|\mathbf{x}+\mathbf{y}\|^2 &= (\mathbf{x}+\mathbf{y},\mathbf{x}+\mathbf{y}) \leqslant (\mathbf{x},\mathbf{x})+2|(\mathbf{x},\mathbf{y})|+(\mathbf{y},\mathbf{y}) \\ &\leqslant \left[\sqrt{(\mathbf{x},\mathbf{x})}+\sqrt{(\mathbf{y},\mathbf{y})}\right]^2 = [\|\mathbf{x}\|+\|\mathbf{y}\|]^2.\end{aligned}$$

The following identity, known as the parallelogram law, holds in inner-product spaces.

Theorem 7.1.2. *If V is an inner-product space, then* $\forall \mathbf{x},\mathbf{y} \in V$,

$$\|\mathbf{x}+\mathbf{y}\|^2 + \|\mathbf{x}-\mathbf{y}\|^2 = 2\|\mathbf{x}\|^2 + 2\|\mathbf{y}\|^2.$$

PROOF:

$$\begin{aligned}(\mathbf{x}+\mathbf{y},\mathbf{x}+\mathbf{y})+(\mathbf{x}-\mathbf{y},\mathbf{x}-\mathbf{y}) &= (\mathbf{x},\mathbf{x})+2(\mathbf{x},\mathbf{y})+(\mathbf{y},\mathbf{y}) \\ &\quad +(\mathbf{x},\mathbf{x})-2(\mathbf{x},\mathbf{y})+(\mathbf{y},\mathbf{y}) \\ &= 2(\mathbf{x},\mathbf{x})+2(\mathbf{y},\mathbf{y}).\end{aligned}$$ □

This theorem simply states that in an inner-product space with its natural norm the sum of the squares of the lengths of the diagonals of a parallelogram is equal to the sum of the squares of the lengths of the sides. An immediate corollary is the identity

$$\tfrac{1}{2}[(\mathbf{x},\mathbf{y})+(\mathbf{y},\mathbf{x})]=\tfrac{1}{4}(\|\mathbf{x}+\mathbf{y}\|^2-\|\mathbf{x}-\mathbf{y}\|^2).$$

We remind the reader that the situation in inner-product spaces with regard to norm is similar to the situation in normed linear spaces with regard to metric. Every normed linear space with its natural topology can be made into a metric space, with the metric generating the same topology as the original norm topology, but the converse is not always true. Analogously, every inner-product space can be normed in the way we have described, and the topology of this norm is what we call the natural topology of the given inner product. On the other hand, there are normed linear spaces which will not admit an inner product that give rise to the original norm function.

The space $\mathbb{R}^n$ admits the following inner product:

$$\forall \mathbf{x},\mathbf{y}\in\mathbb{R}^n,\qquad \mathbf{x}=(x_1,\ldots,x_n),\quad \mathbf{y}=(y_1,\ldots,y_n).$$

Define

$$(\mathbf{x},\mathbf{y})=\sum_{i=1}^{n} x_i y_i.$$

It is a simple matter to verify that this functional on $\mathbb{R}^n\times\mathbb{R}^n$ is bilinear and satisfies the axioms I1–I4. Moreover, the associated norm is precisely the Euclidean norm, so E^n is not only a normed linear space but an inner-product space as well.

Let $\mathcal{C}[a,b]$ be the linear space of continuous functions from $[a,b]$ into $\mathbb{R}$. Define

$$\forall f,g\in\mathcal{C}[a,b],\qquad (f,g)=\int_a^b f(x)g(x)\,dx.$$

This is an inner product which gives rise to the norm

$$\|f\|^2=\int_a^b[f(x)]^2\,dx.$$

7.2. ORTHOGONALITY AND ORTHONORMAL SETS

The Schwarz inequality for real inner-product spaces, and these are the only kind we shall be considering until Section 7.7, asserts that

$$\forall \mathbf{x},\mathbf{y}\in V,\qquad -1\leqslant(\mathbf{x},\mathbf{y})/\|\mathbf{x}\|\,\|\mathbf{y}\|\leqslant 1.$$

This suggests that we might define abstractly the cosine of the angle θ determined by two vectors $\mathbf{x}$ and $\mathbf{y}$ to be

$$\cos\theta=(\mathbf{x},\mathbf{y})/\|\mathbf{x}\|\,\|\mathbf{y}\|.$$

You know that this formula is valid in E^n with its usual inner product, so there is no reason for not defining, for example, the angle θ between e^x and $\cos x$ in the space $\mathcal{C}[0,1]$ by the rule

$$\theta = \cos^{-1}\left[\int_0^1 e^x \cos x\,dx \Big/ \sqrt{\int_0^1 e^{2x}\,dx}\sqrt{\int_0^1 \cos^2 x\,dx}\right].$$

We shall so define the angle between two vectors in an inner-product space. Hence we have

$$\|\mathbf{x}-\mathbf{y}\|^2 \equiv \|\mathbf{x}\|^2 + \|\mathbf{y}\|^2 - 2\|\mathbf{x}\|\,\|\mathbf{y}\|\cos\theta,$$

where θ is the angle between $\mathbf{x}$ and $\mathbf{y}$. This is just the law of cosines from trigonometry.

Definition 7.2.1. If V is an inner-product space and $\mathbf{x},\mathbf{y} \in V$, we say that $\mathbf{x}$ and $\mathbf{y}$ are *orthogonal*, and write $\mathbf{x}\perp\mathbf{y}$, if $(\mathbf{x},\mathbf{y}) = 0$. More generally, we define the angle determined by $\mathbf{x}$ and $\mathbf{y}$, $\mathbf{x} \neq \mathbf{0}$, $\mathbf{y} \neq \mathbf{0}$, by the formula $\|\mathbf{x}\|\,\|\mathbf{y}\|\cos\theta = (\mathbf{x},\mathbf{y})$. If S is a set of vectors from an inner-product space V and if for each pair of distinct vectors $\mathbf{x},\mathbf{y} \in S$ we have $\mathbf{x}\perp\mathbf{y}$, we call S an *orthogonal set*. If, moreover, $\forall \mathbf{x} \in S$, $\|\mathbf{x}\| = 1$, we call S an *orthonormal set*.

The word "orthogonal" is used as a generalization of "perpendicular." Sometimes "normal" is used to denote perpendicularity; for example, we talk about the normal vector to a surface, meaning a vector perpendicular to the plane tangent to the surface. On the other hand, if $\mathbf{x}$ is a nonzero vector, then we "normalize" $\mathbf{x}$ when we divide $\mathbf{x}$ by its norm; i.e., we shrink or magnify $\mathbf{x}$ to a unit vector.

Definition 7.2.2. If $A \subset V$, V an inner product space, the set $A^\perp$, called the *orthogonal complement* of A, consists of those vectors of V which are orthogonal to every vector of A.

$$A^\perp = \{\mathbf{x} \in V : \forall \mathbf{a} \in A,\ (\mathbf{x},\mathbf{a}) = 0\}.$$

Theorem 7.2.3. *Let V be an inner-product space and $A \subset V$ a nonempty subset. Then $A^\perp$ is a subspace of V, and $A \cap A^\perp \subset \{\mathbf{0}\}$. Moreover, $A^\perp = (\operatorname{sp} A)^\perp$.*

PROOF: Let $\mathbf{a} \in A$, $\alpha,\beta \in \mathbb{R}$, $\mathbf{x},\mathbf{y} \in A^\perp$. Then $(\alpha\mathbf{x}+\beta\mathbf{y},\mathbf{a}) = \alpha(\mathbf{x},\mathbf{a}) + \beta(\mathbf{y},\mathbf{a}) = 0$, so $A^\perp$ is closed with respect to linear combinations. Therefore $A^\perp$ is a subspace of V. If $\mathbf{x} \in A \cap A^\perp$, then $(\mathbf{x},\mathbf{x}) = 0$, and we have $\mathbf{x} = \mathbf{0}$. Hence the zero vector is the only vector A and $A^\perp$ can share. To prove the last statement we need only show that if $\mathbf{x} \in A^\perp$, then $\mathbf{x} \in (\operatorname{sp} A)^\perp$. Let $\Sigma_{i=1}^m \alpha_i \mathbf{a}_i$ be an arbitrary element from $\operatorname{sp} A$ and $\mathbf{x} \in A^\perp$. $(\Sigma_{i=1}^m \alpha_i\mathbf{a}_i, \mathbf{x}) = \Sigma_{i=1}^m \alpha_i(\mathbf{a}_i,\mathbf{x}) = 0$. Since this is true for every possible linear combination of elements of A, we have that $\mathbf{x} \in (\operatorname{sp} A)^\perp$. □

As a consequence of this theorem the set $(A^{\perp})^{\perp}$, which we shall write $A^{\perp\perp}$, is always a subspace. It is evident that $A \subset A^{\perp\perp}$ and that $A^{\perp} \subset A^{\perp\perp\perp}$. If $\mathbf{x} \in A^{\perp\perp\perp}$, then $\mathbf{x} \perp A^{\perp\perp}$, so $\mathbf{x} \perp A$. Thus we have $\mathbf{x} \in A^{\perp}$. We conclude that

$$\forall A \subset V, \qquad A^{\perp} = A^{\perp\perp\perp}.$$

Suppose the set $A \subset V$ is a subspace of V. Then the direct sum $A \oplus A^{\perp}$ is the set of all vectors of V of the form $\mathbf{x} = \mathbf{a} + \mathbf{a}'$, $\mathbf{a} \in A$, $\mathbf{a}' \in A^{\perp}$. We assert that if $\mathbf{x} \in A \oplus A^{\perp}$, and $\mathbf{x}$ has the representation just given, then this representation is unique. For suppose $\mathbf{x} = \mathbf{a} + \mathbf{a}' = \mathbf{b} + \mathbf{b}'$, where $\mathbf{a}, \mathbf{b} \in A$, $\mathbf{a}', \mathbf{b}' \in A^{\perp}$. Then $\mathbf{a} + \mathbf{a}' - \mathbf{b} - \mathbf{b}' = \mathbf{0} \Rightarrow \mathbf{a} - \mathbf{b} = \mathbf{b}' - \mathbf{a}'$. The left-hand side of this equation is a vector of A, the right-hand side a vector of $A^{\perp}$. The only vector these spaces share is the zero vector; hence $\mathbf{a} = \mathbf{b}$, $\mathbf{a}' = \mathbf{b}'$, and the representation is unique.

It is apparent that if $A \subset V$ is a subspace, then $A \oplus A^{\perp}$ is also a subspace of V. Under certain circumstances, we have the equality $V = A \oplus A^{\perp}$. We postpone consideration of this question until later.

Theorem 7.2.4. *If S is an orthonormal set in V, where V is an n-dimensional inner-product space, then S contains at most n elements.*

PROOF: Let $\mathbf{x}, \mathbf{y} \in S$. Since $\mathbf{x} \neq \mathbf{0}, \mathbf{y} \neq \mathbf{0}$, the fact that $(\mathbf{x}, \mathbf{y}) = 0 < 1 = \|\mathbf{x}\|\,\|\mathbf{y}\|$ implies that the vectors of S are pairwise linearly independent. Choose any m distinct elements from S, say $\{\mathbf{x}_1, \ldots, \mathbf{x}_m\}$, and suppose some linear combination of these is zero: $\Sigma_{i=1}^m \alpha_i \mathbf{x}_i = 0$. Then for each $i = 1, \ldots, m$, we have $(\Sigma_{i=1}^m \alpha_i \mathbf{x}_i, \mathbf{x}_i) = \alpha_i = 0$, which means the only vanishing linear combination of any m of the elements of S is the trivial one.

Since $\dim V = n$, S can contain at most n elements since any m of the elements of S are linearly independent. ☐

Theorem 7.2.5. *Let S be an orthonormal set in an inner product space V. Then for each finite subset of S, $\{\mathbf{u}_1, \ldots, \mathbf{u}_m\}$, and for every $\mathbf{x} \in V$, $\Sigma_{i=1}^m |(\mathbf{x}, \mathbf{u}_i)|^2 \leqslant \|\mathbf{x}\|^2$ (Bessel's inequality), and $\forall \mathbf{x}, \mathbf{y} \in V$, $\Sigma_{i=1}^m |(\mathbf{x}, \mathbf{u}_i)(\mathbf{y}, \mathbf{u}_i)| \leqslant \|\mathbf{x}\|\,\|\mathbf{y}\|$.*

PROOF: We have $0 \leqslant \|\mathbf{x} - \Sigma_{i=1}^m (\mathbf{x}, \mathbf{u}_i)\mathbf{u}_i\|^2 = \|\mathbf{x}\|^2 - 2\,\Sigma_{i=1}^m |(\mathbf{x}, \mathbf{u}_i)|^2 + (\Sigma_{i=1}^m (\mathbf{x}, \mathbf{u}_i)\mathbf{u}_i, \Sigma_{i=1}^m (\mathbf{x}, \mathbf{u}_i)\mathbf{u}_i) = \|\mathbf{x}\|^2 - 2\Sigma_{i=1}^m |(\mathbf{x}, \mathbf{u}_i)|^2 + \Sigma_{i=1}^m (\mathbf{x}, \mathbf{u}_i)^2 = \|\mathbf{x}\|^2 - \Sigma_{i=1}^m |(\mathbf{x}, \mathbf{u}_i)|^2$.

This proves the Bessel inequality. To prove the second statement, put $a_i = |(\mathbf{x}, \mathbf{u}_i)|$ and $b_i = |(\mathbf{y}, \mathbf{u}_i)|$ and note that for any real number t

$$0 \leqslant \sum_{i=1}^{m} (a_i t + b_i)^2 = \left(\sum_{i=1}^{m} a_i^2 \right) t^2 + 2\left(\sum_{i=1}^{m} a_i b_i \right) t + \left(\sum_{i=1}^{m} b_i^2 \right),$$

so the discriminant must satisfy

$$\left(\sum_{i=1}^{m} a_i b_i \right)^2 - \left(\sum_{i=1}^{m} a_i^2 \right)\left(\sum_{i=1}^{m} b_i^2 \right) \leqslant 0.$$

This is Cauchy's inequality: if we substitute in this inequality for the as and the bs and use the Bessel inequality, we get

$$\left(\sum_{i=1}^{m} |(\mathbf{x},\mathbf{u}_i)|\,|(\mathbf{y},\mathbf{u}_i)|\right)^2 \leqslant \|\mathbf{x}\|^2\|\mathbf{y}\|^2.$$

The proof is complete. □

Definition 7.2.6. If $S=\{\mathbf{u}_\alpha\}$ is an orthonormal set in V, the scalars $(\mathbf{x},\mathbf{u}_\alpha)$ are called the *Fourier coefficients* of $\mathbf{x}$ with respect to S.

Theorem 7.2.7. *If V is a separable inner-product space and $S\subset V$ is an orthonormal set, then S is at most countable.*

PROOF: Let $\mathbf{x},\mathbf{y}$ be distinct elements of S; then $\|\mathbf{x}-\mathbf{y}\|^2=2$. Let $\{\mathbf{z}_i\}$ be a countable dense subset of V. Then for each $\mathbf{x}\in S$ there corresponds at least one n, depending on $\mathbf{x}$, such that $\|\mathbf{x}-\mathbf{z}_n\|<\sqrt{2}/3$. To show that this $\mathbf{z}_n$ can correspond to no other element of S, we note that if $\mathbf{y}\in S$, $\mathbf{y}\neq\mathbf{x}$, and if $\mathbf{z}_m$ is an element of the countable dense set $\{\mathbf{z}_i\}$ which satisfies $\|\mathbf{y}-\mathbf{z}_m\|<\sqrt{2}/3$, then

$$\sqrt{2}=\|\mathbf{x}-\mathbf{y}\|\leqslant\|\mathbf{x}-\mathbf{z}_n\|+\|\mathbf{z}_n-\mathbf{z}_m\|+\|\mathbf{z}_m-\mathbf{y}\|$$

implies that $\sqrt{2}/3<\|\mathbf{z}_n-\mathbf{z}_m\|$, so $\mathbf{z}_n\neq\mathbf{z}_m$, and $n\neq m$. This shows that $\operatorname{card}\{\mathbf{z}_i\}\geqslant\operatorname{card} S$, and the proof is complete. □

We can extend this result somewhat by showing that in an arbitrary inner-product space V with an orthonormal set S, although S may be uncountable, the set of nonzero Fourier coefficients for any $\mathbf{x}\in V$ will be at most countable.

Theorem 7.2.8. *If S is an orthonormal set in an inner-product space V, then for any $\mathbf{x}\in V$, the set of those $\mathbf{u}\in S$ for which $(\mathbf{x},\mathbf{u})\neq 0$ is at most countable.*

PROOF: Let $\mathbf{x}\in V$. Then $\|\mathbf{x}\|<\infty$, and for each finite subset of S we have $\sum_{i=1}^{m}|(\mathbf{x},\mathbf{u}_i)|^2\leqslant\|\mathbf{x}\|^2$. It follows then that for each positive integer n, the number of elements $\mathbf{u}\in S$ for which $|(\mathbf{x},\mathbf{u}_i)|\geqslant 1/n$ cannot exceed $n^2\|\mathbf{x}\|^2$. If this were not so, we would have a sum of more than $n^2\|\mathbf{x}\|^2$ terms, each of which is greater than or equal to $1/n^2$, and the Bessel inequality would not hold. Hence the set of all $\mathbf{u}\in S$ for which $(\mathbf{x},\mathbf{u})\neq 0$ is the union of the sets T_n of those finitely many $\mathbf{u}$s for which

$$\frac{1}{n}\leqslant|(\mathbf{x},\mathbf{u})|<\frac{1}{n-1},\qquad n=1,2,3,\ldots.$$

This is a countable union of finite sets and thus countable. □

As a consequence of Theorem 7.2.8, we can extend the results of Theorem 7.2.5 by letting the sums range over all of the orthonormal set S, since these sums are just ordinary infinite series of positive terms whose partial sums are bounded.

Theorem 7.2.9. *Let S be an orthonormal set in an inner-product space V. Then for every $\mathbf{x} \in V$, $\mathbf{y} \in V$, we have*

$$\sum_{\mathbf{u} \in S} |(\mathbf{x},\mathbf{u})|^2 \leqslant \|\mathbf{x}\|^2 \qquad (\textit{Bessel}),$$

$$\sum_{\mathbf{u} \in S} |(\mathbf{x},\mathbf{u})(\mathbf{y},\mathbf{u})| \leqslant \|\mathbf{x}\|\,\|\mathbf{y}\|.$$

Definition 7.2.10. An orthonormal set S in V is said to be *complete* if there is no other orthonormal set of which S is a proper subset. This is to say, a complete orthonormal set is maximal with respect to orthonormality.

Suppose that $\{\mathbf{u}_1,\ldots,\mathbf{u}_n\}$ is a finite orthonormal set spanning an n-dimensional subspace $M \subset V$. If $\mathbf{x} \in M$, then in terms of this orthonormal set, which is a basis for M, $\mathbf{x}$ has a unique representation of the form

$$\mathbf{x} = \sum_{i=1}^{n} \xi_i \mathbf{u}_i.$$

Observe that the ξ_i are precisely the values of $(\mathbf{x},\mathbf{u}_i)$, the Fourier coefficients of $\mathbf{x}$ with respect to the orthonormal set $\{\mathbf{u}_i\}_{i=1}^n$. We would like to extend this idea to infinite-dimensional spaces; that is, if S is an infinite orthonormal set in V, then $\overline{\operatorname{sp}}\, S$ is a closed linear subspace of V, and it would be very nice if we could express each $\mathbf{x} \in \overline{\operatorname{sp}}\, S$ in terms of a unique countable linear combination of the vectors of S, in which the coefficients were just the Fourier coefficients of $\mathbf{x}$ with respect to S. Things would be even better if we could be sure that V admits a complete orthonormal set S in terms of which each $\mathbf{x} \in V$ could be uniquely represented as an infinite series, i.e., a countable linear combination of vectors from S, the coefficients being the Fourier coefficients with respect to S.

At this point, experience suggests a possible problem with which we might have to cope. If indeed we can represent an $\mathbf{x} \in V$ by $\mathbf{x} = \Sigma_{\mathbf{u} \in S}(\mathbf{x},\mathbf{u})\mathbf{u}$, what does it mean to say a vector $\mathbf{x}$ is equal to an "infinite series" of vectors? We had better present a consistent definition for this concept right now.

Definition 7.2.11. Let $\Sigma_{i=1}^{\infty} \xi_i \mathbf{u}_i$ be an infinite series of vectors of an inner-product space V, or of a normed linear space V, i.e., a countable linear combination of the vectors $\{\mathbf{u}_i\} \subset V$. We say that $\Sigma_{i=1}^{\infty} \xi_i \mathbf{u}_i$ *converges to* $\mathbf{x}$ iff the sequence of partial sums $\mathbf{x}_n$ converges to $\mathbf{x}$, where for each n, $\mathbf{x}_n = \Sigma_{i=1}^{n} \xi_i \mathbf{u}_i$, the convergence being in the norm topology of V.

This is to say, we must have $\lim_n \|\mathbf{x}-\mathbf{x}_n\| = 0$.

Now that we know what we mean when we say an infinite series of vectors converges to a vector, it could well be that each vector in V has an infinite series representation, but that some series of vectors of V may have sequences of partial sums which are Cauchy sequences but not convergent. We had better restrict our attention to inner-product spaces V which are complete as metric spaces, i.e., spaces V in which every Cauchy sequence converges to a point of V. (We have mentioned earlier that every incomplete metric space may be completed. The completion of an inner product space V to a complete inner product space $\hat{V}$ is an analogous process.)

We are ready to push on: We have a direction and a goal. Unless we specifically say something to the contrary, all the spaces under consideration will be complete.

Definition 7.2.12. An infinite-dimensional inner-product space which is complete is called a *Hilbert space*. A complete normed linear space is called a *Banach space*.

Thus every Hilbert space is a Banach space, but not conversely.

Suppose B is the collection of all real sequences $\{x_n\}_{n=1}^{\infty}$; we make B a real linear space by defining addition and multiplication by a scalar in the obvious way

$$\{x_n\}+\{y_n\}=\{x_n+y_n\}, \qquad \lambda\{x_n\}=\{\lambda x_n\}.$$

Define $\|\{x_n\}\| = \sup_n\{|x_n|\}$ for each $\{x_n\}\in B$. Now B contains a normed linear space, usually denoted by the symbol l^∞, the space of bounded real sequences.

Note that the space l^∞ is not separable. Let $\{\xi_n\}$ be any countable collection of points of l^∞, where for each n, ξ_n is the sequence $\{x_k^{(n)}\}_{k=1}^{\infty}$. Let $\xi \in l^\infty$ be the sequence $\{x_k\}_{k=1}^{\infty}$, where for each k, $x_k = x_k^{(k)}+1$ if $|x_k^{(k)}| \leqslant 1$ and $x_k = 0$ if $|x_k^{(k)}|>1$. Then the kth component of $\xi-\xi_k$ is $x_k - x_k^{(k)}$, and $|x_k - x_k^{(k)}| \geqslant 1$. This is to say, $\|\xi-\xi_k\| \geqslant 1$. Hence, regardless of what the countable collection $\{\xi_n\}$ might be, there is an element $\xi \in l^\infty$ and a neighborhood of that point which contains none of the points of $\{\xi_n\}$. Thus $\{\xi_n\}$ is not dense in l^∞, and l^∞ is not separable.

Suppose $\{\xi_n\}_{n=1}^{\infty}$ is a Cauchy sequence in l^∞. It is apparent that the kth components, $k=1,2,3,\ldots$, of the terms ξ_n make up a Cauchy sequence of real numbers $\{x_n^{(k)}\}$, which sequence converges to the real number $x^{(k)}$. Hence $\{\xi_n\}$ converges to $\xi=(x^{(1)},x^{(2)},x^{(3)},\ldots)$. This is to say, l^∞ is complete, i.e., l^∞ is a Banach space.

Let $\xi=(1,0,0,\ldots)$ and $\eta=(0,1,0,\ldots)$. Both ξ and η belong to l^∞; each is a unit vector. Note that

$$\|\xi+\eta\|^2+\|\xi-\eta\|^2 = 2 \neq 2\|\xi\|^2+2\|\eta\|^2 = 4.$$

Since the parallelogram law does not hold in l^∞, l^∞ is not an inner-product space and hence not a Hilbert space. This is to say, the subset of B consisting of all bounded real sequences does not admit an inner product which will give rise to the l^∞ norm.

Suppose we consider the subset of B consisting of those sequences $\{x_n\}$ for which $\Sigma_{n=1}^{\infty}|x_n|^2<\infty$. We can make this set into an inner-product space by defining $\forall\boldsymbol{\xi}=\{x_n\},\forall\boldsymbol{\eta}=\{y_n\}$,

$$(\boldsymbol{\xi},\boldsymbol{\eta})=\sum_{n=1}^{\infty}x_n y_n.$$

That this really is an inner product is easy to verify, and the reader should have no difficulty in proving that this inner-product space, denoted l^2, is complete and hence a Hilbert space. l^2 becomes a Banach space once we define $\|\boldsymbol{\xi}\|_2=(\boldsymbol{\xi},\boldsymbol{\xi})^{1/2}=(\Sigma_{n=1}^{\infty}x_n^2)^{1/2}$. As linear spaces $l^2\subset l^\infty$, but as Banach spaces this inclusion does not make sense.

Theorem 7.2.13. *Let V be a Hilbert space, $\{\xi_i\}$ a sequence of scalars, and $\{\mathbf{u}_i\}$ a countable orthonormal set in V. The infinite series $\Sigma_{i=1}^{\infty}\xi_i\mathbf{u}_i$ converges iff $\Sigma_{i=1}^{\infty}|\xi_i|^2$ converges.*

PROOF: Suppose the series converges; this is to say the sequence of partial sums $\mathbf{S}_n$ is a Cauchy sequence. Hence $\forall\varepsilon>0$, $\exists n_0$ s.t. $\forall m>n\geqslant n_0$, $\|\mathbf{S}_n-\mathbf{S}_m\|^2=\|\Sigma_{n+1}^{m}\xi_i\mathbf{u}_i\|^2=(\Sigma_{n+1}^{m}\xi_i\mathbf{u}_i,\Sigma_{n+1}^{m}\xi_i\mathbf{u}_i)=\Sigma_{n+1}^{m}|\xi_i|^2<\varepsilon^2$. Since this holds for all $m>n\geqslant n_0$, we can conclude that $\Sigma_{i=n_0+1}^{\infty}|\xi_i|^2\leqslant\varepsilon^2$, so $\Sigma_{i=1}^{\infty}|\xi_i|^2<\infty$.

Conversely, if $\Sigma_{i=1}^{\infty}|\xi_i|<\infty$ and $\varepsilon>0$, we have for all m,n sufficiently large, $m>n$, $\Sigma_{n+1}^{m}|\xi_i|^2<\varepsilon^2$, so $\|\mathbf{S}_m-\mathbf{S}_n\|^2<\varepsilon^2$. This means that $\{\mathbf{S}_n\}$ is a Cauchy sequence, and since V is complete, the series converges. □

Corollary 7.2.14. *If the series $\Sigma_{i=1}^{\infty}\xi_i\mathbf{u}_i$ does indeed converge and $\mathbf{x}$ is the point to which it converges, then for each i, $\xi_i=(\mathbf{x},\mathbf{u}_i)$.*

PROOF: Let $\mathbf{S}_n=\Sigma_{i=1}^{n}\xi_i\mathbf{u}_i$. For each $i\leqslant n$, $(\mathbf{S}_n,\mathbf{u}_i)=\xi_i$; if $i>n$, this inner product is zero. Hence, for each fixed i, $\lim_{n\to\infty}(\mathbf{S}_n,\mathbf{u}_i)=\xi_i$. Now, the inner product is a continuous function of each of its variables, and certainly the first one; it has to be, for it determines the topology on V. Hence $\lim_n(\mathbf{S}_n,\mathbf{u}_i)=(\lim_n\mathbf{S}_n,\mathbf{u}_i)=(\mathbf{x},\mathbf{u}_i)$. We conclude that for each i, $\xi_i=(\mathbf{x},\mathbf{u}_i)$. This concludes the proof. □

We now step back and see how far we've come. If V is a Hilbert space, S an orthonormal set, and $\mathbf{x}\in V$, we know that there are at most a countable number of vectors $\mathbf{u}\in S$ for which $(\mathbf{x},\mathbf{u})\neq 0$. Suppose these are indexed: $\{\mathbf{u}_i\}_{i=1}^{\infty}$. Theorem 7.2.9 ensures that the series $\Sigma_{i=1}^{\infty}|(\mathbf{x},\mathbf{u}_i)|^2$ converges, and Theorem 7.2.13 then guarantees that the infinite series $\Sigma_{i=1}^{\infty}(\mathbf{x},\mathbf{u}_i)\mathbf{u}_i$ will converge to some point $\mathbf{y} \in V$. The corollary to this theorem says that for each i, $(\mathbf{x},\mathbf{u}_i)=(\mathbf{y},\mathbf{u}_i)$. Does this mean $\mathbf{x}=\mathbf{y}$? No, it simply means that for each i, $(\mathbf{x}-\mathbf{y},\mathbf{u}_i)=0$, which in turn means that $\mathbf{x}-\mathbf{y}$ is in the orthogonal complement of the set $\{\mathbf{u}_i\}_{i=1}^{\infty}$. Now if this set were complete (as an orthonormal set), we could conclude that $\mathbf{y}=\mathbf{x}$. But at the moment we don't even know that $\{\mathbf{u}_i\}_{i=1}^{\infty}$ is all of S.

We now show that the inner product is indeed continuous in its first (and hence also the second) variable. Let $\mathbf{y}$ be fixed and $\mathbf{x}$ a variable. Let $\varepsilon>0$; we use the Schwarz inequality to show that $(\mathbf{x},\mathbf{y})$ is continuous at $\mathbf{x}_0$. Let $\|\mathbf{x}-\mathbf{x}_0\|<\varepsilon/\|\mathbf{y}\|$. (If $\mathbf{y}=\mathbf{0}$, there is nothing to show; the inner product is constant.) Then $|(\mathbf{x},\mathbf{y})-(\mathbf{x}_0,\mathbf{y})|=|(\mathbf{x}-\mathbf{x}_0,\mathbf{y})|\leqslant\|\mathbf{x}-\mathbf{x}_0\|\cdot\|\mathbf{y}\|<\varepsilon$.

We use this fact to show that if $\mathbf{x}\perp A$, a nonempty set in V, then $\mathbf{x}\perp\bar{A}$. Let $\{\mathbf{a}_n\}\subset A$ be a sequence converging to $\mathbf{a}\in\bar{A}$. Then, because of the continuity of the inner product in its first variable, we have $\lim_n(\mathbf{a}_n,\mathbf{x})=(\lim_n\mathbf{a}_n,\mathbf{x})=0$. Hence $\mathbf{x}\perp A\Rightarrow\mathbf{x}\perp\bar{A}$. Of course the implication may be reversed, so we conclude $A^{\perp}=\bar{A}^{\perp}$. This allows us to improve Theorem 7.2.3 slightly; we could have concluded that $A^{\perp}=\bar{A}^{\perp}=(\operatorname{sp}\bar{A})^{\perp}=(\overline{\operatorname{sp}}\,\bar{A})^{\perp}$. We use this fact in the proof of the next theorem, known as the projection theorem.

Theorem 7.2.15. *Let S be an orthonormal set in a Hilbert space V. For each $\mathbf{x}\in V$, there is a unique element $P_S\mathbf{x}\in\overline{\operatorname{sp}}\,S$ defined by*

$$P_S\mathbf{x}=\sum_{\mathbf{u}\in S}(\mathbf{x},\mathbf{u})\mathbf{u}.$$

Moreover, $\mathbf{x}\in\overline{\operatorname{sp}}\,S\Leftrightarrow\mathbf{x}=P_S\mathbf{x}$, and $(\mathbf{x}-P_S\mathbf{x})\perp\overline{\operatorname{sp}}\,S$. Finally, $\|\mathbf{x}\|^2=\|P_S\mathbf{x}\|^2+\|\mathbf{x}-P_S\mathbf{x}\|^2$.

PROOF: Let $\mathbf{x}\in V$. From Theorem 7.2.8 we know that for at most a countable number of $\mathbf{u}\in S$, $(\mathbf{x},\mathbf{u})\neq 0$, so index these **u**s in some order $\mathbf{u}_1,\mathbf{u}_2,\mathbf{u}_3,\ldots$. The series $\sum_{i=1}^{\infty}(\mathbf{x},\mathbf{u}_i)\mathbf{u}_i$ is convergent, as we have already mentioned, to some point, say $\mathbf{x}_1$.

Suppose we had indexed the **u**s in some other way; let $\{\mathbf{v}_i\}$ be a rearrangement of the sequence $\{\mathbf{u}_i\}$. Then the series $\sum_{i=1}^{\infty}(\mathbf{x},\mathbf{v}_i)\mathbf{v}_i$ converges to some point $\mathbf{x}_2$. We can write

$$\begin{aligned}\|\mathbf{x}_1-\mathbf{x}_2\|^2&=\left(\mathbf{x}_1-\mathbf{x}_2,\sum(\mathbf{x},\mathbf{u}_i)\mathbf{u}_i-\sum(\mathbf{x},\mathbf{v}_i)\mathbf{v}_i\right)\\&=\left(\mathbf{x}_1-\mathbf{x}_2,\sum(\mathbf{x},\mathbf{u}_i)\mathbf{u}_i\right)-\left(\mathbf{x}_1-\mathbf{x}_2,\sum(\mathbf{x},\mathbf{v}_i)\mathbf{v}_i\right)\\&=\sum(\mathbf{x},\mathbf{u}_i)(\mathbf{x}_1-\mathbf{x}_2,\mathbf{u}_i)-\sum(\mathbf{x},\mathbf{v}_i)(\mathbf{x}_1-\mathbf{x}_2,\mathbf{v}_i).\end{aligned}$$

Now for each i we have $(\mathbf{x}_1,\mathbf{u}_i)=(\mathbf{x},\mathbf{u}_i)$ and $(\mathbf{x}_2,\mathbf{v}_i)=(\mathbf{x},\mathbf{v}_i)$. Moreover, for each $\mathbf{u}_i$ there is an index j such that $\mathbf{u}_i=\mathbf{v}_j$ and exactly one such j, call it j_i. Then for each i, $(\mathbf{x}_1-\mathbf{x}_2,\mathbf{u}_i)=(\mathbf{x}_1,\mathbf{u}_i)-(\mathbf{x}_2,\mathbf{u}_i)=(\mathbf{x},\mathbf{u}_i)-(\mathbf{x}_2,\mathbf{v}_{j_i})=(\mathbf{x},\mathbf{u}_i)-(\mathbf{x},\mathbf{v}_{j_i})=(\mathbf{x},\mathbf{u}_i)-(\mathbf{x},\mathbf{u}_i)=0$. Analogously, for each j, $(\mathbf{x}_1-\mathbf{x}_2,\mathbf{v}_j)=0$.

Hence $\|\mathbf{x}_1 - \mathbf{x}_2\|^2 = 0$, and $\mathbf{x}_1 = \mathbf{x}_2$. This is to say,

$$P_S\mathbf{x} = \sum_{\mathbf{u} \in S} (\mathbf{x}, \mathbf{u})\mathbf{u}$$

defines $P_S\mathbf{x}$ unambiguously.

It is clear that $P_S\mathbf{x} \in \overline{\text{sp}}\, S$ since each term of the sequence of partial sums for $P_S\mathbf{x}$ belongs to $\text{sp}\, S$. Hence $\mathbf{x} = P_S\mathbf{x} \Rightarrow \mathbf{x} \in \overline{\text{sp}}\, S$. To show that $(\mathbf{x} - P_S\mathbf{x}) \perp \overline{\text{sp}}\, S$, we show that $(\mathbf{x} - P_S\mathbf{x}) \perp \text{sp}\, S$. Let $\mathbf{u}_0 \in S$; then

$$(\mathbf{x} - P_S\mathbf{x}, \mathbf{u}_0) = (\mathbf{x}, \mathbf{u}_0) - (P_S\mathbf{x}, \mathbf{u}_0) = (\mathbf{x}, \mathbf{u}_0) - (\mathbf{x}, \mathbf{u}_0) = 0.$$

From the observation made just before this theorem, we have $(\mathbf{x} - P_S\mathbf{x}) \perp \overline{\text{sp}}\, S$. Next, observe that $\|\mathbf{x}\|^2 = (\mathbf{x}, \mathbf{x}) = (P_S\mathbf{x} + [\mathbf{x} - P_S\mathbf{x}], P_S\mathbf{x} + [\mathbf{x} - P_S\mathbf{x}]) = (P_S\mathbf{x}, P_S\mathbf{x}) + (\mathbf{x} - P_S\mathbf{x}, \mathbf{x} - P_S\mathbf{x}) = \|P_S\mathbf{x}\|^2 + \|\mathbf{x} - P_S\mathbf{x}\|^2$. All that remains to do is to show that if $\mathbf{x} \in \overline{\text{sp}}\, S$, then $\mathbf{x} = P_S\mathbf{x}$. This is easy; if $\mathbf{x} \in \overline{\text{sp}}\, S$, then the fact that $(\mathbf{x} - P_S\mathbf{x}) \in \overline{\text{sp}}\, S$ and $(\mathbf{x} - P_S\mathbf{x}) \perp \overline{\text{sp}}\, S$ implies that $\mathbf{x} - P_S\mathbf{x} = \mathbf{0}$, and the theorem is proved. □

The point $P_S\mathbf{x}$ is called the projection of $\mathbf{x}$ on $\overline{\text{sp}}\, S$. Note that $P_S(P_S\mathbf{x}) = P_S^2\mathbf{x} = P_S\mathbf{x}$, and that as an operator on V, P_S is linear and continuous and equal to the identity on $\overline{\text{sp}}\, S$. The first property we called attention to is expressed by saying that P_S is *idempotent*.

Theorem 7.2.16. *If V is an inner-product space containing an orthonormal set S, then there is a complete orthonormal set $\hat{S}$ in V which contains S.*

PROOF: Refer to the proof of Theorem 1.4.3; the proof of this theorem is most easily accomplished by applying Zorn's lemma. We leave the details to the reader. □

Theorem 7.2.17. *Let V be a Hilbert space and $S \subset V$ an orthonormal set. If $V = \overline{\text{sp}}\, S$, then S is complete, and conversely.*

PROOF: Assume S is not complete. Then there is a nonzero element $\mathbf{x} \in V$, with $\mathbf{x} \perp S$ and $\|\mathbf{x}\| = 1$. But $\mathbf{x} \perp S \Rightarrow \mathbf{x} \perp \overline{\text{sp}}\, S$; since $\overline{\text{sp}}\, S = V$, we would have $\mathbf{x} \perp V$, which happens iff $\mathbf{x} = \mathbf{0}$.

Now assume S is complete and that $\overline{\text{sp}}\, S \neq V$. Let $\mathbf{x} \in V \setminus \overline{\text{sp}}\, S$; then

$$\frac{\mathbf{x} - P_S\mathbf{x}}{\|\mathbf{x} - P_S\mathbf{x}\|}$$

is a unit vector orthogonal to $\overline{\text{sp}}\, S$ and hence to S. This means S is not complete, and the proof is finished. □

We make the remark that every inner-product space, complete or not, contains a complete orthonormal set, but it is possible that in some incomplete inner-product space the closed linear span of no complete orthonormal set is the whole space (Dixmier, 1953).

We have come a long way; we now have shown, with the last bevy of theorems, that if $\mathbf{x}$ is a point in a Hilbert space V which has a complete orthonormal set S and such an S does exist in V, then $\mathbf{x}$ has a unique Fourier representation in terms of S. There is yet one more theorem in the set, which tells us how to obtain the norm of such a point $\mathbf{x}$ when $\mathbf{x}$ is given in the form of its Fourier expansion.

Theorem 7.2.18. *If S is an orthonormal set in a Hilbert space V and if for each $\mathbf{x} \in V$ we have*

$$\|\mathbf{x}\|^2 = \sum_{\mathbf{u} \in S} |(\mathbf{x},\mathbf{u})|^2,$$

then S is complete. Conversely, if S is complete, the equality holds for every $\mathbf{x} \in V$.

PROOF: If S is not complete, then there is a nonzero $\mathbf{x}$, with $\mathbf{x} \perp S$, and we would have $(\mathbf{x},\mathbf{u}) = 0$ for each $\mathbf{u} \in S$. This proves the first part.

If S is complete, we have $\mathbf{x} = P_S\mathbf{x}$ for each $\mathbf{x}$, so $\mathbf{x} = \Sigma_{\mathbf{u} \in S}(\mathbf{x},\mathbf{u})\mathbf{u}$. Then $\|\mathbf{x}\|^2 = (\Sigma_{\mathbf{u} \in S}(\mathbf{x},\mathbf{u})\mathbf{u},\mathbf{x}) = \Sigma_{\mathbf{u} \in S}|(\mathbf{x},\mathbf{u})|^2$, and the proof is complete. □

The equality $\|\mathbf{x}\|^2 = \Sigma_{\mathbf{u} \in S}|(\mathbf{x},\mathbf{u})|^2$ is called *Parseval's formula*. Compare this to the *Bessel inequality* $\Sigma_{\mathbf{u} \in S}|(\mathbf{x},\mathbf{u})|^2 \leqslant \|\mathbf{x}\|^2$, which holds for arbitrary orthonormal sets S. We proved Bessel's inequality for finite sets S, and since every such set S is contained in a maximal (complete) orthonormal set, Parseval's formula proves that the Bessel inequality holds for arbitrary orthonormal sets as well. It is also well to note that we have a *generalized Parseval formula* in a complete real inner product space V with a complete orthonormal set S:

$$\forall \mathbf{x},\mathbf{y} \in V, \qquad (\mathbf{x},\mathbf{y}) = \sum_{\mathbf{u} \in S} (\mathbf{x},\mathbf{u})(\mathbf{y},\mathbf{u}).$$

We ask the reader to prove this formula by computing

$$(\mathbf{x},\mathbf{y}) = \left(\sum_{\mathbf{u} \in S} (\mathbf{x},\mathbf{u})\mathbf{u},\ \sum_{\mathbf{u} \in S} (\mathbf{y},\mathbf{u})\mathbf{u} \right).$$

We may summarize the results of the last four theorems in one grand statement.

Theorem 7.2.19. *Let V be a real Hilbert space with an orthonormal set S. Then there exists a complete orthonormal set $\hat{S}$ in V, with $S \subset \hat{S}$. Furthermore, the following statements are equivalent.*

a. *S is a complete orthonormal set.*
b. *If $\mathbf{x} \in V$ and $(\mathbf{x},\mathbf{u}) = 0$ for each $\mathbf{u} \in S$, then $\mathbf{x} = 0$.*
c. $\overline{\mathrm{sp}}\, S = V$.

d. *If* $\mathbf{x} \in V$, *then* $\mathbf{x} = \sum_{\mathbf{u} \in S} (\mathbf{x}, \mathbf{u})\mathbf{u}$.
e. *If* $\mathbf{x} \in V$, *then* $\|\mathbf{x}\|^2 = \sum_{\mathbf{u} \in S} |(\mathbf{x}, \mathbf{u})|^2$.
f. *If* $\mathbf{x}, \mathbf{y} \in V$, *then* $(\mathbf{x}, \mathbf{y}) = \sum_{\mathbf{u} \in S} (\mathbf{x}, \mathbf{u})(\mathbf{y}, \mathbf{u})$.

7.3 AN EXAMPLE: THE SPACE $L^2(0,2\pi)$

Let V be a space we shall designate by $L^2(0,2\pi)$. This is a function space, a set of real-valued functions having $[0,2\pi] \subset \mathbb{R}$ as domain. We know how to make a real linear space out of this set. We characterize the members of $L^2(0,2\pi)$ in this way: $f \in L^2(0,2\pi)$ iff f is a "measurable, square integrable," real-valued function on the real interval $[0,2\pi]$. What "measurable" means will be made clearer in the next chapter. Now put an inner product on this linear space by defining

$$\forall f, g \in L^2(0,2\pi), \qquad (f, g) = \int_0^{2\pi} f(t) g(t)\, dt.$$

That this is an inner product is hard to verify since it often happens that $(f, f) = 0$ without f being the zero function. We get out of this difficulty by declaring two functions of $L^2(0,2\pi)$ to be equivalent if they differ on a set of at most measure zero. This too will make more sense after Chapter 8. Hence $L^2(0,2\pi)$ is really a space of equivalence classes of functions. We say that two functions of the same equivalence class are *essentially equal.*

Is this space a Hilbert space? We may not be able to say for sure, but let's conjecture that it is and attempt to test whether our conjecture is reasonable. We have a norm given by

$$\forall f \in L^2(0,2\pi), \qquad \|f\|^2 = \int_0^{2\pi} |f(t)|^2\, dt.$$

Let $\{f_n\}$ be a Cauchy sequence in $L^2(0,2\pi)$, which is to say, given $\varepsilon > 0$, for all m, n sufficiently large,

$$\|f_n - f_m\|^2 = \int_0^{2\pi} |f_n(t) - f_m(t)|^2\, dt < \varepsilon^2.$$

Now for any norm function we always have the inequality

$$|\|a\| - \|b\|| \leq \|a - b\| \leq \|a\| + \|b\|,$$

so the sequence of norms $\{\|f_n\|\}$ is a Cauchy sequence and has a limit $L \in \mathbb{R}$. Next, observe that inequality (3.1.12) yields, for all m, n sufficiently large,

$$\int_0^{2\pi} |f_n(x) - f_m(x)|\, dx \leq \|f_n - f_m\| \sqrt{2\pi} < \sqrt{2\pi}\, \varepsilon,$$

where the function g in (3.1.12) is the characteristic function of $[0,2\pi]$.

Definition 7.3.1. Let A be a set in a space X. The real function χ_A, defined on X by

$$\chi_A(x)=\begin{cases}1, & x\in A,\\ 0, & x\notin A,\end{cases}$$

is called the *characteristic function* of A.

Since we have $\forall n$, $\int_0^{2\pi}|f_n(x)|^2\,dx<\infty$, if $E_n=\{x\in[0,2\pi]:|f_n(x)|$ is unbounded$\}$, the measure of E_n, $\mu(E_n)$, must be zero. Similarly, since for all m,n sufficiently large,

$$\int_0^{2\pi}|f_n(x)-f_m(x)|\,dx<\sqrt{2\pi}\,\varepsilon,$$

if $E_{mn}=\{x\in[0,2\pi]:|f_n(x)-f_m(x)|\geqslant 1/k\}$, k fixed, then

$$\sqrt{2\pi}\,\varepsilon>\int_{E_{mn}}|f_n(x)-f_m(x)|\,dx\geqslant\mu(E_{mn})/k,$$

so $\mu(E_{mn})<k\sqrt{2\pi}\,\varepsilon$ for all sufficiently large m,n. Hence it seems that regardless of how large we fix k, since $\varepsilon>0$ can be arbitrarily small, $\mu(E_{mn})\to 0$ as $m,n\to\infty$. We conclude then that except for a subset of measure zero, $\{f_n(x)\}$ will "converge" to a real number on $[0,2\pi]$. Thus we define a "limit" f for $\{f_n\}$ by the formula

$$f(x)=\frac{1}{2}\left[\bigwedge_{n=1}^{\infty}\bigvee_{k=n}^{\infty}f_k(x)+\bigvee_{n=1}^{\infty}\bigwedge_{k=n}^{\infty}f_k(x)\right]$$

for each $x\in[0,2\pi]\setminus E_0$, where

$$E_0=\left(\lim_{m,n\to\infty}E_{mn}\right)\cup\bigcup_{n=1}^{\infty}E_n.$$

The norm of this f will equal L, and $\lim_n\|f_n-f\|=0$. What $f(x)$ equals on E_0 is of no consequence; $f\in L^2(0,2\pi)$.

This is certainly a very plausible argument that $L^2(0,2\pi)$ is a complete space: you may rest assured that $L^2(0,2\pi)$ is indeed a Hilbert space.

Remember, this function f, our candidate for the limit of the Cauchy sequence $\{f_n\}$, is not the only possible candidate, but all other candidates will differ from f on sets of at most measure zero.

Now let $\{u_1,u_2,u_3,u_4,u_5,\ldots\}$ be the functions in $L^2(0,2\pi)$, $1/\sqrt{2\pi}$, $\cos t/\sqrt{\pi}$, $\sin t/\sqrt{\pi}$, $\cos 2t/\sqrt{\pi}$, $\sin 2t/\sqrt{\pi},\ldots$, respectively. It is an elementary calculus problem to verify that these functions make up an orthonormal set S; S is countable, and $L^2(0,2\pi)$ is infinite dimensional.

Suppose we have the sequences of scalars $\{a_0,a_1,a_2,\ldots\}$ and $\{b_1,b_2,b_3,\ldots\}$ such that

$$a_0^2+\sum_{i=1}^{\infty}\left(a_i^2+b_i^2\right)<\infty.$$

Then the series

$$\frac{a_0}{\sqrt{2\pi}}+\frac{1}{\sqrt{\pi}}\sum_{n=1}^{\infty}(a_n\cos nt+b_n\sin nt)$$

converges to a function $x(t)\in L^2(0,2\pi)$. Moreover,

$$a_0=\frac{1}{\sqrt{2\pi}}\int_0^{2\pi}x(t)\,dt,$$

$$a_n=\frac{1}{\sqrt{\pi}}\int_0^{2\pi}x(t)\cos nt\,dt,$$

$$b_n=\frac{1}{\sqrt{\pi}}\int_0^{2\pi}x(t)\sin nt\,dt.$$

Remember that $x(t)$ is determined only up to a set of measure zero by the scalars $\{a_i\}$ and $\{b_i\}$.

For convenience, let's introduce new constants $\{\alpha_0,\alpha_1,\alpha_2,\dots\}$ and $\{\beta_1,\beta_2,\beta_3,\dots\}$, where

$$\alpha_0=a_0\sqrt{2/\pi}\,,\qquad \alpha_n=a_n/\sqrt{\pi}\,,\qquad \beta_n=b_n/\sqrt{\pi}\,,$$

so that we can write

$$x(t)=\frac{\alpha_0}{2}+\sum_{n=1}^{\infty}(\alpha_n\cos nt+\beta_n\sin nt),$$

and $\forall n=0,1,2,\dots,$

$$\alpha_n=\frac{1}{\pi}\int_0^{2\pi}x(t)\cos nt\,dt,\qquad \beta_n=\frac{1}{\pi}\int_0^{2\pi}x(t)\sin nt\,dt.$$

These last coefficients are the ones generally called the Fourier coefficients of $x(t)$ with respect to the orthonormal set S, and the corresponding series is what is generally called the Fourier series for $x(t)$.

If we could show that S is a complete orthonormal set for $L^2(0,2\pi)$ we could then rest assured that for any function in $L^2(0,2\pi)$ we could produce its Fourier series. Indeed, the set S, usually called the trigonometric sequence, is a complete orthonormal set for $L^2(0,2\pi)$. The proof is lengthy, and we choose to omit it.

Note that $x(t)$ is periodic, with period 2π, if we extend $x(t)$ to a larger interval of $\mathbb{R}$.

The following remark is in order. Suppose $f\in L^2(0,2\pi)$ is continuous on $[0,2\pi]$, except for a jump discontinuity at $x=c$, and is such that at each $x\neq c$, the Fourier series for f converges to $f(x)$. Then the Fourier series converges at $x=c$ to the value $\frac{1}{2}[f(c^+)+f(c^-)]$ (cf. Theorem 7.4.2). Hence such a function f should perhaps be redefined at points of jump discontinuities so that at these points it will agree with its Fourier series representation.

For example, define

$$f(t)=\begin{cases} 1 & \text{if} \quad t\in[0,\pi), \\ \frac{3}{4} & \text{if} \quad t=\pi, \\ 2 & \text{if} \quad t\in(\pi,2\pi]. \end{cases}$$

Then

$$\alpha_0=3,$$

$$\alpha_n=\frac{1}{\pi}\int_0^{\pi}\cos nt\,dt+\frac{2}{\pi}\int_{\pi}^{2\pi}\cos nt\,dt=0, \qquad \forall n=1,2,\ldots,$$

$$\beta_n=\frac{1}{\pi}\int_0^{\pi}\sin nt\,dt+\frac{2}{\pi}\int_{\pi}^{2\pi}\sin nt\,dt=-\frac{\cos nt}{n\pi}\bigg|_0^{\pi}-\frac{2}{n\pi}\cos nt\bigg|_{\pi}^{2\pi}$$

$$=\frac{1}{n\pi}[-\cos n\pi+1-2\cos 2n\pi+2\cos n\pi]$$

$$=\frac{1}{n\pi}[\cos n\pi-1]=\begin{cases} 0 & \text{if } n \text{ is even,} \\ -\dfrac{2}{n\pi} & \text{if } n \text{ is odd.} \end{cases}$$

Thus

$$\frac{3}{2}-\frac{2}{\pi}\sum_{k=0}^{\infty}\frac{1}{2k+1}\sin(2k+1)t$$

is the Fourier series for $f(t)$. Then, $f(t)$ should be redefined so that it is periodic if extended beyond $[0,2\pi]$, and so that $f(0)=f(2\pi)=\frac{3}{2}$ and $f(\pi)=\frac{3}{2}$. Note that $f(\pi/2)=1$, so we can write

$$f\left(\frac{\pi}{2}\right)=1=\frac{3}{2}-\frac{2}{\pi}\sum_{k=0}^{\infty}\frac{1}{2k+1}(-1)^k,$$

which is to say

$$\pi/4=[1-\tfrac{1}{3}+\tfrac{1}{5}-\tfrac{1}{7}+\cdots].$$

Another point should not be overlooked. Theorem 7.2.19 states that in $L^2(0,2\pi)$, the trigonometric sequence S is a complete orthonormal set iff for all $x\in L^2(0,2\pi)$, $\|x\|^2$ equals the sum of the squares of the Fourier coefficients. This means

$$\|x\|^2=a_0^2+\sum_{n=1}^{\infty}(a_n^2+b_n^2)$$

and

$$\|x\|^2\neq\alpha_0^2+\sum_{n=1}^{\infty}(\alpha_n^2+\beta_n^2)=\frac{2}{\pi}a_0^2+\frac{1}{\pi}\sum_{n=1}^{\infty}(a_n^2+b_n^2).$$

These latter coefficients are simpler to work with and have been introduced for convenience. We have, of course,

$$\alpha_0^2+\sum_{n=1}^{\infty}(\alpha_n^2+\beta_n^2)=\frac{1}{\pi}(\|x\|^2+a_0^2),$$

or

$$\|x\|^2=\frac{\pi}{2}\alpha_0^2+\pi\sum_{n=1}^{\infty}(\alpha_n^2+\beta_n^2).$$

Hence we can say S is complete iff $\forall x\in L^2(0,2\pi)$

$$\pi\int_0^{2\pi}[x(t)]^2\,dt=\tfrac{1}{2}\left[\int_0^{2\pi}x(t)\,dt\right]^2 + \sum_{n=1}^{\infty}\left[\left(\int_0^{2\pi}x(t)\cos nt\,dt\right)^2+\left(\int_0^{2\pi}x(t)\sin nt\,dt\right)^2\right].$$

Now let's start with the series

$$\sum_{n=1}^{\infty}\frac{\sin nt}{n}.$$

In this case, each $\alpha_n=0$ and $\beta_n=1/n$. Since $\Sigma|\beta_n|^2<\infty$, the function

$$x(t)=\sum_{n=1}^{\infty}\frac{\sin nt}{n},\qquad 0\leqslant t\leqslant 2\pi,$$

is a function in $L^2(0,2\pi)$, unique up to a set of measure zero, and determined by the constants β_n.

Now let $y(t)=\frac{1}{2}(\pi-t)$, $t\in[0,2\pi]$. Since $y\in L^2(0,2\pi)$, let's write its Fourier series with respect to the trigonometric orthonormal sequence.

$$\alpha_n=\frac{1}{\pi}\int_0^{2\pi}\frac{\pi-t}{2}\cos nt\,dt,\qquad \beta_n=\frac{1}{\pi}\int_0^{2\pi}\frac{\pi-t}{2}\sin nt\,dt.$$

Integration by parts yields $\alpha_n=0$, $\beta_n=1/n$. Hence the Fourier series for $y(t)$ is given by

$$x(t)=\sum_{n=1}^{\infty}\frac{\sin nt}{n}.$$

$x(t)$ and $y(t)$ are essentially equal on $[0,2\pi]$.

As a final example, let

$$f(x)=\begin{cases}\sqrt{x} & \text{if}\quad 0\leqslant x\leqslant\pi,\\ \pi^{-3/2}(x-2\pi)^2 & \text{if}\quad \pi<x\leqslant 2\pi.\end{cases}$$

This function certainly belongs to $L^2(0,2\pi)$. What is its Fourier series, up to several terms? We had better leave this to the most diligent reader (who will meet a virtually insurmountable impasse).

7.4 FOURIER SERIES AND CONVERGENCE

A series of the form

$$\frac{\alpha_0}{2}+\sum_{n=1}^{\infty}(\alpha_n \cos nx+\beta_n \sin nx)$$

is called a trigonometric series; it is not called a Fourier series unless there is an integrable function f which gives rise to the coefficients α_n, β_n by the formulas

$$\alpha_n=\frac{1}{\pi}\int_0^{2\pi} f(t)\cos nt\,dt, \qquad \beta_n=\frac{1}{\pi}\int_0^{2\pi} f(t)\sin nt\,dt.$$

For example, $\sum_{n=1}^{\infty}\cos nx$ is a trigonometric series, but it is not a Fourier series, for one cannot find an integrable function f such that $\beta_n=0$, $\alpha_n=1$ for all $n\geqslant 1$. The proof of this depends upon the following lemma, which we do not prove.

Lemma 7.4.1 (Riemann–Lebesgue). *If f and $|f|$ are integrable on the interval $[a,b]$, then*

$$\lim_{t\to\infty}\int_a^b f(x)\cos tx\,dx=\lim_{t\to\infty}\int_a^b f(x)\sin tx\,dx=0,$$

or, equivalently,

$$\forall b, \qquad \lim_{t\to\infty}\int_a^b f(x)\sin(tx+b)\,dx=0.$$

Thus a Fourier series may not converge, but the terms of $\alpha_n\cos nx$ and $\beta_n \sin nx$ do tend to zero as $n\to\infty$. We have seen that any function f which belongs to $L^2(0,2\pi)$ gives rise to a convergent Fourier series, and that this series defines a function F, but that we do not necessarily have that $F(x)=f(x)$ for all $x\in[0,2\pi]$. F and f are "essentially" equal, in that they differ on at most a set of measure zero. If we start with a function $f\in L^2(0,2\pi)$ and change its values for certain points $x\in[0,2\pi]$ to get a new function g, the resulting Fourier series will be the same, as long as the set of points at which we change the value has measure zero.

The choice of the interval $[0,2\pi]$ was really quite arbitrary; we could just as well have considered the interval $[-\pi,\pi]$. The important thing is that the interval have length 2π. The Hilbert space $L^2(-\pi,\pi)$ is frequently more convenient to work with.

There are functions f which do not belong to $L^2(0,2\pi)$ but which are integrable on $[0,2\pi]$, as we shall see, and so give rise to Fourier series (for example, $f(x)=x^{-1/2}$, $x\neq 0$, $f(0)=0$). However, there is no assurance that the resulting series will even converge at some $x\in[0,2\pi]$, let alone converge to the value $f(x)$. On the other hand, it turns out that in many cases an integrable function f does give rise to a Fourier series which converges to $f(x)$. Hence we shall present a theorem without proof which gives sufficient conditions for the Fourier series of an integrable function f (in terms of the trigonometric sequence) to converge to the function at a point x.

Theorem 7.4.2 (Fourier). *Suppose f and $|f|$ are integrable on $[0,2\pi]$ and f is extended outside this interval so as to be periodic with period 2π. If α_n, β_n are defined by*

$$\alpha_n = \frac{1}{\pi}\int_0^{2\pi} f(t)\cos nt\,dt, \quad \beta_n = \frac{1}{\pi}\int_0^{2\pi} f(t)\sin nt\,dt, \qquad n=0,1,2,\ldots,$$

then the series

$$\frac{\alpha_0}{2} + \sum_{n=1}^{\infty} (\alpha_n \cos nx + \beta_n \sin nx)$$

converges to $f(x)$ if f is continuous at x, or to $\frac{1}{2}[f(x^+)+f(x^-)]$, where $f(x^+)$, $f(x^-)$ denote the right- and left-hand limits of f at x, respectively, when at least one of the following conditions are satisfied.

i. *f is bounded and has only a finite number of maxima and minima and a finite number of discontinuities on $[0,2\pi]$.*
ii. *There is an open interval I centered at x such that f is bounded on I and monotonic on each of the open halves of I.*
iii. *There is a neighborhood of x on which f is of bounded variation (see Definition 8.9.3).*
iv. *There is a punctured neighborhood of x on which f satisfies a Lipschitz condition; i.e., there exists a constant $M>0$ such that for all t, t' in this neighborhood, $|f(t)-f(t')| \leq M|t-t'|$.*
v. *f is differentiable on the right and on the left at x.*
vi. *Both $f(x^+)$, $f(x^-)$ exist, and there is a positive number $\varepsilon>0$ such that*

$$\left|\frac{f(x+t)-f(x^+)}{t} + \frac{f(x-t)-f(x^-)}{t}\right|$$

is integrable on $[-\varepsilon, \varepsilon]$.

vii. *Both $f(x^+)$, $f(x^-)$ exist, and both limits*

$$\lim_{t\uparrow x}\frac{f(t)-f(x^-)}{t-x}, \qquad \lim_{t\downarrow x}\frac{f(t)-f(x^+)}{t-x}$$

exist. (In this case we say that f is quasi-differentiable from both the left and the right at x.)

7.5 THE GRAM–SCHMIDT PROCESS

If V is an inner-product space, there is an algorithm which makes it possible for us to construct an orthonormal set, at least a finite, and by extension, a countable one. It presumes the constructor has the time and patience to pick and choose vectors from V, one at a time, verifying that each one chosen is independent of the ones previously chosen.

Theorem 7.5.1. *Let $\{\mathbf{x}_i\} \subset V$ be a finite or countable collection of linearly independent vectors from an inner product space V. Then there is an orthonormal set S having the same cardinality as the given set, as well as the same linear span. Moreover, the elements of the orthonormal set can be arranged in an order $\{\mathbf{u}_1, \mathbf{u}_2, \ldots\}$ such that for each $n = 1, 2, 3, \ldots$, $\mathrm{sp}\{\mathbf{x}_1, \ldots, \mathbf{x}_n\} = \mathrm{sp}\{\mathbf{u}_1, \ldots, \mathbf{u}_n\}$.*

PROOF: To prove this theorem, we need only show the construction of the set S. None of the $\mathbf{x}_i$ are zero. We define recursively $\mathbf{y}_1, \mathbf{u}_1, \mathbf{y}_2, \mathbf{u}_2, \mathbf{y}_3, \mathbf{u}_3, \ldots,$ as follows:

$$\mathbf{y}_1 = \mathbf{x}_1, \qquad \mathbf{u}_1 = \mathbf{y}_1/\|\mathbf{y}_1\|,$$

$$\mathbf{y}_2 = \mathbf{x}_2 - (\mathbf{x}_2, \mathbf{u}_1)\mathbf{u}_1, \qquad \mathbf{u}_2 = \mathbf{y}_2/\|\mathbf{y}_2\|,$$

$$\mathbf{y}_3 = \mathbf{x}_3 - (\mathbf{x}_3, \mathbf{u}_1)\mathbf{u}_1 - (\mathbf{x}_3, \mathbf{u}_2)\mathbf{u}_2, \quad \mathbf{u}_3 = \mathbf{y}_3/\|\mathbf{y}_3\|,$$

and in general

$$\mathbf{y}_{n+1} = \mathbf{x}_{n+1} - \sum_{i=1}^{n} (\mathbf{x}_{n+1}, \mathbf{u}_i)\mathbf{u}_i, \qquad \mathbf{u}_{n+1} = \mathbf{y}_{n+1}/\|\mathbf{y}_{n+1}\|.$$

If S is finite, the process stops with the nth step. Evidently, $\mathrm{sp}\{\mathbf{u}_1\} = \mathrm{sp}\{\mathbf{x}_1\}$. Assume that for any n, $\mathrm{sp}\{\mathbf{u}_i\}_{i=1}^n = \mathrm{sp}\{\mathbf{x}_i\}_{i=1}^n$; then $\mathbf{y}_{n+1}$ is in the span of $\mathbf{x}_{n+1}$ and the span of the first n **u**s, which is tantamount to saying $\mathbf{y}_{n+1}$ and hence $\mathbf{u}_{n+1}$ belong to $\mathrm{sp}\{\mathbf{x}_i\}_{i=1}^{n+1}$. Clearly, $\mathbf{x}_{n+1}$ is in the span of $\{\mathbf{u}_1, \ldots, \mathbf{u}_n, \mathbf{y}_{n+1}\}$, so it follows that $\mathrm{sp}\{\mathbf{x}_i\}_{i=1}^{n+1} = \mathrm{sp}\{\mathbf{u}_i\}_{i=1}^{n+1}$; that the **u**s so defined comprise an orthonormal set is fairly clear. □

It is certainly true that if an orthonormal set S is finite, it is a basis for its span. On the other hand, if a linear space V is infinite dimensional, an orthonormal set S contained in it consists of linearly independent vectors, as we have seen. If S is complete, then S might serve as a Hamel basis for V. However, such is not always the case. Consider that if $\mathbf{u}, \mathbf{v} \in S$, then $\|\mathbf{u}-\mathbf{v}\|^2 = (\mathbf{u}-\mathbf{v}, \mathbf{u}-\mathbf{v}) = (\mathbf{u}, \mathbf{u}) + (\mathbf{v}, \mathbf{v}) = 2$. Now, if V is complete, let $\mathbf{x}$ be an adherent point of $\mathrm{sp}\, S$, $\mathbf{x} \neq 0$, such that no finite linear combination of points of S is equal to $\mathbf{x}$. Then $\mathbf{x}$ is not orthogonal to S, since S is complete, yet $S \cup \{\mathbf{x}\}$ is a linearly independent set. The Gram–Schmidt (G–S) procedure is illustrated in Figure 7.1.

The following theorem is a restatement of Theorem 7.2.7. The proof is quite different.

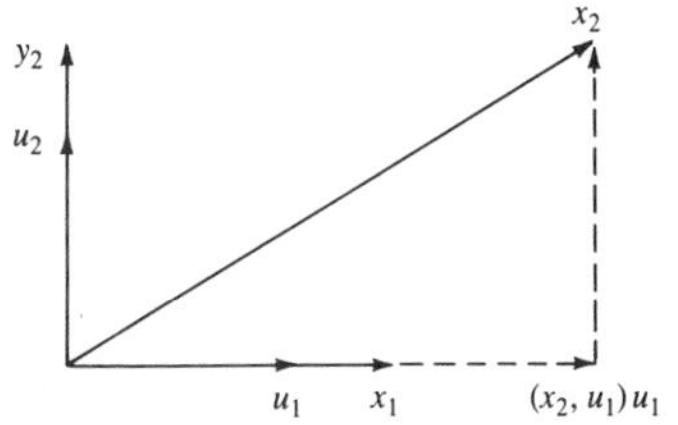

$$y_2 = x_2 - (x_2, u_1)u_1, \qquad u_2 = y_2/\|y_2\|$$

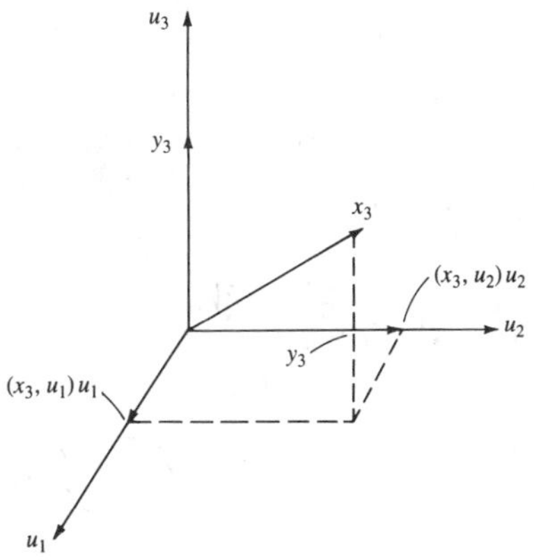

$$y_3 = x_3 - (x_3, u_1)u_1 - (x_3, u_2)u_2, \qquad u_3 = y_3/\|y_3\|$$

FIGURE 7.1 The Gram–Schmidt process.

Theorem 7.5.2. *Let V be a separable Hilbert space, and S a complete orthonormal set in V. Then S is countable.*

PROOF: Let $\{\mathbf{v}_n\}_{n=1}^{\infty}$ be a countable dense subset of V; let $\mathbf{x}_1$ be the first nonzero term in $\{\mathbf{v}_n\}$. Let $\mathbf{x}_2$ be the next element in the sequence independent of $\mathbf{x}_1$, and let $\mathbf{x}_3$ be the first of the terms following $\mathbf{x}_2$ which is independent of $\mathbf{x}_1$ and $\mathbf{x}_2$. In general, let $\mathbf{x}_{n+1}$ be the first element in the sequence beyond the term $\mathbf{x}_n$ such that $\mathbf{x}_{n+1}$ is not in the span of $\{\mathbf{x}_1,\dots,\mathbf{x}_n\}$. The set $\{\mathbf{x}_i\}_{i=1}^{\infty}$ is a set of linearly independent vectors, $\{\mathbf{x}_i\} \subset \{\mathbf{v}_n\}$, and $\mathrm{sp}\{\mathbf{x}_i\} = \mathrm{sp}\{\mathbf{v}_n\}$, so $\overline{\mathrm{sp}}\{\mathbf{x}_i\} = V$. Apply the Gram–Schmidt process to $\{\mathbf{x}_i\}_{i=1}^{\infty}$ to get an orthonormal set $\{\mathbf{u}_i\}_{i=1}^{\infty}$. Evidently, $\overline{\mathrm{sp}}\{\mathbf{u}_i\}_{i=1}^{\infty} = V$, so $\{\mathbf{u}_i\}$ is a complete orthonormal set and is countable.

Now let S be any arbitrary complete orthonormal set in V. card S is clearly infinite; we assert that card $S = \aleph_0$. Now each $\mathbf{u}_i \in \{\mathbf{u}_i\}$ is expressible as a countable linear combination of elements of S. Each $\mathbf{x} \in V$ is expressible in terms of a countable number of the $\mathbf{u}_i$s. Since a countable union of countable sets is countable (Theorem 1.1.5), we conclude that any $\mathbf{x} \in V$ has a unique expression in terms of a countable number of elements of S. Hence S must be at most countable, and the proof is complete. □

The fact that many Hilbert spaces, and indeed Banach spaces, are separable suggests that there may frequently be Banach spaces which contain countable subsets of linearly independent vectors such that the closed linear spans of the subsets equal the containing spaces. In anticipation that this is indeed the case, we give the following definition.

Definition 7.5.3. Let B be a Banach space. A set $\{\mathbf{b}_n\}_{n=1}^{\infty}$ of linearly independent vectors of B is called a *Schauder basis* for B if each $\mathbf{x} \in B$ has a unique representation of the form $\mathbf{x} = \Sigma_{n=1}^{\infty} \xi_n \mathbf{b}_n$, the equality meaning that $\lim_{n\to\infty} \|\mathbf{x} - \Sigma_{i=1}^{n} \xi_i \mathbf{b}_i\| = 0$.

Not every separable Banach space admits a Schauder basis, but the common ones do, and because of Theorem 7.5.2, we can conclude that every separable Hilbert space admits a Schauder basis, specifically some complete orthonormal set.

7.6 APPROXIMATION BY PROJECTION

In analysis or applied calculus we are always interested in ways of approximating a given function, and once we have found a way, it is of paramount importance to know how good the approximation is. The next theorem tells us that if $\mathbf{x}$ is a function in a Hilbert space, and S is some orthonormal set in that space, then $P_S\mathbf{x}$ might be a pretty good approximation for $\mathbf{x}$, good in the sense that $\|\mathbf{x} - P_S\mathbf{x}\|$ might be small.

Theorem 7.6.1. *Let S be an orthonormal set in a Hilbert space V and $\mathbf{x}$ an arbitrary point of V. Then $P_S\mathbf{x}$ is the best approximation that can be found in $\overline{\mathrm{sp}}\, S$ in that if $\mathbf{y} \in \overline{\mathrm{sp}}\, S$, then $\|\mathbf{x} - P_S\mathbf{x}\| \leqslant \|\mathbf{x} - \mathbf{y}\|$.*

PROOF: As a consequence of Theorem 7.2.14 we can write

$$\|\mathbf{x}-\mathbf{y}\|^2 = \|P_S(\mathbf{x}-\mathbf{y})\|^2 + \|(\mathbf{x}-\mathbf{y}) - P_S(\mathbf{x}-\mathbf{y})\|^2.$$

Since $\mathbf{y} \in \overline{\mathrm{sp}}\, S$, $\mathbf{y} = P_S\mathbf{y}$, and since P_S is a linear operator, we have $\|\mathbf{x}-\mathbf{y}\|^2 = \|P_S(\mathbf{x}-\mathbf{y})\|^2 + \|\mathbf{x} - P_S\mathbf{x}\|^2$ and $\|\mathbf{x}-\mathbf{y}\| \geqslant \|\mathbf{x} - P_S\mathbf{x}\|$. □

How well $P_S\mathbf{x}$ approximates $\mathbf{x}$ depends upon the size of $\sqrt{\|\mathbf{x}\|^2 - \|P_S\mathbf{x}\|^2}$. If $\overline{\text{sp}}\,S$ comes close to "filling up" V, then the approximation will be good. If V is a separable Hilbert space and S a complete orthonormal set, we can approximate $\mathbf{x} \in V$ as closely as we like by projecting $\mathbf{x}$ down onto the first n vectors of S, for n sufficiently large. If the complete orthonormal set S for a nonseparable Hilbert space is uncountable, we can't tell which countable subset of S to use to base the finite span upon which to project $\mathbf{x}$ to get a good approximation. The best we can do in this case is to look for an orthonormal set T such that, for a given $\mathbf{x} \in V$, $\|\mathbf{x}\|$ and $\|P_T\mathbf{x}\|$ are very close in value.

For the record, the spaces $L^2(a,b)$, $a, b \in \mathbb{R}$, are separable Hilbert spaces. Some orthonormal sets for these spaces are

1. the trigonometric sequence on $[0, 2\pi]$,
2. the Legendre polynomials on $[-1, 1]$,
3. the Chebyshev functions on $[-1, 1]$,
4. the Jacobi functions on $[-1, 1]$,
5. the Hermite functions on $(-\infty, \infty)$,
6. the Laguerre functions on $[0, \infty)$,
7. the Haar functions on $[0, 1]$, and
8. the Rademacher functions on $[0, 1]$.

All but the last are complete. We get the Legendre polynomials by applying the Gram–Schmidt process to the set $\{t^n\}_{n=0}^{\infty}$ on $[-1, 1]$. The Chebyshev functions are obtained when we apply the G–S process to the set $\{(1-t^2)^{-1/2}t^n\}_{n=0}^{\infty}$ on $[-1,1]$ or (a second variety) to the set $\{(1-t^2)^{1/2}t^n\}_{n=0}^{\infty}$ on $[-1, 1]$. An application of the G–S process to the set $\{(1-t)^{\alpha}(1+t)^{\beta}t^n\}_{n=0}^{\infty}$ on $[-1,1]$ yields the set of Jacobi functions. If we apply the G–S process to the set $\{e^{-t^2/2}t^n\}_{n=0}^{\infty}$ over $(-\infty, \infty)$, we get the set of Hermite functions for $L^2(\mathbb{R})$, and if we orthonormalize the set $\{e^{-t/2}t^n\}_{n=0}^{\infty}$ on $[0, \infty)$, we obtain the Laguerre functions on $[0, \infty)$.

In general, if $f(t)$ is such that $t^n f(t) \in L^2(a,b)$ for each n, one can apply the G–S process to the sequence $\{t^n f(t)\}_{n=0}^{\infty}$ and make one's own orthonormal set. It might not be complete, but it will be one's own! We omit a description of the Haar and Rademacher functions; you can look them up if you like.

We now consider a relatively simple function and approximate it. Let $y = 2^t + 1$ on $[-1, 1]$. The interested reader could construct a Fourier approximation for this function over the interval $[-\pi, \pi]$, but we shall use the Legendre polynomials as a basis. The first six of these are $1/\sqrt{2}$, $\sqrt{\frac{3}{2}}\,t$, $\frac{3}{2}\sqrt{\frac{5}{2}}\,(t^2 - \frac{1}{3})$, $\frac{5}{2}\sqrt{\frac{7}{2}}\,(t^3 - \frac{3}{5}t)$, $\frac{35}{8}\sqrt{\frac{9}{2}}\,(t^4 - \frac{6}{7}t^2 + \frac{3}{35})$, and $\frac{63}{8}\sqrt{\frac{11}{2}}\,(t^5 - \frac{10}{9}t^3 + \frac{5}{21}t)$. We shall find $P_S y$ on the span of the first three of these; to go further appears so formidable that even the most dedicated blanch at the prospect.

$$P_S y(t) = \frac{1}{\sqrt{2}} \int_{-1}^{1} (2^t+1) \frac{1}{\sqrt{2}}\, dt + \sqrt{\frac{3}{2}}\, t \int_{-1}^{1} (2^t+1) \sqrt{\frac{3}{2}}\, t\, dt$$

$$+ \frac{3}{2}\sqrt{\frac{5}{2}}\left(t^2 - \frac{1}{3}\right) \int_{-1}^{1} (2^t+1) \frac{3}{2}\sqrt{\frac{5}{2}}\left(t^2 - \frac{1}{3}\right) dt$$

$$= \frac{1}{2}\left[\frac{3}{2\ln 2} + 2\right] + \frac{3}{2}t\left[\frac{5}{2\ln 2} - \frac{3}{2(\ln 2)^2}\right]$$

$$+ \frac{45}{8}\left(t^2 - \frac{1}{3}\right)\left[\frac{3}{(\ln 2)^3} - \frac{5}{(\ln 2)^2} + \frac{1}{\ln 2}\right]$$

$$\approx .249t^2 + .727t + 1.999.$$

Next, let's approximate $y = 2^t + 1$, not by projection, but simply by a quadratic that agrees with y at $t = -1, 0, 1$. The standard way to find such a polynomial is to use the Lagrange interpolation formula.

Theorem 7.6.2 (Lagrange Interpolation Formula). *Let $x_1 < x_2 < \cdots < x_{n+1}$ be $n+1$ distinct real numbers, and let $\{y_1, y_2, \ldots, y_{n+1}\}$ be $n+1$ values, not necessarily distinct. Then*

$$y = \frac{(x-x_2)\cdots(x-x_{n+1})}{(x_1-x_2)\cdots(x_1-x_{n+1})} y_1 + \frac{(x-x_1)(x-x_3)\cdots(x-x_{n+1})}{(x_2-x_1)(x_2-x_3)\cdots(x_2-x_{n+1})} y_2$$

$$+ \cdots + \frac{(x-x_1)\cdots(x-x_n)}{(x_{n+1}-x_1)\cdots(x_{n+1}-x_n)} y_{n+1}$$

is a polynomial of degree $\leqslant n$ which assumes the value y_i at each x_i, and it is the only such polynomial of degree $\leqslant n$.

PROOF: Obviously, $\deg(y) \leqslant n$, and $y = y_i$ at each x_i. That this polynomial is unique is clear since if $P(x)$ is any other polynomial of degree $\leqslant n$ having the prescribed values at $x_1, \ldots, x_{n+1}$, then $P(x) - y$ is a polynomial of degree $\leqslant n$ having $n+1$ zeros, so $P(x) - y$ must be identically zero. □

Our Lagrange approximation of degree 2 for $y = 2^t + 1$ is obtained easily; $\{x_1, x_2, x_3\} = \{-1, 0, 1\}$, and $\{y_1, y_2, y_3\} = \{\frac{3}{2}, 2, 3\}$. Then

$$L(t) = \frac{t(t-1)}{(-1)(-2)} \cdot \frac{3}{2} + \frac{(t+1)(t-1)}{(1)(-1)} \cdot 2 + \frac{(t+1)(t)}{(2)(1)} \cdot 3$$

$$= \frac{t^2}{4} + \frac{3}{4}t + 2 = .250t^2 + .750t + 2.$$

This is not much different from the quadratic $P_S y(t)$ that we computed.

Now we approximate $y=2^t+1$ by a quadratic which agrees with y at $t=0$; moreover, its first two derivatives agree with the first two derivatives of y at $t=0$. This is to say, we approximate by using the Taylor polynomial of degree 2 centered at the midpoint of the interval $[-1,1]$. This polynomial is most easy to get.

$$T(t)=2+(\ln 2)t+\tfrac{1}{2}(\ln 2)^2t^2\approx .240t^2+.693t+2.$$

These polynomials are reasonably close in value, but Theorem 7.6.1 assures us that $P_S y(t)$ is the best in the sense that the integral $\int_{-1}^{1}[2^t+1-P_S y(t)]^2\,dt$ will be less than or equal to either of the integrals $\int_{-1}^{1}[2^t+1-L(t)]^2\,dt$ or $\int_{-1}^{1}[2^t+1-T(t)]^2\,dt$—or for that matter, the integral $\int_{-1}^{1}[2^t+1-(at^2+bt+c)]^2\,dt$, where a, b, c are arbitrary real constants.

The Lagrange interpolation formula is really quite useful; frequently, one wants a polynomial of minimum degree that will assume certain prescribed values at particular values of its argument. On the other hand, one might want a periodic function, rather than a polynomial, which will assume the prescribed values. We can obtain such a function having period 2π by using a trigonometric analog of the Lagrange interpolation formula.

Definition 7.6.3. Let $x_1<\cdots<x_n$ be n distinct points and $\{y_1,\ldots,y_n\}$ be n (not necessarily distinct) values to be associated with the x_is, respectively. The function

$$f(x)=\frac{y_1\sin(x-x_2)\cdots\sin(x-x_n)}{\sin(x_1-x_2)\cdots\sin(x_1-x_n)}+\cdots+\frac{y_n\sin(x-x_1)\cdots\sin(x-x_{n-1})}{\sin(x_n-x_1)\cdots\sin(x_n-x_{n-1})}$$

is a periodic function, period 2π, that takes the value y_i at x_i for each $i=1,2,\ldots,n$. This expression is known as the *Hermite interpolation formula*.

7.7 COMPLEX INNER-PRODUCT SPACES

It often happens that the scalar field associated with a linear space is $\mathbb{C}$, the field of complex numbers. For example, $V(\mathbb{C})$ may be the complex linear space of n-tuples of complex numbers $\mathbb{C}^n$ or a function space of square-integrable complex-valued functions. A Banach space may be either real or complex, and the same is true for a Hilbert space. We feel that it is appropriate at this time to recall a few facts about complex numbers and to bring to the reader's attention the fact that if we intend to endow a complex linear space with an inner product, we need to rephrase our inner-product axioms so that they may apply to complex as well as real vector spaces. First, a few words about complex numbers.

If z is a complex number, z has a unique representation of the form $z=x+iy$, where $x, y \in \mathbb{R}$ and $i=\sqrt{-1}$ or $i^2=-1$. We call x the real part of z and y the imaginary part, and write $x=\mathrm{Re}(z)$, $y=\mathrm{Im}(z)$, respectively. The number $\bar{z}=x-iy$ is called the *conjugate* of $z=x+iy$. Invariably

$$z\bar{z}=\bar{z}z=x^2+y^2.$$

We call $|z|=\sqrt{x^2+y^2}=\sqrt{z\bar{z}}$ the *modulus* of z. Since $|\ |$ obeys the very special properties of a valuation, we also call $|z|$ the absolute value of the complex number z. One observes immediately that $\forall z, w \in \mathbb{C}$,

$$|z+w| \leqslant |z|+|w|, \qquad |zw|=|z||w|, \qquad z+\bar{z}=2\,\mathrm{Re}(z),$$
$$z-\bar{z}=2\,\mathrm{Im}(z), \qquad z^{-1}=\bar{z}/|z|^2.$$

There are two other ways to represent the complex number $z=x+iy$. They are

$$z=|z|e^{i\theta}, \qquad z=|z|(\cos\theta+i\sin\theta).$$

The first of these is called the *polar form*, the other the *trigonometric form*. In both of these the real number θ is called the *argument* of z, written $\arg(z)$, and we generally assume that $-\pi<\theta\leqslant\pi$. More specifically, $\arg(z)$ is the principal argument of z; θ may equal $\arg(z)+2k\pi$, where k is any integer.

If we write $z=e^{\ln z}=|z|e^{i\arg(z)}$, we can then extend the definition for the natural logarithm function by extending its domain to all of $\mathbb{C}\setminus\{0\}$.

$$\ln(z)=\ln|z|+i\arg(z).$$

Note that $\arg(z)$ must be the principal argument of z. It follows from this that $\ln(-1)=\pi i$, or $e^{\pi i}=-1$.

The reader may easily verify that conjugation obeys the following rules:

$$\overline{z+w}=\bar{z}+\bar{w}, \qquad \overline{zw}=\bar{z}\bar{w}, \qquad \bar{\bar{z}}=z.$$

In particular, for any real number θ,

$$\overline{e^{i\theta}}=\overline{\cos\theta+i\sin\theta}=\cos\theta-i\sin\theta=e^{-i\theta}=e^{\overline{i\theta}}.$$

From this follows

$$|e^{i\theta}|^2=e^{i\theta}\cdot e^{-i\theta}=e^0=1=|e^{i\theta}|.$$

If $z=|z|e^{i\theta}$, we define $\mathrm{sgn}[z]=\mathrm{signum}[z]=e^{-i\theta}$; thus, $z(\mathrm{sgn}[z])=|z|$.

Definition 7.7.1. If V is a complex linear space and if there is a conjugate bilinear function $\mu: V\times V\to\mathbb{C}$ satisfying the following axioms $\forall \mathbf{x},\mathbf{y},\mathbf{z}\in V$, $\forall\alpha\in\mathbb{C}$, we call V, together with μ, a complex inner-product space:

CI1. $\mu(\mathbf{x}+\mathbf{y},\mathbf{z})=\mu(\mathbf{x},\mathbf{z})+\mu(\mathbf{y},\mathbf{z})$.
CI2. $\mu(\mathbf{x},\mathbf{y})=\overline{\mu(\mathbf{y},\mathbf{x})}$; $\mu(\mathbf{x},\mathbf{x})\in\mathbb{R}$.
CI3. $\mu(\alpha\mathbf{x},\mathbf{y})=\alpha\mu(\mathbf{x},\mathbf{y})$; $\mu(\mathbf{x},\alpha\mathbf{y})=\bar{\alpha}\mu(\mathbf{x},\mathbf{y})$.
CI4. $\mu(\mathbf{x},\mathbf{x})\geqslant 0$; $\mu(\mathbf{x},\mathbf{x})=0\Leftrightarrow\mathbf{x}=\mathbf{0}$.

In case V is a real linear space, then the conjugation bar has no effect, and these axioms are the same as the I-axioms given earlier. Note that the inner product for $\mathbb{C}^n$ is given $\forall \mathbf{x},\mathbf{y} \in \mathbb{C}^n$, $\mathbf{x}=(x_1,\ldots,x_n)$, $\mathbf{y}=(y_1,\ldots,y_n)$, by

$$\mu(\mathbf{x},\mathbf{y}) \equiv (\mathbf{x},\mathbf{y}) = \sum_{i=1}^{n} x_i \bar{y}_i, \qquad (\mathbf{x},\mathbf{x}) = \sum_{i=1}^{n} x_i \bar{x}_i = \|\mathbf{x}\|^2.$$

If $L^2(\Omega)$ is a complex Hilbert space analogous to $L^2(a,b)$,

$$(f,g) = \int_\Omega f(z)\overline{g(z)}\,dz,$$

$$(f,f) = \int_\Omega f(z)\overline{f(z)}\,dz = \int_\Omega |f(z)|^2\,dz = \|f\|^2.$$

In general, $(\mathbf{x},\mathbf{y})+(\mathbf{y},\mathbf{x}) = 2\,\mathrm{Re}(\mathbf{x},\mathbf{y})$, so in a complex Hilbert space

$$\|\mathbf{x}+\mathbf{y}\|^2 = \|\mathbf{x}\|^2 + \|\mathbf{y}\|^2 + 2\,\mathrm{Re}(\mathbf{x},\mathbf{y}),$$
$$\|\mathbf{x}-\mathbf{y}\|^2 = \|\mathbf{x}\|^2 + \|\mathbf{y}\|^2 - 2\,\mathrm{Re}(\mathbf{x},\mathbf{y}),$$

and hence the parallelogram law still holds.

The only change we have to make in moving from real inner-product spaces to complex inner-product spaces is to point out that the generalized Parseval relations must be written

$$\sum_{\mathbf{u}} |(\mathbf{x},\mathbf{u})\overline{(\mathbf{y},\mathbf{u})}| \leqslant \|\mathbf{x}\|\,\|\mathbf{y}\|$$

and

$$(\mathbf{x},\mathbf{y}) = \sum_{\mathbf{u}} (\mathbf{x},\mathbf{u})\overline{(\mathbf{y},\mathbf{u})}.$$

In the complex Hilbert space $L^2(0,2\pi)$ the functions

$$\varphi_n(t) = \frac{1}{\sqrt{2\pi}} e^{int}, \qquad t \in [0,2\pi], \quad n = 0, \pm 1, \pm 2, \ldots,$$

make up a complete orthonormal set.

7.8 THE GRAM DETERMINANT AND MEASURES OF k-PARALLELOTOPES

This section is important in that its result is one of the keystones for our calculus. The material contained here is often omitted from first courses in linear algebra.

Definition 7.8.1. Let $\{\mathbf{x}_j\}_{j=1}^{k}$ be k arbitrary vectors in E^n, with its standard inner product. The *Gram determinant*, or *Gramian*, of these k vectors is the determinant

$$G(\mathbf{x}_1,\ldots,\mathbf{x}_k)=\det\begin{bmatrix}(\mathbf{x}_1,\mathbf{x}_1) & (\mathbf{x}_1,\mathbf{x}_2) & \ldots & (\mathbf{x}_1,\mathbf{x}_k)\\ (\mathbf{x}_2,\mathbf{x}_1) & (\mathbf{x}_2,\mathbf{x}_2) & \ldots & (\mathbf{x}_2,\mathbf{x}_k)\\ \vdots & \vdots & & \vdots\\ (\mathbf{x}_k,\mathbf{x}_1) & (\mathbf{x}_k,\mathbf{x}_2) & \ldots & (\mathbf{x}_k,\mathbf{x}_k)\end{bmatrix}.$$

If necessary, the reader may refer back to Theorem 4.3.3, which lists the basic properties of determinants that one derives from the definition. Also, we remind the reader that the Gram–Schmidt process produces an orthonormal set of vectors from an arbitrary set; this slight modification of the process gives rise to a particular orthogonal set.

Let

$$\mathbf{y}_1=\mathbf{x}_1,\qquad \mathbf{y}_2=\mathbf{x}_2-\frac{(\mathbf{x}_2,\mathbf{y}_1)}{(\mathbf{y}_1,\mathbf{y}_1)}\mathbf{y}_1,\qquad \ldots,$$

$$\mathbf{y}_{i+1}=\mathbf{x}_{i+1}-\frac{(\mathbf{x}_{i+1},\mathbf{y}_1)}{(\mathbf{y}_1,\mathbf{y}_1)}\mathbf{y}_1-\cdots-\frac{(\mathbf{x}_{i+1},\mathbf{y}_i)}{(\mathbf{y}_i,\mathbf{y}_i)}\mathbf{y}_i.$$

(We do not normalize the vectors $\mathbf{y}_1,\mathbf{y}_2,\ldots$.) In G, replace $\mathbf{x}_1$ by $\mathbf{y}_1$, and let α_1 be such that $\mathbf{y}_2=\alpha_1\mathbf{y}_1+\mathbf{x}_2$ is orthogonal to $\mathbf{y}_1$ [i.e., let $\alpha_1=-(\mathbf{x}_2,\mathbf{y}_1)/(\mathbf{y}_1,\mathbf{y}_1)$]. Now multiply the first column of G by α_1 and add it to the second column, noting that $\alpha_1(\mathbf{x}_i,\mathbf{y}_1)=(\mathbf{x}_i,\alpha_1\mathbf{y}_1)$ and $(\mathbf{x}_i,\alpha_1\mathbf{y}_1)+(\mathbf{x}_i,\mathbf{x}_2)=(\mathbf{x}_i,\mathbf{x}_2+\alpha_1\mathbf{y}_1)$:

$$G=\det\begin{bmatrix}(\mathbf{y}_1,\mathbf{y}_1) & (\mathbf{y}_1,\mathbf{x}_2+\alpha_1\mathbf{y}_1) & \ldots\\ (\mathbf{x}_2,\mathbf{y}_1) & (\mathbf{x}_2,\mathbf{x}_2+\alpha_1\mathbf{y}_1) & \ldots\\ \vdots & \vdots & \\ (\mathbf{x}_k,\mathbf{y}_1) & (\mathbf{x}_k,\mathbf{x}_2+\alpha_1\mathbf{y}_1) & \ldots\end{bmatrix}.$$

Next, multiply the first row by α_1, and add to the second row, in the way indicated, and replace $\mathbf{x}_2+\alpha_1\mathbf{y}_1$ by $\mathbf{y}_2$.

$$G=\det\begin{bmatrix}(\mathbf{y}_1,\mathbf{y}_1) & (\mathbf{y}_1,\mathbf{y}_2) & \ldots\ \ldots\\ (\mathbf{x}_2+\alpha_1\mathbf{y}_1,\mathbf{y}_1) & (\mathbf{x}_2+\alpha_1\mathbf{y}_1,\mathbf{x}_2+\alpha_1\mathbf{y}_1) & \ldots\ \ldots\\ \vdots & \vdots & \vdots\ \ \vdots\\ (\mathbf{x}_k,\mathbf{y}_1) & (\mathbf{x}_k,\mathbf{x}_2+\alpha_1\mathbf{y}_1) & \ldots\ \ldots\end{bmatrix},$$

$$G=\det\begin{bmatrix}(\mathbf{y}_1,\mathbf{y}_1) & (\mathbf{y}_1,\mathbf{y}_2) & \ldots\\ (\mathbf{y}_2,\mathbf{y}_1) & (\mathbf{y}_2,\mathbf{y}_2) & \ldots\\ \vdots & \vdots & \vdots\end{bmatrix}.$$

Now let $\mathbf{y}_3=\beta_1\mathbf{y}_1+\beta_2\mathbf{y}_2+\mathbf{x}_3$, where β_1, β_2 are such that $\mathbf{y}_3$ is orthogonal to

$\mathbf{y}_1$ and $\mathbf{y}_2$ [i.e., let $\beta_1 = -(\mathbf{x}_3,\mathbf{y}_1)/(\mathbf{y}_1,\mathbf{y}_1)$ and $\beta_2 = -(\mathbf{x}_3,\mathbf{y}_2)/(\mathbf{y}_2,\mathbf{y}_2)$]. Multiply the first column by β_1 and the second column by β_2, and add to the third column, associating the βs with the second terms of the inner products, and then do the same operation on the rows associating the βs now with the first elements of the inner products. Replace $\beta_1\mathbf{y}_1 + \beta_2\mathbf{y}_2 + \mathbf{x}_3$ by $\mathbf{y}_3$, and G will have been reduced to

$$G = \det\begin{bmatrix} (\mathbf{y}_1,\mathbf{y}_1) & 0 & 0 & \cdots \\ 0 & (\mathbf{y}_2,\mathbf{y}_2) & 0 & \cdots \\ 0 & 0 & (\mathbf{y}_3,\mathbf{y}_3) & \cdots \\ \vdots & \vdots & \vdots & \end{bmatrix}.$$

None of the operations we have been doing changes the value of G. Eventually, we have the above matrix reduced to a diagonal matrix, with $G = \prod_{i=1}^{k}(\mathbf{y}_i,\mathbf{y}_i)$.

Note that for each $i \geqslant 1$ we have constructed $\mathbf{y}_{i+1}$ by the procedure

$$\mathbf{y}_1 = \mathbf{x}_1,$$

$$\mathbf{y}_{i+1} = \mathbf{x}_{i+1} - \frac{(\mathbf{x}_{i+1},\mathbf{y}_1)}{(\mathbf{y}_1,\mathbf{y}_1)}\mathbf{y}_1 - \frac{(\mathbf{x}_{i+1},\mathbf{y}_2)}{(\mathbf{y}_2,\mathbf{y}_2)}\mathbf{y}_2 - \cdots - \frac{(\mathbf{x}_{i+1},\mathbf{y}_i)}{(\mathbf{y}_i,\mathbf{y}_i)}\mathbf{y}_i.$$

Hence

$$\|\mathbf{y}_{i+1}\|^2 = (\mathbf{y}_{i+1},\mathbf{y}_{i+1}) = (\mathbf{x}_{i+1},\mathbf{y}_{i+1}) \leqslant \|\mathbf{x}_{i+1}\| \cdot \|\mathbf{y}_{i+1}\|,$$

and we have, for each $i = 1,\ldots,k$,

$$\|\mathbf{y}_i\|^2 = (\mathbf{y}_i,\mathbf{y}_i) \leqslant (\mathbf{x}_i,\mathbf{x}_i) = \|\mathbf{x}_i\|^2.$$

This is an important remark: the set of vectors $\{\mathbf{x}_1,\ldots,\mathbf{x}_k\}$ gives rise to a set $\{\mathbf{y}_1,\ldots,\mathbf{y}_k\}$ of orthogonal vectors iff the $\mathbf{x}_i$s are linearly independent. This is clear; if the $\mathbf{x}_i$s are linearly dependent, they span something less than a k-dimensional space. Hence the number of orthogonal vectors, which of course span the same space as the $\mathbf{x}_i$s, must be the same as the dimension of that space since they are linearly independent. We are now ready to state

Theorem 7.8.2. *If $\{\mathbf{x}_1,\ldots,\mathbf{x}_k\}$ is a set of vectors in E^n and G is the Gram determinant of these vectors, then*

$$0 \leqslant G(\mathbf{x}_1,\ldots,\mathbf{x}_k) \leqslant \prod_{i=1}^{k} \|\mathbf{x}_i\|^2.$$

$G = 0$ iff the vectors $\{\mathbf{x}_i\}$ are linearly dependent. The equality on the right-hand side holds iff the $\mathbf{x}_i$ are mutually orthogonal. □

There is a perfectly natural way to define the Euclidean measure of a k-parallelotope, $k > 1$. We simply define it to be the measure of a

$(k-1)$-dimensional face times the measure of the one-dimensional altitude to that face. Starting with $k=2$, we proceed inductively; we now have a reasonable definition for the k-dimensional measure of a k-parallelotope in E^n.

If you agree that this is a perfectly natural way to measure such figures (it is consistent with the way we do it in 2- and 3-space, and with the idea that the measure of the unit k-cube most certainly should be 1), we have, *mirabile dictu*, that the value of the Gram determinant of $\{\mathbf{x}_1,\ldots,\mathbf{x}_k\}$ is the square of the k-dimensional measure of the k-parallelotope determined by these vectors. This is clear. $\mathbf{y}_1$ is the first edge; $\mathbf{y}_2$ is an altitude to that edge, so $(\mathbf{y}_1,\mathbf{y}_1)(\mathbf{y}_2,\mathbf{y}_2)$ is the square of the measure of the two-dimensional face. $\mathbf{y}_3$ is an altitude to that face, and so on.

Theorem 7.8.3. *If $\{\mathbf{x}_1,\ldots,\mathbf{x}_k\}$ are k vectors in E^n, the Gram determinant $G(\mathbf{x}_1,\ldots,\mathbf{x}_k)$ is the square of the k-dimensional measure of the k-parallelotope determined by these vectors. If $G=0$, the figure is degenerate, i.e., a parallelotope of less than k dimensions.*

There is more to follow. Each vector $\mathbf{x}_j$ in E^n has n components; we can represent each $\mathbf{x}_j$ of the set $\{\mathbf{x}_1,\ldots,\mathbf{x}_k\}$ in terms of its components by

$$\mathbf{x}_j=\left(\xi_1^{(j)},\ldots,\xi_n^{(j)}\right).$$

Let's express $G(\mathbf{x}_1,\ldots,\mathbf{x}_k)$ in terms of the components of these vectors,

$$G=\det\begin{bmatrix} \sum_{i=1}^{n}\xi_i^{(1)}\xi_i^{(1)} & \cdots & \sum_{i=1}^{n}\xi_i^{(1)}\xi_i^{(k)} \\ \sum_{i=1}^{n}\xi_i^{(2)}\xi_i^{(1)} & \cdots & \sum_{i=1}^{n}\xi_i^{(2)}\xi_i^{(k)} \\ \vdots & & \vdots \\ \sum_{i=1}^{n}\xi_i^{(k)}\xi_i^{(1)} & \cdots & \sum_{i=1}^{n}\xi_i^{(k)}\xi_i^{(k)} \end{bmatrix}.$$

Now this vast array can be expressed as a linear combination of simpler determinants, each one of the form (with its coefficient)

$$\xi_{i_1}^{(1)}\xi_{i_2}^{(2)}\cdots\xi_{i_k}^{(k)}\det\begin{bmatrix} \xi_{i_1}^{(1)} & \xi_{i_2}^{(1)} & \cdots & \xi_{i_k}^{(1)} \\ \xi_{i_1}^{(2)} & \xi_{i_2}^{(2)} & \cdots & \xi_{i_k}^{(2)} \\ \vdots & \vdots & & \vdots \\ \xi_{i_1}^{(k)} & \xi_{i_2}^{(k)} & \cdots & \xi_{i_k}^{(k)} \end{bmatrix},$$

where each column is made up of the i_jth terms of the n-nomials in the jth column of the original determinant and the common factor has then been factored out. If any of the i_j are equal, these simple determinants vanish, the i_j being numbers from 1 to n.

We can disregard the simple determinants in which two of the subscripts of the ξs are equal; from the rest, group together all the summands having the same k subscripts $i_1,\ldots,i_k$, and add these together; there should be $k!$ of them since that's the number of permutations of the ordered k-tuple there are. Now think hard! Each of these $k!$ determinants has the same absolute value, and each is multiplied by the number $\xi_{i_1}^{(1)}\xi_{i_2}^{(2)}\cdots\xi_{i_k}^{(k)}$. Now the value of

$$\det\begin{bmatrix} \xi_{i_1}^{(1)} & \cdots & \xi_{i_k}^{(1)} \\ \vdots & & \vdots \\ \xi_{i_1}^{(k)} & \cdots & \xi_{i_k}^{(k)} \end{bmatrix}$$

shall be denoted $M(i_1,\ldots,i_k)$, and we do not lose any generality if we assume $1 \leqslant i_1 < i_2 < \cdots < i_k \leqslant n$, as you will see. If we interchange two columns of M, we change its sign; more generally, if we make an odd permutation of the columns, the sign of M changes, but it does not change if we make an even permutation. Hence we shall express the sum of these $k!$ terms

$$M(i_1,\ldots,i_k)\sum(-1)^{\sigma}\xi_{i_1}^{(1)}\xi_{i_2}^{(2)}\cdots\xi_{i_k}^{(k)},$$

where the sum is over all permutations of $(i_1,\ldots,i_k)$, and σ is 0 or 1 according as the permutation is even or odd. But this sum is by definition equal to the value of the determinant. This is to say, the value of the sum of all the simple determinants having the same set of k subscripts is $M^2(i_1,\ldots,i_k)$. This means that $G(\mathbf{x}_1,\ldots,\mathbf{x}_k)$ is equal to the sum of the squares of all the expressions of the form $M(i_1,\ldots,i_k)$, with $1 \leqslant i_1 < i_2 < \cdots < i_k \leqslant n$.

The linear mapping that sends the unit k-cube in E^k to the k-parallelotope in E^n determined by the vectors, in order, has as its matrix

$$T=\begin{bmatrix} \xi_1^{(1)} & \xi_1^{(2)} & \cdots & \xi_1^{(k)} \\ \xi_2^{(1)} & \xi_2^{(2)} & \cdots & \xi_2^{(k)} \\ \vdots & \vdots & & \vdots \\ \xi_n^{(1)} & \xi_n^{(2)} & \cdots & \xi_n^{(k)} \end{bmatrix}.$$

Hence $M^2(i_1,\ldots,i_k)$ is just the square of the determinant of the $k \times k$ minor consisting of the $i_1, i_2,\ldots,i_k$ rows.

Theorem 7.8.4. *Let P be a k-parallelotope in E^n and T the linear part of the affine map that sends the unit cube in E^k to P. The k-dimensional Euclidean measure of P is equal to the square root of the sum of the squares of the determinants of the $\binom{n}{k}$ distinct principal minors of T. In particular, if $k=n$, the Euclidean measure of P is just the absolute value of* $\det T$.

Moreover, the absolute value of the determinant of each minor $M(i_1,\ldots,i_k)$ is the Euclidean measure of the projection of P onto the

$(i_1 i_2 \ldots i_k)$-coordinate subspace of E^n. This means that we have in E^n the generalized Pythagorean theorem; namely, the square of the measure of a k-dimensional parallelotope P is equal to the sum of the squares of the measures of the projections of P on each of the k-dimensional coordinate subspaces.

One immediate consequence of the results of this section is Hadamard's inequality, listed in Chapter 3, useful in estimating an upper bound for a determinant.

Theorem 7.8.5. *If*

$$D = \det\begin{bmatrix} \xi_{11} & \cdots & \xi_{1n} \\ \xi_{21} & \cdots & \xi_{2n} \\ \vdots & & \vdots \\ \xi_{n1} & \cdots & \xi_{nn} \end{bmatrix},$$

then

$$D^2 \leqslant \prod_{i=1}^{n} \left(\sum_{j=1}^{n} \xi_{ij}^2 \right).$$

PROOF: Let $\{\mathbf{x}_1, \ldots, \mathbf{x}_n\}$ be n vectors having components comprising the rows of the matrix shown. Then

$$D^2 = G(\mathbf{x}_1, \ldots, \mathbf{x}_n) \leqslant \prod_{i=1}^{n} \|\mathbf{x}_i\|^2$$

$$= \left(\sum_{j=1}^{n} \xi_{1j}^2 \right)\left(\sum_{j=1}^{n} \xi_{2j}^2 \right) \cdots \left(\sum_{j=1}^{n} \xi_{nj}^2 \right) = \prod_{i=1}^{n} \left(\sum_{j=1}^{n} \xi_{ij}^2 \right).$$

Since equality holds iff $\{\mathbf{x}_1, \ldots, \mathbf{x}_n\}$ is an orthogonal set, we can't in general improve on this inequality. Note that since D is also the determinant of the transpose, we also have $D^2 \leqslant \prod_{j=1}^{n} (\sum_{i=1}^{n} \xi_{ij}^2)$. □

7.9 VECTOR PRODUCTS IN E^3

We have defined an inner product in E^n; one might conjecture that the term "inner" is used to suggest that the product of two vectors in E^n is a vector in E^1, a subspace of E^n. We have met the term exterior product as well; such a product, a vector in the Grassmann algebra, was an element of some "outer" space containing the space of the factors. Hence neither of these products is an element of the same space as its factors.

In E^3 there is a useful "true" product. This is to say, if **A** and **B** are vectors in E^3, we can define a vector product $\mathbf{A} \times \mathbf{B}$ which is itself a vector in E^3.

Definition 7.9.1. Let **A**, **B** be nonzero vectors in E^3. The *vector product* $\mathbf{A}\times\mathbf{B}$ is a vector orthogonal to both **A** and **B**, having length equal to $\|\mathbf{A}\times\mathbf{B}\| = \|\mathbf{A}\|\|\mathbf{B}\|\sin\theta$, where θ is the principal arccosine of $(\mathbf{A},\mathbf{B})/\|\mathbf{A}\|\|\mathbf{B}\|$. Equivalently, $\|\mathbf{A}\times\mathbf{B}\| = \sqrt{\|\mathbf{A}\|^2\|\mathbf{B}\|^2-(\mathbf{A},\mathbf{B})^2}$.

Since there are two such vectors, one the opposite of the other, we specify that the *right-hand rule* will be brought into play: To determine the direction of $\mathbf{A}\times\mathbf{B}$, let the fingers of the right hand point from the terminal end of **A** to the terminal end of **B**; the direction of the outstretched thumb is the direction of $\mathbf{A}\times\mathbf{B}$. Hence $\mathbf{A}\times\mathbf{B} = -(\mathbf{B}\times\mathbf{A})$, for $\mathbf{A},\mathbf{B}\in E^3$.

Now E^3 has a well-defined scalar product already defined. To find $\mathbf{A}\times\mathbf{B}$, we first look for a vector **X** such that

$$(\mathbf{A},\mathbf{X}) = (\mathbf{B},\mathbf{X}) = 0.$$

That is, we solve the system of equations

$$a_1x_1+a_2x_2+a_3x_3=0, \qquad b_1x_1+b_2x_2+b_3x_3=0.$$

We can assume **A** and **B** are linearly independent. Solutions are

$$x_1=(a_2b_3-a_3b_2)k, \qquad x_2=(a_3b_1-a_1b_3)k, \qquad x_3=(a_1b_2-a_2b_1)k,$$

where k is an arbitrary parameter. Putting $x_1^2+x_2^2+x_3^2=\|\mathbf{A}\|^2\|\mathbf{B}\|^2-(\mathbf{A},\mathbf{B})^2$ yields $k=1$.

Theorem 7.9.2. *If* $\mathbf{A}=(a_1,a_2,a_3)$, $\mathbf{B}=(b_1,b_2,b_3)$ *are two vectors in* E^3, *then* $\mathbf{A}\times\mathbf{B}$ *is the vector*

$$\mathbf{A}\times\mathbf{B}=(a_2b_3-a_3b_2,\, a_3b_1-a_1b_3,\, a_1b_2-a_2b_1).$$

PROOF: All we need to verify is that $\mathbf{A}\times\mathbf{B}$ points in the correct direction. The linear map T that sends the positively oriented unit cube $K\subset E^3$ into the parallelopiped determined by the edges, in order, **A**, **B**, and $\mathbf{A}\times\mathbf{B}$, has for its matrix

$$\begin{bmatrix} a_1 & a_2 & a_3 \\ b_1 & b_2 & b_3 \\ a_2b_3-a_3b_2 & a_3b_1-a_1b_3 & a_1b_2-a_2b_1 \end{bmatrix}^{t},$$

the determinant of which is

$$(a_2b_3-a_3b_2)^2+(a_3b_1-a_1b_3)^2+(a_1b_2-a_2b_1)^2\geqslant 0.$$

Hence the parallelopiped is positively oriented, as required. □

Observe that $\|\mathbf{A}\times\mathbf{B}\|$ is numerically equal to the area of the parallelogram determined by **A** and **B**.

Theorem 7.9.3. *The vector product satisfies the following properties, and these properties, together with the right-hand rule stipulation, characterize the vector product in* E^3.

V1. $\forall \mathbf{A},\mathbf{B} \in E^3$, $(\mathbf{A}\times\mathbf{B}) = -(\mathbf{B}\times\mathbf{A})$.
V2. $\forall \mathbf{A},\mathbf{B},\mathbf{C} \in E^3$, $\mathbf{A}\times(\mathbf{B}+\mathbf{C}) = (\mathbf{A}\times\mathbf{B})+(\mathbf{A}\times\mathbf{C})$.
V3. $\forall \mathbf{A},\mathbf{B} \in E^3$, $\forall \alpha \in \mathbb{R}$, $\alpha(\mathbf{A}\times\mathbf{B}) = \alpha\mathbf{A}\times\mathbf{B} = \mathbf{A}\times\alpha\mathbf{B}$.
V4. $\forall \mathbf{A},\mathbf{B} \in E^3$, $(\mathbf{A},\mathbf{A}\times\mathbf{B}) = 0$.
V5. $\forall \mathbf{A},\mathbf{B} \in E^3$, $\|\mathbf{A}\times\mathbf{B}\|^2 + (\mathbf{A},\mathbf{B})^2 = \|\mathbf{A}\|^2\|\mathbf{B}\|^2$.

We omit the proof.

Definition 7.9.4. By the *scalar triple product* of three vectors $\mathbf{A},\mathbf{B},\mathbf{C} \in E^3$ we mean the scalar $(\mathbf{A},\mathbf{B}\times\mathbf{C})$.

For convenience, the inner product in E^n, $\mathbb{R}^n$, $\mathbb{C}^n$, and more general sequence spaces is often written with a dot, and it is occasionally referred to as the *dot product*. The scalar triple product is thus usually written

$$\mathbf{A}\cdot\mathbf{B}\times\mathbf{C}.$$

Theorem 7.9.5

$$\mathbf{A}\cdot\mathbf{B}\times\mathbf{C} = \det\begin{bmatrix} a_1 & a_2 & a_3 \\ b_1 & b_2 & b_3 \\ c_1 & c_2 & c_3 \end{bmatrix}.$$

As a consequence of this, we see that the volume of the parallelopiped determined by $\mathbf{A}$, $\mathbf{B}$, and $\mathbf{C}$ is $|\mathbf{A}\cdot\mathbf{B}\times\mathbf{C}|$.

Theorem 7.9.6

$$\forall \mathbf{A},\mathbf{B},\mathbf{C} \in E^3, \qquad \mathbf{A}\times(\mathbf{B}\times\mathbf{C}) = (\mathbf{A}\cdot\mathbf{C})\mathbf{B} - (\mathbf{A}\cdot\mathbf{B})\mathbf{C}.$$

From this follows the Jacobi identity

$$\mathbf{A}\times(\mathbf{B}\times\mathbf{C}) + \mathbf{C}\times(\mathbf{A}\times\mathbf{B}) + \mathbf{B}\times(\mathbf{C}\times\mathbf{A}) \equiv \mathbf{0},$$

and the fact that in general the vector product is nonassociative.

We omit the proofs; both are straightforward.

PROBLEMS

1. Prove Theorem 7.2.16.

2. Write a Fourier series for x^2 on $[-\pi,\pi]$ and then on $[0,2\pi]$.

3. Try to obtain a Fourier series for

$$f(x) = \begin{cases} \sqrt{x}, & 0 \leqslant x \leqslant \pi, \\ \pi^{-3/2}(x-2\pi)^2, & \pi \leqslant x \leqslant 2\pi. \end{cases}$$

4. Suppose T is a symmetric linear operator on an inner-product space. This is to say, for every pair of vectors $\mathbf{x},\mathbf{y}$ in the space, $(T\mathbf{x},\mathbf{y})=(\mathbf{x},T\mathbf{y})$. Let $\mathbf{x},\mathbf{y}$ be two distinct eigenvectors of T corresponding to distinct eigenvalues of T. Show that $\mathbf{x}\perp\mathbf{y}$.

5. Let $\Omega(a,b)$ be the space of twice differentiable functions on $[a,b]\subset E^1$ such that for each $x\in\Omega(a,b)$, $x(a)=x(b)$, and $x'(a)=x'(b)$. A differential operator L over this space having the form

$$Lx(t)=\frac{d}{dt}\left[p(t)\frac{d(x(t))}{dt}\right]+q(t)x(t),$$

where p, q are given functions, is called a *Sturm–Liouville* differential operator. Show that L is a symmetric operator [i.e., $\forall x,y\in\Omega(a,b)$, $(Lx,y)=(x,Ly)$] and that if $p(t)=t^2-1$, $q(t)=0$, $a=-1$, $b=1$, the Legendre polynomials are eigenfunctions of this linear operator. The Legendre polynomials on $[a,b]$ are given by the formula

$$P_n(t)=\frac{1}{2^n n!}\left[(t^2-1)^n\right]^{(n)},\qquad n=0,1,2,\ldots.$$

However, it must be noted that these polynomials have not been normalized. (The superscript in parentheses of course indicates nth derivative.)

6. Suppose P is a parallelopiped in E^6 determined by the vectors:

$$P:\quad\begin{bmatrix}0&4&1\\2&2&-3\\1&3&-1\\1&3&-1\\2&1&4\\2&1&4\end{bmatrix}.$$

It is a bit of a chore to find an orthogonal matrix (the columns are the elements of an orthonormal set in E^6), such as

$$T=\begin{bmatrix}\frac{1}{2}&\frac{1}{2}&\frac{1}{2}&\frac{1}{2}&0&0\\0&0&0&0&\frac{1}{\sqrt{2}}&\frac{1}{\sqrt{2}}\\\frac{1}{\sqrt{2}}&-\frac{1}{\sqrt{2}}&0&0&0&0\\0&0&\frac{1}{\sqrt{2}}&-\frac{1}{\sqrt{2}}&0&0\\0&0&0&0&\frac{1}{\sqrt{2}}&-\frac{1}{\sqrt{2}}\\\frac{1}{2}&\frac{1}{2}&-\frac{1}{2}&-\frac{1}{2}&0&0\end{bmatrix},$$

which will map P to the parallelopiped P', a figure in the particular coordinate

E^3 subspace indicated by the matrix for P'

$$\begin{bmatrix} 2 & 6 & -2 \\ 2\sqrt{2} & \sqrt{2} & 4\sqrt{2} \\ -\sqrt{2} & \sqrt{2} & 2\sqrt{2} \\ 0 & 0 & 0 \\ 0 & 0 & 0 \\ 0 & 0 & 0 \end{bmatrix}.$$

Use Theorem 7.9.5 to find the volume of P'. Since T is a congruence map, you have the volume of P. Now find the volume of P by using Theorem 7.8.4, and the result had better be the same! (If P had not been conveniently "fixed up" by the author, such a matrix T would be hard to find.)

7. Suppose $\mathbf{x}_1=(0,2,1,1,2,2)$, $\mathbf{x}_2=(4,2,3,3,1,1)$, $\mathbf{x}_3=(1,-3,-1,-1,4,4)$ are three vectors in E^6 which determine a parallelopiped. Use the G–S process to obtain three vectors $\mathbf{u}_1$, $\mathbf{u}_2$, $\mathbf{u}_3$ such that $\mathbf{u}_1=\mathbf{x}_1/\|\mathbf{x}_1\|$, $\mathbf{u}_2\in \mathrm{sp}\{\mathbf{x}_1,\mathbf{x}_2\}$, $\mathbf{u}_3\in \mathrm{sp}\{\mathbf{x}_1,\mathbf{x}_2,\mathbf{x}_3\}$, and $\{\mathbf{u}_1,\mathbf{u}_2,\mathbf{u}_3\}$ is an orthonormal set. Let $\mathbf{u}_1$, $\mathbf{u}_2$, $\mathbf{u}_3$ be the first three columns of a matrix A; get the last three columns by letting $\mathbf{x}_4=\mathbf{e}_3$, $\mathbf{x}_5=\mathbf{e}_4$, $\mathbf{x}_6=\mathbf{e}_5$, or the simplest vectors you can find such that $\mathrm{sp}\{\mathbf{x}_1,\mathbf{x}_2,\ldots,\mathbf{x}_6\}=E^6$. Or, if by inspection you can find three additional columns for A such that all six columns will comprise an orthonormal set, do so, and you will have an orthogonal matrix A. If you can't do this, continue with the G–S process to get an orthonormal set $\{\mathbf{u}_1,\mathbf{u}_2,\ldots,\mathbf{u}_6\}$ from $\{\mathbf{x}_1,\mathbf{x}_2,\ldots,\mathbf{x}_6\}$ and let these $\mathbf{u}_i$s be the columns for A.

   Next, compute A^{-1} by the well-known process of transforming the 6×12 matrix

$$\begin{bmatrix} a_{11} & \cdots & a_{16} & 1 & \cdots & 0 \\ \vdots & A & \vdots & \vdots & \ddots & \vdots \\ a_{61} & \cdots & a_{66} & 0 & \cdots & 1 \end{bmatrix}$$

into

$$\begin{bmatrix} 1 & \cdots & 0 & b_{11} & \cdots & b_{16} \\ \vdots & \ddots & \vdots & \vdots & B & \vdots \\ 0 & \cdots & 1 & b_{61} & \cdots & b_{66} \end{bmatrix}$$

by means of elementary row operations. The 6×6 matrix B is A^{-1}. The matrix A^{-1} will play the role of T in the previous problem.

   Do all this, and then prove the following: If T is an orthogonal matrix, T preserves lengths, angles, and measures, but not necessarily orientations. In short, T is a congruence mapping.

8. Suppose B is a reflexive Banach space, which implies $B = B^{**}$ or more correctly, B is linearly and isometrically isomorphic with B^{**}, its bidual. Let $T\colon B\to B$ be a continuous linear endomorphism. We define the adjoint of T, denoted T^*, by $T^*\colon B^*\to B^*$. For each $F\in B^*$, T^*F is the linear functional in

B^* defined by the equation $(T^*F)f = F(Tf)$, $\forall f \in B$. We define the norm on B^* by

$$\|F\| = \sup\{|Ff| : f \in B, \|f\| = 1\},$$

so $\|T^*F\|$ is defined. Then logically we would define

$$\|T^*\| = \sup\{\|T^*F\| : F \in B^*, \|F\| = 1\}.$$

Show that T^* is a linear endomorphism on B^*, with $\|T^*\| = \|T\|$, and for any scalar α and any continuous endomorphisms S and T on B,

$$(S+T)^* = S^* + T^*, \qquad (\alpha T)^* = \alpha T^*, \qquad (ST)^* = T^*S^*.$$

Next, let N be the null space of T, R the closure of the range of T, N^* the null space of T^*, and R^* the closure of the range of T^*. Let A be a closed subspace of B. Define, and this is somewhat ambiguous if B happens to be an inner product space, $A^\perp$, the orthogonal complement of A, to be the set in B^*:

$$A^\perp = \{F \in B^* : Ff = 0, \forall f \in A\}.$$

Thus $N^\perp$ and $R^\perp$ are defined. Show that $N^\perp = R^*$ and $R^\perp = N^*$.

Finally, suppose now that T is such that for a fixed positive constant k, $\|T^n\| \leqslant k$, $n = 1,2,3\ldots$. This is to say, $\{T^n\}_{n=1}^\infty$ is a uniformly bounded sequence of linear endomorphisms on B. Define

$$T_n = (1/n)[I + T + T^2 + \cdots + T^{n-1}],$$

where $I = T^0$ is the identity mapping. Prove that $\{T_n\}$ converges pointwise to a projector P on B whose range is the null space of $I - T$ and whose null space is the closure of the range of $I - T$, and that $\|P\| \leqslant k$. (To say that $\{T_n\}$ converges pointwise to P is to say that for each fixed $f \in B$, $\{T_n f\} \to Pf$.)

9. Let A be a normed linear space and A_0 a proper closed subspace of A. Prove that for each real number t, $0 < t < 1$, there is a point $x_t \in A$ with $\|x_t\| = 1$, such that $\|x - x_t\| \geqslant t$ for all $x \in A_0$. (Compare this with Lemma 6.5.18.)

This lemma is by F. Riesz. It says that we can always find a point on the unit sphere of A which is at a distance from A_0 as close to 1 as we like. Of course, each point of the unit sphere of A is a distance 1 from $\mathbf{0} \in A_0$, but it might be that for no $\mathbf{x} \in A$, $\|\mathbf{x}\| = 1$, is $\rho(\mathbf{x}, A_0) = 1$.

As a hint, start with $\mathbf{x}_1 \in A \setminus A_0$ and take a point $\mathbf{x}_0 \in A_0$, with $\|\mathbf{x}_0 - \mathbf{x}_1\| \leqslant d/t$, where $d = \inf\{\|\mathbf{x} - \mathbf{x}_1\| : \mathbf{x} \in A_0\}$. (Why is $d > 0$?) Put $h = \|\mathbf{x}_0 - \mathbf{x}_1\|^{-1}$, and consider the point $h(\mathbf{x}_1 - \mathbf{x}_0)$.

10. Refer to Problem 2. With the aid of your results, obtain sums for

(a) $\displaystyle\sum_{n=1}^{\infty} \frac{1}{n^2}$,

(b) $\displaystyle\sum_{n=1}^{\infty} (-1)^{n-1} n^{-2}$,

(c) $\displaystyle\sum_{n=1}^{\infty} (2n-1)^{-2}$.

11. Let $f(x)=(\pi^2-x^2)^2$ on $[-\pi,\pi]$. Are f, f', f'', f''' of period 2π on this interval (i.e., does $f^{(i)}(\pi)=f^{(i)}(-\pi)$, $i=0,1,2,3$)? Expand $f(x)$ as a Fourier series on $[-\pi,\pi]$. Compute the sum $\Sigma_{n=1}^{\infty}(-1)^{n-1}n^{-4}$.

12. Prove Theorems 7.9.3 and 7.9.5.

13. Prove Theorem 7.9.6.

14. Find the first three Hermite functions H_1, H_2, H_3. Let $f(t)=(1+t^2)^{-1}$. If $S=\text{sp}\{H_1,H_2,H_3\}$, express $P_S f$.

15. If $\mathbf{A}=(3,1,2)$ and $\mathbf{B}=(-1,1,4)$ and $\alpha=3$, calculate $\mathbf{A}\times(\alpha\mathbf{B})+(\alpha\mathbf{A})\times\mathbf{B}$.

16. If $z=3-3\sqrt{-3}$, find $\ln z$. Now suppose $z=x+iy$. Can you express $\arg(z^2)$ in terms of x and y? Is $\ln z^2=2\ln z$?

17. Prove that in an inner-product space, the inner product $(\mathbf{x},\mathbf{y})$ is continuous in both variables, as well as in each separately.

18. Prove that if V is a Hilbert space and S a complete orthonormal set, then

$$\mathbf{x}\cdot\mathbf{y}=(\mathbf{x},\mathbf{y})=\sum_{\mathbf{u}\in S}(\mathbf{x},\mathbf{u})(\mathbf{y},\mathbf{u}).$$

19. Prove that the sequence space l^2, as defined in the text, is a Hilbert space.

20. Prove that l^2 is separable.

21. Prove that a real normed linear space L in which the parallelogram law holds is an inner-product space. This is to say, the parallelogram law characterizes inner-product spaces. Go at the problem like this: Start with a normed linear space L in which $\forall\mathbf{x},\mathbf{y}$, $\|\mathbf{x}+\mathbf{y}\|^2+\|\mathbf{x}-\mathbf{y}\|^2=2\|\mathbf{x}\|^2+2\|\mathbf{y}\|^2$. Can you define an inner product $(\mathbf{x},\mathbf{y})$ such that $\forall\mathbf{x}$, $(\mathbf{x},\mathbf{x})=\|\mathbf{x}\|^2$? The inner product must satisfy the four inner-product axioms; the corollary statement immediately after Theorem 7.1.2 tells how you must define $(\mathbf{x},\mathbf{y})$.

Having defined $(\mathbf{x},\mathbf{y})$ to be $\frac{1}{4}(\|\mathbf{x}+\mathbf{y}\|^2-\|\mathbf{x}-\mathbf{y}\|^2)$, show that the axioms I2 and I4 are satisfied, that $(\mathbf{x},\mathbf{x})=\|\mathbf{x}\|^2$, and that $(\mathbf{x},\mathbf{0})=(\mathbf{0},\mathbf{y})=0$. By applying the parallelogram law to $\mathbf{x}+\mathbf{z}$ and $\mathbf{y}+\mathbf{z}$, $\mathbf{x}-\mathbf{z}$ and $\mathbf{y}-\mathbf{z}$, respectively, see if you can obtain

$$\begin{aligned}(\mathbf{x},\mathbf{z})+(\mathbf{y},\mathbf{z})&=\tfrac{1}{4}\left[\|\mathbf{x}+\mathbf{z}\|^2+\|\mathbf{y}+\mathbf{z}\|^2\right]-\tfrac{1}{4}\left[\|\mathbf{x}-\mathbf{z}\|^2+\|\mathbf{y}-\mathbf{z}\|^2\right]\\&=\tfrac{1}{8}\left[\|\mathbf{x}+\mathbf{z}+\mathbf{y}+\mathbf{z}\|^2+\|\mathbf{x}+\mathbf{z}-\mathbf{y}-\mathbf{z}\|^2\right]\\&\quad-\tfrac{1}{8}\left[\|\mathbf{x}-\mathbf{z}+\mathbf{y}-\mathbf{z}\|^2+\|\mathbf{x}-\mathbf{z}-\mathbf{y}+\mathbf{z}\|^2\right]\\&=\tfrac{1}{8}\left[\|\mathbf{x}+\mathbf{y}+2\mathbf{z}\|^2-\|\mathbf{x}+\mathbf{y}-2\mathbf{z}\|^2\right]=\tfrac{1}{2}(\mathbf{x}+\mathbf{y},2\mathbf{z}).\end{aligned}$$

Set $\mathbf{y}=\mathbf{0}$ to get, $\forall\mathbf{x},\mathbf{z}$, $(\mathbf{x},\mathbf{z})=\frac{1}{2}(\mathbf{x},2\mathbf{z})$. Now replace $\mathbf{x}$ by $\mathbf{x}+\mathbf{y}$ to get $(\mathbf{x}+\mathbf{y},\mathbf{z})=(\mathbf{x},\mathbf{z})+(\mathbf{y},\mathbf{z})$, and axiom I1 is satisfied. Use a continuity argument to prove $(t\mathbf{x},\mathbf{y})=t(\mathbf{x},\mathbf{y})$. Start by letting t be rational, $t=m/n$. Put $\mathbf{x}'=\mathbf{x}/n$, so that

$(t\mathbf{x},\mathbf{y})=(m\mathbf{x}',\mathbf{y})$. For all $\mathbf{x}'$, $\mathbf{y}$, we have $(m\mathbf{x}',\mathbf{y})=m(\mathbf{x}',\mathbf{y})$, by additivity. Show $n(\mathbf{x}',\mathbf{y})=(\mathbf{x},\mathbf{y})$, and it follows that $(m\mathbf{x}/n,\mathbf{y})=(m/n)(\mathbf{x},\mathbf{y})$. Now apply the continuity argument.

22. Prove that for nonzero vectors $\mathbf{x}$, $\mathbf{y}$ in an inner-product space with its natural norm, $\|\mathbf{x}+\mathbf{y}\|=\|\mathbf{x}\|+\|\mathbf{y}\|$ iff $\mathbf{x}$ is a positive multiple of $\mathbf{y}$.

23. Prove that if $(\mathbf{x},\mathbf{z})=(\mathbf{y},\mathbf{z})$ for every $\mathbf{z}$ in an inner-product space, then $\mathbf{x}=\mathbf{y}$.

24. Prove that in an inner-product space

$$\mathbf{x}\perp\mathbf{y} \quad\Leftrightarrow\quad \|\mathbf{x}+t\mathbf{y}\|=\|\mathbf{x}-t\mathbf{y}\|$$
$$\Leftrightarrow\quad \|\mathbf{x}+t\mathbf{y}\|\geqslant\|\mathbf{x}\|$$

for each scalar t.

25. Let $\{(1,1,-2,0,1),\ (-3,1,2,1,4),\ (0,1,1,1,0),\ (-1,0,1,0,1),\ (4,1,-1,-2,-3)\}$ be an ordered set of vectors in E^5. Find an orthonormal set $(\mathbf{e}_1,\ldots,\mathbf{e}_5)$ such that the span of the first k, $k=1,\ldots,5$, of the unit vectors is the same as the span of the first k given vectors. (Are the given vectors independent?)

26. Write a Fourier series for

$$f(x)=\begin{cases}-1, & -\pi<x<0,\\ 1, & 0<x<\pi,\\ 0, & x=-\pi,0,\pi.\end{cases}$$

27. Find the Fourier series for $f(x)=e^x$, $-\pi<x<\pi$.

28. Do the same for the function 2^x+1, $-\pi<x<\pi$.

29. If $F(x)$ is an L^2-function on the interval $[-m,m]$, put $t=\pi x/m$ and define $f(t)=F(x)$. Obtain a Fourier series for $f(t)$, $-\pi\leqslant t\leqslant\pi$, and then replace t by $\pi x/m$; you will have a Fourier series for $F(x)$ on $[-m,m]$. Suppose $F(x)=x^2-x+\frac{1}{6}$ if $0<x<1$ and $F(x+1)=F(x)$ for all x. Show that F is an even function. Write a Fourier series for F on $[-1,1]$.

30. Find the Fourier series on $[0,3]$ for

$$F(x)=\begin{cases}x/2, & 0\leqslant x\leqslant 2,\\ -x+3, & 2\leqslant x\leqslant 3.\end{cases}$$

31. Suppose a is not an integer and let $f(x)=\cos(ax)$, $x\in[-\pi,\pi]$. Find the Fourier series for $\cos(ax)$. You should get

$$\cos ax=\frac{2a\sin a\pi}{\pi}\left[\frac{1}{2a^2}+\sum_{n=1}^{\infty}(-1)^n\frac{\cos nx}{a^2-n^2}\right].$$

Now put $x=0$, and derive the formula

$$\frac{\pi}{\sin \pi a}=\frac{1}{a}+2a\sum_{n=1}^{\infty}\frac{(-1)^{n+1}}{n^2-a^2}.$$

Finally, set $x=\pi$, and obtain a formula for $\cot(a\pi)$, $a\notin\mathbb{Z}$.

32. Compute $\int_\gamma(\ln z+e^z)\,dz$, where γ is the circle in the complex plane given by $|z|=1$.

33. Let $f(x)=1$ if x is irrational and $f(x)=-1$ if x is rational. You will see in the next chapter that f is integrable. Can you write its Fourier series?

34. If z is the complex number $x+iy$, show that e^z is never zero.

35. Suppose that z is complex and $|z|<1$. Show that

$$\ln(1+z)=\tfrac{1}{2}\ln(1+2r\cos\theta+r^2)+i\arctan[(r\sin\theta)/(1+r\cos\theta)],$$

where $r=|z|$ and $\theta=\arg(z)$.

36. Show that

$$\tfrac{1}{2}+\cos\theta+\cos 2\theta+\cdots+\cos n\theta=\frac{\sin[(n+\frac{1}{2})\theta]}{2\sin(\theta/2)}.$$

[*Hint*: $\cos k\theta=\frac{1}{2}(e^{ik\theta}+e^{-ik\theta})$, $\sin k\theta=(1/2i)(e^{ik\theta}-e^{-ik\theta})$. Convert the left-hand side to

$$\tfrac{1}{2}[e^{-in\theta}+e^{-i(n-1)\theta}+\cdots+e^{-i\theta}+1+e^{i\theta}+\cdots+e^{in\theta}]$$
$$=\tfrac{1}{2}[(e^{i(n+1)\theta}-e^{-in\theta})/(e^{i\theta}-1)].]$$

37. Write a polynomial of degree $\leqslant 5$ whose values at $\{-2,-1,0,1,2,3\}$ are, respectively, $\{1,4,-1,2,1,0\}$.

38. Let $(1,-3,0,-1,4)$, $(1,1,1,0,1)$, $(-2,2,1,1,-1)$, $(0,1,1,0,-2)$ be four vectors in E^5. Evaluate the Gram determinant of these four vectors, and find the associated diagonal determinant. Find the four-dimensional measure of the parallelotope spanned by these four vectors.

39. Evaluate $(1,1,1)\times(1,2,3)+(1,-2,-1)\times(1,-3,4)$.

40. Find the area of the parallelogram spanned by $(1,-1,4)$ and $(2,3,8)$.

41. Let $\mathbf{A}$, $\mathbf{B}$, $\mathbf{C}$ be noncoplanar vectors in E^3 with endpoints A, B, C. Why is, or is not, the vector $\mathbf{V}=\frac{1}{2}(\mathbf{B}-\mathbf{A})\times(\mathbf{C}-\mathbf{A})$ perpendicular to the plane of the triangle ABC, and of length $\|\mathbf{V}\|=\text{area}(ABC)$?

42. Find the area of the triangle ABC in E^3, where $\mathbf{A}=(1,4,1)$, $\mathbf{B}=(2,3,5)$, $\mathbf{C}=(-7,9,-1)$.

43. Let $\mathbf{A}=(a_1,a_2,a_3)$, $\mathbf{B}=(b_1,b_2,b_3)$, $\mathbf{C}=(c_1,c_2,c_3)$, and $\mathbf{D}=\mathbf{A}\times\mathbf{B}+\mathbf{B}\times\mathbf{C}+\mathbf{C}\times\mathbf{A}$. Find $\mathbf{D}\cdot(\mathbf{B}-\mathbf{A})$ and $\mathbf{D}\cdot(\mathbf{C}-\mathbf{A})$.

44. Show that $(\mathbf{A}\times\mathbf{B})\cdot\mathbf{C}=\mathbf{A}\cdot(\mathbf{B}\times\mathbf{C})$ and that $(\mathbf{A}\times\mathbf{B})\cdot(\mathbf{B}\times\mathbf{C})\times(\mathbf{C}\times\mathbf{A})=(\mathbf{A}\cdot\mathbf{B}\times\mathbf{C})^2$.

45. Write a Fourier series for $f(x)=x$, $0\leqslant x\leqslant\pi$.
[*Hint*: Let $F(t)=f(x)$, where $t=2x$, and expand $F(t)$ over $[0,2\pi]$; then replace t by $2x$. In this case, $F(t)=t/2$.]

46. Show that $\ln|\sin(x/2)|=-\ln 2-\Sigma_{n=1}^{\infty}(1/n)\cos nx$, $x\neq 2n\pi$.

47. If $f\in L^2(0,2\pi)$, then $\|f\|^2/\pi=\alpha_0^2/2+\Sigma_{n=1}^{\infty}(\alpha_n^2+\beta_n^2)$, where these α_i and β_i are the usual Fourier coefficients for f. (This is Parseval's formula.) Consider the series $\Sigma_{n=1}^{\infty}1/n^2$; find an appropriate Fourier series and a function f to go with the series so that you can evaluate

$$\sum_{n=1}^{\infty}\frac{1}{n^2}=\frac{1}{\pi}\int_0^{2\pi}|f(x)|^2\,dx=\frac{1}{\pi}\|f\|^2.$$

Then $\Sigma_{n=1}^{\infty}1/n^4$ and $\Sigma_{n=1}^{\infty}1/n^6$ should be evaluated in a similar fashion.

48. Find a proof, or create your own, for Lemma 7.4.1.

49. Find an infinite-dimensional inner-product space V with a complete orthonormal set S such that S will also be a Hamel basis for V. (Consider the set of sequences all but a finite number of terms of which are zero. Find an inner product for this set; is the resulting space complete?)

50. Suppose $\mathfrak{F}$ is a set of vectors, and we have made $\mathfrak{F}$ into a partially ordered normed linear space with a positive cone. For example, $\mathfrak{F}$ could be the collection of real functions integrable on $[a,b]$, $a,b\in\mathbb{R}$, normed by $\|f\|=\int_a^b|f(x)|\,dx$. (We call this the L^1-norm.) If $\mathfrak{F}$ is such that for every pair of elements $\mathbf{f},\mathbf{g}$ in the positive cone we have $\|\mathbf{f}+\mathbf{g}\|=\|\mathbf{f}\|+\|\mathbf{g}\|$, $\mathfrak{F}$ is called an L-space. Give some examples of L-spaces. Is $L^2(a,b)$ and L-space? Show that if an inner-product space is an L-space, then for all $\mathbf{x},\mathbf{y}$ in the positive cone C, we have $\|\mathbf{x}-\mathbf{y}\|=|\|\mathbf{x}\|-\|\mathbf{y}\||$. Deduce that in such an inner-product space, if $\mathbf{x},\mathbf{y}\in C$, then $\mathbf{x}$ is a positive scalar multiple of $\mathbf{y}$, so that the positive cone is at most a ray.

51. Let p be the linear space of real polynomials in x. What is the dimension of p? What is the "Hamel dimension" of p? Can you find a norm for p? If you can norm p, is p a Banach space under this norm? Can you find an inner product for p? If you can, is p a Hilbert space with this inner product? Finally, try to prove that if B is an infinite-dimensional Banach space with a Hamel basis H, then $\text{card}(H)>\aleph_0$.

52. Refer to Theorem 7.4.2 and try to find an integrable function f defined on $[-\pi,\pi]$ for which the Fourier series does not converge; you may have to search the literature to find one. You will see in a later chapter that the trigonometric series $\Sigma_{n=2}^{\infty}\sin nx/\ln n$ converges for all $x\in[-\pi,\pi]$ to a function $f(x)$. Suppose f is integrable. Then $F(x)=\int_0^x f(t)\,dt$ is absolutely continuous and periodic, and hence there is a convergent Fourier series for $F(x)$. Since $f(x)$ is odd, $F(x)$ is even, and $F'(x)=f(x)$, except possibly for a set of measure zero. Find the Fourier coefficients for $F(x)=\alpha_0/2+\Sigma_{n=1}^{\infty}\alpha_n\cos nx$. This series converges in particular at $x=0$; does $\Sigma_{n=1}^{\infty}\alpha_n$ converge? This shows that f must not have been integrable. It can be shown that for no integrable function is this convergent trigonometric series the Fourier series.

CHAPTER

8

MEASURE AND INTEGRATION

The basic theory of the Riemann integral in E^n is an old story; what we would like to do in this chapter is to give a somewhat extended notion of the integral of a real-valued function f over a set $A \subset E^n$. As you push on in mathematics you will meet many different kinds of integrals, several of which we shall describe for you, but none is so fundamental as the Riemann integral.

Just what is the integral of a function? At the risk of oversimplification we shall say that the integral of a real function over a set A is the average value that the function assumes on A multiplied by the measure of A. If a body B with mass m has a measure μ, then the average density of B is m/μ. If a distance d is covered in time t by a particle moving somewhat erratically, the average rate of the particle is d/t. Thus, if the density of a body is a function of position in the body, then the integral over B of the density function is m, the value of the mass, and if the rate of a particle is a function of time, the integral of the rate function over a time interval is the value d of the distance covered by the particle during that time interval. Hence integration involves an averaging process.

Suppose a body B has mass m and measure μ and, at each point $p \in B$, $\delta(p)$ is a function which gives the density of B at p. Now, the reader should realize that the term density is actually defined to be the ratio of mass to measure, so the value $\delta(p)$, the density at p, is really an "instantaneous" density. This is to say, $\delta(p)$ is the limit of the ratios of mass to measure of a decreasing sequence of increments of B, $\Delta_n B$, each of which contains p, as the measures of these increments tend to zero.

$$\delta(p) = \lim\{m(\Delta_n B)/\mu(\Delta_n B) : p \in \Delta_n B, \forall n, \text{ and } \{\Delta_n B\} \downarrow \{p\}\}.$$

Hence a density function is a kind of derivative.

Suppose B has been partitioned into the collection of subsets $\{B_i\}_{i=1}^n$. Then the average density δ_i of each B_i is m_i/μ_i, and we have $m_i=\delta_i\mu_i$. It follows that $m=\sum_{i=1}^n m_i=\sum_{i=1}^n \delta_i\mu_i$. Note that the average density δ of B is equal to the "weighted average" of the average densities δ_i:

$$\sum_{i=1}^{n} \delta_i\cdot(\mu_i/\mu)\Big/\sum_{i=1}^{n}(\mu_i/\mu)=\frac{1}{\mu}\sum_{i=1}^{n}\delta_i\mu_i\Big/1$$
$$=m/\mu=\delta.$$

Now, if the mass of B is known, we already know the integral of the density function over B; it is simply the mass. But if the mass is unknown, what we have said suggests that we might be able to approximate the value of the integral of $\delta(p)$ over B by partitioning B into n subsets B_i, each subset being small enough in measure so that $\delta(p)$ never strays too far from its average value δ_i on B_i, and evaluating the sum

$$\sum_{i=1}^{n}\delta(p_i)(\text{measure of } B_i),\qquad \forall i,\ p_i\in B_i.$$

Analogously, we ought to be able to approximate the integral of a rate function over a time interval by the same sort of process.

Unfortunately, this heuristic approach to the notion of integral is neither precise enough nor sufficiently general for us, but it does show the way. We shall begin like this: Suppose f is a real-valued function defined on a set $A\subset E^n$. By a partition of A we mean a finite collection of nonempty subsets of A, $\{A_i\}_{i=1}^n$, such that $A=\bigcup A_i$ and $A_i\cap A_j=\varnothing$, $i\neq j$. Suppose P is the class of all partitions of A; P may be partially ordered in the following way. If $\pi_1,\pi_2\in P$, we say π_2 is *finer* than π_1, $\pi_1\prec\pi_2$, if every set of the partition π_1 is a union of sets of the partition π_2. Generally, two partitions of A are not comparable, but there is always a common refinement $\pi=\pi_1\vee\pi_2$, where π is the coarsest partition of A such that each set of π_1 and each set of π_2 is a union of sets of π. Actually, P is a lattice with this ordering, but it is enough for us to note that P is an upward-directed set.

If $\pi_\alpha\in P$ is an arbitrary partition of A, let $S_\alpha=\sum_{i=1}^n f(p_i)$ (measure of A_i), where $\pi_\alpha=\{A_i\}_{i=1}^n$ and $p_i\in A_i$ for each i. Now S_α is not well defined because we have not specified how the points p_i are chosen once the partition has been made, but this will not matter. We assume that for each partition, the points are determined by some rule. Now the collection $\{S_\alpha\}$ is a net (Definition 2.2.1) and this net converges iff for each $\varepsilon>0$, there is an index α_0 such that for every β, $\alpha_0\prec\beta$, $|S_{\alpha_0}-S_\beta|<\varepsilon$. ($\alpha\prec\beta\Leftrightarrow\pi_\alpha\prec\pi_\beta$.)

Hence, we make this preliminary definition: if the limit of the net $\{S_\alpha\}$ exists, is finite, and is independent of the choice of points p_i, we define the integral of f over A to be this limiting value.

This is a rather general definition for the Riemann integral of a real function f over a set A; it is consistent with the notion that the value of the integral divided by the measure of the set A is the average value of the function f over A. However, it relies entirely on the fact that we can measure A and all the subsets of A. Unfortunately, we cannot always do this, as we shall see, so we need to investigate the notion of measure in order to put matters on firmer ground.

8.1 MEASURE

Almost everyone may have an intuitive idea what is meant by the measure of a set, but rest assured that the notion of measure is much more complicated than you might suspect. You have already met the most basic kind of measure, distance measure, when we introduced a "distance assigner" ρ to make a metric space out of a set. The function ρ assigned to each point pair $\{x, y\}$ in the set a certain real number in such a way as to satisfy a particular set of axioms, which axioms were chosen so that this abstract distance between the points would be consistent with our idea of real distance between two points. We spoke of ρ, the metric for a set V, as a real-valued function having $V \times V$ as a domain, but we could have called ρ a set function, i.e., a function having a class of sets as domain and values in $\mathbb{R}$, where specifically the domain of ρ would be the class of all doublets and singletons in $P(V)$.

Now we want to measure more general sets; a real-valued function μ, to be a measure, must be a set function which assigns real number values to sets, and it must do it in such a way as to be consistent with our idea of what the measure of a set should be. Certainly we should insist that the following axioms hold.

μ1. $\forall A \in \operatorname{dom} \mu$, $0 \leqslant \mu(A) \leqslant \infty$. $\varnothing \in \operatorname{dom} \mu$, and $\mu(\varnothing) = 0$.

μ2. $\forall A, B \in \operatorname{dom} \mu$, if $A \cup B \in \operatorname{dom} \mu$, $\mu(A \cup B) \leqslant \mu(A) + \mu(B)$. Moreover, if $A \cup B \in \operatorname{dom} \mu$ and $A \cap B = \varnothing$, $\mu(A \cup B) = \mu(A) + \mu(B)$. If $A \subset B$, then $\mu(A) \leqslant \mu(B)$.

These two axioms stipulate that $\operatorname{dom} \mu$ is not empty and that μ is actually an "extended" real-valued function which is additive on disjoint sets, but suppose A, B are nonempty sets in $\operatorname{dom} \mu$ with $A \subset B$ properly. Then B is the disjoint union of A and $B \setminus A$, but if $B \setminus A$ is not in the domain of μ, we cannot say that $\mu(B) = \mu(A) + \mu(B \setminus A)$, nor can we write, in case $A \cup B \neq B$, $\mu(A \cup B) = \mu(A) + \mu(B \setminus A)$. This is a real defect; hence we introduce an additional axiom.

μ3. The domain of μ is a ring of sets (Definition 1.2.3).

If we line up a countable set $\{B_n\}$ of boxes side by side and the measure of each box B_n is 2^{-n}, then the union of these boxes ought to have measure

1. Hence, we augment our set of axioms to include

μ4. $\forall\{A_i\}_{i=1}^{\infty} \subset \operatorname{dom}\mu$, if $A_i \cap A_j = \varnothing$, $i \neq j$, and $\bigcup_{i=1}^{\infty} A_i \in \operatorname{dom}\mu$, then $\mu(\bigcup_{i=1}^{\infty} A_i) = \Sigma_{i=1}^{\infty}\mu(A_i)$.

We are now ready to define what we mean by a measure.

Definition 8.1.1. Let V be a set and $R \subset P(V)$ a ring of sets. A set function μ defined on R which satisfies axioms $\mu 1-\mu 4$ is a *measure* on R.

As an example of a measure, we let f be an extended real-valued nonnegative function defined on E^1 and let R be the ring of all finite subsets of E^1. For each $\{x_1,\ldots,x_n\} \in R$, define $\mu(\{x_1,\ldots,x_n\}) = \Sigma_{i=1}^{n} f(x_i)$, and $\mu(\varnothing) = 0$.

Another rather trivial example is this one. Let V be a set and $R = P(V)$. For each $A \in R$, define $\mu(A) = \operatorname{card} A$.

It is clear that if μ is a measure on a ring R, then μ is a nonnegative increasing function in the sense that if $A, B \in R$ with $A \subset B$, then $\mu(A) \leqslant \mu(A) + \mu(B \setminus A) = \mu(B)$. Moreover, $\mu(B\setminus A) = \mu(B) - \mu(A)$, provided that $\mu(A) < \infty$.

Theorem 8.1.2. *If μ is a measure on a ring R, if $A \in R$, and if $\{A_i\}_{i=1}^{\infty}$ is any sequence of sets of R such that $A \subset \bigcup_{i=1}^{\infty} A_i$, then $\mu(A) \leqslant \Sigma_{i=1}^{\infty}\mu(A_i)$. On the other hand, if the sets of the sequence are mutually disjoint and $\bigcup A_i \subset A$, then $\Sigma_{i=1}^{\infty}\mu(A_i) \leqslant \mu(A)$.*

PROOF: Define $B_1 = A_1$ and $B_i = A_i \setminus \bigcup\{A_j : 1 \leqslant j < i\}$. Then the sets of the sequence $\{B_i\}$ are mutually disjoint, $B_i \subset A_i$ for each i, and $\bigcup B_i = \bigcup A_i$. Now $\mu(A) = \mu(\bigcup\{B_i \cap A\}) = \Sigma\mu(B_i \cap A)$. Since for each i, $\mu(B_i \cap A) \leqslant \mu(B_i) \leqslant \mu(A_i)$, it follows that $\mu(A) \leqslant \Sigma\mu(A_i)$.

On the other hand, if the A_i are disjoint and $\bigcup A_i \subset A$, for each n we have $\mu(\bigcup_{i=1}^{n} A_i) = \Sigma_{i=1}^{n}\mu(A_i) \leqslant \mu(A)$. The series $\Sigma_{i=1}^{\infty}\mu(A_i)$ is a series of nonnegative terms whose partial sums are bounded above by $\mu(A)$; hence it converges if $\mu(A) < \infty$, and we have $\Sigma_{i=1}^{\infty}\mu(A_i) \leqslant \mu(A)$. If $\mu(A) = \infty$, the inequality still holds. □

Theorem 8.1.3. *If μ is a measure on a ring R and $\{A_n\}_{n=1}^{\infty}$ is an increasing sequence of sets of R such that $\bigcup A_n \in R$, then $\mu(\bigcup A_n) = \mu(\lim_n A_n) = \lim_n \mu(A_n)$. On the other hand, if $\{A_n\}_{n=1}^{\infty} \subset R$ is a decreasing sequence for which $\mu(A_n) < \infty$ for at least one index n and $\lim_n A_n = \bigcap A_n \in R$, then $\mu(\bigcap A_n) = \mu(\lim_n A_n) = \lim_n \mu(A_n)$. This is to say, a measure μ on R is continuous from below and from above at each set $A \in R$.*

PROOF: Put $A_0 = \varnothing$ and write $\mu(\lim A_n) = \mu(\bigcup A_n) = \mu(\bigcup[A_n \setminus A_{n-1}]) = \Sigma\mu(A_n \setminus A_{n-1}) = \lim_n \Sigma_{i=1}^{n}\mu(A_i \setminus A_{i-1}) = \lim_n \mu(\bigcup_{i=1}^{n}[A_i \setminus A_{i-1}]) = \lim_n \mu(A_n)$.

In the event that $\{A_n\} \subset R$ is decreasing and for some m, $\mu(A_m) < \infty$, and $\bigcap A_n \in R$, we have $\mu(A_n) \leqslant \mu(A_m)$ for all $n \geqslant m$, so $\mu(\bigcap A_n) < \infty$. Now $\{A_m \setminus A_n\}$ is an increasing sequence of sets of R, and $\bigcup_{n=1}^{\infty}(A_m \setminus A_n) = A_m \setminus \bigcap_{n=1}^{\infty} A_n \in R$, so we can write

$$\mu\left(\bigcup_{n=1}^{\infty}[A_m \setminus A_n]\right) = \mu\left(A_m \setminus \bigcap_{n=1}^{\infty} A_n\right) = \mu(A_m) - \mu(\lim_n A_n)$$
$$= \mu(A_m \setminus \lim_n A_n) = \mu(\lim_n [A_m \setminus A_n])$$
$$= \lim_n \mu(A_m \setminus A_n) = \lim_n [\mu(A_m) - \mu(A_n)]$$
$$= \mu(A_m) - \lim_n \mu(A_n).$$

Hence $\mu(\lim_n A_n) = \lim_n \mu(A_n)$. □

Suppose $\mathcal{C}$ is a nonempty class of subsets of a universal set V. $\mathcal{C}$ may not be a ring, but it is easy to see that there is a unique ring (σ-ring) $\mathfrak{R}(\mathcal{C})$ ($\mathcal{S}(\mathcal{C})$) generated by $\mathcal{C}$, where $\mathfrak{R}(\mathcal{C})$ ($\mathcal{S}(\mathcal{C})$) is the smallest ring (σ-ring) of sets contained in $P(V)$ which contains $\mathcal{C}$. $P(V)$ contains $\mathcal{C}$ and is a σ-ring; consider the collection of all rings (σ-rings) in $P(V)$ which contain $\mathcal{C}$. The intersection of all these rings (σ-rings) is itself a ring (σ-ring) which contains $\mathcal{C}$, and it is clearly the smallest such ring. The reader may have to ponder over this a bit.

If we have a set function μ defined on a nonempty class $\mathcal{C}$ of sets, not necessarily a ring, which satisfies axioms $\mu 1$, $\mu 2$, and $\mu 4$, we can extend μ to the class $\mathfrak{R}(\mathcal{C})$ directly if $\mathcal{C}$ is a special class of sets called a semiring.

Definition 8.1.4. Suppose $\mathcal{C}$ is a nonempty class of sets having the following properties.

1. $\forall A, B \in \mathcal{C}$, $A \cap B \in \mathcal{C}$. In particular, $\varnothing \in \mathcal{C}$.
2. If $A, B \in \mathcal{C}$ and $A \subset B$, then there is a finite subcollection $\{A_0, A_1, \ldots, A_n\} \subset \mathcal{C}$ such that $A = A_0 \subset A_1 \subset \cdots \subset A_n = B$, and $\forall i$, $(A_i \setminus A_{i-1}) \in \mathcal{C}$.

Then $\mathcal{C}$ is a *semiring* of sets.

Now suppose $\mathcal{C}$ is a semiring of sets from $P(V)$. Let $\mathfrak{R}$ be the collection of sets which can be formed by taking unions of arbitrary finite collections of disjoint sets of $\mathcal{C}$. Then $\mathfrak{R}$ is a ring, and $\mathfrak{R} = \mathfrak{R}(\mathcal{C})$. This is far from obvious, so consider the following remarks carefully, and work out some of the details for yourself.

$\mathfrak{R}$ is closed with respect to finite intersections and disjoint unions. This is easy to see. Now refer to the definition above and note that if $A, B \in \mathcal{C}$ and $A \subset B$, then $B \setminus A \in \mathfrak{R}$. Next, it will follow that if $A \in \mathcal{C}$ and $B \in \mathfrak{R}$, and $A \subset B$, then $B \setminus A \in \mathfrak{R}$, and finally, if $A, B \in \mathfrak{R}$ and $A \subset B$, then $B \setminus A \in \mathfrak{R}$.

This means that $\mathcal{R}$ is closed with respect to differences. If $A, B \in \mathcal{R}$, then $A \cup B = (A \setminus B) \cup (A \cap B) \cup (B \setminus A) \in \mathcal{R}$, so $\mathcal{R}$ is a ring, and we must have $\mathcal{R} \supset \mathcal{R}(\mathcal{C})$. It is clear from the definition of $\mathcal{R}$ that $\mathcal{R} \subset \mathcal{R}(\mathcal{C})$, so we have the equality $\mathcal{R} = \mathcal{R}(\mathcal{C})$.

Having a set function μ defined on a semiring $\mathcal{C}$ satisfying axioms $\mu 1$, $\mu 2$, and $\mu 4$, we extend μ to $\mathcal{R}(\mathcal{C})$ in the obvious way. If $A \in \mathcal{R}(\mathcal{C})$, then A is a disjoint union of sets of $\mathcal{C}$

$$A = \bigcup_{i=1}^{n} \{A_i : A_i \in \mathcal{C}, A_i \cap A_j = \varnothing, i \neq j\},$$

so we define $\mu(A) = \sum_{i=1}^{n} \mu(A_i)$.

You will meet shortly an important semiring from $P(E^n)$, namely, the class $\mathcal{B}$ of "half-open" rational boxes. By a half-open rational box B we mean the set $B \subset E^n$ defined by

$$B = \bigtimes_{i=1}^{n} \{[a_i, b_i) : \forall i, a_i, b_i \in \mathbf{Q}, a_i < b_i\}.$$

If we can measure half-open boxes in E^n, it seems that we are in a position to measure any set in E^n that is a union of finitely many such boxes. But this is not enough; we want to be able to measure sets that are countable unions of half-open boxes. More generally, if we have a measure μ on a ring $\mathcal{R}$, we want to be able to extend μ at least to the σ-ring $\mathcal{S}(\mathcal{R})$ generated by $\mathcal{R}$. Recall that a σ-ring is a ring which is closed with respect to countable unions (and intersections as well). We proceed as follows.

We call a collection $\mathcal{H}$ of sets *hereditary* if $A \in \mathcal{H}$ and $B \subset A$ implies that $B \in \mathcal{H}$. Certainly the power set of a set S is hereditary. We denote $\mathcal{H}(\mathcal{R})$ the smallest hereditary σ-ring which contains $\mathcal{R}$, where $\mathcal{R}$ is a ring on which a measure μ is defined. If $\mathcal{S}(\mathcal{R})$ is the σ-ring generated by $\mathcal{R}$, it is clear that $\mathcal{S}(\mathcal{R}) \subset \mathcal{H}(\mathcal{R})$. That these σ-rings exist and are unique is not hard to see; each is the intersection of all σ-rings, or hereditary σ-rings, as the case may be, which contain $\mathcal{R}$.

We call an extended real-valued set function μ^* defined on a class of sets $\mathcal{C}$ *subadditive* if whenever $A, B, A \cup B \in \mathcal{C}$,

$$\mu^*(A \cup B) \leqslant \mu^*(A) + \mu^*(B),$$

equality not necessarily holding if $A \cap B = \varnothing$. We call μ^* *countably subadditive* if $\mu^*(\bigcup_{n=1}^{\infty} A_n) \leqslant \sum_{n=1}^{\infty} \mu^*(A_n)$ whenever $\{A_n\}$, $\bigcup A_n \in \mathcal{C}$. We say μ^* is increasing if $\mu^*(A) \leqslant \mu^*(B)$ whenever $A, B \in \mathcal{C}$ and $A \subset B$. Finally, we give

Definition 8.1.5. An extended real-valued nonnegative increasing countably subadditive set function μ^* defined on a hereditary σ-ring $\mathcal{H}$ such that $\mu^*(\varnothing) = 0$ is called an *outer measure* on $\mathcal{H}$.

Consider the following example: let X be a real line and $\mathcal{H} = P(X)$. If $A \subset X$, the *characteristic function* of A is that function χ defined on X by

$\chi_A(x)=1$ if $x\in A$, and 0 otherwise. Let $x_0\in X$ be fixed, and define for each $A\subset X$,

$$\mu^*(A)=\chi_A(x_0).$$

Then μ^* is nonnegative on $\mathcal{H}$, $\mu^*(\varnothing)=0$, μ^* is certainly increasing, and finally, it is countably subadditive. Hence μ^* is an outer measure. Is it also a measure? Now define ν^* on $\mathcal{H}$ as follows. For each $A\in P(X)$, let $\nu^*(A)=1$ if A contains a rational number, and $\nu^*(A)=0$ otherwise. Is ν^* an outer measure? Is it a measure?

Theorem 8.1.6. *Let μ be a measure on a ring $\mathcal{R}$ and define*

$$\forall A\in\mathcal{H}(\mathcal{R}),\qquad \mu^*(A)=\inf\left\{\sum_{n=1}^{\infty}\mu(A_n): A_n\in\mathcal{R}, A\subset\bigcup_{n=1}^{\infty}A_n\right\}.$$

Then μ^* is an outer measure on $\mathcal{H}(\mathcal{R})$, and $\mu^*=\mu$ on $\mathcal{R}$. We say that μ^* is the outer measure induced by μ.

PROOF: It is a fact that the class of all sets which can be covered by countably many sets of $\mathcal{R}$ is a σ-ring, in fact, a hereditary σ-ring, which is to say, it contains $\mathcal{H}(\mathcal{R})$. Hence, if $A\in\mathcal{H}(\mathcal{R})$, there does exist a countable collection $\{A_n\}_{n=1}^{\infty}\subset\mathcal{R}$ such that $A\subset\bigcup_{n=1}^{\infty}A_n$.

If $A\in\mathcal{R}$, then $A=A\cup\varnothing\cup\varnothing\cup\cdots$, and $\mu^*(A)\leqslant\mu(A)+\mu(\varnothing)+\cdots=\mu(A)$. On the other hand, if $A\subset\bigcup A_n$, where each $A_n\in\mathcal{R}$, $\mu(A)\leqslant\Sigma\mu(A_n)$, and in particular,

$$\mu(A)\leqslant\inf\left\{\sum_{n=1}^{\infty}\mu(A_n): A\subset\bigcup A_n\right\}=\mu^*(A),$$

so we have $\mu(A)=\mu^*(A)$ on $\mathcal{R}$, and $\mu^*(\varnothing)=0$.

If $A,B\in\mathcal{H}(\mathcal{R})$, with $A\subset B$, it is clear that $\mu^*(A)\leqslant\mu^*(B)$ and that for any $A\in\mathcal{H}(\mathcal{R})$, $\mu^*(A)\geqslant 0$. Now suppose $A\in\mathcal{H}(\mathcal{R})$ and $\{A_n\}_{n=1}^{\infty}\subset\mathcal{H}(\mathcal{R})$ such that $A\subset\bigcup A_n$. Let $\varepsilon>0$. For each n, choose a sequence of sets $\{A_{nj}\}_{j=1}^{\infty}$ from $\mathcal{R}$ such that $A_n\subset\bigcup_{j=1}^{\infty}A_{nj}$ and $\Sigma_j\mu(A_{nj})\leqslant\mu^*(A_n)+\varepsilon/2^n$. This can be done since $\mu^*(A_n)$ is defined to be the infimum of such sums $\Sigma_j\mu(A_{nj})$. Since the collection $\{A_{nj}\}$ is a countable cover for A, we have

$$\mu^*(A)\leqslant\sum_{n=1}^{\infty}\sum_{j=1}^{\infty}\mu(A_{nj})\leqslant\sum_{n=1}^{\infty}\left[\mu^*(A_n)+\varepsilon/2^n\right]\leqslant\sum_{n=1}^{\infty}\mu^*(A_n)+\varepsilon.$$

Since $\varepsilon>0$ was arbitrary, we conclude that $\mu^*(A)\leqslant\Sigma_{n=1}^{\infty}\mu^*(A_n)$ and μ^* is an outer measure on $\mathcal{H}(\mathcal{R})$. $\square$

This theorem assures us that a measure μ on a ring $\mathcal{R}$ can be extended to an outer measure μ^* on $\mathcal{H}(R)$. Now in E^n, the semiring $\mathcal{B}$ of half-open boxes generates a σ-ring which contains E^n itself since E^n is contained in a countable union of boxes of $\mathcal{B}$. Hence $\mathcal{H}(\mathcal{B})$ is actually the power set of

E^n, so every subset of E^n has an outer measure, assuming we have already assigned a measure to boxes. But an outer measure may not be a measure since it need not be additive on disjoint unions; hence we must push on.

Definition 8.1.7. Let μ^* be an outer measure on a hereditary σ-ring $\mathcal{H}$. A set $M \in \mathcal{H}$ is said to be *μ^*-measurable* if for every set $A \in \mathcal{H}$,

$$\mu^*(A) = \mu^*(A \cap M) + \mu^*(A \cap \tilde{M}).$$

What this definition says is that the μ^*-measurable sets of $\mathcal{H}$ are those sets M which with their respective complements partition each set of $\mathcal{H}$ into a pair of subsets for which the outer measure is additive.

The following theorem is an important one for our development, and the proof is a marvelous exercise in manipulation. The author is indebted to Paul Halmos for this proof; it is too good not to pass on.

Theorem 8.1.8. *Let μ^* be an outer measure on a hereditary σ-ring $\mathcal{H}$ and $\mathcal{S}$ the class of all μ^*-measurable sets of $\mathcal{H}$. Then $\mathcal{S}$ is a σ-ring which contains all the sets of $\mathcal{H}$ having outer measure* 0. *If $A \in \mathcal{H}$ and $\{M_n\}_{n=1}^{\infty}$ is a collection of mutually disjoint sets in $\mathcal{S}$ with $M = \cup M_n$, then*

$$\mu^*(M \cap A) = \sum_{n=1}^{\infty} \mu^*(M_n \cap A).$$

PROOF: We first show that $\mathcal{S}$ is a ring. For any $A \in \mathcal{H}$,

$$\mu^*(A) = \mu^*(A \cap \varnothing) + \mu^*(A \cap \tilde{\varnothing}),$$

so $\varnothing \in \mathcal{S}$. If $M, N \in \mathcal{S}$ and $A \in \mathcal{H}$, then

$$\mu^*(A) = \mu^*(A \cap M) + \mu^*(A \cap \tilde{M}), \tag{8.1.1}$$

$$\mu^*(A \cap M) = \mu^*(A \cap M \cap N) + \mu^*(A \cap M \cap \tilde{N}), \tag{8.1.2}$$

$$\mu^*(A \cap \tilde{M}) = \mu^*(A \cap \tilde{M} \cap N) + \mu^*(A \cap \tilde{M} \cap \tilde{N}). \tag{8.1.3}$$

Substituting (8.1.2) and (8.1.3) into (8.1.1) we have

$$\mu^*(A) = \mu^*(A \cap M \cap N) + \mu^*(A \cap M \cap \tilde{N}) + \mu^*(A \cap \tilde{M} \cap N) + \mu^*(A \cap \tilde{M} \cap \tilde{N}). \tag{8.1.4}$$

Now, replace A by $A \cap (M \cup N)$ in (8.1.4) to get

$$\mu^*(A \cap (M \cup N)) = \mu^*(A \cap M \cap N) + \mu^*(A \cap M \cap \tilde{N}) + \mu^*(A \cap \tilde{M} \cap N) + \mu^*(\varnothing). \tag{8.1.5}$$

Now put (8.1.5) into (8.1.4) and obtain

$$\begin{aligned} \mu^*(A) &= \mu^*(A \cap (M \cup N)) + \mu^*(A \cap \tilde{M} \cap \tilde{N}) \\ &= \mu^*(A \cap (M \cup N)) + \mu^*(A \cap (\widetilde{M \cup N})). \end{aligned} \tag{8.1.6}$$

This proves that $M \cup N \in \mathcal{S}$. Now, if we replace A in (8.1.4) by $A \cap (\widetilde{M \setminus N}) = A \cap (\tilde{M} \cup N)$ we get

$$\mu^*(A \cap (\widetilde{M \setminus N})) = \mu^*(A \cap (\tilde{M} \cup N)) = \mu^*(A \cap M \cap N) + \mu^*(\varnothing) + \mu^*(A \cap \tilde{M} \cap N) + \mu^*(A \cap \tilde{M} \cap \tilde{N}). \tag{8.1.7}$$

Since $A \cap M \cap \tilde{N} = A \cap (M \setminus N)$, we can put (8.1.7) into (8.1.4) to get

$$\mu^*(A) = \mu^*(A \cap (M \setminus N)) + \mu^*(A \cap (\widetilde{M \setminus N})), \tag{8.1.8}$$

which proves that $M \setminus N \in \mathcal{S}$. Thus $\mathcal{S}$ is a ring.

If $\{M_i\}_{i=1}^{\infty}$ is a sequence of disjoint sets of $\mathcal{S}$ with $M = \cup M_i$, then for each n we have

$$\mu^*\left(A \cap \bigcup_{i=1}^{n} M_i\right) = \sum_{i=1}^{n} \mu^*(A \cap M_i).$$

To see this, note that $M_1 \cap M_2 = \varnothing$, $M_1 \cap \tilde{M}_2 = M_1$, and $M_2 \cap \tilde{M}_1 = M_2$, so from (8.1.5) we get

$$\mu^*(A \cap (M_1 \cup M_2)) = \mu^*(A \cap M_1) + \mu^*(A \cap M_2).$$

The equation above follows by induction. Now put $N_n = \cup_{i=1}^{n} M_i$, so $N_n \in \mathcal{S}$, and we can write

$$\mu^*(A) = \mu^*(A \cap N_n) + \mu^*(A \cap \tilde{N}_n) \geqslant \sum_{i=1}^{n} \mu^*(A \cap M_i) + \mu^*(A \cap \tilde{M}).$$

Since this inequality holds for every n, we can write

$$\mu^*(A) \geqslant \sum_{i=1}^{\infty} \mu^*(A \cap M_i) + \mu^*(A \cap \tilde{M}) \geqslant \mu^*(A \cap M) + \mu^*(A \cap \tilde{M}). \tag{8.1.9}$$

Since $A \cap M$, $A \cap \tilde{M} \in \mathcal{H}$ and $A = (A \cap M) \cup (A \cap \tilde{M})$, we always have $\mu^*(A) \leqslant \mu^*(A \cap M) + \mu^*(A \cap \tilde{M})$, so equality must hold. Hence we have that $M \in \mathcal{S}$, so $\mathcal{S}$ is closed under countable disjoint unions, and it follows that

$$\sum_{i=1}^{\infty} \mu^*(A \cap M_i) + \mu^*(A \cap \tilde{M}) = \mu^*(A \cap M) + \mu^*(A \cap \tilde{M}) = \mu^*(A). \tag{8.1.10}$$

Now replace A by $M \cap A$ in (8.1.10) to get

$$\mu^*(M \cap A) = \sum_{i=1}^{\infty} \mu^*(M \cap A \cap M_i) + \mu^*(M \cap A \cap \tilde{M}) = \sum_{i=1}^{\infty} \mu^*(M_i \cap A). \tag{8.1.11}$$

Suppose $N \in \mathcal{H}$ and that $\mu^*(N)=0$. Then for every $A \in \mathcal{H}$ we have

$$\mu^*(A)=\mu^*(N)+\mu^*(A) \geqslant \mu^*(A \cap N)+\mu^*(A \cap \tilde{N}).$$

Since the reverse inequality must hold, we have equality, and $N \in \mathcal{S}$.

Since every countable union of sets of $\mathcal{S}$ may be written as countable union of disjoint sets of $\mathcal{S}$, we conclude that $\mathcal{S}$ is indeed a σ-ring, and the proof of the theorem is complete. □

As a consequence of the fact that $\mathcal{S}$ is a σ-ring, the last statement of the theorem tells us that if $\{M_n\}_{n=1}^{\infty}$ is any sequence of disjoint sets of $\mathcal{S}$, then $M=\bigcup_{n=1}^{\infty} M_n \in \mathcal{S} \subset \mathcal{H}$, and $\mu^*(M)=\sum_{n=1}^{\infty}\mu^*(M_n)$. Hence the following corollary is immediate.

Corollary 8.1.9. *μ^* restricted to $\mathcal{S}$ is a measure on $\mathcal{S}$.*

Consider this example. Let A be the set of algebraic numbers in the interval $[0,1]$. (An algebraic number is one that is a zero, or a root, of a polynomial having rational coefficients.) A is a countable set, so each point of A can be indexed: $A=\{a_n\}_{n=1}^{\infty}$. Let $\mathcal{B}$ be the semiring of half-open intervals in E^1; if $[a,b)$ is a typical interval, let $\mu([a,b))=b-a$. Let $\varepsilon>0$. A can be covered by a countable union of intervals,

$$A \subset \bigcup_{n=1}^{\infty} \{[a_n, b_n) : a_n \in A, a_n < b_n\}.$$

Suppose that for each n, $b_n - a_n = \varepsilon/2^n$. Then $\mu^*(A) \leqslant \sum \mu([a_n, b_n)) = \varepsilon$. This implies that $\mu^*(A)=0$, so A is μ^*-measurable. Moreover, every subset of A is μ^*-measurable.

We have remarked that every subset of E^n has an outer measure induced by the natural measure of the half-open rational boxes, but we must state here and now that not every set is measurable. This is to say, the class $\mathcal{S}$ of μ^*-measurable sets of E^n is not all of $P(E^n)$. We shall give an example later on of a set in E^1 which is not measurable.

This section is not meant to be a rigorous treatment of measure theory. Our purpose has been simply to give the reader an idea of how a theory of measure can be developed. Let's see how far we have come. Suppose we have a set, say E^n, from which we can pick out a certain class of sets $\mathcal{B}$ which has the structure of a semiring of sets, and we have a well-defined set function μ on $\mathcal{B}$ which assigns a measure to each set of $\mathcal{B}$ in such a way that axioms $\mu 1$, $\mu 2$, and $\mu 4$ are satisfied. We can extend μ to a ring $\mathcal{R}=\mathcal{R}(\mathcal{B})$, so that μ is now a measure on $\mathcal{R}$. This measure μ induces an outer measure μ^* on the hereditary σ-ring $\mathcal{H}(\mathcal{R})$ generated by $\mathcal{R}$, but μ^* is not in general countably additive on $\mathcal{H}(\mathcal{R})$. There is, however, a σ-ring $\mathcal{S} \subset \mathcal{H}(\mathcal{R})$, with $\mathcal{R} \subset \mathcal{S}(\mathcal{R}) \subset \mathcal{S}$, such that on $\mathcal{S}$, μ^* is countably additive. On $\mathcal{R}$, $\mu^*=\mu$, and $\mathcal{S}$ contains all the sets of $\mathcal{H}(\mathcal{R})$ which have outer measure zero. If we denote

μ^*, restricted to $\mathcal{S}$, by $\bar{\mu}$, $\bar{\mu}$ is a measure on $\mathcal{S}$, and $\bar{\mu}$ agrees with μ on $\mathcal{R}$, so $\bar{\mu}$ is indeed an extension of μ to all of $\mathcal{S}$.

We shall henceforth call $\mathcal{S}$ the class of measurable sets of our original set, and $\bar{\mu}$, the measure on $\mathcal{S}$, will be denoted simply μ.

Definition 8.1.10. Suppose μ is a measure on a ring $\mathcal{R}$, where $\mathcal{R}$ may be a σ-ring. We say μ is *finite* if $\forall A \in \mathcal{R}$, $\mu(A) < \infty$. We call μ *σ-finite* if $\forall A \in \mathcal{R}$, $\exists \{A_n\}_{n=1}^{\infty} \subset \mathcal{R}$ s.t. $\mu(A_n) < \infty$, $n = 1, 2, 3, \ldots$, and $A \subset \bigcup A_n$. We call the measure μ *complete* if $P(A) \subset \mathcal{R}$ whenever $A \in \mathcal{R}$ and $\mu(A) = 0$.

We bring this section to a close by stating a theorem which we do not prove; it is given simply for reference.

Theorem 8.1.11. (a) *If μ is a measure on a ring $\mathcal{R}$, then there is an extension $\bar{\mu}$ of μ to the σ-ring $\mathcal{S}(\mathcal{R})$. If μ is σ-finite, the extension $\bar{\mu}$ is unique, and $\bar{\mu}$ is also σ-finite.*

(b) *If μ is a measure on a σ-ring $\mathcal{S}$, the the class $\bar{\mathcal{S}}$ of all sets of the form $S \triangle N$, where $S \in \mathcal{S}$ and N is a subset of a set of $\mathcal{S}$ having measure zero, is also a σ-ring which contains $\mathcal{S}$, and the set function $\bar{\mu}$ on $\bar{\mathcal{S}}$ defined by $\bar{\mu}(S \triangle N) = \mu(S)$ is a complete measure on $\mathcal{S}$ and an extension of μ.*

(c) *If μ is a σ-finite measure on a ring $\mathcal{R}$ and if μ^* is the outer measure induced by μ, then the completion of the extension of μ to $\mathcal{S}(\mathcal{R})$ is identical with μ^* on the class of μ^*-measurable sets.*

Corollary 8.1.12. *Suppose μ is a σ-finite measure on a ring $\mathcal{R}$ and μ is extended and completed to a measure on $\bar{\mathcal{S}}(\mathcal{R})$. Then for each $A \in \bar{\mathcal{S}}(\mathcal{R})$ having finite measure and each $\varepsilon > 0$, there exists a set $A_0 \in \mathcal{R}$ such that $\mu(A \triangle A_0) < \varepsilon$.*

PROOF: By Theorem 8.1.11 the set A differs from some set $B \in \mathcal{S}(\mathcal{R})$ by at most a set of measure zero. Recall that if $B \in \mathcal{S}(\mathcal{R})$, then B can be covered by a countable union of sets of $\mathcal{R}$. [The class of all sets which can be covered by countable unions of sets of $\mathcal{R}$ is a σ-ring which contains $\mathcal{R}$; hence it contains $\mathcal{S}(\mathcal{R})$.] Since μ was extended to $\mathcal{S}(\mathcal{R})$, we can write

$$\mu(B) = \inf\left\{ \sum_{i=1}^{\infty} \mu(B_i) : B \subset \bigcup B_i,\ B_i \in \mathcal{R},\ i = 1, 2, 3, \ldots \right\}.$$

Hence for some $\{B_i\}_1^{\infty} \subset \mathcal{R}$, we have $B \subset \bigcup B_i$, and $\mu(B) \leqslant \mu(\bigcup B_i) \leqslant \mu(B) + \varepsilon/3$.

There is an integer n such that

$$\mu\left(\bigcup_{i=1}^{n} B_i \right) \geqslant \mu\left(\bigcup_{i=1}^{\infty} B_i \right) - \varepsilon/3.$$

Let $A_0 = \bigcup_{i=1}^{n} B_i \in \mathfrak{R}$. We have

$$\mu(B \setminus A_0) \leqslant \mu\left(\bigcup_{i=1}^{\infty} B_i \setminus A_0\right) = \mu\left(\bigcup_{i=1}^{\infty} B_i\right) - \mu(A_0) \leqslant \varepsilon/3,$$

and

$$\mu(A_0 \setminus B) \leqslant \mu\left(\bigcup_{i=1}^{\infty} B_i \setminus B\right) = \mu\left(\bigcup_{i=1}^{\infty} B_i\right) - \mu(B) \leqslant \varepsilon/3,$$

so $\mu(A_0 \triangle B) < \varepsilon$.

Since $\mu(B \triangle A) = 0$, we conclude that $\mu(A_0 \triangle A) < \varepsilon$. □

8.2 MEASURE SPACES AND A DARBOUX INTEGRAL

Definition 8.2.1. Suppose V is a nonempty set, $\mathbb{S} \subset P(V)$ a class of subsets on which a set function μ is defined. We call $(V, \mathbb{S}, \mu)$ a *measure space* in the event that

1. $\mathbb{S}$ is a σ-ring of sets;
2. $\mu(\varnothing) = 0$; $\forall A \in \mathbb{S}$, $0 \leqslant \mu(A) \leqslant \infty$;
3. $\forall \{A_n\}_{n=1}^{\infty} \subset \mathbb{S}$, $\mu(\bigcup A_n) \leqslant \Sigma \mu(A_n)$; if $A_i \cap A_j = \varnothing$, $i \neq j$, then $\mu(\bigcup A_n) = \Sigma \mu(A_n)$; and
4. each point of V belongs to at least one set of $\mathbb{S}$.

If, in addition, $A \in \mathbb{S}$ and $\mu(A) = 0 \Rightarrow P(A) \subset \mathbb{S}$, we call $(V, \mathbb{S}, \mu)$ a *complete measure space*.

We do not insist that the set V of a measure space $(V, \mathbb{S}, \mu)$ itself be measurable, but if it is, then $\mathbb{S}$ is a σ-algebra, closed with respect to complementation. However, V is the union of all the sets of $\mathbb{S}$, even if no countable collection of sets of $\mathbb{S}$ covers V.

Let A be a measurable set in a measure space $(V, \mathbb{S}, \mu)$. By a measurable partition of A we simply mean a partition of A each set of which is measurable. Now, if P is the class of measurable partitions of A, P is a partial order, the order being defined for all $\pi_1, \pi_2 \in P$, by $\pi_1 \prec \pi_2$ iff π_2 is finer than π_1 iff every set of the partition π_1 is a union of sets of the partition π_2. It is not hard to see that since $\mathbb{S}$ is a σ-ring, there always exists a measurable partition π which is the common refinement of π_1 and π_2, $\pi = \pi_1 \vee \pi_2$ in the event that π_1 and π_2 are not comparable.

Suppose f is a real-valued function defined on A, and further suppose that $\mu(A) < \infty$ and that f is bounded on A. Let $\pi = \{A_i\}_{i=1}^{n}$ be an arbitrary measurable partition of A, and define

$$\underline{f}_i = \inf\{f(x) : x \in A_i\}, \qquad \bar{f}_i = \sup\{f(x) : x \in A_i\},$$

$$\underline{S}(f, \pi, A) = \sum_{i=1}^{n} \underline{f}_i \mu(A_i), \qquad \bar{S}(f, \pi, A) = \sum_{i=1}^{n} \bar{f}_i \mu(A_i).$$

It is easy to show that if π_1, π_2 are two measurable partitions of A and $\pi = \pi_1 \vee \pi_2$ is the common refinement, then

$$\underline{S}(f, \pi_1, A) \leqslant \underline{S}(f, \pi, A) \leqslant \bar{S}(f, \pi, A) \leqslant \bar{S}(f, \pi_2, A).$$

This is to say, as a partition is refined, lower sums increase and upper sums decrease. Moreover, any upper sum majorizes any lower sum. Consequently, each of the following limits exist since f is bounded on A and $\mu(A) < \infty$:

$$\sup_{\pi \in P} \underline{S}(f, \pi, A) = \underline{\int}_A f\,d\mu = \text{the lower integral of } f \text{ over } A,$$

$$\inf_{\pi \in P} \bar{S}(f, \pi, A) = \bar{\int}_A f\,d\mu = \text{the upper integral of } f \text{ over } A,$$

and

$$\underline{\int}_A f\,d\mu \leqslant \bar{\int}_A f\,d\mu.$$

Definition 8.2.2. If the lower and upper integrals of f over A are equal, we say that f is *μ-integrable* over A, and we denote the common value of these integrals by $\int_A f\,d\mu$. We refer to this integral as the *Darboux integral* of f over A with respect to the measure μ.

We can improve this notion of integrability slightly by postulating that if a real-valued function f is unbounded on a set N for which $\mu(N) = 0$, then the integral of f over N is zero. Hence, if f is defined on a set A, and N is a subset of A having measure zero such that f is bounded on $A \setminus N$, then the integral of f over A will be taken to be the same as the integral of f over $A \setminus N$ if this latter integral exists. Thus we can relax the restriction that f be bounded on A; it will suffice that f be "essentially bounded" on A, meaning that f is bounded everywhere on A except for a subset of A having measure zero. (To say that f is essentially bounded on a set A is to say that f is bounded "almost everywhere" on A.)

Note that we can regard the integral over A as a linear functional on the space of real-valued functions which are integrable over A; this is to say, if $\alpha, \beta \in \mathbb{R}$ and f, g are integrable functions, then

$$\int_A (\alpha f + \beta g)\,d\mu = \alpha \int_A f\,d\mu + \beta \int_A g\,d\mu.$$

Moreover, the integral of f may be thought of as an additive set function in that if S, T are disjoint measurable sets over which f is integrable, then $\int_S f\,d\mu$ maps S into $\mathbb{R}$, and

$$\int_{T \cup S} f\,d\mu = \int_T f\,d\mu + \int_S f\,d\mu.$$

Consider the following example. Let $V=\mathbb{Z}^+$, $\mathfrak{S}=P(V)$, and $\mu(A)=\Sigma_{a\in A}a$ for every $A\subset V$. μ is finite only on the finite subsets of V, but μ is σ-finite on $\mathfrak{S}$. Suppose f is defined on V by $f(a)=a^{-1}\cdot 2^{-a}$. Since $(V,\mathfrak{S},\mu)$ is a measure space, we can consider the integral of f over any measurable set $A\in\mathfrak{S}$, which is to say, over any subset of V. Let A be the set $\{1,2,3,4,\ldots,10\}$; what is $\int_A f\,d\mu$? For this set the finest possible partition is the one consisting of ten singletons. Evidently, for this partition there is only one sum, namely $\Sigma_{a=1}^{10}af(a)=\Sigma_{a=1}^{10}2^{-a}=\frac{1023}{1024}$, so this is the value of the integral.

Those of you who have had an introduction to statistics will recognize that the function $f(a)$ that we have defined is a "probability density" function for the measure space $(V,\mathfrak{S},\mu)$. This is to say, if X is a "random variable" that takes values in V, not quite randomly, but in a way that can best be described by saying the probability that X will assume the value $a\in V$ is $a^{-1}2^{-a}$, then $\int_A f\,d\mu$ gives to probability that X will assume a value in the set A. Note that $\int_V f\,d\mu=1$.

As another example we consider the interval $[0,1]=V$. Let S be the σ-ring generated by the semiring of half-open intervals in $[0,1]$ of the form $[a,b)$ and μ be the measure on $\mathfrak{S}$ obtained by extending the set function μ defined on the semiring by $\mu([a,b))=b^3-a^3$ (μ is indeed a measure). Let f be defined on V by $f(x)=(ax+b)^{-1/2}/3$. What is $\int_V f\,d\mu$ $(a,b>0)$?

We partition $[0,1]$ as indicated: $0=x_0<x_1<\cdots<x_n=1$. A lower sum is, since $f(x)\searrow$

$$\sum_{i=1}^{n} f(x_i)(x_i^3-x_{i-1}^3)=\sum_{i=1}^{n} f(x_i)(x_i^2+x_ix_{i-1}+x_{i-1}^2)(x_i-x_{i-1}).$$

As $\max_i\{x_i-x_{i-1}\}\to 0$, this lower sum approaches the limit $\int_0^1(x^2/\sqrt{ax+b}\,)\,dx$, an elementary Riemann integral whose value is

$$2\Big[(3a^2-4ab+8b^2)\sqrt{a+b}-8b^{5/2}\Big]/15a^2.$$

Thus $\int_V f\,d\mu=\int_V f\,d(x^3)=\int_0^1 3x^2f(x)\,dx$, which suggests that in this case $d\mu=3x^2dx$.

Suppose V is a Hausdorff space and we are able to make V into a complete measure space $(V,\mathfrak{S},\mu)$, where $\mathfrak{S}$ includes the topology for V. Then $\mathfrak{S}$ also includes the closed sets of V. In a Hausdorff space any closed set F is the intersection of all open sets G which contain F $(F=\cap[\{\tilde{p}\}:p\notin F])$ and any open set G is the union of all closed sets F contained in G (see Problem 6.7). Suppose $A\subset V$ is a set, and for each $\varepsilon>0$ we can find a closed set F_ε and an open set G_ε with $F_\varepsilon\subset A\subset G_\varepsilon$ such that $\mu(G_\varepsilon\setminus F_\varepsilon)<\varepsilon$. This is tantamount to saying that there exists an increasing sequence of closed sets $\{F_n\}$ and a decreasing sequence of open sets $\{G_n\}$ such that for each n, $F_n\subset A\subset G_n$ and $\mu(G_n\setminus F_n)<1/n$. Hence both limits $\lim\mu(G_n)$ and $\lim\mu(F_n)$ exist and are equal, or both are infinite. This means that A differs from each of the measurable sets $\cup F_n$ and $\cap G_n$ by a set of measure zero. Hence $A\in\mathfrak{S}$, and we have the following useful theorem.

Theorem 8.2.3. *Suppose $(V, \mathcal{S}, \mu)$ is a complete measure space and V is a Hausdorff space as well. Assume further that $\mathcal{S}$ includes the open sets of V. If $A \subset V$ is such that for each $\varepsilon > 0$ there exist sets F, G, with F closed, G open, $F \subset A \subset G$, and $\mu(G \setminus F) < \varepsilon$, then $A \in \mathcal{S}$.*

8.3 THE MEASURE SPACE $(E^n, \mathfrak{M}, \mu)$ AND LEBESGUE MEASURE

Clearly, we are far more interested in the space E^n than in any others, for it is in E^n that we are going to do integration. Hence we want to establish a particular measure for this space. Now E^n is a linear metric space, and the metric is *translation invariant*, meaning that for all $x, y, z \in E^n$, $\rho(x, y) = \rho(x+z, y+z)$. Thus we might want our measure for E^n to be translation invariant in the same sense.

Definition 8.3.1. Let $(V, \mathcal{S}, \mu)$ be a linear measure space. The measure μ is said to be *invariant* if for each nonempty set $M \in \mathcal{S}$ and each $\mathbf{x} \in V$, $M + \mathbf{x} \in \mathcal{S}$ and $\mu(M + \mathbf{x}) = \mu(M)$.

This is a perfectly reasonable property for a measure to have; after all, we would expect congruent figures in E^n to have equal measures. However, not all measures are invariant!

Let $\mathfrak{B} \subset P(E^n)$ be the family of half-open rational boxes, a typical box B being of the form

$$B = \bigtimes_{i=1}^{n} \{[a_i, b_i) : a_i, b_i \in \mathbf{Q}, a_i < b_i, i = 1, \dots, n\}.$$

We define a set function μ on $\mathfrak{B}$ by the rule $\mu(B) = \prod_{i=1}^{n}(b_i - a_i)$, and we further stipulate that $\mu(\varnothing) = 0$. Now $\mathfrak{B} \cup \{\varnothing\}$ is a semiring of sets, so μ can be extended to a measure on $\mathfrak{R}(\mathfrak{B})$, and as we have seen, this measure can be further extended via the outer measure induced by μ to a complete measure on a σ-ring of sets $\mathfrak{M} \subset P(E^n)$ which contains $\mathcal{S}(\mathfrak{B})$ and all subsets of sets of $\mathcal{S}(\mathfrak{B})$ which have measure zero. Since the original set function μ on $\mathfrak{B}$ is evidently translation invariant, it follows that the extended measure on $\mathfrak{M}$ is invariant. It is clear that our complete measure μ on $\mathfrak{M}$ is also σ-finite.

Definition 8.3.2. The complete σ-finite invariant measure μ that we have just defined for E^n is called n-dimensional *Lebesgue measure*. We denote by $\mathfrak{M}$ the class of Lebesgue-measurable sets in E^n.

Observe that if $B \in \mathfrak{B}$ is the half-open unit cube in E^n, $B = [0,1)^n$, and $B_k = [0, 1+1/k)^n$, $k = 1, 2, 3, \dots$, then $\overline{B} = \bigcap_{k=1}^{\infty} B_k$. Since $\lim \mu(B_k) = 1$, we have that $\mu(\overline{B}) = 1 = \mu(B)$. This is clearly the case for every $B \in \mathfrak{B}$. This, together with the invariance of μ, implies that the Lebesgue measure of singletons and all k-dimensional boxes in E^n, $k<n$, are zero. It follows immediately that $\mu(\mathring{B})=\mu(B)$. The collection of the interiors of the sets of $\mathfrak{B}$ is a countable base for the topology for E^n; hence every open set $G \subset E^n$ is a countable union of sets from this base. Suppose $G = \bigcup(\mathring{B}_n : B_n \in \mathfrak{B})$. Now, $\bigcup_{n=1}^{\infty} B_n$ is measurable, and we know that $\bigcup_{n=1}^{\infty} \partial B_n$ is a set of measure zero (a countable union of sets of measure zero has measure zero), so G differs from $\bigcup B_n$ by a set of measure zero. Thus G is measurable, so $\mathfrak{M}$ contains all open sets of E^n. Since $E^n \in \mathfrak{M}$ all closed sets are also measurable.

As a consequence of Theorem 8.2.3, a set $A \subset E^n$ is Lebesgue measurable if for each $\varepsilon > 0$, there is a closed set F and an open set G with $F \subset A \subset G$ such that $\mu(G \setminus F) < \varepsilon$. Now the ring $\mathfrak{R}(\mathfrak{B})$ generated by $\mathfrak{B}$ is the class of sets which are unions of finitely many disjoint sets of $\mathfrak{B}$. If $S \in \mathfrak{R}(\mathfrak{B})$, it is clear that $\mu(\mathring{S}) = \mu(S) = \mu(\bar{S})$. We are assured by Corollary 8.1.12 that if $A \in \mathfrak{M}$ and $\mu(A)<\infty$, then for each $\varepsilon>0$ there is a set $S \in \mathfrak{R}(\mathfrak{B})$ such that $\mu(S \Delta A)<\varepsilon$. Since S is a finite union of disjoint boxes of $\mathfrak{B}$, it should not be too difficult for the reader to complete the details of the proof that there is a closed set F and an open set G with $F \subset A \subset G$ and that $\mu(G \setminus F)<\varepsilon$. We summarize:

Theorem 8.3.3. *Suppose μ is Lebesgue measure for E^n and $\mathfrak{M}$ is the class of Lebesgue-measurable sets. Then $\mathfrak{M}$ contains the open sets and closed sets of E^n. Every finite or countable set is in $\mathfrak{M}$ and has measure zero. Each compact set is measurable and has finite measure. If $A \in \mathfrak{M}$, then $\tilde{A} \in \mathfrak{M}$. Finally, $A \in \mathfrak{M}$* iff *for each $\varepsilon > 0$ there is a closed set F and an open set G with $F \subset A \subset G$, such that $\mu(G \setminus F) < \varepsilon$.*

We close this section with the following comment. A great many students make the hasty conjecture that if a set $A \subset E^n$ has positive measure, then A contains an open set. Don't you believe it!

8.4 THE LEBESGUE INTEGRAL IN E^n

The integral that we defined in Section 8.2 is a perfectly good one, but we have something even better in mind. We admit, however, that when one actually gets down to evaluating an integral, it is something very close to Riemann integration that is brought into play.

An obvious difficulty we face with our integral is that it seems impossible to determine all the measurable partitions of a measurable set A; another problem is finding the various suprema and infima of the function values over the various subsets. Here is a method for alleviating many of the problems inherent in our first definition of the integral; any function integrable by our first definition will be integrable under the new scheme, and we shall pick up some more functions which are integrable.

Definition 8.4.1. By an *extended real-valued function* f on E^n we mean a function f which may assume the values $+\infty$ or $-\infty$. If f is such a function then by the *support* of f we mean the closed set

$$\operatorname{supp} f = \operatorname{cl}\{\mathbf{x} \in E^n : f(\mathbf{x}) \neq 0\},$$

and by the *zero-set* of f we mean the set

$$Z_f = \{\mathbf{x} \in E^n : f(\mathbf{x}) = 0\}.$$

Definition 8.4.2. Let $(V, \mathbb{S}, \mu)$ be a measure space and f an extended real-valued function on V. f is called a *μ-measurable function* iff

$$\forall a, b, \quad 0 \leqslant a < b \leqslant +\infty, \qquad f^{-1}((a, b]) \in \mathbb{S},$$

and

$$\forall c, d, \quad -\infty \leqslant c < d \leqslant 0, \qquad f^{-1}([c, d)) \in \mathbb{S}.$$

Sets of the form $(a, b]$ or $[c, d)$ are called basic Borel sets of the real line; these sets generate the σ-ring of Borel sets of $\mathbb{R}$, and they are, of course, measurable sets. Notice the direct connection with a continuous function. f is continuous iff the preimage of each basic open set is open. Here we have $f > 0$ is measurable iff the preimage of each basic measurable set is measurable. We are intentionally ignoring the set Z_f, because whether that set is measurable or not, we can with every justification simply define

$$\int_{Z_f} f = 0.$$

Moreover, if f assumes the value $+\infty$ on a set A, and $\mu(A) = 0$, we define $\int_A f$ to be zero.

Now if we have a μ-measurable function f, we can partition the set $[-\infty, 0) \cup (0, \infty]$ and proceed. But wait—a simplifying idea presents itself.

Definition 8.4.3. Let f be an extended real-valued function. Then f can be decomposed into the difference of two nonnegative functions: $f = f^+ - f^-$. We define f^+ and f^- by their values

$$\forall \mathbf{x} \in \operatorname{dom} f \quad \begin{cases} f^+(\mathbf{x}) = f(\mathbf{x}) \vee 0 = \max\{f(\mathbf{x}), 0\}, \\ f^-(\mathbf{x}) = -(f(\mathbf{x}) \wedge 0) = -\min\{f(\mathbf{x}), 0\}. \end{cases}$$

It is clear that for all $\mathbf{x} \in \operatorname{dom} f$

$$f(\mathbf{x}) = f^+(\mathbf{x}) - f^-(\mathbf{x}), \qquad |f(\mathbf{x})| = f^+(\mathbf{x}) + f^-(\mathbf{x}),$$

$$f^+(\mathbf{x}) \geqslant 0, \qquad f^-(\mathbf{x}) \geqslant 0, \qquad f^+(\mathbf{x}) \cdot f^-(\mathbf{x}) = 0,$$

$$Z_f = Z_{f^+} \cap Z_{f^-}.$$

Moreover, it is fairly apparent that f is measurable iff each of f^+, f^- is measurable. f^+ and f^- are called the positive and negative parts of f, respectively.

We shall define the integrals of f^+ and f^-. Having done that we shall define the integral of f to be the difference of the integrals of f^+ and f^-.

Let π be a simple partition of $[0,\infty]$ of the form $\{0=a_0, a_1, a_2, \ldots, a_n, a_{n+1}=\infty\}$, f^+ be the positive part of a measurable, extended real-valued function f and $A \subset \operatorname{dom} f$.

Let

$$\underline{S}(f^+,\pi,A)=\sum_{i=0}^{n} a_i\mu\left(A\cap(f^+)^{-1}((a_i,a_{i+1}])\right),$$

$$\bar{S}(f^+,\pi,A)=\sum_{i=1}^{n+1} a_i\mu\left(A\cap(f^+)^{-1}((a_{i-1},a_i])\right),$$

where it is to be understood that $\bar{S}=\infty$ if $\mu(A\cap(f^+)^{-1}((a_n,\infty]))>0$, but that if this measure is zero, the last term of the sum is zero. The set A is such that $A\setminus Z_{f^+}$ is measurable.

$\bar{S}$ is an upper sum and $\underline{S}$ is a lower sum for f^+ over A, relative to the partition π. Define $\|\pi\|$ to be the maximum of the lengths of the finite intervals of π.

We can make the same observations that we made before concerning the upper sums and lower sums. For any two partitions π_1, π_2, with $\pi=\pi_1\vee\pi_2$, we have

$$0\leqslant\underline{S}(f^+,\pi_1,A)\leqslant\underline{S}(f^+,\pi,A)\leqslant\bar{S}(f^+,\pi,A)\leqslant\bar{S}(f^+,\pi_2,A).$$

Hence if an upper sum is finite, the lower sums must approach a common limit as the partitions are refined, i.e., as $\|\pi\|\to 0$. On the other hand, if for all partitions π, the upper sums are infinite, we shall consider the limit as $\|\pi\|\to 0$, and $a_n\to\infty$, of the lower sums $\underline{S}(f^+,\pi,A)$. If this limit is finite, it will be the value of the integral. We combine this in the following definition.

Definition 8.4.4. Let f be an extended real-valued function on a measure space $(V,\mathbb{S},\mu)$ which is nonnegative and measurable. Let A be a set in the domain of f such that $A\setminus Z_f$ is measurable. Then

$$\int_A f=\lim_{\substack{\|\pi\|\to 0\\ a_n\to\infty}}\underline{S}(f,\pi,A),$$

provided the limit on the right-hand side is finite. If such is the case, we say f is μ-integrable over A.

If f is an arbitrary μ-measurable extended real-valued function on a measure space $(V,\mathbb{S},\mu)$ the integral of f over A is defined to be

$$\int_A f=\int_A f^+-\int_A f^-,$$

provided both integrals on the right-hand side exist, where f^+, f^- are the positive and negative parts of f, respectively.

Because the measure μ plays such a role in this whole process, we indicate the integral of f over A with respect to the measure μ by the more complete symbol $\int_A f\,d\mu$. We refer to this integral as the *Lebesgue integral* of f over A in case $V = E^n$, $\mathcal{S} = \mathfrak{M}$, and μ is Lebesgue measure.

What functions are μ-integrable? Certainly, if f is not measurable and a measurable set A lies outside the zero-set of f, there is some interval, say $(a, b]$, such that $f^{-1}((a, b])$ is not measurable, and thus an approximating sum cannot exist for $\int_A f\,d\mu$. If f is measurable, it still might not be μ-integrable in that one of $\int_A f^+\,d\mu$ or $\int_A f^-\,d\mu$ is infinite. If f is measurable and bounded on a set A and $A \setminus Z_f$ is measurable with finite measure, then it seems clear that f is integrable. A measurable function f is μ-integrable iff $|f|$ is μ-integrable. Thus the function

$$f(t) = (\sin t)/t$$

is not μ-integrable on $[0, \infty)$ since

$$\int_0^\infty \frac{|\sin t|}{t}\,dt = \infty.$$

On the other hand,

$$\int_0^\infty \frac{\sin t}{t}\,dt = \frac{\pi}{2},$$

as we shall see in Section 11.2, so $f(t)$ is integrable in some sense over $[0, \infty]$.

Consider for a moment the real line E^1, with its standard topology. The Lebesgue measure on E^1 is the measure generated by the Borel sets of the line that we have been talking about. Now think of the symbol X as varying at random over E^1. Suppose you are given absolute assurance that the relative probability that the random variable X will take the value x is given by the formula

$$f(x) = \begin{cases} \dfrac{x^{\alpha-1}}{\Gamma(\alpha)\beta^\alpha} e^{-x/\beta}, & x > 0, \\ 0, & \text{otherwise} \end{cases}$$

where α and β are positive real constants and Γ is the gamma function, defined by

$$\Gamma(\alpha) = \int_0^\infty t^{\alpha-1} e^{-t}\,dt.$$

We call f the probability density function for the random variable X, and in this case, we say that X has a gamma distribution.

The integral $\int_a^b f(x)\,dx$ gives the probability that X will be in the interval $[a, b]$; this is to say, the integral assigns a probability measure to the set $[a, b]$.

Let us quickly point out that $f(x)$ does not give the probability that X take the value x; that probability is given by $\int_x^x f(t)\,dt$, which is zero. However, $f(x)$ is a probability density; i.e., a limiting ratio of the probability measure of a neighborhood of x to the ordinary measure of that neighborhood.

Thus we can obtain new measures from old. If $F(x)=\int_{-\infty}^{x} f(t)\,dt$, where f is the density function described above and A is a μ-measurable set in E^1, then $\mu(A)=\int_A d\mu=\int_A dx$, and the gamma measure of A is

$$\int_A dF=\int_A f(x)\,dx=\int_A f\,d\mu.$$

In general, if f is a positive μ-integrable function over the sets of a σ-ring $\mathfrak{M}$, then $\int f\,d\mu$ is a new measure for the old measure space. Note that in general this new measure is not invariant.

8.5 SIGNED MEASURES

The last statement in the previous section suggests that there might be a place for negative measure in analysis, and indeed there is. Therefore, we shall lay down a bit of theory to cover the possibility.

Suppose $(V,\mathbb{S},\mu)$ is a complete measure space and ν is another complete measure on V over the same σ-ring $\mathbb{S}$.

It might be well to emphasize that in generating a σ-ring of measurable sets, we started with a ring $\mathfrak{R}$ of sets upon which we have a measure function already defined. The class $\mathbb{S}(\mathfrak{R})$ is the smallest σ-ring containing $\mathfrak{R}$; we get $\mathbb{S}(\mathfrak{R})$ by forming all the sets possible from $\mathfrak{R}$ by taking countable unions and differences. The σ-ring $\mathbb{S}$ is, strictly speaking, what we should call an "enriched" σ-ring. We adjoin to $\mathbb{S}(\mathfrak{R})$ all the subsets of sets of measure zero to obtain $\mathbb{S}$, the completed σ-ring.

Consider the functional σ defined over $\mathbb{S}$ by

$$\forall A\in\mathbb{S}\qquad \sigma(A)=\mu(A)-\nu(A).$$

It is not hard to see that the basic measure axioms σ fails to satisfy are the ones involving inequalities, and it is also not true in general that if $A\in\mathbb{S}$ and $\sigma(A)=0$, then $\sigma(B)=0$ for every $B\subset A$. Hence we modify our set of axioms so that we may refer to such a functional σ as a complete signed measure.

First, observe that if for some A, $B\in\mathbb{S}$, $\mu(A)=\infty=\nu(B)$, then $\sigma(A\cup B)=\infty-\infty$, an undefined quantity. Therefore, we must restrict the values of σ slightly. If σ is to be a signed measure over a σ-ring $\mathbb{S}$, σ and $\mathbb{S}$ must satisfy the following "signed measure" axioms.

- $\sigma 1$. $\mathbb{S}$ must be a σ-ring of sets.
- $\sigma 2$. $\sigma(\varnothing)=0$; σ may assume any real value over $\mathbb{S}$, and at most one of the values $+\infty$ or $-\infty$.

$\sigma 3$. $\forall A, B \in \mathbb{S}$, $\sigma(A \cup B) + \sigma(A \cap B) = \sigma(A) + \sigma(B)$.
$\sigma 4$. If $\{A_i\}_{i=1}^{\infty}$ is a sequence of mutually disjoint sets of $\mathbb{S}$,

$$\sigma\left(\bigcup_{i=1}^{\infty} A_i\right) = \sum_{i=1}^{\infty} \sigma(A_i).$$

There is yet another axiom to follow, but first we must lay a bit of groundwork.

Suppose $(V, \mathbb{S}, \sigma)$ is a signed measure space, and assume that for no $A \in \mathbb{S}$ is $\sigma(A) = -\infty$ and that σ does indeed assume both positive and negative values over $\mathbb{S}$.

Theorem 8.5.1 (Hahn Decomposition Theorem). *Let A be a measurable set in the signed measure space $(V, \mathbb{S}, \sigma)$. Then A can be decomposed into two disjoint parts $A^+, A^- \in \mathbb{S}$, with $A = A^+ \cup A^-$, such that for every measurable subset $S \subset A^+$, $\sigma(S) \geqslant 0$ and for every measurable subset $S \subset A^-$, $\sigma(S) \leqslant 0$. This decomposition is essentially unique in that if (B^+, B^-) is another such decomposition, $\sigma(A^+ \triangle B^+) = \sigma(A^- \triangle B^-) = 0$. Such a decomposition is called a Hahn decomposition with respect to σ.*

We omit the proof; the interested reader may refer to any standard text in measure theory.

We are now ready to enrich, or complete, the σ-ring of sets. Suppose each set of $\mathbb{S}$ has been Hahn decomposed.

$\sigma 5$. If (A^+, A^-) is a Hahn decomposition of a set $A \in \mathbb{S}$ and $\sigma(A^+) = \sigma(A^-) = 0$, then $P(A) \subset \mathbb{S}$, and for every $S \in P(A)$, $\sigma(S) = 0$.

Definition 8.5.2. Let V be a nonempty set, $\mathbb{S}$ a nonempty class of subsets of V, and σ an extended real-valued set function whose domain is $\mathbb{S}$. If $\mathbb{S}$ and σ satisfy axioms $\sigma 1$–$\sigma 5$, we call $(V, \mathbb{S}, \sigma)$ a *complete signed measure space*.

Now suppose $(V, \mathbb{S}, \sigma)$ is a signed measure space. We can define two functionals σ^+, σ^- on $\mathbb{S}$ as follows: for each $A \in \mathbb{S}$, with (A^+, A^-) its Hahn decomposition, put

$$\sigma^+(A) = \sigma(A^+), \qquad \sigma^-(A) = -\sigma(A^-).$$

It is not hard to verify that σ^+ and σ^- are ordinary measures on V, having $\mathbb{S}$ as their common domain, and that $\sigma = \sigma^+ - \sigma^-$. If σ never takes on the value $-\infty$, σ^- is finite valued. We call $\sigma = \sigma^+ - \sigma^-$ the *Jordan representation* for the signed measure σ.

Let $\mathfrak{M}$ be the Lebesgue-measurable sets of E^1. For each $A \in \mathfrak{M}$, let

$$\varphi(A)=\int_A \frac{2x}{(1+x^2)^2}\,dx.$$

Then φ is a signed measure for E^1.

If $\mu_1(A)=\int_A e^{-t}\,dt$ and $\mu_2(A)=\int_A (1+t^2)^{-1}\,dt$ define two measures on $([0,\infty),\mathfrak{M},\mu)$, then $\sigma=\mu_1-\mu_2$ is a signed measure on $[0,\infty)$. What is the σ-measure of the interval $[0,1]$? Why, $1-e^{-1}-\pi/4$, of course.

If $\mathcal{K}$ is a class of compact oriented k-manifolds in E^n and μ is a measure on $\mathcal{K}$, then the measure σ,

$$\forall M \in \mathcal{K} \begin{cases} \sigma(M)=\mu(M) & \text{if } M \text{ is positively oriented,} \\ \sigma(M)=-\mu(M) & \text{if } M \text{ is negatively oriented,} \end{cases}$$

is a signed measure over $\mathcal{K}$.

8.6 AFFINE MAPS ON $(E^n, \mathfrak{M}, \mu)$

Recall that if K is the unit cube in E^k and P is a k-parallelotope in E^n, $n \geqslant k$, then P is the image of K under a nonsingular affine map φ whose linear part is T, and the n-dimensional measure of P is zero when $k<n$. The k-dimensional measure of P is obtained from the generalized Pythagorean theorem; the measure of the projection of P on the $(i_1 \ldots i_k)$-coordinate subspace is given by

$$|dx_{i_1}\cdots dx_{i_k}\varphi(K)|=|\det T'|,$$

where T' is the appropriate submatrix of T. Note that the expression on the left-hand side of the equation is (see Section 5.8) the absolute value of the integral of the k-form $dx_{i_1}\cdots dx_{i_k}$ over $\varphi(K)$; that is,

$$dx_{i_1}\cdots dx_{i_k}(\varphi(K))=\int_{\varphi(K)} dx_{i_1}\cdots dx_{i_k}.$$

Lemma 8.6.1. *Let $T\colon E^n \to E^n$ be linear and $A \subset E^n$ be μ-measurable. Then TA is μ-measurable, with $\mu(TA)=\mu(A)|\det T|$, and thus any affine image of A is μ-measurable.*

PROOF: If A is measurable with $\mu(A)<\infty$, then, $\varepsilon>0$ having been given, we can find a closed set F and an open set G, with $F \subset A \subset G$ and $\mu(G\setminus F)<\varepsilon/6$. If T has rank less than n, the statement of the theorem is trivially true, since in that case $\mu(TA)=0$, so we assume T is bijective and hence a homeomorphism. This is to say, we have $TF \subset TA \subset TG$, with TF closed and TG open. It is not hard to see that if B is any box in E^n, then $\mu(TB)=|\det T|\mu(B)$. Hence, if F' is any countable union of disjoint (or perhaps abutting) boxes, $\mu(TF')=|\det T|\mu(F')$. It follows fairly easily that if Z is a set of measure zero, $\mu(TZ)=0$. Finally, we note that we can "approximate" F, as well as G, by countable unions F' and G' of disjoint (perhaps abutting) boxes, in the sense that $\mu(F \triangle F') < \varepsilon/6$, $\mu(G \triangle G') < \varepsilon/6$, and matters are arranged so that $F' \subset G \subset G'$. We now have $\mu(G' \setminus F') < \varepsilon/2$, and $\mu(TG' \setminus TF') < |\det T|\varepsilon/2$.

Since $\mu(F \triangle F') < \varepsilon/6$, we can cover $F \triangle F'$ with a countable collection of disjoint or abutting boxes whose union F'' has a measure less than $\varepsilon/2$. Since $F' \subset F \cup (F \triangle F') \subset F \cup F''$, we have $TF' \subset TF \cup TF''$, and $\mu(TF') \leqslant \mu(TF) + \mu(TF'')$. Thus

$$\tfrac{1}{2}\varepsilon|\det T| > \mu(TG') - \mu(TF') > \mu(TG') - \mu(TF) - \tfrac{1}{2}\varepsilon|\det T|$$
$$> \mu(TG) - \mu(TF) - \tfrac{1}{2}\varepsilon|\det T|,$$

and finally,

$$\mu(TG \setminus TF) < \varepsilon|\det T|.$$

Since $\varepsilon > 0$ was arbitrary, we conclude that TA is measurable. Because of the invariance of μ, the affine image of a measurable set is measurable. It follows rather easily that $\mu(TA) = \mu(A)|\det T|$.

In the event that $\mu(A) = \infty$, we can argue this way. Since $\mu(A) = \infty$, A is not bounded. Let $\{B_n\}$ be an increasing sequence of boxes centered at a common point of A such that $A \subset \bigcup_{n=1}^{\infty} B_n$ and $\forall n$, $\mu(B_n) = n$. Then $A = A \cap \bigcup_{n=1}^{\infty}(B_n \setminus B_{n-1})$, where $B_0 = \varnothing$. Each set $A \cap (B_n \setminus B_{n-1})$ is measurable with finite measure, and if T is nonsingular or singular, we can write $TA = \bigcup_{n=1}^{\infty} T(A \cap (B_n \setminus B_{n-1}))$, a disjoint union. Thus,

$$\mu(TA) = \sum_{n=1}^{\infty} \mu[T(A \cap (B_n \setminus B_{n-1}))] = |\det T| \cdot \sum_{n=1}^{\infty} \mu(A \cap (B_n \setminus B_{n-1}))$$
$$= |\det T| \cdot \mu(A) = \infty,$$

if $\det T \neq 0$. If T is singular, each finite measure is zero, and we conclude that $\mu(TA) = 0$. □

Theorem 8.6.2. *Let $\varphi: E^n \to E^m$, $m \geqslant n$, be a nonsingular affine map with linear part T, and let $A \subset E^n$ be μ-measurable. Then $\varphi(A)$ is μ-measurable, where μ is n-dimensional Lebesgue measure, and $\mu(\varphi(A)) = \mu(A) \cdot g$, where g^2 is the sum of the squares of the* nth *order determinants in T.*

PROOF: If $m = n$, the previous lemma tells the whole story. Let $m > n$. We can find the measure of the projection of $\varphi(A)$ on each n-dimensional coordinate subspace of E^m; since the rank of φ is n, at least one of these measures will be nonzero if $\mu(A)$ is positive. This is all a consequence of the previous lemma.

Suppose B_i is a box in E^n, say one of the boxes of an approximating countable union $\bigcup_{i=1}^{\infty} B_i$ of disjoint or abutting boxes for A. Evidently, for each i, $\mu(\varphi(B_i)) = \mu(B_i) \cdot g$, where g^2 is the sum of the squares of all the nth order determinants in T. Hence,

$$\mu[\varphi(\cup B_i)] = \mu[\cup \varphi(B_i)] = \sum \mu[\varphi(B_i)] = \sum g\mu(B_i) = g\mu[\cup B_i].$$

As $\cup_{i=1}^{\infty} B_i \to A$, say as a decreasing sequence of open sets, it then must follow that $\mu[\varphi(\cup B_i)] \to \mu[\varphi(A)] = \mu(A)\cdot g$. $\square$

We remark that φ is a homeomorphism from E^n to $\varphi(E^n) \subset E^m$. The set $\varphi(A)$ should be regarded as a subset of the subspace $\varphi(E^n)$, endowed with the relative topology. If A is measurable and F and G are two "sandwiching" sets, F closed and G open, then $\varphi(F)$, $\varphi(G)$ is a sandwiching pair for $\varphi(A)$, with $\mu(\varphi(G)\backslash\varphi(F))$ small.

If f is a real-valued measurable function on E^m, we can now consider the integral of f over $\varphi(A)$, where $A \subset E^n$ is measurable and φ: $E^n \to E^m$ is a nonsingular affine mapping. First, remember that if ω is a constant n-form on E^m, the symbol $\int_{\varphi(A)}\omega$ is a particular number which is a certain linear combination of the signed measures of the projections of $\varphi(A)$ on certain n-dimensional coordinate subspaces of E^m, the coefficients of the linear combination being the coefficients of ω and the subspaces being the ones corresponding to the basic n-forms in ω. The symbol $\int_{\varphi(A)} f\omega$ is now something quite different; it is now the integral of a nonconstant n-form over $\varphi(A)$. To evaluate this, we would have to integrate f over the various projections multiplied by the respective constant coefficients of ω. However, to integrate f over the projections is not simple to do. We get around the difficulty by calculating the pullback of the form and integrating the pullback. Thus

$$\int_{\varphi(A)} f\omega = \int_A f\varphi\omega^* = \int_A f^*\omega^*.$$

On the other hand, $\mu(\varphi(A))$ is always representable by the symbol $\int_{\varphi(A)} d\mu$, and the value of this is obtained by evaluating $\int_A g\,d\mu$. The μ in the first integral refers to the n-dimensional measure in the space of $\varphi(A)$. (If φ were something more complicated than an affine mapping, the space of $\varphi(A)$ is not generally congruent to E^n, and we might choose a different symbol for measure in such a space.)

Finally, the integral $\int_{\varphi(A)} f\,d\mu$ is not the integral of an n-form; it is a Lebesgue integral of a measurable function with respect to the measure μ. To evaluate such an integral, we would use the pullback. That is,

$$\int_{\varphi(A)} d\mu = \int_A g\,d\mu,$$

and

$$\int_{\varphi(A)} f\,d\mu = \int_A f^*g\,d\mu,$$

where f^* is the pullback of f with respect to φ.

Physically, such an integral might represent the total charge on the surface $\varphi(A)$ if f is the charge density.

8.7 INTEGRATION BY PULLBACKS; THE AFFINE CASE

Suppose T is the linear part of an affine map $\varphi: E^n \to E^m$, $m \geqslant n$, and we don't distinguish between T and the matrix representing T, relative to the standard orthonormal bases for Euclidean space. It has been reasonably demonstrated that the pullbacks for basic k-forms, $k=1,\ldots,n$, may be computed methodically from the matrix T (Section 5.9). Moreover, we have shown by several typical examples that if one applies a basic k-form ω (as a functional) on E^m to an oriented k-parallelotope $P \subset E^m$, the result is precisely what one gets by applying the pullback ω^*, as we have defined the pullback, to the pullback of P in E^k:

$$\omega(P)=\omega^*(P^*)=\omega^*(\varphi^{-1}(P)).$$

Hence the definition for the pullback of a basic form is consistent.

If T is nonsingular and f is a nonconstant 0-form, for each $y \in \operatorname{im}\varphi$, f applied to y is the same as f^* applied to the pullback of y, under φ. This is to say, if $x=\varphi^{-1}(y)$, then $f(y)=f^*(x)=f\varphi(x)$. Thus the pullback of a nonconstant 0-form f under φ is simply $f\varphi$.

Suppose ω is a basic k-form, $k=1,\ldots,n$, and f is an arbitrary 0-form. We assert that the pullback of the product $(f\omega)^*$ is equal to the product of the pullbacks $f^*\omega^*$ under φ. Let $y \in \operatorname{im}\varphi$ be arbitrary. For each such y, $f(y)\omega$ is a constant k-form, and its pullback is $(f(y)\omega)^*$. But $y=\varphi(x)$, for some $x \in \operatorname{dom}\varphi$, so we have

$$\forall y \in \operatorname{im}\varphi, \qquad (f(y)\omega)^*=f(y)\omega^*=f\varphi(x)\omega^*=f^*(x)\omega^*.$$

We conclude that since the expressions $(f\omega)^*$ and $f^*\omega^*$ agree pointwise, they are equal.

This last result allows us to conclude that under φ, the pullback of a nonconstant k-form $f\omega$ is $f^*\omega^*$. That the pullback of a sum is the same as the sum of the pullbacks seems apparent, so we sum up what we have said in the following.

Theorem 8.7.1. *Let φ be a nonsingular affine map $E^n \to E^m$, $m \geqslant n$, and ω an arbitrary measurable n-form on E^m, $n=0,1,\ldots,m$. For any measurable set $A \subset E^n$,*

$$\int_{\varphi(A)} \omega = \int_A \omega^*.$$

PROOF: Since A is measurable, so is $\varphi(A)$, and A can be approximated from above by a decreasing sequence of measurable sets $\{G_i\}_{i=1}^{\infty}$, each G_i being the union of a countable number of disjoint or abutting positively oriented boxes. $\varphi(G_i)$ is a similar union of parallelotopes covering $\varphi(A)$. For each such G_i we have

$$\omega(\varphi(G_i)) = \int_{\varphi(G_i)} \omega = \int_{G_i} \omega^* = \omega^*(G_i).$$

The integral $\int_{G_i} \omega^*$ is an ordinary integral which converges to $\int_A \omega^*$ as $\mu(G_i) \searrow \mu(A)$. Since the left-hand integral must converge to $\int_{\varphi(A)} \omega$, the equality stated in the theorem holds. The possibility that the values might be infinite is, of course, real. □

We give a short example. Let B be the quadrilateral whose vertices in order are $(0,0,0,0)$, $(1,-1,0,2)$, $(7,1,2,4)$, and $(3,1,1,1)$. The rank of the matrix of the three nonzero vectors is 2, so B is a planar figure. Let A be the quadrilateral in whose vertices in order are $(0,0)$, $(1,0)$, $(1,2)$, $(0,1)$; then the linear map whose matrix is

$$T = \begin{bmatrix} 1 & 3 \\ -1 & 1 \\ 0 & 1 \\ 2 & 1 \end{bmatrix}, \qquad T: E^2 \to E^4,$$

maps A onto B and maintains the orientation.

Let $f(x, y, z, w) = x^2 - 2yz + 3w - 1$. We want to calculate the two integrals

$$\int_B f\,dy\,dw \qquad \text{and} \qquad \int_B f\,d\mu.$$

$f(x, y, z, w)$ pulls back to $f^*(u, v) = u^2 + 8uv + 7v^2 + 6u + 3v - 1$. This follows from the direct substitution

$$T\begin{bmatrix} u \\ v \end{bmatrix} = \begin{bmatrix} x \\ y \\ z \\ w \end{bmatrix} = \begin{bmatrix} u+3v \\ -u+v \\ v \\ 2u+v \end{bmatrix} = \begin{bmatrix} 1 & 3 \\ -1 & 1 \\ 0 & 1 \\ 2 & 1 \end{bmatrix} \begin{bmatrix} u \\ v \end{bmatrix}.$$

$(dy\,dw)^* = -3\,du\,dv$, so

$$\int_B f\,dy\,dw = \int_A (-3)(u^2 + 8uv + 7v^2 + 6u + 3v - 1)\,du\,dv$$
$$= -3\int_0^1 \int_0^{u+1} f^*(u, v)\,dv\,du = -66.$$

Since A is positively oriented,

$$\int_A f^*\,du\,dv = \int\int_A f^*\,du\,dv = \int\int_A f^*\,dv\,du.$$

To evaluate $\int_B f\,d\mu$, we need to calculate the g factor; it is equal to

$$\left[4^2 + 1^2 + (-5)^2 + (-1)^2 + (-3)^2 + (-2)^2\right]^{1/2} = \sqrt{56} = 2\sqrt{14}.$$

Then

$$\int_B f\,d\mu = \int_A f^* 2\sqrt{14}\,d\mu = 2\sqrt{14}\int\int_A f^*\,du\,dv = 44\sqrt{14}.$$

The theory we have developed so far allows us to integrate measurable n-forms in E^m over affine images of measurable sets in E^n. Unfortunately, these image sets $\varphi(A)$ are "flat." We need to extend our theory to deal with mappings φ which are nonlinear. Now, every C^1-mapping φ has a "linear part," i.e., its derivative, and we can make good approximations to such mappings φ by "almost affine" mappings as long as we approximate over sufficiently tiny measurable regions and the linear part of φ is nonsingular on these small regions. We shall develop the foundations which we need to do this in the next chapter.

The reader should be curious about what a nonmeasurable set looks like. Indeed, are there any?

8.8 A NONMEASURABLE SET IN E^1

You may skim through this section, or work out the details for yourself. In any event, think about what you do read.

Let $E \subset [0,1)$ be any set and $x \in [0,1)$ a point. By $H = E + x$ modulo 1, we mean the set

$$H = \{y : y = e + x,\ e \in E,\ e + x < 1\} \cup \{y : y = e + x - 1,\ e \in E,\ e + x \geqslant 1\}.$$

H is a disjoint union of two parts H_1 and H_2, $H = H_1 \cup H_2$, and it is clear that if E is measurable, then $\mu(E) = \mu(H)$. Now let R be the set of rational numbers in $[0,1)$ and define

$$E_x = R + x \bmod 1, \qquad x \in [0,1).$$

Let $x_1, x_2 \in [0,1)$, and let

$$y \in E_{x_1} \cap E_{x_2}.$$

Take any $z \in E_{x_1}$ and write

$$z - x_2 = (z - x_1) + (x_1 - y) + (y - x_2).$$

Then $z \in E_{x_2}$. (Show this.) This means that $E_{x_1} = E_{x_2}$, or they are disjoint.

Now, let $\mathfrak{D}$ be the class of all distinct sets of the form E_x; $\mathfrak{D}$ is a class of mutually disjoint sets. Using the axiom of choice, form a set Q_0, consisting of exactly one point of each set of $\mathfrak{D}$. Since $\mathfrak{D}$ is nonempty and each E_x contains x, Q_0 is nonempty.

Let r be the sequence of rational numbers in $(0,1)$, $r = \{r_n\}_{n=1}^{\infty}$, and for each n put

$$Q_n = Q_0 + r_n \bmod 1.$$

The sets Q_n are mutually disjoint. (Show that this is true by showing that if $y \in Q_{n_1} \cap Q_{n_2}$, then $y - r_{n_1}$ and $y - r_{n_2}$ both belong to Q_0 mod 1, and their difference is rational, so both belong to E_0. This contradicts the fact that Q_0 contains only one point from each E_x.)

Now we have a countable collection of disjoint sets $\{Q_n\}$, since each Q_n is a translate of Q_0, $\mu(Q_n) = \mu(Q_0)$ for each n.

Each $x \in [0, 1)$ belongs to Q_n for some n. (Show this to be true.) This means that $\bigcup_{n=1}^{\infty} Q_n = [0, 1)$, and since the Q_n are disjoint we have

$$\mu\left(\bigcup_{n=1}^{\infty} Q_n\right) = \sum_{n=1}^{\infty} \mu(Q_n) = \lim_{n \to \infty} n\mu(Q_0).$$

If $\mu(Q_0) = 0$, $\mu([0, 1)) = 0$. If $\mu(Q_0) = a > 0$, $\mu([0, 1)) = \infty$. The only conclusion we can draw is that Q_0 is a nonmeasurable set.

Theorem 8.8.1. *There do exist sets in E^1, and by extension in E^n, which are not measurable.*

8.9 THE RIEMANN–STIELTJES INTEGRAL IN E^1

We have introduced the Darboux integral with respect to a measure and the Lebesgue integral, and now we shall present a third type of integral. We shall restrict ourselves to the real line, and the measure we shall use is one called Jordan content.

Definition 8.9.1. Let $A \subset E^1$ be a set. The *inner content* of A is the supremum of the sums of the lengths of finitely many nonoverlapping intervals whose union is contained in A. The *outer content* of A is the infimum of the sum of the lengths of finitely many nonoverlapping intervals whose union contains A. If the inner and outer contents are equal, we call the common value the *Jordan content* of A.

Note the distinction between outer content and outer measure. Outer content is the infimum of finite sums, whereas outer measure is the infimum of countable sums of measures.

The reason for using content instead of measure is that the theory of the Riemann integral is generally presented this way. The Riemann–Stieltjes integral is an extension of the Riemann integral. There is a Lebesgue–Stieltjes integral; it is not too far removed from the Riemann–Stieltjes integral that we are going to talk about. The Riemann integral can be extended to higher dimensions; the notion of content is clearly extendable to E^n, an interval extending to a box. However, we won't try to extend the Stieltjes integral beyond E^1.

Definition 8.9.2. Let B be a closed box in E^n,

$$B=[a_1,b_1]\times\cdots\times[a_n,b_n],\qquad \forall i=1,\ldots,n,\quad a_i<b_i.$$

Suppose B is partitioned into a finite number of abutting closed sub-boxes R_k, each with nonempty interior. A scalar-valued function s defined on B is called a *step function* on B if s is constant in the interiors of each of the sub-boxes R_k. The values of s on the boundaries of the R_k are of no concern. If $\operatorname{supp} s\subset B$, s is a step function over all of E^n.

Note that a step function is differentiable almost everywhere (i.e., except for a set of measure zero) and that the differential is zero almost everywhere.

Definition 8.9.3. A function $f\colon E^1\to\mathbb{R}$ is said to be of *bounded variation* on the interval $[a,b]$, and we write $f\in BV(a,b)$, if there exists a constant M such that for any partition π of $[a,b]$, $\pi=\{a=x_0,\ldots,x_n=b\}$, we have $\sum_{i=1}^n|f(x_i)-f(x_{i-1})|\leqslant M$. If $f\in BV(a,b)$ for all intervals $[a,b]$, we say $f\in BV$.

Theorem 8.9.4. *If f is monotonic on $[a,b]$, or if f is continuous on $[a,b]$ and differentiable on (a,b) with f' bounded on (a,b), then $f\in BV(a,b)$. On the other hand, if $f\in BV(a,b)$, then f is bounded on $[a,b]$.*

PROOF: If $f\nearrow$ on $[a,b]$, then for every partition $\pi=\{a=x_0,x_1,\ldots,x_n=b\}$, $\sum_{i=1}^n|f(x_i)-f(x_{i-1})|=f(b)-f(a)$. If $f\searrow$ on $[a,b]$, the sum will be $f(a)-f(b)$. If f satisfies the next set of conditions, then for any such partition π we have, using the mean value theorem,

$$\sum|f(x_i)-f(x_{i-1})|=\sum|f'(\xi_i)(x_i-x_{i-1})|\leqslant M(b-a).$$

Thus, in all these cases, $f\in BV(a,b)$.

Conversely, if $f\in BV(a,b)$, let M be a constant such that for any partition of $[a,b]$, $\sum|f(x_i)-f(x_{i-1})|\leqslant M$. In particular, for each $x\in(a,b)$ we have

$$M\geqslant|f(x)-f(a)|+|f(b)-f(x)|\geqslant|f(x)-f(a)|\geqslant|f(x)|-|f(a)|,$$

so that

$$|f(x)|\leqslant|f(a)|+M.$$

Thus f is bounded by $|f(a)|+|f(b)|+M$ on $[a,b]$. □

Definition 8.9.5. The *total variation* of f on $[a,b]$ is the supremum of the set of numbers $\sum|f(x_i)-f(x_{i-1})|$ over the set of all partitions of $[a,b]$. We denote this possibly infinite number $V_f(a,b)$.

It is evident that $V_f(a,b)<\infty \Leftrightarrow f\in BV(a,b)$ and that the function $V_f(a,x)$ is a nonnegative, increasing function on $[a,b]$. We define $V_f(a,a)=0$.

Theorem 8.9.6. *Suppose $f, g, h \in BV(a,b)$. Let λ, μ be the suprema, respectively of $|f|,|g|$ on $[a,b]$, and let ν be the infimum of $|h|$ on $[a,b]$. Then*

$$(f\pm g)\in BV(a,b), \qquad (fg)\in BV(a,b),$$
$$V_{f\pm g}(a,b)\leqslant V_f(a,b)+V_g(a,b),$$
$$V_{fg}(a,b)\leqslant \mu V_f(a,b)+\lambda V_g(a,b).$$

If $\nu>0$,

$$1/h\in BV(a,b),$$
$$V_{1/h}(a,b)\leqslant V_h(a,b)/\nu^2.$$

If $a\leqslant c\leqslant b$, $f\in BV(a,c)\cap BV(c,b)$, and

$$V_f(a,c)+V_f(c,b)=V_f(a,b).$$

Finally, $V_f(a,x)$ and $V_f(a,x)-f(x)$ are both nonnegative increasing functions on $[a,b]$, so that f can always be expressed as a difference of two such functions.

We leave the proof of this theorem as an exercise: we shall however, point out that if $a\leqslant x<y\leqslant b$, then

$$[V_f(a,y)-f(y)]-[V_f(a,x)-f(x)]=V_f(x,y)-[f(y)-f(x)],$$

and the right-hand side of this equation is nonnegative.

We are now ready to define the Riemann–Stieltjes (R–S) integral of a bounded real-valued function of a single real variable.

Definition 8.9.7. Let $\pi=\{a=x_0<x_1<\cdots<x_n=b\}$ be an arbitrary partition of $[a,b]$, t_i an arbitrary point of $[x_{i-1},x_i]$, and f, α two bounded real functions on $[a,b]$. A number

$$S(\pi,f,\alpha)=\sum_{i=1}^{n} f(t_i)[\alpha(x_i)-\alpha(x_{i-1})]$$

is called a *Riemann–Stieltjes* sum of f with respect to α.

We say that f is *Riemann-integrable* with respect to α on $[a,b]$ if the numbers $S(\pi,f,\alpha)$ converge in the following sense: there exists a number A such that for any $\varepsilon>0$, there is a partition π_ε such that for all partitions π finer than π_ε and any choice of the t_i in the ith intervals of the partition π, we have $|A-S(\pi,f,\alpha)|<\varepsilon$. If such an A exists, we call that number the *value of the integral of f* with respect to α over $[a,b]$ and denote it by $\int_a^b f\,d\alpha$.

If f is integrable in this sense, we say $f \in R(\alpha)$ on $[a, b]$. It is to be understood that if $a<b$, $\int_b^a f\,d\alpha = -\int_a^b f\,d\alpha$ and if $a=b$, the value of the integral is zero. The parts f, α, and $[a, b]$ of the integral $\int_a^b f\,d\alpha$ are called, respectively, the *integrand*, the *integrator*, and the *interval of integration*.

Theorem 8.9.8. *Suppose $f, g \in R(\alpha)$ on $[a, b]$, and c_1, c_2 are constants. Then $(c_1 f + c_2 g) \in R(\alpha)$ on $[a, b]$. If $f \in R(\beta)$ on $[a, b]$ as well, then $f \in R(c_1\alpha + c_2\beta)$ on $[a, b]$. In fact, the R–S integral is linear in its integrand and its integrator. Moreover, if $c \in (a, b)$, then $f \in R(\alpha)$ on $[a, c]$ and $[c, b]$, and we have $\int_a^c f\,d\alpha + \int_c^b f\,d\alpha = \int_a^b f\,d\alpha$.*

Again we leave the busy work to the reader. The proof is straightforward, a direct consequence of the definition of the R–S integral.

The next theorem is not quite so basic.

Theorem 8.9.9. *If $f \in R(\alpha)$ on $[a, b]$, then $\alpha \in R(f)$ on $[a, b]$, and we have the famous integration by parts formula:*

$$\int_a^b f\,d\alpha + \int_a^b \alpha\,df = f(b)\alpha(b) - f(a)\alpha(a).$$

PROOF: Let $\varepsilon > 0$. Let π_ε be a partition such that for all partitions π finer than π_ε we have $|S(\pi, f, \alpha) - \int_a^b f\,d\alpha| < \varepsilon$. Consider an R–S sum for $\int_a^b \alpha\,df$:

$$S(\pi', \alpha, f) = \sum_{i=1}^{n} \alpha(t_i)(f(x_i) - f(x_{i-1})) = \sum_i \alpha(t_i) f(x_i) - \sum_i \alpha(t_i) f(x_{i-1}),$$

where π' is finer than π_ε. We can write

$$f(b)\alpha(b) - f(a)\alpha(a) = \sum_i f(x_i)\alpha(x_i) - \sum_i f(x_{i-1})\alpha(x_{i-1}),$$

and subtracting these two equations, we obtain

$$\begin{aligned} &f(b)\alpha(b) - f(a)\alpha(a) - S(\pi', \alpha, f) \\ &\quad = \sum_i f(x_i)[\alpha(x_i) - \alpha(t_i)] + \sum f(x_{i-1})[\alpha(t_i) - \alpha(x_{i-1})]. \end{aligned}$$

Suppose π is the partition whose points consist of the points of π' as well as all the points t_i in the above expression; $\pi_\varepsilon \prec \pi' \prec \pi$. In terms of the finer partition π, we can write the expression on the right-hand side of the above equation in the form

$$S(\pi, f, \alpha) = \sum_{i=1}^{2n} f(x_i)[\alpha(x_i) - \alpha(x_{i-1})].$$

So we have

$$f(b)\alpha(b) - f(a)\alpha(a) - S(\pi', \alpha, f) = S(\pi, f, \alpha).$$

As $\|\pi'\| \to 0$, $S(\pi', \alpha, f)$ must then converge to $f(b)\alpha(b) - f(a)\alpha(a) - \int_a^b f\,d\alpha$, and we have our formula. □

Suppose we have that $f \in R(\alpha)$ on $[a, b]$ and g is a strictly monotonic continuous function which maps an interval $[c, d]$ onto $[a, b]$. Assume $g\uparrow$, so that $g(c) = a$, $g(d) = b$. If $x = g(y)$, we might replace $f(x)$ and $\alpha(x)$ by $f(g(y)) = h(y)$, and $\alpha(g(y)) = \beta(y)$. This is to say, we make a change of variable. Does $h \in R(\beta)$ on $[c, d]$? Of course. If π is a suitably fine partition so that $S(\pi, f, \alpha)$ is within ε of $\int_a^b f\,d\alpha$ for all choices of the points t_i, then $g^{-1}(\pi)$ is a partition of $[c, d]$, $\{c = y_0, y_1, \ldots, y_n = d\}$, with $y_{i-1} = g^{-1}(x_{i-1}) \leq g^{-1}(t_i) = \tau_i \leq y_i = g^{-1}(x_i)$ for each i, and an R–S sum for h with respect to β on $[c, d]$ is forthcoming which is precisely equal to $S(\pi, f, \alpha)$. We conclude that $\int_c^d h\,d\beta$ exists and is equal to $\int_a^b f\,d\alpha$.

It is reasonably clear that had g been strictly decreasing, we would have found that $\int_c^d h\,d\beta = -\int_a^b f\,d\alpha$. We now summarize this result.

Theorem 8.9.10. *If $f \in R(\alpha)$ on $[a, b]$ and g is a homeomorphism mapping an interval $[c, d]$ onto $[a, b]$, and if $h = f \circ g$, and $\beta = \alpha \circ g$, then $h \in R(\beta)$ on $[c, d]$, and*

$$\int_c^d h\,d\beta = \int_{g(c)}^{g(d)} f\,d\alpha.$$

Theorem 8.9.11. *Suppose $f \in R(\alpha)$ and α is C^1 on $[a, b]$. Then*

$$\int_a^b f\,d\alpha = \int_a^b f(x)\alpha'(x)\,dx.$$

This theorem shows that the Riemann–Stieltjes integral is indeed an extension of the ordinary Riemann integral.

PROOF: Let $\varepsilon > 0$. $\alpha'(x)$ is uniformly continuous on $[a, b]$, so if t_1 and t_2 are sufficiently close together, $\alpha'(t_1)$ and $\alpha'(t_2)$ can't be very far apart. Choose a partition π fine enough so that

$$\left| S(\pi, f, \alpha) - \int_a^b f\,d\alpha \right| < \varepsilon/2$$

and, if M is an upper bound for $|f(x)|$ on $[a, b]$, $|\alpha'(t_1) - \alpha'(t_2)| < \varepsilon/2M(b-a)$ whenever $|t_1 - t_2| < \|\pi\|$. Thanks to the mean value theorem we can write

$$\begin{aligned} S(\pi, f, \alpha) &= \sum_i f(t_i)[\alpha(x_i) - \alpha(x_{i-1})] \\ &= \sum_i f(t_i)\alpha'(t_i')(x_i - x_{i-1}), \qquad t_i' \in (x_{i-1}, x_i). \end{aligned}$$

Let $\underline{t}_i, \bar{t}_i$ be points in $[x_{i-1}, x_i]$ where $f(t_i)\alpha'(x)$ assumes its minimum and maximum values, respectively, on that closed subinterval, so that we have

$$\begin{aligned}\sum_i f(t_i)\alpha'(\underline{t}_i)(x_i - x_{i-1}) &\leqslant \sum_i f(t_i)\alpha'(\tau_i)(x_i - x_{i-1})\\ &\leqslant \sum_i f(t_i)\alpha'(\bar{t}_i)(x_i - x_{i-1})\end{aligned}$$

for any choice of $\tau_i \in [x_{i-1}, x_i]$. Since $|\bar{t}_i - \underline{t}_i| < \|\pi\|$, we can write

$$\begin{aligned}0 &\leqslant \sum_i f(t_i)(x_i - x_{i-1})\left[\alpha'(\bar{t}_i) - \alpha'(\underline{t}_i)\right]\\ &\leqslant \sum_i |f(t_i)|(x_i - x_{i-1})|\alpha'(\bar{t}_i) - \alpha'(\underline{t}_i)|\\ &\leqslant M\frac{\varepsilon}{2M(b-a)}\sum_i (x_i - x_{i-1}) = \varepsilon/2.\end{aligned}$$

It follows that if we take τ_i to be t_i, we have an R–S sum $S(\pi, f\alpha', x)$, and if we take τ_i to be t_i', we have our original R–S sum $S(\pi, f, \alpha)$. Since these two sums differ by less than $\varepsilon/2$, we conclude that

$$\left|S(\pi, f\alpha', x) - \int_a^b f\,d\alpha\right| < \varepsilon,$$

and, finally, that $\int_a^d f\,d\alpha = \int_a^b f(x)\alpha'(x)\,dx$. □

One situation that occurs frequently in applications is the case where the integrator is a step function. The following theorem shows when we can handle such cases.

Theorem 8.9.12. *Let $\alpha(x)$ be a step function on $[a, b]$ with jumps of α_i at each x_i in the partition determined by the step function α. If f is a function defined on $[a, b]$ such that both f and α are not discontinuous from the same side at each x_i, then $f \in R(\alpha)$ and*

$$\int_a^b f\,d\alpha = \sum_{i=1}^n f(x_i)\alpha_i.$$

PROOF: It suffices to consider the simplest case where $a < c < b$, $\alpha(x) = \alpha(a)$ on $[a, c)$, and $\alpha(x) = \alpha(b)$ on $(c, b]$, and at least one of f or α is continuous at c from the left, and the other from the right. Without loss of generality, assume $\alpha(c) = \alpha(b)$, so that f is continuous from the left at c. Denote by $\alpha(c^-)$ the left limit of α at c; $\alpha(c^-) = \alpha(a)$. Let π be any partition which includes c; then $S(\pi, f, \alpha) = f(t)[\alpha(c) - \alpha(c^-)]$, where t is in the subinterval whose right endpoint is c. The limiting value of $S(\pi, f, \alpha)$ is $\alpha_c f(c)$, where α_c is the jump of α at c.

If f is continuous on the right at c and α continuous from the left, the limiting value of $S(\pi, f, \alpha)$ is the same. If c is an endpoint of $[a, b]$, it is a simple matter to extend the proof to include these two cases. Finally, we can extend this result to arbitrary step functions by making use of the additivity of the integral as a set function. □

Theorem 8.9.13. *If $c \in [a, b]$, and f and α are both discontinuous from the same side at c, then $f \notin R(\alpha)$ on $[a, b]$.*

PROOF: Assume f and α are discontinuous from the left at c, and let π be any partition which includes c as a partition point. Two R–S sums which differ only in the choice of a point $t_i \in [x_{i-1}, c]$, where $t_i = c$ for one and $t_i \in [x_{i-1}, c)$ for the other, will differ by an amount

$$|f(c) - f(t_i)||\alpha(c) - \alpha(x_{i-1})|.$$

If the norm of the partition is fine enough, we necessarily have two positive numbers δ_1, δ_2 such that $|f(c) - f(t_i)| > \delta_1$ and $|\alpha(c) - \alpha(x_{i-1})| > \delta_2$, regardless of how $t_i \in [x_{i-1}, c)$ is chosen.

Now suppose $\varepsilon > 0$ is given with $\varepsilon < \delta_1\delta_2$. Then no matter how fine a partition π may be, it can be refined to include c, and then there will always be an R–S sum (the one with $t_i = c$) which differs from other R–S sums by more than ε. Hence the integral $\int_a^b f\,d\alpha$ cannot exist. □

Note that if $\alpha(x) = [\![x]\!]$, where $[\![x]\!]$ is defined to be the greatest integer which is less than or equal to x, and if $\sum_{i=1}^n a_i$ is any sum of numbers, then we can define a function f on $[0, n]$ by

$$f(x) = a_i, \qquad \text{if} \quad i-1 < x \leq i, \quad i = 1, 2, \ldots, n,$$

and

$$f(0) = 0,$$

so that f is continuous from the left at each integer of $[0, n]$. Since $\alpha(x)$ is continuous from the right at each integer, $\int_0^n f\,d\alpha$ exists, and by Theorem 8.9.12 its value is $\sum_{i=1}^n f(i)\alpha_i = \sum_{i=1}^n f(i)$. This is to say

$$\int_0^n f\,d[\![x]\!] = \sum_{i=1}^n a_i,$$

so that every finite sum can be written as a Riemann–Stieltjes integral.

There is a famous formula known as Euler's summation formula which we can get by the use of Stieltjes integration.

Theorem 8.9.14. *Suppose f is C^1 on $[a, b]$. Then we have Euler's summation formula*

$$\sum_{n=[\![a]\!]+1}^{[\![b]\!]} f(n) = \int_a^b f(x)\,dx + \int_a^b f'(x)(x - [\![x]\!])\,dx$$
$$+ f(a)(a - [\![a]\!]) - f(b)(b - [\![b]\!]).$$

PROOF: We have

$$\int_a^b f\,d(x-[\![x]\!])+\int_a^b (x-[\![x]\!])\,df=f(b)(b-[\![b]\!])-f(a)(a-[\![a]\!]),$$

so

$$\int_a^b f'(x)(x-[\![x]\!])\,dx+f(a)(a-[\![a]\!])-f(b)(b-[\![b]\!])+\int_a^b f(x)\,dx$$

$$=\int_a^b f\,d[\![x]\!]=\sum_{n=[\![a]\!]+1}^{[\![b]\!]} f(n). \qquad \square$$

In case a and b are both integers, the conclusion of the theorem reduces to

$$\sum_{n=a+1}^{b} f(n)=\int_a^b f(x)\,dx+\int_a^b f'(x)(x-[\![x]\!])\,dx.$$

Thus

$$\sum_{n=a}^{b} f(n)=\int_a^b f(x)\,dx+\int_a^b f'(x)(x-[\![x]\!])\,dx+f(a),$$

or

$$\sum_{n=a}^{b} f(n)=\int_a^b f(x)\,dx+\int_a^b f'(x)(x-[\![x]\!]-\tfrac{1}{2})\,dx+\tfrac{1}{2}(f(a)+f(b)).$$

We have shown that every function $\alpha\in BV(a,b)$ can be expressed as the difference of two nonnegative increasing functions $\alpha_1(x)=V_\alpha(a,x)$ and $\alpha_2(x)=V_\alpha(a,x)-\alpha(x)$. Since the integral is linear in both integrand and integrator, it suffices to assume in developing R–S theory for functions of bounded variation that these functions are increasing functions. It is perfectly straightforward, when the integrator α is increasing on $[a,b]$, to define upper and lower Darboux sums, just as we did at the beginning of this chapter, and if these sums converge to upper and lower integrals, respectively, we can state that if $\alpha\nearrow$ on $[a,b]$, then $f\in R(\alpha)$ if the upper and lower integrals are equal, which is tantamount to saying that given $\varepsilon>0$, there exists a partition π_ε, such that for all partitions π finer than π_ε we have

$$\bar{S}(\pi,f,\alpha)-\underline{S}(\pi,f,\alpha)<\varepsilon.$$

Theorem 8.9.15. *If f,α are real-valued functions defined on $[a,b]$, one continuous and the other of bounded variation, then $f\in R(\alpha)$ and $\alpha\in R(f)$. In particular, if $f\in BV(a,b)$ or if f is continuous on $[a,b]$, then $\int_a^b f(x)\,dx$ exists.*

PROOF: Because of the integration by parts theorem it suffices to show that just one of the integrals exists. Assume that f is continuous and $\alpha\in BV(a,b)$.

We can assume $\alpha \nearrow$ on $[a, b]$. Since f is bounded and uniformly continuous on $[a, b]$, choose a partition as follows. $\varepsilon > 0$ having been given, a $\delta > 0$ can be found such that $|t_1 - t_2| < \delta \Rightarrow |f(t_1) - f(t_2)| < \varepsilon/2(\alpha(b) - \alpha(a))$; take a partition π_ε with $\|\pi_\varepsilon\| < \delta$.

If

$$M_i = \sup\{f(t) : t \in [x_{i-1}, x_i]\}$$

and

$$m_i = \inf\{f(t) : t \in [x_{i-1}, x_i]\},$$

then for any π finer than π_ε we have

$$\bar{S}(\pi, f, \alpha) - \underline{S}(\pi, f, \alpha) = \sum_{i=1}^{n} (M_i - m_i)(\alpha(x_i) - \alpha(x_{i-1})) < \varepsilon.$$

Thus $f \in R(\alpha)$. The last statement is an obvious corollary to the general statement of the theorem. □

There are some useful mean value theorems for R–S integration; we now present two of them.

Theorem 8.9.16. *Let $f \in R(\alpha)$, and assume $\alpha \nearrow$ on $[a, b]$. There exists a real number c with* $\inf f(x) \leq c \leq \sup f(x)$, $x \in [a, b]$, *such that*

$$\int_a^b f\,d\alpha = c\int_a^b d\alpha = c(\alpha(b) - \alpha(a)),$$

and if f is continuous on $[a, b]$, then for some $x_0 \in [a, b]$, $f(x_0) = c$.

PROOF: If $\alpha(b) = \alpha(a)$, there is nothing to prove. Otherwise, we have, for any partition π,

$$(\inf f(x))(\alpha(b) - \alpha(a)) \leq \underline{S}(\pi, f, \alpha) \leq \bar{S}(\pi, f, \alpha)$$
$$\leq (\sup f(x))(\alpha(b) - \alpha(a)),$$

so $\int_a^b f\,d\alpha$ lies between these outer bounds. The completeness of $\mathbb{R}$ guarantees the existence of c, and if f is continuous on $[a, b]$, f assumes all values between and including its infimum and supremum on $[a, b]$. □

Theorem 8.9.17. *Assume α continuous on $[a, b]$ and $f \nearrow$ on $[a, b]$. Then for some point $c \in [a, b]$ we have*

$$\int_a^b f\,d\alpha = f(a)\int_a^c d\alpha + f(b)\int_c^b d\alpha.$$

PROOF: $f \in BV(a, b)$, so $f \in R(\alpha)$, and we get, using the previous theorem and integration by parts,

$$\begin{aligned}\int_a^b f\,d\alpha &= f(b)\alpha(b) - f(a)\alpha(a) + \int_a^b \alpha\, df \\ &= f(b)\alpha(b) - f(a)\alpha(a) - \alpha(c)(f(b) - f(a)) \\ &= f(a)\int_a^c d\alpha + f(b)\int_c^b d\alpha. \quad \square\end{aligned}$$

What follows is an extremely important application of Theorems 8.9.9 and 8.9.16.

If $f: \mathbb{R} \to \mathbb{R}$ is a C^n-function, we can write for a fixed x and fixed a,

$$f(x) - f(a) = \int_a^x f'(y)\, dy.$$

Let $t = x - y$, and write

$$\int_a^x f'(y)\, dy = -\int_{x-a}^0 f'(x-t)\, dt = \int_0^{x-a} f'(x-t)\, dt.$$

Using Theorem 8.9.9 repeatedly, we get

$$\begin{aligned}\int_0^{x-a} f'(x-t)\, dt &= f'(x-t)\cdot t\Big|_0^{x-a} - \int_0^{x-a}(-1)f''(x-t)t\, dt \\ &= f'(a)(x-a) + \frac{f''(x-t)t^2}{2}\Big|_0^{x-a} \\ &\quad - \int_0^{x-a}(-1)f'''(x-t)\frac{t^2}{2}\, dt \\ &= f'(a)(x-a) + \frac{f''(a)(x-a)^2}{2} + \int_0^{x-a} f^{(3)}(x-t)\frac{t^2}{2}\, dt \\ &= \sum_{k=1}^{n-1}\frac{f^{(k)}(a)(x-a)^k}{k!} + \int_0^{x-a}\frac{f^{(n)}(x-t)t^{n-1}}{(n-1)!}\, dt.\end{aligned}$$

Since $f^{(n)}/(n-1)!$ is continuous, apply Theorem 8.9.16 to get

$$f(x) - f(a) = \sum_{k=1}^{n-1}\frac{f^{(k)}(a)(x-a)^k}{k!} + \frac{f^{(n)}(\tau)}{n!}(x-a)^n,$$

where τ is some number between a and x. We have the following formula known as Taylor's formula.

Theorem 8.9.18. *Suppose $f: \mathbb{R} \to \mathbb{R}$ is C^n; i.e., f and its first n derivatives are continuous in a neighborhood of a point $a \in \mathbb{R}$. Then we can represent $f(x)$*

for each x sufficiently close to a by the formula

$$f(x)=\sum_{k=0}^{n-1}\frac{f^{(k)}(a)(x-a)^k}{k!}+\frac{f^{(n)}(\tau_x)}{n!}(x-a)^n,$$

where τ_x is a number depending upon x which lies between a and x.

We have given a rather lengthy presentation of the basic theory of Riemann–Stieltjes integration in E^1. Of course, as we mentioned earlier, this theory can be modified by making the notion of measure more sophisticated. Although this book has been designed to give the reader a useful introduction to many different areas of mathematics which are intimately connected with advanced calculus, it is, alas, not able to tell the whole story.

We mentioned at the outset of this section that there is a Lebesgue–Stieltjes integral, so we shall at least tell you what it is. Recall that if the integrator in a Stieltjes integral is a function of bounded variation on an interval $[a, b]$, it could be expressed as the difference of two nonnegative increasing functions, so we might as well consider integrators that are of that more simple type. Suppose now that we replace Jordan content by Lebesgue measure, and consider integrands f which are Lebesgue-measurable functions along with integrators φ which are increasing. We call the integral $\int_a^b f\,d\varphi$ a Lebesgue–Stieltjes integral; its value is the value of the Lebesgue integral $\int_{\varphi(a)}^{\varphi(b)} F(\xi)\,d\xi$ whenever this latter value exists. The function F is defined on the interval $[\varphi(a), \varphi(b)]$ by the following relations:

1. $\forall\xi\in[\varphi(a),\varphi(b)]$, if $\exists x\in[a,b]$ s.t. $\xi=\varphi(x)$, then $F(\xi)=f(x)$.
2. If $\xi_0\in[\varphi(a),\varphi(b)]$ is such that for no $x\in[a,b]$ is $\varphi(x)=\xi_0$, there is a unique $x_0\in[a,b]$ at which φ has a jump discontinuity, with $\varphi(x_0^-)\leqslant\xi_0\leqslant\varphi(x_0^+)$; in this case, define $F(\xi_0)=f(x_0)$.

If it happens to be the case that F as defined is measurable on $[\varphi(a), \varphi(b)]$ and f is measurable on $[a, b]$ and that $\varphi(x)=\int_a^x\psi(t)\,dt$ for some measurable function ψ, then

$$\int_a^b f\,d\varphi=\int_a^b f(x)\,d\varphi(x)=\int_a^b f(x)\psi(x)\,dx,$$

and the integral on the right-hand side is an ordinary Lebesgue integral.

We bring this section to a close by presenting this last little theorem.

Theorem 8.9.19. *Let $\alpha\in BV(a,b)$ and $f\in R(\alpha)$ on $[a,b]$. Define*

$$F(x)=\int_a^x f\,d\alpha\qquad\forall x\in[a,b].$$

Then

1. $F\in BV(a,b)$;
2. *every point of continuity of α is a point of continuity of F; and*
3. *$F'(x)$ exists at each $x\in(a,b)$ where $\alpha'(x)$ exists and f is continuous, and for such x, $F'(x)=f(x)\alpha'(x)$.*

PROOF: Again, we may assume that α is increasing on $[a, b]$. Theorem 8.9.16 assures the existence of a number c such that for any pair $x_1, x_2 \in [a, b]$, $x_1 < x_2$,

$$F(x_2) - F(x_1) = c(\alpha(x_2) - \alpha(x_1)),$$

where $\inf\{f(x): x \in [x_1, x_2]\} \leqslant c \leqslant \sup\{f(x): x \in [x_1, x_2]\}$. Now f is assumed to be bounded on $[a, b]$ since $f \in R(\alpha)$ on $[a, b]$, so if $|f(x)| \leqslant M$, we can write $|F(x_2) - F(x_1)| \leqslant M(\alpha(x_2) - \alpha(x_1))$. It follows immediately that $F \in BV(a, b)$. Moreover, the above equation shows that if x_1 is a point of continuity of α, it must necessarily be such a point for F.

Finally, suppose x_1 is a point of continuity of f and $\alpha'(x_1)$ exists. Divide both sides of the above equation by $x_2 - x_1$. As $x_2 \to x_1$, the number c is forced to approach $f(x_1)$, the quotient $[\alpha(x_2) - \alpha(x_1)]/[x_2 - x_1]$ approaches $\alpha'(x_1)$, and we are left with

$$\lim_{x_2 \to x_1} \frac{F(x_2) - F(x_1)}{x_2 - x_1} = f(x_1)\alpha'(x_1).$$

If $x_2 < x_1$, we get the same result, so the third part of the theorem is proved. □

Corollary 8.9.20. *Suppose $\alpha \in BV(a, b)$, $f \in R(\alpha)$ on $[a, b]$, and α is absolutely continuous on $[a, b]$. Then $F(x) = \int_a^x f\,d\alpha$ is absolutely continuous.*

PROOF: The theorem assures that F is continuous and hence uniformly continuous on $[a, b]$. Assume that $\alpha \nearrow$ on $[a, b]$, and let M be an upper bound for $|f(x)|$ on $[a, b]$. Then for any finite set of (not necessarily distinct) points $\{x_1, \ldots, x_{2n}\}$ such that $\sum_{i=1}^n |x_{2i} - x_{2i-1}| < \delta$, δ to be determined later, we have

$$\sum_{i=1}^{n} |F(x_{2i}) - F(x_{2i-1})| \leqslant M \sum_{i=1}^{n} |\alpha(x_{2i}) - \alpha(x_{2i-1})|.$$

Suppose $\varepsilon > 0$. Since α is absolutely continuous on $[a, b]$, we can choose δ small enough so that $\sum_{i=1}^n |\alpha(x_{2i}) - \alpha(x_{2i-1})| < \varepsilon/M$. □

If a real-valued function $f(x)$ is absolutely continuous on an interval $[a, b]$, does it follow that $f \in BV(a, b)$? What about the converse statement? We leave these two questions for the reader to ponder over while we move on to the next section.

8.10 FUBINI'S THEOREM

We have presented a fairly comprehensive treatment of integration in this chapter, and the reader who is not completely numbed by it all might suddenly realize at this moment that we have failed to tell you one rather important thing: how to evaluate an integral. Of course, Theorem 5.4.2 tells how to evaluate $\int_a^b f'(x)\,dx$, but then this has long been well known. If the integrand of this simple integral is of bounded variation, an approximate value can be obtained by various methods. But we are concerned with the evaluation of integrals of functions over sets in E^n.

Generally, an integral is evaluated by using Darboux approximation or something closely related to it. What we are going to do is show how an integral in E^2 can be evaluated by an iteration process. We start by presenting this theorem:

Theorem 8.10.1. *Suppose $f(t,x)$ is a (uniformly) continuous function on the closed rectangle $[a,b]\times[c,d]$ and that $D_2 f$ is also (uniformly) continuous on this rectangle. Let*

$$g(x)=\int_a^b f(t,x)\,dt.$$

Then g is differentiable, and

$$g'(x)=\frac{d}{dx}\int_a^b f(t,x)\,dt=\int_a^b D_2 f(t,x)\,dt.$$

PROOF: Let $\varepsilon>0$. For all h sufficiently small we have

$$\left|\frac{g(x+h)-g(x)}{h}-\int_a^b D_2 f(t,x)\,dt\right|$$
$$\leqslant\int_a^b\left|\frac{f(t,x+h)-f(t,x)}{h}-D_2 f(t,x)\right|dt<\varepsilon(b-a)$$

since for each $t\in[a,b]$ there is a constant c_t between x and $x+h$ such that $f(t,x+h)-f(t,x)=hD_2 f(t,c_t)$ and the uniform continuity of $D_2 f$ ensures that for all $t\in[a,b]$, if h is sufficiently small, $|D_2 f(t,c_t)-D_2 f(t,x)|<\varepsilon$. Hence

$$\lim_{h\to 0}\frac{g(x+h)-g(x)}{h}=g'(x)=\int_a^b D_2 f(t,x)\,dt. \quad \square$$

Suppose R is the (positively oriented) rectangle $[a,b]\times[c,d]$. The next theorem tells how to evaluate $\int_R f\,d\mu$ by an iterated integration process.

Theorem 8.10.2 (Fubini—the Simple Version). *Let $f\colon R\to\mathbb{R}$ be continuous; then the maps*

$$\varphi\colon x\mapsto\int_a^b f(t,x)\,dt,\qquad \theta\colon t\mapsto\int_c^d f(t,x)\,dx$$

are continuous and

$$\int_R f(t,x)\,d\mu=\int\int_R f(t,x)\,dt\,dx=\int_c^d\left[\int_a^b f(t,x)\,dt\right]dx$$
$$=\int_a^b\left[\int_c^d f(t,x)\,dx\right]dt.$$

PROOF: We already know that $\int_R f\,d\mu$ exists and equals $\iint_R f\,d\mu$ since R is positively oriented. Put $\varphi(x)=\int_a^b f(t,x)dt$. Then for each $x\in[c,d]$,

$$\varphi(x+h)-\varphi(x)=\int_a^b [f(t,x+h)-f(t,x)]\,dt.$$

Since f is uniformly continuous on R, $\varphi(x+h)-\varphi(x)\to 0$ as $h\to 0$, so φ is continuous on $[c,d]$. Analogously, $\theta(t)=\int_c^d f(t,x)\,dx$ is continuous on $[a,b]$.

Now put

$$\psi(t,x)=\int_c^x f(t,v)\,dv, \qquad c\leqslant x\leqslant d.$$

Then $D_2\psi(t,x)=f(t,x)$, so $D_2\psi$ is uniformly continuous on R. It is easy to see that $\psi(t,x)$ is also continuous on R, and hence uniformly continuous, since for any $(t_0,x_0)\in R$,

$$\psi(t,x)-\psi(t_0,x_0)=\int_c^{x_0}[f(t,v)-f(t_0,v)]\,dv+\int_{x_0}^x f(t,v)\,dv.$$

and if (t,x) is sufficiently close to (t_0,x_0), each of the integrals on the right-hand side is small.

Now put $g(x)=\int_a^b \psi(t,x)\,dt$. Then

$$g'(x)=\int_a^b D_2\psi(t,x)\,dt=\int_a^b f(t,x)\,dt,$$

and

$$g(d)-g(c)=\int_c^d\left[\int_a^b f(t,x)\,dt\right]dx.$$

However,

$$g(d)-g(c)=\int_a^b \psi(t,d)\,dt-\int_a^b \psi(t,c)\,dt=\int_a^b\left[\int_c^d f(t,v)\,dv\right]dt,$$

since $\psi(t,c)=0$, $\forall t\in[a,b]$. Hence we have

$$\int_a^b\left[\int_c^d f(t,x)\,dx\right]dt=\int_c^d\left[\int_a^b f(t,x)\,dt\right]dx.$$

Finally, $\iint_R f(t,x)\,d\mu$ can be approximated as closely as we like by a Riemann sum of the form

$$\sum_{i,j}^{n,m} f(P_{ij})(x_j-x_{j-1})(t_i-t_{i-1}),$$

where P_{ij} is a point in $[t_{i-1},t_i]\times[x_{j-1},x_j]$. But this is the sum

$$\sum_{j=1}^{m}\left[\sum_{i=1}^{n} f(P_{ij})(t_i-t_{i-1})\right](x_j-x_{j-1}),$$

which is an approximation for $\int_c^d[\int_a^b f(t,x)\,dt]\,dx$. This completes the proof. □

This is the simplest possible form of Fubini's theorem. Note that it says that not only can you integrate f over R by a repeated integration, but that the order in which you do the repeated integration doesn't matter. Needless to say, the theorem extends from E^2 to E^n; the rectangle R becomes a box in E^n. An induction proof for this will work.

We need a further extension of this theorem if it is to be of much use, and we describe briefly what it is. Suppose A is a compact subset of E^n and f is a real-valued function to be integrated over A. In theory, we look for a C^1 injection $E^n \to E^n$ that will map a box B onto A and then integrate over B the product of the pullback of f and the g factor of the injection. The simple form of Fubini's theorem tells us that we can do this by performing n repeated one-dimensional integrations, in various orders, at least when f is continuous. But if A is rather special (or can be partitioned into "special" subsets), we can dispense with the pullback operation. Here is what we mean by "special."

Suppose A is such that we can find two continuous boundary functions

$$\bar{x}_n = \bar{x}_n(x_1,\ldots,x_{n-1}), \qquad \underline{x}_n = \underline{x}_n(x_1,\ldots,x_{n-1})$$

such that for all $(x_1,\ldots,x_{n-1},0)$ in A_{n-1}, the projection of A onto the $(x_1\ldots x_{n-1})$-coordinate subspace, the line segment $[\underline{x}_n(x_1,\ldots,x_{n-1}), \bar{x}_n(x_1,\ldots,x_{n-1})]$ is contained in A. A is the union of all such line segments.

Now suppose A_{n-1} is such that we can find two continuous boundary functions

$$\bar{x}_{n-1} = \bar{x}_{n-1}(x_1,\ldots,x_{n-2}), \qquad \underline{x}_{n-1} = \underline{x}_{n-1}(x_1,\ldots,x_{n-2})$$

such that for each point $(x_1,\ldots,x_{n-2},0,0)$ in A_{n-2}, the projection of A onto the $(x_1\ldots x_{n-2})$-coordinate subspace, the line segment $[\underline{x}_{n-1}(x_1,\ldots,x_{n-2}), \bar{x}_{n-1}(x_1,\ldots,x_{n-2})]$ is contained in A_{n-1} and A_{n-1} is the union of all such line segments.

Continue in this way until we get to A_1, the projection of A onto the x_1 axis. Suppose this region is bounded by $\bar{x}_1 = b$ and $\underline{x}_1 = a$, and $[a,b]$ is contained in A_1.

The order in which we do the projecting is immaterial, as long as we go from E^n to an $(n-1)$-dimensional coordinate subspace to a one-dimensional coordinate subspace. When this can be done, we call A a special region.

Note that if A is convex, all the projections are convex, and there will be no question about the particular line segments being contained in the projections.

Theorem 8.10.3 (Fubini). *Let $A \subset E^n$ be a special compact set as described above and f a continuous real function on A. Then*

$$\int \cdots \int_A f\,d\mu = \int_a^b \int_{\underline{x}_2(x_1)}^{\bar{x}_2(x_1)} \cdots \int_{\underline{x}_n(x_1,\ldots,x_{n-1})}^{\bar{x}_n(x_1,\ldots,x_{n-1})} f\,dx_n \cdots dx_2\,dx_1,$$

where the expression on the right-hand side is the iterated integral (with all the parentheses omitted).

A more general statement of the theorem is the following.

Suppose A is a μ-measurable set in E^n, f a positive μ-measurable, extended real-valued function on A, X and Y are subsets of E^n such that $A = X \times Y$, μ_X, μ_Y are Lebesgue measures on the Euclidean subspaces spanned by X, Y, respectively, and X, Y are μ_X- and μ_Y-measurable, respectively. Then

$f(x, y)$ is μ_X-measurable for each $y \in Y$;

$f(x, y)$ is μ_Y-measurable for each $x \in X$;

$\int_X f(x, y)\, d\mu_X$ is μ_Y-measurable;

$\int_Y f(x, y)\, d\mu_Y$ is μ_X-measurable; and

$$\int_A f\, d\mu = \int_Y \int_X f(x, y)\, d\mu_X d\mu_Y = \int_X \int_Y f(x, y)\, d\mu_Y d\mu_X.$$

We omit the proof; it may be found in more advanced texts on real analysis.

8.11 APPROXIMATE CONTINUITY

Suppose f is a real-valued function on the measure space $(E^n, \mathfrak{M}, \mu)$. Let $\mathbf{p} \in E^n$. We want to give the reader one additional type of continuity that a real function might enjoy: approximate continuity at $\mathbf{p}$.

Let $S \in \mathfrak{M}$ be a measurable set and B an open box in E^n. The quotient $\mu(S \cap B)/\mu(B)$ expresses the *relative measure* of S in B; since $\mu(B) > 0$, this quotient is always a real number between 0 and 1, inclusive. Now think of B "shrinking" continuously to a point $\mathbf{x}$; the relative measure of S in B converges perhaps to a value we might reasonably call the metric density of S at $\mathbf{x}$. To make the description more detailed, we write

$$\bar{\varphi}_n(\mathbf{x}) = \sup\left[\frac{\mu(S \cap B)}{\mu(B)} : \mathbf{x} \in B,\ 0 < \operatorname{diam} B < \frac{1}{n}\right],$$

$$\underline{\varphi}_n(\mathbf{x}) = \inf\left[\frac{\mu(S \cap B)}{\mu(B)} : \mathbf{x} \in B,\ 0 < \operatorname{diam} B < \frac{1}{n}\right].$$

Then let $\bar{\varphi}(\mathbf{x}) = \lim_{n \to \infty} \bar{\varphi}_n(\mathbf{x})$ and $\underline{\varphi}(\mathbf{x}) = \lim_n \underline{\varphi}_n(\mathbf{x})$. Since $\{\bar{\varphi}_n(\mathbf{x})\} \searrow$ and $\{\underline{\varphi}_n(\mathbf{x})\} \nearrow$, the limits exist, and we obviously have

$$0 \leq \underline{\varphi}(\mathbf{x}) \leq \bar{\varphi}(\mathbf{x}) \leq 1.$$

Definition 8.11.1. $\underline{\varphi}(\mathbf{x})$ and $\bar{\varphi}(\mathbf{x})$ as described above are called the *lower metric density* and *upper metric density* of the measurable set S at $\mathbf{x}$. If these two values are equal, we denote the common value by $\varphi(\mathbf{x})$, and this value is the metric density of the measurable set S at $\mathbf{x}$.

The reader might make a note of the fact that the notion of the metric density of a measurable set S at $\mathbf{x}$ is somewhat like the notion of a derivative at $\mathbf{x}$.

We simply state, but do not prove, a theorem called the Lebesgue density theorem.

Theorem 8.11.2 (Lebesgue). *If S is a measurable set, the metric density of S exists and is equal to 1 at every point of S except for a set of measure zero.*

Definition 8.11.3. A measurable function $f\colon E^n \to \mathbb{R}$ is said to be *approximately continuous* at $\mathbf{p}$ iff for every open interval (a, b) such that $f(\mathbf{p}) \in (a, b)$, the set $S = f^{-1}((a, b))$ has metric density 1 at $\mathbf{p}$.

Theorem 8.11.4. *If f is continuous at $\mathbf{p}$, then f is approximately continuous at $\mathbf{p}$, but not conversely.*

PROOF: Since f is continuous at $\mathbf{p}$, any interval (a, b), with $a < f(\mathbf{p}) < b$, gives rise to a preimage $S = f^{-1}((a, b))$, which is an open neighborhood of $\mathbf{p}$, and hence for all n sufficiently large we would have $\underline{\varphi_n}(\mathbf{p}) = 1$, so the metric density of S at $\mathbf{p}$ is 1.

Consider the characteristic function χ_{T^+} where the set $T^+ \subset [0,1]$ is described as follows: For each $n = 1, 2, 3, \ldots$, let

$$T_n = \left(\frac{1}{n+1} + \frac{n-1}{2n^2(n+1)}, \frac{1}{n} - \frac{n-1}{2n^2(n+1)} \right)$$

and $T^+ = \bigcup_{n=1}^{\infty} T_n$. T^+ is a union of disjoint open intervals, $T_1 = (\frac{1}{2}, 1)$, $T_2 = (\frac{9}{24}, \frac{11}{24})$, $T_3 = (\frac{20}{72}, \frac{22}{72})$, etc., and $\mu(T_n) = 1/n^2(n+1)$ for each n. $\chi_{T^+}(x) = 1$, if $x \in T^+$, and 0, otherwise. Let $f(x)$ be the even function defined on $[-1, 1]$ by $f(-x) = f(x) = \chi_{T^+}(x)$, $0 \leqslant x \leqslant 1$. Let $T^- \subset [-1, 0]$ be the symmetric image of T^+, and define $T = T^+ \cup T^- \subset [-1, 1]$. It is apparent that f is not continuous at $x = 0$. It is not hard to see that f is measurable and that, if (a, b) is any open interval containing $0 = f(0)$, $S = f^{-1}((a, b))$ is a measurable set and S has metric density 1 at $x = 0$. Consider the case when $b > 1$; then $S = [-1, 1]$. If $0 < b \leqslant 1$, $S = [-1, 1] \setminus T$. In the immediate neighborhood of 0 it is clear that the ratio of the relative measures T to $\tilde{T}$ is of the order $1/n$. Hence, for a small interval I about the origin, the relative measure of S in I is of the order $(n-1)/n$, and this approaches 1. We have thus argued convincingly that this function $f(x)$ is approximately continuous at 0, which proves that approximate continuity at a point does not imply continuity. □

Note that approximate continuity is a term reserved for measurable functions. If f is approximately continuous at each point $\mathbf{p}$ of a set D, we say f is approximately continuous on D.

Theorem 8.11.5. *If f is approximately continuous almost everywhere on a bounded measurable set $B \subset E^n$, then f may be approximated "in measure" (see Definition 10.1.1, part 7) by a step function s or a continuous function g on B.*

PROOF: Assume that $\mu(B)=1$. The statement of the theorem means that if f is approximately continuous almost everywhere on the set B, then for each $\varepsilon>0$ and each $\delta>0$, we can find a continuous function g such that the measure of the set in B on which $|f(\mathbf{x})-g(\mathbf{x})|<\varepsilon$ is greater than $1-\delta=\mu(B)-\delta$. Let $A \subset B$ be the set on which f is approximately continuous. For each $\mathbf{y} \in A$ there is a sequence of closed boxes $\{B_{ny}\}_{n=1}^{\infty}$ with $\{B_{ny}\} \searrow \{\mathbf{y}\}$ such that for each n the relative measure

$$\frac{\mu(A \cap B_{ny})}{\mu(B_{ny})} > 1 - \frac{\varepsilon}{2},$$

$\varepsilon>0$ having been arbitrarily fixed. This is tantamount to saying that the relative measure of the set $\{\mathbf{x} \in A : |f(\mathbf{x})-f(\mathbf{y})|<\varepsilon\}$ is greater than $1-\varepsilon/2$. The collection $\mathfrak{B}=\{B_{ny} : \mathbf{y} \in A,\ n=1,2,3,\ldots\}$ covers A, and since A is bounded and has finite measure, it can be shown (with some difficulty) that we can extract from $\mathfrak{B}$ a finite collection of disjoint boxes $\{B_{n_1 y_1},\ldots,B_{n_k y_k}\}$ for which $\sum_{i=1}^{k}\mu(B_{n_i y_i})>1-\varepsilon/2$. The collection $\mathfrak{B}$ is a kind of covering for A known as a Vitali covering. The Vitali covering theorem assures that we can indeed extract such a finite collection.

Define a step function $s(\mathbf{x})$ as follows:

$$s(\mathbf{x}) = \begin{cases} f(\mathbf{y}_i), & \text{for each } \mathbf{x} \in B_{n_i y_i}, \quad i=1,\ldots,k. \\ 0, & \text{elsewhere.} \end{cases}$$

If E is the set $\{\mathbf{x} \in A : |f(\mathbf{x})-s(\mathbf{x})|<\varepsilon\}$, then $\forall i=1,\ldots,k$,

$$\frac{\mu(E \cap B_{n_i y_i})}{\mu(B_{n_i y_i})} > 1 - \frac{\varepsilon}{2}.$$

Hence

$$\mu(E) \geqslant \sum_{i=1}^{k} \mu(E \cap B_{n_i y_i}) > \left(1-\frac{\varepsilon}{2}\right) \sum_{i=1}^{k} \mu(B_{n_i y_i}) > \left(1-\frac{\varepsilon}{2}\right)^2 > (1-\varepsilon).$$

The step function $s(\mathbf{x})$, defined on B, fulfills the requirements, since $\mu(B)=\mu(A)$.

Suppose s is restricted to the closed set $\bigcup_{i=1}^{k} B_{n_i y_i}$. By the Tietze extension theorem (Theorem 6.8.5), the restricted s has a continuous extension g to all of B. This continuous function g is the desired function. □

The Lebesgue density theorem states that a measurable set has metric density 1 almost everywhere, i.e., except for a set of measure zero. The next theorem is an analogous statement about measurable functions.

Theorem 8.11.6. *If f is measurable, then f is approximately continuous almost everywhere.*

We omit the proof. The point of the theorem is simply to give some idea of just how "continuous" or "tame" a measurable function is.

Do not fail to note that as a consequence of Theorem 8.11.5 every measurable function on a bounded measurable set can in a certain sense be approximated by a step function or a continuous function. Hence the Lebesgue integral of a bounded measurable function over a bounded measurable set can be approximated by the Riemann integral of a continuous function.

PROBLEMS

1. Prove that if π_1, π_2 are two measurable partitions of a measurable set A, then
$$\underline{S}(f,\pi_1,A) \leqslant \underline{S}(f,\pi,A) \leqslant \bar{S}(f,\pi,A) \leqslant \bar{S}(f,\pi_2,A),$$
where $\pi = \pi_1 \vee \pi_2$.

2. Prove that if $\{A_n\}$ is a countable collection of measurable sets, $\bigcap_{n=1}^{\infty} A_n$ is measurable.

3. Suppose A is a Lebesgue-measurable set in E^n. Prove that $\forall \varepsilon > 0$, $\exists F, G$, with F closed, G open, and $F \subset A \subset G$, s.t. $\mu(G \setminus F) < \varepsilon$.

4. Find a partition of $[0,\infty]$, $\pi_0 = \{0 = a_0, a_1, \ldots, a_n, a_{n+1} = \infty\}$, which will show clearly that $\underline{S}(f,\pi,[0,\infty])$ tends to ∞ as $a_n \to \infty$, $\|\pi\| \to 0$, where $f(x) = |\sin x|/x$.

5. Suppose W is a "random variable" ranging from 0 to ∞. Specifically, the value of W is to represent the "waiting time" until a manufactured item breaks down once it has been put into use. For appropriate values of parameters α and β, the relative probability, or probability density, of W assuming a particular value w is rather accurately given by the probability density function
$$f(w) = \frac{\alpha w^{\alpha-1}}{\beta^{\alpha}} e^{-(w/\beta)^{\alpha}}, \qquad 0 \leqslant w.$$

What is the probability that a new car will last more than five years, given that for automobiles reasonable values for α and β are 2 and 10, respectively?

This random variable W is said to have a *Weibull distribution* with parameters α and β.

6. Let

$$x_1 = 4u_1 + 4u_2 + 3u_3 - u_4 + 2,$$
$$x_2 = u_1 + 2u_2 + 3u_3 + u_4 - 3,$$
$$x_3 = 4u_1 + 8u_2 + 10u_3 + 2u_4 + 2$$

be a mapping from E^4 to E^3. Calculate $dx_1\,dx_3$, using a matrix method. (Compare with Section 5.9.)

7. Let φ: $E^2 \to E^3$ be given by

$$x_1 = 4u_1^2 + 3u_2,$$
$$x_2 = -u_1 + u_2^2,$$
$$x_3 = 2u_1u_2 + u_1.$$

Let $f(x_1, x_2, x_3) = x_1 + x_2 + x_3$. If A is the unit 2-cube $[0,1]\times[0,1] \subset E^2$, evaluate $\int_{\varphi(A)} f\,dx_1\,dx_3$.

8. Evaluate the following Stieltjes integrals:

(a) $\int_0^1 x\,d(x^2)$,
(b) $\int_0^2 x^2\,d(x^2)$,
(c) $\int_{-\pi}^{\pi} e^{|x|}\,d(\cos x)$,
(d) $\int_{1/2}^{17/4} [\![x]\!]\,d[\![2x]\!]$,
(e) $\int_{-\pi}^{e^2} x^3\,d[\![x]\!]$,
(f) $\int_{-3}^{1} x\,d|x|$,
(g) $\int_0^6 (x^2 + [\![x]\!])\,d|3-x|$,
(h) $\int_0^4 e^x\,d(x + [\![x]\!])$,
(i) $\int_1^5 [\![x]\!]\,d|x|$,
(j) $\int_1^n ([\![x]\!]/x^3)\,d[\![x]\!]$.

9. Let $x = S/\sqrt{2}$, $y = 1 - S/\sqrt{2}$, $0 \leqslant S \leqslant \sqrt{2}$ be a parametric representation for an oriented line segment in E^2. Suppose the density of this segment at each point is equal to the distance of the point from the endpoint (1,0). Attach two point masses, one of mass 2 at $(1/\sqrt{2}, 1 - 1/\sqrt{2})$ and another of mass 4 at $(5/4\sqrt{2}, 1 - 5/4\sqrt{2})$. Find the moment of inertia of this line segment about the origin in E^2.

10. Prove Theorem 8.9.6.

11. Prove Theorem 8.9.8.

12. Write a Taylor polynomial of degree 4 for $f(x) = \arctan x$. Do the same for $g(x) = e^{-1/x^2}$. In both cases let $x = 0$ be the center of the expansion, and for the second function, remove the discontinuities at $x = 0$.

13. Suppose ν is a measure on the measure space $(E^n, \mathfrak{M}, \mu)$ such that $\mathfrak{M}$ is the domain of ν and, for every $\varepsilon > 0$, there exists a $\delta > 0$ such that for every

$M \in \mathfrak{M}$ for which $\mu(M) < \delta$ we have $\nu(M) < \varepsilon$. This is to say, if $\{M_n\} \subset \mathfrak{M}$ is a sequence of sets and $\mu(M_n) \to 0$, then $\nu(M_n) \to 0$. Such a measure ν on $\mathfrak{M}$ is said to be *absolutely continuous* with respect to μ.

Let $\{M_n\} \subset \mathfrak{M}$ be a decreasing sequence of sets converging in measure to a point $\mathbf{p} \in E^n$. Consider the limit $\lim_{n\to\infty} \nu(M_n)/\mu(M_n)$ if this limit exists. It is clearly a number which depends upon $\mathbf{p}$. Suppose for each $\mathbf{p}$, this limit does exist; the set of values determines a function $f(\mathbf{p})$. This function we call the Radon–Nikodym (R–N) derivative of ν with respect to μ, and we can write

$$f = d\nu / d\mu.$$

We remark that if the limit exists, it must be independent of the choice of convergent sequences $\{M_n\} \searrow \{\mathbf{p}\}$.

Suppose X is a random variable in E^1, and suppose π is a probability measure on E^1 which is absolutely continuous with respect to μ, where specifically, for each measurable set $S \subset E^1$, $\pi(S)$ equals the probability that the random variable X will assume a value $x \in S$. In particular, $\pi(E^1) = 1$. Show that the Radon–Nikodym derivative of π with respect to μ, the ordinary measure on E^1, is a real function f, and that $\int_S f(x)\,dx = \pi(S)$.

Next, suppose (X, Y, Z) is a random variable in E^3, and $\pi(S)$ is the probability that $(X, Y, Z) = (x, y, z) \in S$. Suppose that for each set $S \subset E^3$ of the form $S = (-\infty, x) \times (-\infty, y) \times (-\infty, z)$, we have $\pi(S) = (1 - e^{-x})(1 - e^{-y})(1 - e^{-z})$ if $x \geq 0$, $y \geq 0$, $z \geq 0$, and $\pi(S) = 0$, otherwise. Show that the R–N derivative of π with respect to μ at an arbitrary point (x, y, z) is the function

$$f(x, y, z) = \begin{cases} e^{-x}e^{-y}e^{-z}, & x, y, z \geq 0, \\ 0, & \text{otherwise.} \end{cases}$$

and that

$$\frac{\partial^3}{\partial x\, \partial y\, \partial z}\left[(1 - e^{-x})(1 - e^{-y})(1 - e^{-z})\right] = f(x, y, z), \qquad x, y, z > 0.$$

Finally, what is the probability that the random point (X, Y, Z) will be in the region $[1,2] \times [3,4] \times [5,6]$?

14. Evaluate $\int_0^a \int_0^b \int_0^c (x^2 + y^2 + z^2)\, dz\, dy\, dx$.

15. Evaluate $\int_0^a \int_0^x \int_0^{x+y} xyz\, dz\, dy\, dx$, and sketch the region of integration.

16. Find the volume of the solid bounded by

$$x^2 + y^2 + z^2 = a^2, \qquad x^2 + y^2 = ax, \qquad a > 0.$$

17. Find the volume bounded by

$$x^2 + y^2 = a^2, \qquad x^2 + z^2 = a^2, \qquad a > 0.$$

18. Find the mass of the region bounded by $z=0$, $z=x+2y+8$, $\sqrt{x^2+y^2}=1-x(x^2+y^2)^{-1/2}$, if the density function is given by $\delta(x,y,z)=\pi$.

19. Compute

$$\int_0^R \int_0^{\sqrt{R^2-x^2}} \ln(1+x^2+y^2)\,dy\,dx.$$

20. Show that

$$\int_0^{\pi/2} \frac{|ab|\,dx}{a^2\cos^2 x+b^2\sin^2 x}=\frac{\pi}{2}, \qquad a,b\neq 0.$$

Now show that

$$\int_0^{\pi/2} \frac{dx}{[a^2\cos^2 x+b^2\sin^2 x]^2}=\frac{\pi(a^2+b^2)}{4|a^3b^3|}.$$

[*Hint*: Try $u=(b/a)\tan x$.]

21. Prove that if f is a measurable function according to Definition 8.4.2, then for each singleton $y\in[-\infty,\infty]\setminus\{0\}$, $f^{-1}(\{y\})$ is a measurable set.

22. Let R be the set of rational numbers in $[0,1]$. What is the outer content of this set? The inner content? The outer measure? Why is, or is not, $f(x)=x^3$ R–S integrable over R if x is the integrator? Compute $\int_R f\,d\mu$.

23. Prove or disprove each of these:

$$f(x)=x\sin\frac{1}{x}\in BV[0,1],$$

$$g(x)=x^2\sin\frac{1}{x}\in BV[0,1],$$

$$h(x)=\sqrt{x}\sin\frac{1}{x}\in BV[0,1].$$

(Define $f(0)=g(0)=h(0)=0$.)

24. Complete the details of the proof of Lemma 8.6.1; i.e., show that $\mu(TA)=\mu(A)|\det T|$.

25. Let $\varphi:[0,1]\to[0,1]$ be the Cantor function. Let $f(x)=x+\varphi(x)$. Show that $f:[0,1]\to[0,2]$ is a homeomorphism. Let $g=f^{-1}$, so $g:[0,2]\to[0,1]$ is also a homeomorphism. Show that there exists a set $D\subset f(C)$, where C is the Cantor ternary set such that $\mu(D)>0$. Show that D contains a nonmeasurable set N. Since $g(N)$ has measure zero, you will have shown that a homeomorphic image of a measurable set may be nonmeasurable and that the homeomorphic image of a set of measure zero may have positive measure. (Note that f is not a linear homeomorphism.)

26. Use a clever little algebraic maneuver to evaluate

$$\int_1^2 \frac{12x^7-8}{(x^7+x+4)^2}\,dx.$$

27. Let

$$f(x,y)=\begin{cases}\dfrac{x^3}{y^2}e^{-x^2/y}, & \text{if } y>0,\\ 0, & \text{if } y=0,\end{cases}$$

be defined on the upper half-plane $y \geqslant 0$. Is $f(x, y)$ continuous at (0,0)? Find $(d/dx)\int_0^1 f(x,y)\,dy$ and $\int_0^1 (\partial/\partial x) f(x,y)\,dy$. Are they identically equal?

28. Let

$$f(x,y)=\begin{cases} y^{-2}, & \text{if } 0<x<y<1,\\ -x^2, & \text{if } 0<y<x<1,\\ 0, & \text{otherwise on } [0,1]\times[0,1],\end{cases}$$

be defined on $[0,1]\times[0,1]$. Calculate

$$\int_0^1\int_0^1 f(x,y)\,dx\,dy \qquad \text{and} \qquad \int_0^1\int_0^1 f(x,y)\,dy\,dx.$$

29. The half-open boxes in E^n comprise a base for a topology. Is it the same as the ordinary topology? Finer? Coarser?

30. Suppose $(V, \mathbb{S}, \mu)$ is a complete measure space and $\{A_n\}_{n=1}^{\infty}$ is an arbitrary countable collection of measurable sets. Show that

$$\mu\left(\underline{\lim}\, A_n\right) \leqslant \underline{\lim}\, \mu(A_n) \leqslant \overline{\lim}\, \mu(A_n) \leqslant \mu\left(\overline{\lim}\, A_n\right).$$

31. Let $f(x, y)=x^2+y^3$, $R=\{(x,y)\in E^2: x^2+y^2 \leqslant 4\}$. Let π be the partition of R described as follows:

$$A_1=\{(x,y): 0\leqslant x^2+y^2<\tfrac{1}{4}\}, \qquad A_2=\{(x,y): x^2+y^2<1\}\setminus A_1$$
$$A_3=\{(x,y): 1\leqslant x^2+y^2<\tfrac{9}{4}\}, \qquad A_4=R\setminus A_3.$$

Evaluate $\underline{S}(f,\pi,R)$ and $\bar{S}(f,\pi,R)$.

32. Suppose that $V=\mathbb{Z}^+$ and that we define a set function μ as follows. For each nonempty subset $A\subset V$, $A=\{a_k\}$, $\mu(A)=\Sigma_k 1/a_k$, and $\mu(\varnothing)=0$. Thus $\mu(\{1,2,3\})=\frac{11}{6}$. Is μ a measure on $P(V)$? Give an example of an infinite set $A\subset V$ for which $\mu(A)<\infty$.

Suppose $f: V\to\mathbb{R}$ is defined by $f(x)=1/(x+1)$. Compute $\int_V f\,d\mu$. If $A\subset V$, $A=\{x\in V: x=n(n+2), n=1,2,3,\ldots\}$, evaluate $\int_A f\,d\mu$.

33. Define μ on $P([0,\infty))$ by $\forall A\subset[0,\infty)$, $\mu(A)=\text{card}\,A$. Is μ a measure on $[0,\infty)$, and is it translation invariant? If $f(x)=x^2$, and $A=\{7,8,9,\ldots,19\}$, evaluate $\int_A f\,d\mu$.

34. If μ is Lebesgue measure in E^n, the measure of an open box is the same as the measure of its closure. Is this true for any open set?
[*Hint*: If you were able to construct a Cantor set with positive measure, consider its complement.]

35. Show that if $f:[0,1]\to\mathbb{R}$ is continuous, then

$$\int_0^\pi xf(\sin x)\,dx=\frac{\pi}{2}\int_0^\pi f(\sin x)\,dx.$$

Next, evaluate

$$\int_0^\pi \frac{x\sin x}{1+\cos^2 x}\,dx.$$

36. If $\mathbb{S}$ is a σ-ring of sets and $\{S_n\}_{n=1}^\infty\subset\mathbb{S}$, prove that $\limsup S_n$ and $\liminf S_n$ belong to $\mathbb{S}$.

37. Let H be a Hilbert parallelotope, where, specifically, H is the box in the real Hilbert space l^2 whose determining edges are the vectors $\{\mathbf{e}_n/n\}_{n=1}^\infty$. That is, $H=\{\mathbf{x}\in l^2:\mathbf{x}=(x_1,x_2,\ldots)\ \forall i,\ 0\leqslant x_i\leqslant 1/i\}$. What would the Euclidean volume of H be, and what is the length of its main diagonal?

38. If you haven't already done this, prove that a semiring of sets which is closed with respect to unions is a ring.

39. Prove that the class of half-open rational boxes in E^n is a semi-ring of sets.

40. Let $\mathfrak{B}$ be the class of half-open rational boxes in E^1. Show that the outer measure of the set $\mathbb{Q}\subset E^1$ is zero.

41. Suppose μ is Lebesgue measure on E^n. If $G\subset E^n$ is open, prove or disprove that $\mu(G)=\mu(\overline{G})$. What if G is open and dense in E^n? Open and connected? Open and convex?

42. Can you find a function $f:[0,1]\to\mathbb{R}$ which is approximately continuous on $[0,1]$ but discontinuous at each rational number in $[0,1]$?

43. Prove that if $f(x)$ is approximately continuous on $[a,b]$, then f assumes all values between $f(a)$ and $f(b)$ on $[a,b]$.

44. Show that the class of measurable functions on E^n is a ring. Is this class closed with respect to countable sums? Give an example showing that the sum or the product of two nonmeasurable functions might be measurable.
[*Hint*: $f\cdot g=\frac{1}{4}[(f+g)^2-(f-g)^2]$.]
Prove that f is measurable iff f^+ and f^- are measurable implies $|f|$ is measurable, but that $|f|$ being measurable does not imply that f is measurable.

45. f defined on $[a,b]$ satisfies a *uniform Lipschitz condition* of order $\alpha>0$ on $[a,b]$ if $\exists M>0$ s.t. $\forall x,y\in[a,b]$, $|f(x)-f(y)|<M|x-y|^\alpha$. Show that f is uniformly continuous on $[a,b]$ if $\alpha>1$, and in fact that f is constant.

Show that $f \in BV[a,b]$ if $\alpha=1$.

Find an example of a function f satisfying a uniform Lipschitz condition of order $\alpha<1$ such that $f \notin BV[a,b]$.

Find an example of a function $f \in BV[a,b]$ which satisfies no uniform Lipschitz condition on $[a,b]$.

46. Show that if $s \neq 1$,

(a) $\sum_{k=1}^{n} 1/k^s = 1/n^{s-1} + s\int_1^n [\![x]\!]/x^{s+1}\,dx$;

(b) $\sum_{k=1}^{2n}(-1)^k/k^s = s\int_1^{2n}(2[\![x/2]\!]-[\![x]\!])/x^{s+1}\,dx$.

47. If f and g are integrable with respect to α, prove that

$$\frac{1}{2}\int_a^b\left[\int_a^b\left[\det\begin{pmatrix} f(x)g(x)\\ f(y)g(y)\end{pmatrix}\right]^2 d\alpha(y)\right]d\alpha(x)$$
$$=\left(\int_a^b[f(x)]^2\,d\alpha(x)\right)\left(\int_a^b[g(x)]^2\,d\alpha(x)\right)$$
$$-\left(\int_a^b f(x)g(x)\,d\alpha(x)\right)^2.$$

48. If f is integrable on $[a,b]$, show that

$$\int_a^b f(x)\,dx = \lim_{n\to\infty}\frac{b-a}{n}\sum_{k=1}^{n} f\left(a+k\frac{b-a}{n}\right),$$

and deduce that

$$\lim_{n\to\infty}\sum_{k=1}^{n}\frac{n}{k^2+n^2}=\frac{\pi}{4} \quad\text{and}\quad \lim_{n\to\infty}\sum_{k=1}^{n}\frac{1}{\sqrt{n^2+k^2}}=\ln\left(1+\sqrt{2}\right).$$

49. Define $f(x,y)$ on $[0,1]\times[0,1]$ as follows:

$$f(x,y)=\begin{cases}1 & \text{if } x \text{ is rational},\\ 2y & \text{if } x \text{ is irrational}.\end{cases}$$

Compute

(a) $\overline{\int_0^1} f(x,y)\,dx$ and $\underline{\int_0^1} f(x,y)\,dx$.

(b) For fixed x, find: $\overline{\int_0^t} f(x,y)\,dy$ and $\underline{\int_0^t} f(x,y)\,dy$, $0 \leq t \leq 1$.

(c) If $F_t(x)=\int_0^t f(x,y)\,dy$ exists for each $t\in[0,1]$, evaluate $\int_0^1 F_t(x)\,dx$.

(d) What can you say about $\int_0^1\int_0^1 f(x,y)\,dx\,dy$?

50. Define f on $[0,1]$ as follows: $f(0)=0$, and if $2^{-n-1}<x\leq 2^{-n}$, $f(x)=2^{-n}$, $n=0,1,2,\ldots$.

(a) Does $\int_0^1 f(x)\,dx$ exist? Why?

(b) Let $F(x)=\int_0^x f(t)\,dt$. Show that on $(0,1]$

$$F(x)=2^{-[\![-\ln x/\ln 2]\!]}x-\left(\tfrac{1}{3}\right)2^{-2[\![-\ln x/\ln 2]\!]}.$$

51. Suppose f is differentiable on $[a,b]$ and $f'(x)=0$ at $\{x_1<x_2<\cdots<x_n\}\subset[a,b]$. Express the total variation of f on $[a,b]$.

CHAPTER

9

DIFFERENTIABLE MAPPINGS

We have talked about mappings in general and about the meaning of continuity, but we have not yet said very much about the derivative of a map. The time has come; to do the things we want to do with integrals we must develop some differential theory.

9.1 THE DERIVATIVE OF A MAP

We are going to restrict ourselves to mappings from one Euclidean space to another. First, recall what happened way back in the first few weeks of elementary calculus. We fixed a point $p \in E^1$ and considered a random variable $p+h$ in a neighborhood of p. Then we took a continuous real-valued function f and applied it to $p+h$. We might have written its value

$$f(p+h)=f(p)+c_p h+|h|g_p(h),$$

where c_p is a fixed real number depending upon p. $|h|g_p(h)$ is a function of h but also depends on p, which is a "compensator"; it just makes the equation true for each h in a neighborhood of zero. Suppose now that g_p is such that $\lim_{h\to 0} g_p(h)=0$. If this is the case, it is clear that the number c_p is simply $f'(p)$. If f is differentiable at p and we wrote the above equation with $c_p \neq f'(p)$, then no compensating function g_p, with $\lim_{h\to 0} g_p(h)=0$, can be found.

Now the term $c_p h$ is a multiple of h; this is to say, $c_p h$ is a particular linear image of h under the linear map $E^1 \to E^1$ that sends h into $c_p h$. The matrix for this linear map is the one-by-one matrix (c_p), and its determinant is c_p. Let's try to extend this notion to get a definition for differentiability, and the derivative, of a mapping $\varphi\colon E^n \to E^m$.

Definition 9.1.1. $\boldsymbol{\varphi}: E^n \to E^m$ is *differentiable at* $\mathbf{P} \in E^n$ if there exists a linear mapping $\mathbf{L_P}: E^n \to E^m$ and a mapping $\mathbf{G_P}: E^n \to E^m$, both defined in a neighborhood N of the origin, such that

$$\boldsymbol{\varphi}(\mathbf{P}+\mathbf{H}) = \boldsymbol{\varphi}(\mathbf{P}) + \mathbf{L_P}(\mathbf{H}) + \|\mathbf{H}\|\mathbf{G_P}(\mathbf{H}), \qquad \forall \mathbf{H} \in N; \quad (9.1.1)$$

$$\lim_{\mathbf{0}} \mathbf{G_P} = \mathbf{0}. \quad (9.1.2)$$

If these maps exist, we call $\mathbf{L_P}$ the *derivative* of $\boldsymbol{\varphi}$ at $\mathbf{P}$. The linear map $\mathbf{L_P}$ is sometimes said to be tangent to $\boldsymbol{\varphi}$ at $\mathbf{P}$ and the best linear approximation for $\boldsymbol{\varphi}$ at $\mathbf{P}$.

Theorem 9.1.2. *If* $\boldsymbol{\varphi}: E^n \to E^m$ *is differentiable at* $\mathbf{P} \in E^n$, *then the derivative* $\mathbf{L_P}$ *is unique.*

PROOF: Suppose $\mathbf{L}$ and $\mathbf{M}$ are linear maps from E^n into E^m such that we have for all $\mathbf{H}$ in a neighborhood N of $\mathbf{0}$

$$\boldsymbol{\varphi}(\mathbf{P}+\mathbf{H}) - \boldsymbol{\varphi}(\mathbf{P}) = \mathbf{L}(\mathbf{H}) + \|\mathbf{H}\|\mathbf{G}_1(\mathbf{H}) = \mathbf{M}(\mathbf{H}) + \|\mathbf{H}\|\mathbf{G}_2(\mathbf{H}),$$

with $\lim_{\mathbf{0}}\mathbf{G}_1 = \mathbf{0} = \lim_{\mathbf{0}}\mathbf{G}_2$.

Let $\mathbf{x} \in E^n$ be arbitrary, but not zero. For some scalar $t \neq 0$, $t\mathbf{x} \in N$, and we have

$$\mathbf{L}(t\mathbf{x}) - \mathbf{M}(t\mathbf{x}) = \|t\mathbf{x}\|(\mathbf{G}_2(t\mathbf{x}) - \mathbf{G}_1(t\mathbf{x})).$$

Factor out the t, and we have

$$\|\mathbf{L}(\mathbf{x}) - \mathbf{M}(\mathbf{x})\| = \|\mathbf{x}\|\|\mathbf{G}_2(t\mathbf{x}) - \mathbf{G}_1(t\mathbf{x})\|.$$

Letting $t \to 0$, we have $\mathbf{L}(\mathbf{x}) = \mathbf{M}(\mathbf{x})$, $\forall \mathbf{x} \in E^n$, as the conclusion we must draw. □

Definition 9.1.3. If $\boldsymbol{\psi}: E^n \to E^m$ is a mapping such that

$$\lim_{\mathbf{x} \to \mathbf{0}} \frac{\boldsymbol{\psi}(\mathbf{x})}{\|\mathbf{x}\|} = \mathbf{0},$$

we say that $\boldsymbol{\psi}$ is "little oh of $\mathbf{x}$" and write $\boldsymbol{\psi}(\mathbf{x}) = o(\mathbf{x})$.

In terms of the "little oh" notation, we can write the following: $\boldsymbol{\varphi}$ is differentiable at $\mathbf{P}$ if there exists a linear map $\mathbf{L_P}$ such that for all $\mathbf{H}$ sufficiently small in norm,

$$\boldsymbol{\varphi}(\mathbf{P}+\mathbf{H}) = \boldsymbol{\varphi}(\mathbf{P}) + \mathbf{L_P}\mathbf{H} + o(\mathbf{H}),$$

and if such an $\mathbf{L_P}$ exists, we can write $\boldsymbol{\varphi}'(\mathbf{P}) = \mathbf{L_P}$.

If $\boldsymbol{\varphi}'(\mathbf{P})$ exists, there must be a matrix which represents this linear map. How we find this matrix depends upon how we represent the mapping $\boldsymbol{\varphi}$ in the first place. Now, $\boldsymbol{\varphi}: \mathbf{x} \mapsto \mathbf{y}$, where $\mathbf{x} \in E^n$ and $\mathbf{y}$ is a unique point in E^m. For $\boldsymbol{\varphi}$ to be well defined, we must know each coordinate of $\mathbf{y}$. This is to say, for each $i=1,\ldots,m$, here must be a mapping $f_i\colon E^n \to \mathbb{R}$ such that $f_i\colon \mathbf{x} \mapsto y_i$. Hence the mapping $\boldsymbol{\varphi}$ itself must have m components, or otherwise said, $\boldsymbol{\varphi}$ must be a vector-valued function of the vector $\mathbf{x}$, described like this:

$$\begin{aligned} y_1 &= f_1(x_1,\ldots,x_n), \\ y_2 &= f_2(x_1,\ldots,x_n), \\ &\vdots \\ y_m &= f_m(x_1,\ldots,x_n). \end{aligned}$$

More simply, $\boldsymbol{\varphi} = (f_1,\ldots,f_m)$.

Theorem 9.1.4. *For $\boldsymbol{\varphi}\colon E^n \to E^m$ to be continuous at $\mathbf{P}$, it is necessary and sufficient that each component function f_i be continuous at $\mathbf{P}$.*

PROOF: Recall that in E^m, all norm topologies are the same, and the same is true in E^n (Corollary 6.5.14). Hence we shall use what is called the sup norm (uniform norm) in these spaces instead of the Euclidean norm; it will be easier. If $\mathbf{x} \in E^k$, $\mathbf{x} = (x_1,\ldots,x_k)$, the sup norm of $\mathbf{x}$ is the maximum of the absolute values of the x_i. The sup norm in E^k generates a topology in which an open ball centered at $\mathbf{x}$ is an open box centered at $\mathbf{x}$.

Let $\mathbf{P} \in E^n$ be a fixed point, and suppose $\boldsymbol{\varphi}$ is continuous at $\mathbf{P}$. Then for any open neighborhood of $\boldsymbol{\varphi}(\mathbf{P}) \in E^m$, say $N(\boldsymbol{\varphi}(\mathbf{P}))$, there is an open neighborhood of $\mathbf{P}$, call it $N(\mathbf{P})$, such that $\boldsymbol{\varphi}(N(\mathbf{P})) \subset N(\boldsymbol{\varphi}(\mathbf{P}))$. Now $N(\boldsymbol{\varphi}(\mathbf{P}))$ contains an open box B which contains $\boldsymbol{\varphi}(\mathbf{P})$; B is of the form $(a_1,b_1)\times(a_2,b_2)\times\cdots\times(a_m,b_m)$, and $N(\mathbf{P})$ contains an open set B' such that $\mathbf{P} \in B'$ and $\boldsymbol{\varphi}(B') \subset B$. We can now say that each f_i is continuous at $\mathbf{P}$ since for any open neighborhood of $\boldsymbol{\varphi}(\mathbf{P})$, say $N(\boldsymbol{\varphi}(\mathbf{P}))$, there exists an open neighborhood of $\mathbf{P}$, B', such that for each i, $f_i(B') \subset (a_i,b_i)$. Now suppose each f_i is continuous at $\mathbf{P}$. Let B be an arbitrary open box centered at $\boldsymbol{\varphi}(\mathbf{P})$; $B=(a_1,b_1)\times\cdots\times(a_m,b_m)$. For each i, $f_i^{-1}((a_i,b_i))$ is an open set in E^n which contains $\mathbf{P}$; call it A_i. Let $A = \bigcap_{i=1}^m A_i$. A is open and contains $\mathbf{P}$, and it is not hard to see that $\boldsymbol{\varphi}(A) \subset B$. Hence $\boldsymbol{\varphi}$ is continuous at $\mathbf{P}$. □

Suppose now that $f\colon E^n \to E^1$ is differentiable at $\mathbf{P}$; then for all $\mathbf{H}$ sufficiently close to $\mathbf{0}$ we have

$$f(\mathbf{P}+\mathbf{H}) = f(\mathbf{P}) + L_{\mathbf{P}}(\mathbf{H}) + o(\mathbf{H}).$$

In particular, consider the standard basis vectors for E^n, $\{\mathbf{e}_1,\ldots,\mathbf{e}_n\}$; for all sufficiently small scalars t,

$$f(\mathbf{P}+t\mathbf{e}_j) = f(\mathbf{P}) + tL_{\mathbf{P}}(\mathbf{e}_j) + o(t\mathbf{e}_j).$$

Letting $t \to 0$, we see that $L_{\mathbf{P}}$ maps the jth unit vector into the jth partial derivative of f, evaluated at $\mathbf{P}$, so $L_{\mathbf{P}}$ must be the gradient of f, evaluated at $\mathbf{P}$.

Definition 9.1.5. Let $f: E^n \to E^1$. By the *gradient* of f, denoted $\mathbf{grad}\, f$, we mean the n-tuple consisting of the n first partial derivatives of f.

If

$$\lim_{t \to 0} \frac{f(\mathbf{P}+t\mathbf{e}_j)-f(\mathbf{P})}{t}$$

exists, it is called the jth first partial derivative of f at $\mathbf{P}$. We denote the jth first partial derivative of f at $\mathbf{P}$ by the symbols $D_j f(\mathbf{P})$ or $(\partial f/\partial x_j)(\mathbf{P})$. If all the n first partials exist at $\mathbf{P}$, then $\mathbf{grad}\, f(\mathbf{P}) = (D_1 f(\mathbf{P}),\ldots,D_n f(\mathbf{P}))$.

This $1\times n$ matrix, the gradient of f at $\mathbf{P}$, is the matrix that represents $L_{\mathbf{P}}$. If the gradient does not exist at $\mathbf{P}$, f is not differentiable at $\mathbf{P}$. Unfortunately, the existence of $\mathbf{grad}\, f(\mathbf{P})$ does not guarantee the differentiability of f at $\mathbf{P}$.

Suppose, however, that the gradient of f exists in a neighborhood of a point $\mathbf{P}=(p_1,\ldots,p_n)$. If $\mathbf{H}=(h_1,\ldots,h_n)$ is sufficiently close to $\mathbf{0}$, we can write

$$\begin{aligned} &f(p_1,\ldots,p_{i-1},p_i+h_i,\ldots,p_n+h_n)-f(p_1,\ldots,p_i,p_{i+1}+h_{i+1},\ldots,p_n+h_n) \\ &\quad = D_i f(p_1,\ldots,p_i,p_{i+1}+h_{i+1},\ldots,p_n+h_n)h_i+o(h_i) \end{aligned}$$

for each $i=1,\ldots,n$ since each partial exists in a neighborhood of $\mathbf{P}$. Hence

$$\begin{aligned} &f(p_1+h_1,\ldots,p_n+h_n)-f(p_1,\ldots,p_n) \\ &\quad = f(p_1+h_1,\ldots,p_n+h_n)-f(p_1,p_2+h_2,\ldots,p_n+h_n) \\ &\qquad + f(p_1,p_2+h_2,\ldots,p_n+h_n)-f(p_1,p_2,p_3+h_3,\ldots,p_n+h_n)+\cdots \\ &\qquad + f(p_1,\ldots,p_{n-1},p_n+h_n)-f(p_1,\ldots,p_n) \\ &\quad = D_1 f(p_1,p_2+h_2,\ldots,p_n+h_n)h_1+o(h_1) \\ &\qquad + D_2 f(p_1,p_2,p_3+h_3,\ldots,p_n+h_n)h_2 \\ &\qquad + o(h_2)+\cdots+D_n f(p_1,\ldots,p_n)h_n+o(h_n). \end{aligned}$$

Now, $\sum_{i=1}^{n} o(h_i)=o(\mathbf{H})$; this is clear if you just think what "little oh" means. But this last expression can be improved if you also have the fact that each partial derivative is continuous at $\mathbf{P}$, so that for each i

$$D_i f(p_1,\ldots,p_i,p_{i+1}+h_{i+1},\ldots,p_n+h_n)h_i = D_i f(p_1,\ldots,p_n)h_i+o(\mathbf{H}).$$

We can write

$$f(\mathbf{P}+\mathbf{H})-f(\mathbf{P})=\sum_{i=1}^{n} D_i f(p_1,\ldots,p_n)h_i+o(\mathbf{H})+no(\mathbf{H});$$

and this last is simply equal to $(\mathbf{grad}\, f(\mathbf{P}))\cdot\mathbf{H}+o(\mathbf{H})$. Thus we have proved the following theorem.

Theorem 9.1.6. *If $f: E^n \to E^1$ is differentiable at* **P**, *then* **grad** *f exists at* **P** *and $f'(\mathbf{P}) = \mathbf{grad}\, f(\mathbf{P})$. If* **grad** *$f$ exists in a neighborhood of* **P** *and is continuous at* **P**, *then f is differentiable at* **P**.

It is clear that we now have a way to represent the derivative of a mapping $\boldsymbol{\varphi}: E^n \to E^m$. Since $\boldsymbol{\varphi}$ must have m real-valued component functions, a reasonable guess is that the derivative of $\boldsymbol{\varphi}$, if it exists, must be representable, or have a "realization," as some mathematicians say, which is the $m \times n$ matrix whose rows are the gradients of the component functions. More specifically, if $\boldsymbol{\varphi}$ has a derivative at **P**, then there is a linear map $\mathbf{L_P}$: $E^n \to E^m$ such that for all **H** sufficiently close to **0**

$$\boldsymbol{\varphi}(\mathbf{P}+\mathbf{H}) = \boldsymbol{\varphi}(\mathbf{P}) + \mathbf{L_P}(\mathbf{H}) + o(\mathbf{H}).$$

Evidently, if each component function of $\boldsymbol{\varphi}$ is differentiable at **P**, then the matrix referred to is a linear mapping that satisfies this equation, and the uniqueness of $\mathbf{L_P}$ ensures that this matrix is the derivative of $\boldsymbol{\varphi}$ at **P**. Thus the differentiability of each of the component functions at **P** implies the differentiability of $\boldsymbol{\varphi}$ at **P**.

On the other hand, if $\boldsymbol{\varphi}$ is differentiable at **P**, then consider what $\mathbf{L_P}$ does to each unit vector $\mathbf{e}_i \in E^n$.

$$\boldsymbol{\varphi}(\mathbf{P}+t\mathbf{e}_i) = \begin{bmatrix} f_1(\mathbf{P}+t\mathbf{e}_i) \\ \vdots \\ f_m(\mathbf{P}+t\mathbf{e}_i) \end{bmatrix} = \begin{bmatrix} f_1(\mathbf{P}) \\ \vdots \\ f_m(\mathbf{P}) \end{bmatrix} + t\mathbf{L_P}(\mathbf{e}_i) + o(t\mathbf{e}_i).$$

Divide through by t and let $t \to 0$; we have

$$\begin{bmatrix} D_i f_i(\mathbf{P}) \\ \vdots \\ D_i f_m(\mathbf{P}) \end{bmatrix} = \mathbf{L_P}(\mathbf{e}_i).$$

Thus $\mathbf{L_P}$ maps each $\mathbf{e}_i$ into a vector in E^m, which vector is precisely the ith column of the matrix whose rows are the gradients at **P** of the component functions. We can say that if $\boldsymbol{\varphi}$ is differentiable at **P**—its derivative is that matrix—so the gradients of the component functions exist at **P**. But we cannot yet say each component is differentiable since there are examples of functions that are not differentiable, or even continuous, but whose gradients exist at a point. For example, let $f(x, y) = xy(x^4 + y^4)^{-1}$ when $(x, y) \neq (0,0)$, and $f(0,0) = 0$; **grad** $f(0,0) = (0,0)$, but f is not continuous at $(0,0)$.

However, one last observation is all we need. If $\boldsymbol{\varphi}$ is differentiable at **P**, then for each component f_i we have

$$f_i(\mathbf{P}+\mathbf{H}) = f_i(\mathbf{P}) + (\mathbf{grad}\, f_i(\mathbf{P})) \cdot \mathbf{H} + o_i(\mathbf{H}),$$

where $o_i(\mathbf{H})$ is the ith component of $o(\mathbf{H})$, and thus is itself "little oh of $\mathbf{H}$." We now can conclude that f_i is differentiable at $\mathbf{P}$.

Theorem 9.1.7. *Let $\varphi: E^n \to E^m$, $\varphi = (f_1,\ldots,f_m)$. Then φ is differentiable at* $\mathbf{P}$ *iff each component function f_i is differentiable at* $\mathbf{P}$. *Moreover, differentiability at* $\mathbf{P}$ *implies continuity at* $\mathbf{P}$ *for φ and the components of φ.*

Definition 9.1.8. If $\varphi: E^n \to E^m$ is a differentiable mapping, the $m \times n$ matrix whose rows are the gradients of the component functions is called the *Jacobian matrix* for φ and is denoted J_φ. The determinant of this matrix, when $n=m$, is referred to as the *Jacobian* of φ. Thus the derivative of φ at $\mathbf{P}$ is the matrix $J_\varphi(\mathbf{P})$.

Note that if $\varphi = (f_1,\ldots,f_m)$ there are mn first partial derivatives of φ. The ijth element of J_φ is $D_j f_i$, $i=1,\ldots,m$ and $j=1,\ldots,n$. We call φ a C^1-mapping if φ and all mn first partials exist and are continuous. Since, as we have seen, C^1-mappings are always differentiable, we almost invariably restrict our attention to mappings of this class. Of course, it is possible for a map φ to be differentiable, but not C^1, but rather than carefully stating "φ is continuously differentiable" when φ is C^1, we shall simply say φ is a differentiable map; C^1-ness being implied unless the contrary is specifically stated.

Definition 9.1.9. If φ is a differentiable map whose derivative at $\mathbf{P}$ is $J_\varphi(\mathbf{P})$, then a *differential* of φ at $\mathbf{P}$ is a vector $J_\varphi(\mathbf{P})\mathbf{H}$. We denote "the" differential of φ by $d\varphi$; its "value" at a point $\mathbf{P}$ in terms of an arbitrary vector $\mathbf{H}$ is $d\varphi(\mathbf{P}) = J_\varphi(\mathbf{P})\mathbf{H}$.

For example, if $\varphi: E^n \to E^1$ and we denote symbolically a vector $\mathbf{H}$ by $(dx_1,\ldots,dx_n)$, then

$$d\varphi(\mathbf{P}) = (\mathbf{grad}\,\varphi(\mathbf{P})) \cdot (dx_1,\ldots,dx_n)$$

is how one would usually write "the value" of the differential of φ at $\mathbf{P}$.

Theorem 9.1.10 (Chain Rule). *If $\varphi: E^k \to E^n$ and $\psi: E^n \to E^m$ are C^1 maps, then the composite $\psi\circ\varphi: E^k \to E^m$ is C^1, and we have $\forall \mathbf{P} \in E^k$*

$$J_{\psi\varphi}(\mathbf{P}) = J_\psi(\varphi(\mathbf{P}))J_\varphi(\mathbf{P}).$$

PROOF: Certainly, $\psi\circ\varphi$ is continuous since each of φ and ψ is. Let $\mathbf{P}$, $\mathbf{H} \in E^k$, and let $\mathbf{K} = \varphi(\mathbf{P}+\mathbf{H}) - \varphi(\mathbf{P})$, so that $\mathbf{H} \to \mathbf{0} \Rightarrow \mathbf{K} \to \mathbf{0}$. Then

$$\psi\varphi(\mathbf{P}+\mathbf{H}) - \psi\varphi(\mathbf{P}) = \psi(\varphi(\mathbf{P})+\mathbf{K}) - \psi\varphi(\mathbf{P}) = J_\psi(\varphi(\mathbf{P}))\mathbf{K} + o(\mathbf{K}).$$

Now, $\mathbf{K}$ can also be written $J_\varphi(\mathbf{P})\mathbf{H}+o(\mathbf{H})$, so we have

$$\psi\varphi(\mathbf{P}+\mathbf{H})-\psi\varphi(\mathbf{P})=J_\psi(\varphi(\mathbf{P}))J_\varphi(\mathbf{P})\mathbf{H}+J_\psi(\varphi(\mathbf{P}))o(\mathbf{H})+o(\mathbf{K}).$$

Since $J_\psi(\varphi(\mathbf{P}))$ is a bounded linear mapping and $\mathbf{H}\to\mathbf{0}\Rightarrow\mathbf{K}\to\mathbf{0}$, $J_\psi(\varphi(\mathbf{P})o(\mathbf{H})=o(\mathbf{H})=o(\mathbf{K})$, so we finally have

$$\psi\varphi(\mathbf{P}+\mathbf{H})=\psi\varphi(\mathbf{P})+J_\psi(\varphi(\mathbf{P}))J_\varphi(\mathbf{P})\mathbf{H}+o(\mathbf{H}).$$

Since each entry in the matrices J_ψ and J_φ is continuous at $\varphi(\mathbf{P})$ and $\mathbf{P}$, respectively, so are the mk entries in the matrix product continuous at $\mathbf{P}$, which is to say, $\psi\circ\varphi$ is C^1 at $\mathbf{P}$. □

We have talked about derivatives and partial derivatives. If f is real valued on E^n, then the ith first partial derivative is an ordinary derivative of f if we hold all the other variables constant. This is to say, if we imagine an $(n+1)$th coordinate axis orthogonal to all the others, and then intersect the hypersurface $x_{n+1}=f(x_1,\ldots,x_n)$ with a plane through the point $\mathbf{P}$ parallel to the x_ix_{n+1}-coordinate plane, we get a plane curve; the slope of the tangent line at $(\mathbf{P}, f(\mathbf{P}))$ is the value of $D_i f(\mathbf{P})$. For this reason, we sometimes call this ith first partial the derivative of f in the direction of the ith unit vector.

We now generalize this idea.

Definition 9.1.11. Let $f\colon E^n\to E^1$ and $\mathbf{u}\in E^n$ be a unit vector. If $\lim_{t\to 0}\{[f(\mathbf{P}+t\mathbf{u})-f(\mathbf{P})]/t\}$ exists, we call this limit the *directional derivative* of f at $\mathbf{P}$ in the direction $\mathbf{u}$ and denote it by $f_{\mathbf{u}}(\mathbf{P})$.

Theorem 9.1.12. *If $f\colon E^n\to E^1$ has a continuous gradient at $\mathbf{P}$ and* **grad** *f is defined in a neighborhood of $\mathbf{P}$, f has a directional derivative at $\mathbf{P}$ in any direction $\mathbf{u}$, and the value of the directional derivative $f_{\mathbf{u}}(\mathbf{P})=(\mathbf{grad}\,f(\mathbf{P}))\cdot\mathbf{u}$. On the other hand, if f has a directional derivative at $\mathbf{P}$ in any direction $\mathbf{u}$, then* **grad** *$f(\mathbf{P})$ exists.*

PROOF: If **grad** f is continuous at $\mathbf{P}$ and exists in a neighborhood of $\mathbf{P}$, then f is differentiable at $\mathbf{P}$, and we can write, for any unit vector $\mathbf{u}$ and a small scalar t,

$$f(\mathbf{P}+t\mathbf{u})=f(\mathbf{P})+(\mathbf{grad}\,f(\mathbf{P}))\cdot t\mathbf{u}+o(t\mathbf{u}).$$

Divide by t to get

$$\frac{f(\mathbf{P}+t\mathbf{u})-f(\mathbf{P})}{t}=(\mathbf{grad}\,f(\mathbf{P}))\cdot\mathbf{u}+\frac{o(t\mathbf{u})}{t}.$$

Since $o(t\mathbf{u})/t$ goes to zero as $t\to 0$, we have that $f_{\mathbf{u}}(\mathbf{P})$ exists and equals $(\mathbf{grad}\,f(\mathbf{P}))\cdot\mathbf{u}$.

On the other hand, if f has a directional derivative at $\mathbf{p}$ in any direction $\mathbf{u}$, it has such a derivative at $\mathbf{P}$ in the n directions of the basic unit vectors, so $\mathbf{grad}\, f(\mathbf{P})$ exists. □

Note that the derivative $f_{\mathbf{u}}(\mathbf{P})$ is maximal when $\mathbf{u} = (\mathbf{grad}\, f(\mathbf{P}))/\|\mathbf{grad}\, f(\mathbf{P})\|$ and is zero if $\mathbf{u} \perp \mathbf{grad}\, f(\mathbf{P})$. In E^n, if $f\colon E^n \to E^1$ is C^1, $f(\mathbf{x}) = 0$ describes a "smooth" hypersurface. Suppose $\mathbf{P}$ is a point on this surface. It is not obvious, in fact it is difficult to see, that the vector $\mathbf{grad}\, f(\mathbf{P})$ is normal to the hypersurface $f(\mathbf{x}) = 0$ at $\mathbf{x} = \mathbf{P}$. Since f is C^1 and $f(\mathbf{P}) = 0$, write

$$f(\mathbf{P}+\mathbf{H}) = (\mathbf{grad}\, f(\mathbf{P})) \cdot \mathbf{H} + o(\mathbf{H})$$

for all $\mathbf{H}$ sufficiently small. Put $\mathbf{x} = \mathbf{P}+\mathbf{H}$, and write

$$f(\mathbf{x}) = (\mathbf{grad}\, f(\mathbf{P})) \cdot (\mathbf{x}-\mathbf{P}) + o(\mathbf{x}-\mathbf{P}).$$

Thus the equation

$$f(\mathbf{x}) = (\mathbf{grad}\, f(\mathbf{P})) \cdot (\mathbf{x}-\mathbf{P}) + o(\mathbf{x}-\mathbf{P}) = 0$$

is the equation of our hypersurface in E^n, at least in a neighborhood of $\mathbf{P}$. Now, $l(\mathbf{x}) = (\mathbf{grad}\, f(\mathbf{P})) \cdot (\mathbf{x}-\mathbf{P}) = 0$ is clearly the equation of a hyperplane through $\mathbf{P}$ when $\mathbf{grad}\, f(\mathbf{P}) \neq \mathbf{0}$. Note that

$$f(\mathbf{x}) - l(\mathbf{x}) = o(\mathbf{x}-\mathbf{P})$$

and $f(\mathbf{P}) = l(\mathbf{P}) = 0$. Assume $\mathbf{grad}\, f(\mathbf{P}) \neq \mathbf{0}$. Since $\lim_{\mathbf{P}}[o(\mathbf{x}-\mathbf{P})/\|\mathbf{x}-\mathbf{P}\|] = 0$, we can conclude that $l(\mathbf{x}) = 0$ must be the equation of the tangent hyperplane to $f(\mathbf{x}) = 0$ at $\mathbf{P}$. The components of $\mathbf{grad}\, f(\mathbf{P})$ are the coefficients of the linear equation, and the fact that $\mathbf{grad}\, f(\mathbf{P})$ is normal to the generating vector $\mathbf{x}-\mathbf{P}$ of this tangent hyperplane implies that $\mathbf{grad}\, f(\mathbf{P})$ is normal to the surface at $\mathbf{P}$. If $\mathbf{grad}\, f(\mathbf{P}) = \mathbf{0}$, this vector is normal to everything.

Theorem 9.1.13. *Let f be a C^1-function on E^n and $\mathbf{P}$ a point on the surface $f(\mathbf{x}) = 0$. Then $\mathbf{grad}\, f(\mathbf{P})$ is normal to the surface at $\mathbf{P}$, and $(\mathbf{grad}\, f(\mathbf{P})) \cdot (\mathbf{x}-\mathbf{P}) = 0$ is the equation of the tangent hyperplane to the surface at $\mathbf{P}$ whenever $\mathbf{grad}\, f(\mathbf{P}) \neq \mathbf{0}$.*

We have defined the Jacobian and the Jacobian matrix of a C^1-mapping $\boldsymbol{\varphi}$; of course, the matrix $J_{\boldsymbol{\varphi}}$ has no determinant if it is not square. As a linear mapping, $J_{\boldsymbol{\varphi}}(\mathbf{x})\colon E^n \to E^m$, $\mathbf{x}$ an arbitrary fixed point of E^n, is continuous, meaning that if $\mathbf{y} \in E^n$ is sufficiently close to $\mathbf{y}_0 \in E^n$, $J_{\boldsymbol{\varphi}}(\mathbf{x})\mathbf{y}$ is necessarily close to $J_{\boldsymbol{\varphi}}(\mathbf{x})\mathbf{y}_0$ in E^m. But $J_{\boldsymbol{\varphi}}(\mathbf{x})$ is a matrix or, if you like, a "matrix-valued" function of $\mathbf{x}$. The Jacobian of a C^1-mapping at $\mathbf{x}$, $\det J_{\boldsymbol{\varphi}}(\mathbf{x})$, is a real-valued function of $\mathbf{x}$, as is the norm of the Jacobian matrix. It is natural to ask if these functions are continuous, or even differentiable. Recall that $\|J_{\boldsymbol{\varphi}}(\mathbf{x})\| = \sup\{\|J_{\boldsymbol{\varphi}}(\mathbf{x})\mathbf{y}\| : \|\mathbf{y}\| = 1\}$.

Theorem 9.1.14. *Let* $\boldsymbol{\varphi}: E^n \to E^m$ *be a* C^1*-mapping and* $\mathbf{x} \in E^n$ *an arbitrary point. Then* $J_{\boldsymbol{\varphi}}(\mathbf{x})$ *is continuous at* $\mathbf{x}_0$ *in the sense that for any fixed vector* $\mathbf{y} \in E^n$, $\lim_{\mathbf{x} \to \mathbf{x}_0} J_{\boldsymbol{\varphi}}(\mathbf{x})\mathbf{y} = J_{\boldsymbol{\varphi}}(\mathbf{x}_0)\mathbf{y}$, *and* $\lim_{\mathbf{x} \to \mathbf{x}_0} \| J_{\boldsymbol{\varphi}}(\mathbf{x}) - J_{\boldsymbol{\varphi}}(\mathbf{x}_0)\| = 0$.

Moreover, if $m = n$, $\det J_{\boldsymbol{\varphi}}(\mathbf{x})$ *is a continuous function of* $\mathbf{x}$.

Finally, the function $J_{\boldsymbol{\varphi}}(\mathbf{x})\mathbf{y}: E^n \times E^n \to E^m$ *is a continuous function of* $(\mathbf{x},\mathbf{y})$ *in the sense that for any* $(\mathbf{x}_0,\mathbf{y}_0) \in E^n \times E^n$

$$\lim_{(\mathbf{x},\mathbf{y}) \to (\mathbf{x}_0,\mathbf{y}_0)} J_{\boldsymbol{\varphi}}(\mathbf{x})\mathbf{y} = J_{\boldsymbol{\varphi}}(\mathbf{x}_0)\mathbf{y}_0.$$

PROOF: Let $\boldsymbol{\varphi}: E^n \to E^m$ be C^1, $\mathbf{x}_0, \mathbf{y}_0$ fixed points in E^n. Each component of $J_{\boldsymbol{\varphi}}(\mathbf{x})\mathbf{y}$, for $\mathbf{x}$, $\mathbf{y}$ arbitrary points of E^n, is given by $(\mathbf{grad}\, f_i(\mathbf{x})) \cdot \mathbf{y}$. Since each component of $\mathbf{grad}\, f_i(\mathbf{x})$ is a continuous function of $\mathbf{x}$ and the inner product is continuous as a function of two variables as well as continuous in each variable separately, it is clear that

$$\lim_{(\mathbf{x},\mathbf{y}) \to (\mathbf{x}_0,\mathbf{y}_0)} (\mathbf{grad}\, f_i(\mathbf{x})) \cdot \mathbf{y} = (\mathbf{grad}\, f_i(\mathbf{x}_0)) \cdot \mathbf{y}_0.$$

It follows that

$$\lim_{(\mathbf{x},\mathbf{y}) \to (\mathbf{x}_0,\mathbf{y}_0)} J_{\boldsymbol{\varphi}}(\mathbf{x})\mathbf{y} = J_{\boldsymbol{\varphi}}(\mathbf{x}_0)\mathbf{y}_0,$$

which is to say, $J_{\boldsymbol{\varphi}}(\mathbf{x})\mathbf{y}$ is a continuous function in both variables as well as a continuous function in each variable separately. This means that

$$\lim_{\mathbf{x} \to \mathbf{x}_0} J_{\boldsymbol{\varphi}}(\mathbf{x})\mathbf{y} = J_{\boldsymbol{\varphi}}(\mathbf{x}_0)\mathbf{y} \qquad \text{for each } \mathbf{y}$$

and

$$\lim_{\mathbf{y} \to \mathbf{y}_0} J_{\boldsymbol{\varphi}}(\mathbf{x})\mathbf{y} = J_{\boldsymbol{\varphi}}(\mathbf{x})\mathbf{y}_0 \qquad \text{for each } \mathbf{x}.$$

That $\det J_{\boldsymbol{\varphi}}(\mathbf{x})$ is continuous is obvious since it is simply a sum of products of continuous functions. To prove that $\lim_{\mathbf{x} \to \mathbf{x}_0} \| J_{\boldsymbol{\varphi}}(\mathbf{x}) - J_{\boldsymbol{\varphi}}(\mathbf{x}_0)\| = 0$ is not quite so easy.

Suppose that $\lim_{\mathbf{x} \to \mathbf{x}_0} \| J_{\boldsymbol{\varphi}}(\mathbf{x}) - J_{\boldsymbol{\varphi}}(\mathbf{x}_0)\| \neq 0$. Then it must follow that there is a positive number $\delta > 0$ such that for each $k = 1,2,3,\ldots$, we can find a point $\mathbf{x}_k \in E^n$, with $\|\mathbf{x}_k - \mathbf{x}_0\| \leq 1/k$ and $\| J_{\boldsymbol{\varphi}}(\mathbf{x}_k) - J_{\boldsymbol{\varphi}}(\mathbf{x}_0)\| > \delta$. For any fixed k, there is a $\mathbf{y}_k \in E^n$, with $\|\mathbf{y}_k\| = 1$ such that

$$\delta/2 < \left\| \left[J_{\boldsymbol{\varphi}}(\mathbf{x}_k) - J_{\boldsymbol{\varphi}}(\mathbf{x}_0) \right] \mathbf{y}_k \right\| \leq \| J_{\boldsymbol{\varphi}}(\mathbf{x}_k) - J_{\boldsymbol{\varphi}}(\mathbf{x}_0)\|.$$

We can assume that the points $\mathbf{y}_k$ are all distinct without loss of any generality. Since the unit sphere in E^n is compact, there is a convergent subsequence $\{\mathbf{y}_{k_i}\}_{i=1}^{\infty}$ converging to $\mathbf{y}_0$, and the corresponding subsequence $\{\mathbf{x}_{k_i}\}$ of course converges to $\mathbf{x}_0$. Now,

$$\lim_{(\mathbf{x}_{k_i},\mathbf{y}_{k_i}) \to (\mathbf{x}_0,\mathbf{y}_0)} J_{\boldsymbol{\varphi}}(\mathbf{x}_{k_i})\mathbf{y}_{k_i} = J_{\boldsymbol{\varphi}}(\mathbf{x}_0)\mathbf{y}_0,$$

and

$$\lim_{\mathbf{y}_{k_i} \to \mathbf{y}_0} J_{\boldsymbol{\varphi}}(\mathbf{x}_0)\mathbf{y}_{k_i} = J_{\boldsymbol{\varphi}}(\mathbf{x}_0)\mathbf{y}_0,$$

so for a sufficiently large index k_i we have

$$\| J_{\boldsymbol{\varphi}}(\mathbf{x}_{k_i})\mathbf{y}_{k_i} - J_{\boldsymbol{\varphi}}(\mathbf{x}_0)\mathbf{y}_0 \| + \| J_{\boldsymbol{\varphi}}(\mathbf{x}_0)\mathbf{y}_{k_i} - J_{\boldsymbol{\varphi}}(\mathbf{x}_0)\mathbf{y}_0 \|$$

less than $\delta/4$. This is a contradiction since we have arrived at the strange inequality

$$\begin{aligned} \delta/2 &< \| J_{\boldsymbol{\varphi}}(\mathbf{x}_{k_i})\mathbf{y}_{k_i} - J_{\boldsymbol{\varphi}}(\mathbf{x}_0)\mathbf{y}_{k_i} \| \\ &\leqslant \| J_{\boldsymbol{\varphi}}(\mathbf{x}_{k_i})\mathbf{y}_{k_i} - J_{\boldsymbol{\varphi}}(\mathbf{x}_0)\mathbf{y}_0 \| + \| J_{\boldsymbol{\varphi}}(\mathbf{x}_0)\mathbf{y}_0 - J_{\boldsymbol{\varphi}}(\mathbf{x}_0)\mathbf{y}_{k_i} \| < \delta/4. \quad \square \end{aligned}$$

9.2 TAYLOR'S FORMULA

We naturally assume that the Taylor formula for real-valued functions of one variable is well known to all, at least reasonably well known. Nevertheless we again put it in view. (See Theorem 8.9.18.)

If $f: \mathbb{R} \to \mathbb{R}$ is of class C^r and $p \in \mathbb{R}$, then the following representation for $f(x)$ in terms of its derivatives is known as Taylor's formula. ($f^{(r)}(x)$ is the rth derivative of f at x.)

$$f(x) = f(p) + f'(p)(x-p) + \cdots + \frac{f^{(r)}(\pi_x)}{r!}(x-p)^r,$$

where π_x is between x and p. In particular, if $p=0$, the formula is sometimes called Maclaurin's formula.

Before we generalize this formula to mappings $\boldsymbol{\varphi}: E^n \to E^m$, we need to establish a preliminary result. This result is important enough to justify theorem status.

Definition 9.2.1. Let $f: E^n \to E^1$. By $D_{ij}f$ we mean the partial derivative with respect to the jth variable of $D_i f$. Thus

$$D_{ij}f = D_j(D_i f) = \frac{\partial^2 f}{\partial x_j \partial x_i} = \frac{\partial}{\partial x_j}\left(\frac{\partial f}{\partial x_i}\right).$$

Theorem 9.2.2. *Suppose f is of class C^2, mapping E^2 into E^1. Let (x_0, y_0) be an arbitrary but fixed point. Then*

$$D_{21}f(x_0, y_0) = D_{12}f(x_0, y_0).$$

This is to say, the "mixed second partials" of f are identically equal.

PROOF: We remind the reader that $D_{12}f(x, y)$ means differentiation first with respect to x followed by differentiation of that first partial with respect to y. In the other common notation,

$$D_{12}f(x,y)=\frac{\partial^2 f(x,y)}{\partial y\,\partial x}=\frac{\partial}{\partial y}\left[\frac{\partial f(x,y)}{\partial x}\right].$$

$$D_{12}f(x_0,y_0)=\lim_{y\to y_0}\frac{D_1 f(x_0,y)-D_1 f(x_0,y_0)}{y-y_0}$$

$$=\lim_{y\to y_0}\left[\lim_{x\to x_0}\left(\frac{\dfrac{f(x,y)-f(x_0,y)}{x-x_0}-\dfrac{f(x,y_0)-f(x_0,y_0)}{x-x_0}}{y-y_0}\right)\right]$$

$$=\lim_{y\to y_0}\left[\lim_{x\to x_0}\left(\frac{\dfrac{f(x,y)-f(x,y_0)}{y-y_0}-\dfrac{f(x_0,y)-f(x_0,y_0)}{y-y_0}}{x-x_0}\right)\right].$$

Because f is C^1 and $D_{12}f(x,y)$ exists and is continuous at (x_0,y_0), we can interchange the order of the limit operations. When we do this, the last expression reduces to $D_{21}f(x_0,y_0)$. To see this, note that, given $\varepsilon>0$, for all (x,y) sufficiently close to (x_0,y_0), the expression

$$\frac{f(x,y)-f(x,y_0)}{(y-y_0)(x-x_0)}-\frac{f(x_0,y)-f(x_0,y_0)}{(y-y_0)(x-x_0)}$$

is necessarily "within ε" of $D_{12}f(x_0,y_0)$. Hence

$$\lim_{y\to y_0}\left[\left(\frac{f(x,y)-f(x,y_0)}{y-y_0}-\frac{f(x_0,y)-f(x_0,y_0)}{y-y_0}\right)\Big/(x-x_0)\right]$$

is within ε of $D_{12}f(x_0,y_0)$.

It follows that

$$\lim_{x\to x_0}\frac{D_2 f(x,y_0)-D_2 f(x_0,y_0)}{x-x_0}=D_{21}f(x_0,y_0)$$

is within ε of $D_{12}f(x_0,y_0)$. Since $\varepsilon>0$ was arbitrary, we have the desired equality. □

The reader should note several things. First, we only used the fact that the mixed partial $D_{12}f(x,y)$ was continuous at (x_0,y_0) and that f was C^1 in a neighborhood of (x_0,y_0). Second, and more important, it immediately follows that if $f\colon E^n\to E^1$ is suitably differentiable, then any pair of mixed partials of the same orders are equal. That is,

$$\begin{aligned}D_1^2D_2^3D_3f(x,y,z,w)&=D_3D_1^2D_2^3f=D_2^3D_1^2D_3f=D_2^3D_3D_1^2f\\&=D_1^2D_3D_2^3f=D_3D_2^3D_1^2f.\end{aligned}$$

Suppose now that $f\colon E^n \to \mathbb{R}$ is C^k. We introduce a bit of notation at this time which will simplify matters for us considerably. Let ∇, pronounced del, be the symbolic vector

$$\nabla = (D_1, \ldots, D_n) = \left(\frac{\partial}{\partial x_1}, \ldots, \frac{\partial}{\partial x_n} \right).$$

Hence $\nabla f = \mathbf{grad}\, f$. If $\boldsymbol{\varphi}\colon E^n \to E^n$, the symbol $\nabla \cdot \boldsymbol{\varphi}$ (the dot product of del and the mapping $\boldsymbol{\varphi} = (f_1, \ldots, f_n)$ would denote

$$\nabla \cdot \boldsymbol{\varphi} = D_1 f_1 + \cdots + D_n f_n = \operatorname{div} \boldsymbol{\varphi}.$$

The *divergence* of $\boldsymbol{\varphi}$, $\operatorname{div}\boldsymbol{\varphi}$, is an expression that frequently appears in applied calculus. For example, if $\mathbf{F}$ is a flow vector of a fluid (see Section 5.6), where $\mathbf{F}(\mathbf{P})$ is the velocity vector of a fluid at the point $\mathbf{P}$, then $\nabla \cdot \mathbf{F}(\mathbf{P}) = \operatorname{div} \mathbf{F}(\mathbf{P})$ is the rate of change of volume per unit volume of an infinitesimal ball of fluid containing the point $\mathbf{P}$.

If $\boldsymbol{\varphi} = (f_1, f_2, f_3)$ and $\nabla = (D_1, D_2, D_3)$, then $\nabla \times \boldsymbol{\varphi}$, the vector product of these two things is what we call the *curl* of $\boldsymbol{\varphi}$:

$$\mathbf{curl}\, \boldsymbol{\varphi} = \nabla \times \boldsymbol{\varphi}.$$

If $\mathbf{F}$ is a flow vector in E^3, then $\frac{1}{2}\,\mathbf{curl}\,\mathbf{F}(\mathbf{P}) = \frac{1}{2}\,(\nabla \times \mathbf{F})(\mathbf{P})$ is the vector angular velocity of an infinitesimal ball of the fluid containing $\mathbf{P}$.

If $\mathbf{H} \in E^n$ and $\nabla = (D_1, \ldots, D_n)$, $\mathbf{H} \cdot \nabla$ is the linear operator

$$\mathbf{H} \cdot \nabla = h_1 D_1 + \cdots + h_n D_n,$$

and $\forall f\colon E^n \to \mathbb{R}$,

$$(\mathbf{H} \cdot \nabla) f = \mathbf{H} \cdot \mathbf{grad}\, f.$$

If $\mathbf{H}$ is a unit vector, then

$$(\mathbf{H} \cdot \nabla) f(\mathbf{x}) = f_{\mathbf{H}}(\mathbf{x}),$$

the directional derivative of f at $\mathbf{x}$ in the direction $\mathbf{H}$.

By $(\mathbf{H} \cdot \nabla)^k$ we mean the operator that we would get by expanding algebraically the kth power of $(\mathbf{H} \cdot \nabla)$. Since we have, for suitably differentiable functions, equality of the mixed partials, we shall assume commutativity of the Ds in the expansion of

$$(h_1 D_1 + \cdots + h_n D_n)^k.$$

The binomial formula used to be something everyone learned in high school, but the multinomial formula was reserved for the more sophisticated students. A typical term of the expansion is of the form

$$C_{i_1 \cdots i_n} h_1^{i_1} \cdots h_n^{i_n} D_1^{i_1} \cdots D_n^{i_n},$$

where $i_1+i_2+\cdots+i_n=k$ and $D_j^{i_j}$ is the i_jth order of the differentiation with respect to the jth variable, and $C_{i_1\cdots i_n}$ is the *multinomial coefficient*

$$C_{i_1\cdots i_n}=\begin{pmatrix} k \\ i_1 i_2\cdots i_n\end{pmatrix}=\frac{k!}{(i_1!)(i_2!)\cdots(i_n!)}.$$

Now we are ready to proceed. Recall that $f\colon E^n\to E^1$ is a C^k-function. We write, for a fixed $\mathbf{P}\in E^n$ and an arbitrarily fixed $\mathbf{H}$, $g(t)=f(\mathbf{P}+t\mathbf{H})$, where t is a real parameter ranging over $[0,1]$. Then by the chain rule

$$g'(t)=[\mathbf{grad}\, f(\mathbf{P}+t\mathbf{H})]\cdot\mathbf{H}=(\mathbf{H}\cdot\nabla)f(\mathbf{P}+t\mathbf{H}),$$

and

$$\begin{aligned} g''(t)&=\frac{d}{dt}[h_1D_1f(\mathbf{P}+t\mathbf{H})+\cdots+h_nD_nf(\mathbf{P}+t\mathbf{H})]\\ &=[h_1\,\mathbf{grad}\,D_1f(\mathbf{P}+t\mathbf{H})]\cdot\mathbf{H}+\cdots+[h_n\mathbf{grad}\,D_nf(\mathbf{P}+t\mathbf{H})]\cdot\mathbf{H}\\ &=h_1^2D_{11}f(\mathbf{P}+t\mathbf{H})+h_1h_2D_{12}f(\mathbf{P}+t\mathbf{H})+\cdots+h_n^2D_{nn}f(\mathbf{P}+t\mathbf{H})\\ &=(\mathbf{H}\cdot\nabla)^2f(\mathbf{P}+t\mathbf{H}),\end{aligned}$$

where we have written D_{ij} for D_jD_i. By induction, it really does work out that

$$g^{(k)}(t)=(\mathbf{H}\cdot\nabla)^kf(\mathbf{P}+t\mathbf{H}).$$

Using Taylor's formula for one variable, we write

$$g(t)=g(0)+g'(0)t+\cdots+\frac{g^{(k-1)}(0)}{(k-1)!}t^{k-1}+\frac{g^{(k)}(\tau)}{k!}t^k.$$

where τ is between 0 and t. Hence we have by substitution, and then putting $t=1$,

$$\begin{aligned} f(\mathbf{P}+\mathbf{H})&=f(\mathbf{P})+(\mathbf{H}\cdot\nabla)f(\mathbf{P})+\frac{(\mathbf{H}\cdot\nabla)^2f(\mathbf{P})}{2!}\\ &\quad+\cdots+\frac{(\mathbf{H}\cdot\nabla)^{k-1}f(\mathbf{P})}{(k-1)!}+\frac{1}{k!}(\mathbf{H}\cdot\nabla)^kf(\mathbf{P}+\tau\mathbf{H})\end{aligned}$$

where τ is between 0 and 1.

In this formula, we can replace $\mathbf{P}+\mathbf{H}$ by $\mathbf{x}$ and $\mathbf{H}$ by $\mathbf{x}-\mathbf{P}$ to get an expansion for $f(\mathbf{x})$ "about a point $\mathbf{P}$." Clearly, this is a complicated and unwieldly expression, but then, what can we say?

Theorem 9.2.3. *If $\varphi: E^n \to E^m$ is a C^k-mapping, we can write a Taylor expression for $\varphi = (f_1, \ldots, f_m)$ about a point $\mathbf{P} \in E^n$,*

$$\varphi(\mathbf{P}+\mathbf{H}) = \begin{bmatrix} \sum_{j=0}^{k-1} \frac{(\mathbf{H}\cdot\nabla)^j f_1(\mathbf{P})}{j!} + \frac{(\mathbf{H}\cdot\nabla)^k}{k!} f_1(\mathbf{P}+\tau_1\mathbf{H}) \\ \vdots \\ \sum_{j=0}^{k-1} \frac{(\mathbf{H}\cdot\nabla)^j f_m(\mathbf{P})}{j!} + \frac{(\mathbf{H}\cdot\nabla)^k}{k!} f_m(\mathbf{P}+\tau_m\mathbf{H}) \end{bmatrix},$$

where for each $i = 1, \ldots, m$, $0 \leqslant \tau_i \leqslant 1$.

There is no assurance that any of the τ_i will be equal; generally they are all different. However, each of the points $\mathbf{P} + \tau_i\mathbf{H}$ lies on the line between $\mathbf{P}$ and $\mathbf{P}+\mathbf{H}$.

9.3 THE INVERSE FUNCTION THEOREM

The famous inverse function theorem is certainly one of the fundamental theorems of calculus. It states that if $\varphi: E^n \to E^n$ is C^1 and, at a point $\mathbf{P} \in E^n$, the Jacobian is nonzero, then there is a neighborhood of $\mathbf{P}$ on which φ is bijective; hence φ has a local inverse at $\varphi(\mathbf{P})$. The proof given here is hard to follow, but so are all the other proofs we have seen.

Theorem 9.3.1 (Inverse Function Theorem). *Let $\varphi: E^n \to E^n$ be C^1, and suppose at a point $\mathbf{P} \in E^n$, $\det J_\varphi(\mathbf{P}) \neq 0$. Then there is a neighborhood A of $\mathbf{P}$ such that $\varphi: A \to \varphi A$ is a* diffeomorphism, *that is, a homeomorphism which is continuously differentiable both ways. Moreover, for each $\mathbf{u} \in A$ we have*

$$J_{\varphi^{-1}}(\varphi(\mathbf{u})) = \left[J_\varphi(\mathbf{u})\right]^{-1}.$$

PROOF: Let φ, $\mathbf{P}$ be the mapping and point satisfying the hypothesis of the theorem. Let D be a compact, convex neighborhood of $\mathbf{P}$ such that $J_\varphi(\mathbf{x})$ is nonsingular on D. Now for each $\mathbf{x} \in D$, $J_\varphi(\mathbf{x})$ is a linear map on E^n and hence bounded, each entry of $J_\varphi(\mathbf{x})$ is continuous in $\mathbf{x}$, and we have shown that $\| J_\varphi(\mathbf{x})\|$ is also a continuous function of $\mathbf{x}$, where this norm is defined:

$$\| J_\varphi(\mathbf{x})\| = \sup_{\mathbf{y} \in E^n} \left\{ \| J_\varphi(\mathbf{x})\mathbf{y}\| : \|\mathbf{y}\| = 1 \right\}.$$

Furthermore, for each $\mathbf{x} \in D$, $0 < \| J_\varphi(\mathbf{x})\| < \infty$, and since D is compact and $\| J_\varphi(\mathbf{x})\|$ is continuous in $\mathbf{x}$, $\| J_\varphi(\mathbf{x})\|$ assumes a maximum and a minimum on D. Let $\underline{\mathbf{x}} \in D$ be a point where $\| J_\varphi(\mathbf{x})\|$ assumes a minimum value,

$$0 < \| J_\varphi(\underline{\mathbf{x}})\| \leqslant \| J_\varphi(\mathbf{x})\|, \qquad \forall \mathbf{x} \in D.$$

Now put

$$\lambda_{\mathbf{x}} = \inf_{\mathbf{y}}\left\{ \left\| \frac{J_\varphi(\mathbf{x})}{\|J_\varphi(\mathbf{x})\|}\mathbf{y} \right\| : \|\mathbf{y}\| = 1 \right\}.$$

Since $J_\varphi(\mathbf{x})$ is nonsingular on D, $\lambda_{\mathbf{x}} > 0$ for all $\mathbf{x} \in D$. Now, $\lambda_{\mathbf{x}}$ is also a continuous function of $\mathbf{x}$, so in D, there is an $\mathbf{x}_0$ where $\lambda_{\mathbf{x}}$ assumes a minimal value λ and $\lambda > 0$. Hence we can write, for any $\mathbf{y} \neq 0$ in E^n and any $\mathbf{x} \in D$,

$$\infty > \|J_\varphi(\mathbf{x})\| \geqslant \left\| J_\varphi(\mathbf{x}) \frac{\mathbf{y}}{\|\mathbf{y}\|} \right\| \geqslant \lambda_{\mathbf{x}} \|J_\varphi(\mathbf{x})\| \geqslant \lambda \|J_\varphi(\underline{\mathbf{x}})\| > 0.$$

We are now ready to move on to the next part of the proof. We ought to emphasize, perhaps, that $\lambda_{\mathbf{x}}$ does indeed assume a minimum value; this simply depends on the fact that

$$\left\| \frac{J_\varphi(\mathbf{x})}{\|J_\varphi(\mathbf{x})\|}\mathbf{y} \right\|$$

is a continuous function of the two variables $\mathbf{x}$ and $\mathbf{y}$ by Theorem 9.1.14, and hence it assumes a minimum value on the compact set $D \times \{\mathbf{y} : \|\mathbf{y}\| = 1\} \subset E^n \times E^n$.

Now, let $\mathbf{u}, \mathbf{v} \in D$ and put $\mathbf{v} = \mathbf{u} + \mathbf{h}$. Since

$$\varphi(\mathbf{u}+\mathbf{h}) - \varphi(\mathbf{u}) - J_\varphi(\mathbf{u})\mathbf{h} = o_{\mathbf{u}}(\mathbf{h})$$

and φ is C^1, it is clear that $o_{\mathbf{u}}(\mathbf{h})$ is a continuous function of $\mathbf{h}$ and of $\mathbf{u}$ (again, we use the fact that if $\mathbf{u}$ makes a sufficiently small change, the effect of $J_\varphi(\mathbf{u})$ on $\mathbf{h}$ will necessarily be small). But because of the way $\mathbf{h}$ and $\mathbf{u}$ appear in the left-hand side, it is apparent that $o_{\mathbf{u}}(\mathbf{h})$ is continuous in both variables simultaneously; this is to say, if $(\mathbf{u}', \mathbf{h}') \to (\mathbf{u}, \mathbf{h})$, then $o_{\mathbf{u}'}(\mathbf{h}') \to o_{\mathbf{u}}(\mathbf{h})$.

Define the function $f(\mathbf{u}, \mathbf{h})$ by

$$f(\mathbf{u},\mathbf{h}) = \left\| \frac{o_{\mathbf{u}}(\mathbf{h})}{\|\mathbf{h}\|} \right\|, \qquad (\mathbf{u},\mathbf{h}) \in D \times B,$$

where B is a compact ball in E^n centered at the origin. $f(\mathbf{u},\mathbf{h}) \to 0$ as $\mathbf{h} \to 0$, for each $\mathbf{u} \in D$, so the discontinuity at $\mathbf{h} = \mathbf{0}$ is removable; we define $f(\mathbf{u}, \mathbf{0}) = 0$ for each $\mathbf{u} \in D$ and remark that since $D \times B$ is compact, $f(\mathbf{u},\mathbf{h})$ is uniformly continuous on $D \times B$.

We assert that the convergence $f(\mathbf{u},\mathbf{h}) \to 0$ as $\mathbf{h} \to \mathbf{0}$ is uniform with respect to $\mathbf{u}$. This is to say, given $\varepsilon > 0$, if $\|\mathbf{h}\|$ is sufficiently small, $|f(\mathbf{u},\mathbf{h})| < \varepsilon$ $\forall \mathbf{u} \in D$. This is fairly clear; since $f(\mathbf{u},\mathbf{h})$ is uniformly continuous on $D \times B$ whenever $\|(\mathbf{u},\mathbf{h}) - (\mathbf{u},\mathbf{0})\| = \|\mathbf{h}\|$ is sufficiently small, $|f(\mathbf{u},\mathbf{h}) - f(\mathbf{u},\mathbf{0})| = |f(\mathbf{u},\mathbf{h})| = f(\mathbf{u},\mathbf{h}) < \varepsilon$. Hence taking for the closed ball B the ball whose radius is sufficiently small so that for every $(\mathbf{u},\mathbf{h})\in D\times B$, we have

$$0\leqslant f(\mathbf{u},\mathbf{h})=\left\|\frac{o_{\mathbf{u}}(\mathbf{h})}{\|\mathbf{h}\|}\right\|<\frac{\lambda}{2}\|J_{\varphi}(\underline{\mathbf{x}})\|.$$

Now, let $A=D\cap(B+\mathbf{P})$, where $B+\mathbf{P}$ is the translation of B by $\mathbf{P}$. A is a compact, convex neighborhood of $\mathbf{P}$, and for any points $\mathbf{u},\mathbf{v}\in A$, with $\mathbf{u}\neq\mathbf{v}$, we have, putting $\mathbf{v}=\mathbf{u}+\mathbf{h}$ and $\mathbf{k}=\boldsymbol{\varphi}(\mathbf{u}+\mathbf{h})-\boldsymbol{\varphi}(\mathbf{u})=J_{\varphi}(\mathbf{u})\mathbf{h}+o_{\mathbf{u}}(\mathbf{h})$,

$$\frac{\|\mathbf{k}\|}{\|\mathbf{h}\|}\geqslant\frac{\|J_{\varphi}(\mathbf{u})\mathbf{h}\|}{\|\mathbf{h}\|}-\frac{\|o_{\mathbf{u}}(\mathbf{h})\|}{\|\mathbf{h}\|}>\lambda\|J_{\varphi}(\underline{\mathbf{x}})\|-\frac{\lambda}{2}\|J_{\varphi}(\underline{\mathbf{x}})\|=\frac{\lambda}{2}\|J_{\varphi}(\underline{\mathbf{x}})\|>0.$$

This implies that $\mathbf{k}\neq 0$. Since $\mathbf{u},\mathbf{v}\in A$ were arbitrary distinct points of A, we have now that $\boldsymbol{\varphi}\colon A\to\boldsymbol{\varphi}A$ is bijective.

It is clear that since $\boldsymbol{\varphi}$ is C^1 on A, $\mathbf{k}\to\mathbf{0}$ as $\mathbf{h}\to\mathbf{0}$. But the inequality

$$\frac{\|\mathbf{k}\|}{\|\mathbf{h}\|}>\frac{\lambda}{2}\|J_{\varphi}(\underline{\mathbf{x}})\|>0\qquad\forall\mathbf{h}\in B$$

implies that $\|\mathbf{h}\|/\|\mathbf{k}\|<2/\lambda\|J_{\varphi}(\underline{\mathbf{x}})\|<\infty$, for all $\mathbf{h}\in B$. Hence, if $\mathbf{k}\to\mathbf{0}$, $\mathbf{h}$ must necessarily approach zero. This is to say that at any point $\boldsymbol{\varphi}(\mathbf{u})\in\boldsymbol{\varphi}(A)$ we have

$$\mathbf{h}=\boldsymbol{\varphi}^{-1}(\boldsymbol{\varphi}(\mathbf{u})+\mathbf{k})-\boldsymbol{\varphi}^{-1}(\boldsymbol{\varphi}(\mathbf{u})).$$

Thus $\boldsymbol{\varphi}^{-1}$ is continuous on $\boldsymbol{\varphi}(A)$. We have now established that $\boldsymbol{\varphi}\colon A\to\boldsymbol{\varphi}A$ is a homeomorphism.

But we can say more. Since $\mathbf{k}=J_{\varphi}(\mathbf{u})\mathbf{h}+o_{\mathbf{u}}(\mathbf{h})$, and $J_{\varphi}(\mathbf{u})$ is nonsingular, we can write

$$\mathbf{h}=\left(J_{\varphi}(\mathbf{u})\right)^{-1}\mathbf{k}-\left(J_{\varphi}(\mathbf{u})\right)^{-1}o_{\mathbf{u}}(\mathbf{h}).$$

Notice that

$$\frac{\left\|-\left(J_{\varphi}(\mathbf{u})\right)^{-1}o_{\mathbf{u}}(\mathbf{h})\right\|}{\|\mathbf{k}\|}=\frac{\|\mathbf{h}\|\left\|\left(J_{\varphi}(\mathbf{u})\right)^{-1}\mathbf{G}(\mathbf{h})\right\|}{\|\mathbf{k}\|},$$

where $\mathbf{G}(\mathbf{h})=o_{\mathbf{u}}(\mathbf{h})/\|\mathbf{h}\|$. The expression on the right-hand side for all $\mathbf{h}\neq\mathbf{0}$, $\mathbf{h}\in B$, is majorized by

$$\frac{2}{\lambda\|J_{\varphi}(\underline{\mathbf{x}})\|}\left\|\left(J_{\varphi}(\mathbf{u})\right)^{-1}\mathbf{G}(\mathbf{h})\right\|.$$

Now $\mathbf{G}(\mathbf{h})\to\mathbf{0}$ as $\mathbf{h}\to\mathbf{0}$, so $\mathbf{G}(\mathbf{h})\to\mathbf{0}$ as $\mathbf{k}\to\mathbf{0}$, which means that

$$\lim_{\mathbf{k}\to\mathbf{0}}\frac{\left\|-\left(J_{\varphi}(\mathbf{u})\right)^{-1}o_{\mathbf{u}}(\mathbf{h})\right\|}{\|\mathbf{k}\|}=0.$$

Thus $-(J_{\varphi}(\mathbf{u}))^{-1}o_{\mathbf{u}}(\mathbf{h})$ is $o(\mathbf{k})$, and we have

$$\mathbf{h}=\left(J_{\varphi}(\mathbf{u})\right)^{-1}\mathbf{k}+o(\mathbf{k})=\boldsymbol{\varphi}^{-1}(\boldsymbol{\varphi}(\mathbf{u})+\mathbf{k})-\boldsymbol{\varphi}^{-1}(\boldsymbol{\varphi}(\mathbf{u})).$$

Since $(J_{\varphi}(\mathbf{u}))^{-1}$ is indeed a linear map on E^n, for each $\mathbf{u}\in D$, we conclude that φ^{-1} is differentiable on φA. The uniqueness of the linear map tells us that $(J_{\varphi}(\mathbf{u}))^{-1}=J_{\varphi^{-1}}(\varphi(\mathbf{u}))$.

Finally, we remark that since all the entries in the matrix $(J_{\varphi}(\mathbf{u}))^{-1}$ are continuous functions of $\mathbf{u}$ and that since φ is bicontinuous if $\mathbf{w}=\varphi(\mathbf{u})$ on A, then the entries in the expression $J_{\varphi^{-1}}(\varphi(\mathbf{u}))$ are continuous functions of $\mathbf{w}$. Thus φ^{-1} is itself C^1. □

After this, we should have a breather, but the next theorem we are going to present is also rather taxing.

Theorem 9.3.2. *Let $\varphi: E^n\to E^m$ be a diffeomorphism on $A\to\varphi A$, with A measurable. We use the symbol g_{φ} to stand for the square root of the sum of the squares of the $n\times n$ determinants in J_{φ}. Then φA has n-dimensional measure $\mu(\varphi A)=\int_A g_{\varphi}\,d\mu$. We call g_{φ} the g factor for the mapping φ.*

PROOF: Assume first that A is bounded, with $0<\mu(A)<\infty$, and let $M=\sup\{g_{\varphi}(\mathbf{x}):\mathbf{x}\in A\}$. Let $\{A_i\}$ be a measurable partition of A; later we shall stipulate how fine the partition must be. Since A is conditionally compact, we can certainly find a measurable partition of A such that $\max\{\operatorname{diam} A_i\}$ is as small as we like. For each i let $\mathbf{P}_i$ be a point in A_i. We can write

$$\begin{aligned}\mu(\varphi(A_i))&=\mu(\{\varphi(\mathbf{P}_i+\mathbf{H}):\mathbf{P}_i+\mathbf{H}\in A_i\})\\&=\mu\big(\{\varphi(\mathbf{P}_i)+J_{\varphi}(\mathbf{P}_i)\mathbf{H}+o(\mathbf{H}):\mathbf{H}\in A_i-\mathbf{P}_i\}\big)\\&=\mu\big(\{J_{\varphi}(\mathbf{P}_i)\mathbf{H}+o(\mathbf{H}):\mathbf{H}\in A_i-\mathbf{P}_i\}\big).\end{aligned}$$

(Note that the set $A_i-\mathbf{P}_i=\{\mathbf{x}:\mathbf{x}=\mathbf{a}-\mathbf{P}_i,\ \mathbf{a}\in A_i\}$.) Assume for the moment that $\mu(A_i)>0$. We can write the inequalities, for $\mu(A_i)$ sufficiently small:

$$\begin{aligned}\mu\big(J_{\varphi}(\mathbf{P}_i)(A_i-\mathbf{P}_i)\big)-o(\mu(A_i-\mathbf{P}_i))&\leqslant\mu(\varphi(A_i))\\&\leqslant\mu\big(J_{\varphi}(\mathbf{P}_i)(A_i-\mathbf{P}_i)\big)+o(\mu(A_i-\mathbf{P}_i))\end{aligned}$$

and

$$\frac{g_{\varphi}(\mathbf{P}_i)\mu(A_i)}{\mu(A_i)}-\frac{o(\mu(A_i))}{\mu(A_i)}\leqslant\frac{\mu(\varphi(A_i))}{\mu(A_i)}\leqslant\frac{g_{\varphi}(\mathbf{P}_i)\mu(A_i)}{\mu(A_i)}+\frac{o(\mu(A_i))}{\mu(A_i)}$$

(see Theorem 8.6.2). So as $\operatorname{diam} A_i\to0$ and $A_i\searrow\mathbf{P}_i$ we have

$$\lim_{A_i\searrow\mathbf{P}_i}\mu(\varphi(A_i))/\mu(A_i)=g_{\varphi}(\mathbf{P}_i).$$

Next, note that g_{φ} is bounded and uniformly continuous on each A_i. Hence there exists an $\hat{\mathbf{x}}_i\in A_i$ such that

$$\int_{A_i}g_{\varphi}\,d\mu=g_{\varphi}(\hat{\mathbf{x}}_i)\mu(A_i).$$

Now we stipulate how fine the partition shall be. Let $\varepsilon > 0$. Let $0 < \delta < 1$ be a number such that $\delta \leqslant \varepsilon/2M\mu(A)$. The partition shall be sufficiently fine so that for each i

$$g_{\boldsymbol{\varphi}}(\mathbf{P}_i)\mu(A_i)(1-\delta) < g(\hat{\mathbf{x}}_i)\mu(A_i) < g_{\boldsymbol{\varphi}}(\mathbf{P}_i)\mu(A_i)(1+\delta)$$

and

$$g_{\boldsymbol{\varphi}}(\mathbf{P}_i)\mu(A_i)(1-\delta) < \mu(\boldsymbol{\varphi}(A_i)) < g_{\boldsymbol{\varphi}}(\mathbf{P}_i)\mu(A_i)(1+\delta).$$

The inequalities simply say that both $\mu(\boldsymbol{\varphi}(A_i))/\mu(A_i)$ and $g_{\boldsymbol{\varphi}}(\hat{\mathbf{x}}_i)$ are within $\varepsilon/\mu(A)$ of $g_{\boldsymbol{\varphi}}(\mathbf{P}_i)$. Cross-subtraction yields

$$0 \leqslant \Big|\sum_i \big(\mu(\boldsymbol{\varphi}A_i) - g_{\boldsymbol{\varphi}}(\hat{\mathbf{x}}_i)\mu(A_i)\big)\Big| \leqslant \sum_i |\mu(\boldsymbol{\varphi}A_i) - g_{\boldsymbol{\varphi}}(\hat{\mathbf{x}}_i)\mu(A_i)|$$

$$< \sum_i 2\delta g_{\boldsymbol{\varphi}}(\mathbf{P}_i)\mu(A_i) \leqslant \frac{\varepsilon}{\mu(A)} \sum_i \mu(A_i) = \varepsilon.$$

Thus we have $|\mu(\boldsymbol{\varphi}A) - \int_A g_{\boldsymbol{\varphi}}\, d\mu| < \varepsilon$. We conclude that $\mu(\boldsymbol{\varphi}A) = \int_A g_{\boldsymbol{\varphi}}\, d\mu$.

If $\mu(A_i) = 0$, there is an open set $B \supset A_i$ such that $\mu(B) > 0$ but $\mu(B) < \varepsilon/M$. Then $\mu(\boldsymbol{\varphi}(B)) < M\mu(B) < \varepsilon$. Since $\varepsilon > 0$ was arbitrary and $\mu(\boldsymbol{\varphi}A) \leqslant \mu(\boldsymbol{\varphi}B)$, we conclude that $\mu(A) = 0 \Rightarrow \mu(\boldsymbol{\varphi}A) = 0$.

If A is unbounded, or has infinite measure, or if $g_{\boldsymbol{\varphi}}$ is unbounded on A, we can consider a sequence of compact sets $\{A_i\}$ increasing to A. For each i we have

$$\mu(\boldsymbol{\varphi}A_i) = \int_{A_i} g_{\boldsymbol{\varphi}}\, d\mu.$$

If $\lim_{A_i \nearrow A} \int_{A_i} g_{\boldsymbol{\varphi}}\, d\mu$ exists and is finite, this is the measure of $\boldsymbol{\varphi}A$. If the limit is ∞, $\mu(\boldsymbol{\varphi}A) = \infty$. □

Suppose we have a measurable set A and a diffeomorphism $\boldsymbol{\varphi}\colon A \to \boldsymbol{\varphi}A$, with $A \subset E^n$ and $\boldsymbol{\varphi}A \subset E^m$, and let f be a measurable real function defined on $\boldsymbol{\varphi}A$. Since f can be written as $f^+ - f^-$, we shall assume without loss of generality that f is positive. Consider $\int_{\boldsymbol{\varphi}(A)} f\, d\mu$.

A fine partition of $\operatorname{im}(f)$ gives rise to measurable partition of $\boldsymbol{\varphi}A$, which in turn establishes a measurable partition $\{A_i\}$ of A via $\boldsymbol{\varphi}^{-1}$. The composite function $f\boldsymbol{\varphi}$ obviously takes on the same value at each $\mathbf{u} \in A_i$ that f assumes at each $\boldsymbol{\varphi}(\mathbf{u}) \in \boldsymbol{\varphi}A_i$. Hence for each i, let m_i, M_i be, respectively, the infimum and the supremum of the values $f\boldsymbol{\varphi}$ assumes on A_i; these will be the infimum and the supremum, respectively, of the values f assumes on $\boldsymbol{\varphi}A_i$.

For each i, we have $m_i\mu(\boldsymbol{\varphi}A_i) = m_i \int_{A_i} g_{\boldsymbol{\varphi}}\, d\mu$ and $M_i\mu(\boldsymbol{\varphi}A_i) = M_i \int_{A_i} g_{\boldsymbol{\varphi}}\, d\mu$. Hence, if $\Sigma_i m_i \int_{A_i} g_{\boldsymbol{\varphi}}\, d\mu$ converges as $\|\pi\| \to 0$, the Lebesgue integral $\int_A f\boldsymbol{\varphi} g_{\boldsymbol{\varphi}}\, d\mu$ exists. But then so must the integral $\int_{\boldsymbol{\varphi}(A)} f\, d\mu$ exist, and have the same value. We summarize this in the following theorem.

Theorem 9.3.3. *Let* φ *be a diffeomorphism on a measurable set* $A \subset E^n$ *to* $\varphi A \subset E^m$ *and* f *a measurable real function on* φA. *Then* $\int_{\varphi A} f\, d\mu$ *exists if and only if* $\int_A f\varphi g_\varphi\, d\mu$ *exists; when they exist (in the Lebesgue sense), they have the same value, i.e.,*

$$\int_{\varphi(A)} f\, d\mu = \int_A f\varphi g_\varphi\, d\mu.$$

The function $f\varphi$ *is the pullback of* f, *relative to* φ, *and we usually denote it by* f^*.

Corollary 9.3.4. *Suppose* φ, f, *and* A *are described in the theorem. Then, if* A *is positively oriented,*

$$\int_{\varphi(A)} f\, dx_{i_1} \cdots dx_{i_n} = \int_A f\varphi \det J'_\varphi\, du_1 \cdots du_n,$$

where J'_φ *is the submatrix of* J_φ *consisting of the* $(i_1, i_2, \dots, i_n)$*th rows.*

PROOF: We are not integrating f over $\varphi(A)$; rather we are integrating f over the projection of $\varphi(A)$ onto the particular coordinate subspace indicated. □

Having this corollary, we are now prepared, in theory at least, to integrate functions over "smooth" surfaces and to integrate k-forms over "smooth" k-surfaces. Loosely said, by a "smooth" k-surface, we mean the C^1-image of a box in E^k under a map φ such that the Jacobian matrix J_φ has rank k at every point of the box. There is one additional notion that we have not yet mentioned, but it follows readily from what we have already done. We have defined the integral of a k-form over a k-dimensional surface; can we extend this idea to get a reasonable notion of the integral of a k-form over an l-dimensional surface? We can certainly try this.

If $k < l$, we shall regard the integral

$$\int_{\varphi(A)} f\, dx_{i_1} \cdots dx_{i_k}, \qquad \dim \varphi(A) = l > k,$$

as the ordinary integral

$$\int_{\pi_{i_1} \cdots i_k \varphi A} f\, dx_{i_1} \cdots dx_{i_k}.$$

This is to say, we project $\varphi(A)$ onto the $(x_{i_1} \dots x_{i_k})$-coordinate subspace and evaluate by pullback.

If $k > l$ in the above integral, we regard the integral as

$$\int_{\pi_{i_1} \cdots i_k \varphi A} f\, d\mu$$

where μ is the l-dimensional measure. We evaluate by pullback. For example, if $A=[0,1]\times[0,1]\subset E^2$ and $\boldsymbol{\varphi}$ is given by

$$\boldsymbol{\varphi}=\begin{bmatrix} 1 & 2 \\ 3 & -1 \\ -1 & 1 \\ 4 & 3 \end{bmatrix},$$

then

$$\int_{\boldsymbol{\varphi}(A)}(x+y+z+w)\,dx\,dz\,dw=\int_{\pi_{xzw}\boldsymbol{\varphi}(A)}(x+y+z+w)\,d\mu$$

$$=\int_A[(u+2v)+(3u-v)+(-u+v)$$

$$+(4u+3v)]\,g'_\varphi d\mu,$$

where g'_φ is the g-factor associated with the xzw submatrix of $\boldsymbol{\varphi}$. This yields

$$\int_0^1\int_0^1(7u+5v)\sqrt{3^2+(-5)^2+(-7)^2}\,du\,dv=6\sqrt{83}\,.$$

The area of $\boldsymbol{\varphi}(A)$ in E^4 is $\sqrt{305}$, and the area of $\pi_{xzw}\boldsymbol{\varphi}(A)$ in the E^3 onto which $\boldsymbol{\varphi}(A)$ was projected is $\sqrt{83}$.

We make a final comment in this section. In all the integration we've been doing, except in Corollary 9.3.4, we have ignored orientation. Perhaps we should have stated that in each case we assumed the measurable set A is positively oriented! However, in many applications we are not concerned with orientation. In the special case where $\boldsymbol{\varphi}$ is a C^1-diffeomorphism on a set $A\to\boldsymbol{\varphi}A$ and $A\subset E^n$, $\boldsymbol{\varphi}A\subset E^n$, the g-factor seems to be just the Jacobian. If the Jacobian is negative on A, $\boldsymbol{\varphi}$ is an orientation reversing mapping, so if orientation is being ignored, remember that the g-factor is really the absolute value of the Jacobian (the square root of the square). If orientation is to be considered, do not use the absolute value of the Jacobian.

9.4 THE IMPLICIT FUNCTION THEOREM

The main theorem of this section is an extension of the inverse function theorem but is much deeper and has some far reaching consequences. Before we state it, we shall tell you a little of what it means.

Suppose we have n C^1-equations in $n+k$ variables. Experience leads one to believe that you should be able to solve for n of the unknowns in terms of the remaining k variables. This is more or less true; however, we need a certain amount of "independence" among the equations. It turns out that if the Jacobian matrix of the system is of rank n, at least in a neighborhood of a "point" $(x_1,\ldots,x_{n+k})$ solution of the system, then we can solve for the n variables associated with the $n \times n$ submatrix with nonvanishing determinant in terms of the other k variables.

If the rank is less than n, we can solve for a lesser number of the variables in terms of the remaining ones.

The theorem not only ensures that under certain conditions, given a system of n equations in $n+k$ variables, we can solve for n of the variables in terms of the remaining k variables (which are then referred to as parameters), but it also guarantees that if we have $n+k$ functions in n variables, equated to $n+k$ parameters, under certain conditions we can solve for each of the n variables in terms of n of the $n+k$ parameters and we can solve for the remaining k parameters in terms of the first n parameters (the parameters being the values of the $n+k$ equations).

Before we tackle the theorem, we would like to introduce an old notation for the Jacobian matrix.

Definition 9.4.1. If φ: $E^n \rightarrow E^m$ is a differentiable mapping where $\varphi = (f_1, f_2, \ldots, f_m)$, the *Jacobian matrix* of φ is the matrix

$$J_\varphi = \frac{\partial(f_1,\ldots,f_m)}{\partial(x_1,\ldots,x_n)} = \begin{bmatrix} \dfrac{\partial f_1}{\partial x_1} & \cdots & \dfrac{\partial f_1}{\partial x_n} \\ \vdots & & \vdots \\ \dfrac{\partial f_m}{\partial x_1} & \cdots & \dfrac{\partial f_m}{\partial x_n} \end{bmatrix}.$$

For $n \leqslant m$, the g factor is most readily obtained from the formula:

$$g_\varphi = \sqrt{|\det[J_\varphi^t \cdot J_\varphi]|}. \quad \text{(the } t \text{ indicates transpose)}$$

Now we are ready to go. If you find this famous theorem hard to digest, you will have lots of company.

Theorem 9.4.2 (Implicit Function Theorem). *Let* $\psi=(f_1,\ldots,f_n)$ *be a mapping from* $E^n\times E^k$ *into* E^n; *suppose that* ψ *is* C^1 *on an open set* $S\subset E^n\times E^k$ *and maps a point* $\mathbf{P}=(p_1,\ldots,p_{n+k})\in S$ *to* $\mathbf{0}\in E^n$. *Suppose further that the determinant*

$$\det\frac{\partial(f_1,\ldots,f_n)}{\partial(x_1,\ldots,x_n)}$$

does not vanish at $\mathbf{P}$. *Then there is a unique mapping* $\boldsymbol{\theta}: E^k\to E^n$, $\boldsymbol{\theta}=(\theta_1,\ldots,\theta_n)$, *defined and* C^1 *on a neighborhood of* $(p_{n+1},\ldots,p_{n+k})\in E^k$ *such that*

$$\boldsymbol{\theta}(p_{n+1},\ldots,p_{n+k})=(p_1,\ldots,p_n)$$

and for each $\mathbf{t}$ *in the neighborhood*

$$\psi(\boldsymbol{\theta}(\mathbf{t}),\mathbf{t})\equiv\mathbf{0}.$$

Specifically, this theorem says that if you have n C^1-functions in $n+k$ variables $\{x_1,\ldots,x_{n+k}\}$, and there is an $(n+k)$-tuple of values $(p_1,\ldots,p_{n+k})$ for these variables which makes the n functions vanish simultaneously, then, assuming that the determinant of the Jacobian matrix involving the first n variables is not zero at the "solution point" $(p_1,\ldots,p_{n+k})$, each of the first n variables can be expressed uniquely as a function of the remaining k variables. Moreover, if these solutions are substituted back into the original n functions, they all vanish identically. This is to say, if the last k unknowns are regarded as independent parameters $x_{n+1}=t_1,\ldots,x_{n+k}=t_k$, then for any $\mathbf{t}=(t_1,\ldots,t_k)$ in a neighborhood of $(p_{n+1},\ldots,p_{n+k})$, the n functions will all vanish simultaneously when t_1 is substituted for x_{n+1}, t_2 for $x_{n+2},\ldots,t_k$ for x_{n+k}, and $\theta_1(t_1,\ldots,t_k)$ is substituted for $x_1,\ldots,\theta_n(t_1,\ldots,t_k)$ is substituted for x_n.

PROOF: The space $E^n\times E^k$ is E^{n+k}, a space of $(n+k)$-tuples. The product topology of $E^n\times E^k$ is the same as the Euclidean topology of E^{n+k}. The space $E^n\oplus E^k$, called the direct sum of E^n and E^k, is the collection of all vectors $\mathbf{x}$ of the form $\mathbf{x}=\mathbf{x}_1+\mathbf{x}_2$, $\mathbf{x}_1\in E^n$, $\mathbf{x}_2\in E^k$, and it is understood that the only vector common to the summands is the zero vector. This last stipulation guarantees that the representation for $\mathbf{x}$ in the form described is unique. The norm topology for this space, whatever norm we use, is the same as the norm topology for E^{n+k}, so we can regard E^{n+k}, $E^n\times E^k$, and $E^n\oplus E^k$ as topologically isomorphic.

The reason for this discussion is that we need to identify E^n, a space of n-tuples, with the particular subspace of the space $E^n\times E^k$ consisting of the $(n+k)$-tuples the last k-entries of which are zeros, and to identify E^k with the particular subspace of $E^n\times E^k$ consisting of the $(n+k)$-tuples which have all zeros in the first n entries. These subspaces we denote by $\bar{E}^n$ and $\bar{E}^k$, respectively.

Define a new mapping $\boldsymbol{\varphi}$: $E^{n+k} \to E^{n+k}$ thus:

$$\boldsymbol{\varphi} = (f_1, \ldots, f_n, \pi_{n+1}, \ldots, \pi_{n+k}),$$

where the f_is are the original components of $\boldsymbol{\psi}$, and π_{n+j} is the projection map on E^{n+k} which maps a point $\mathbf{x}$ to its $(n+j)$th coordinate. Note this important fact:

$$\det J_{\boldsymbol{\varphi}}(\mathbf{x}) = \det \frac{\partial(f_1, \ldots, f_n)}{\partial(x_1, \ldots, x_n)}(\mathbf{x}).$$

These determinants are nonzero on some open set containing $\mathbf{P}$, so that $\boldsymbol{\varphi}$ is a diffeomorphism on a neighborhood A of $\mathbf{P}$. Furthermore,

$$\boldsymbol{\varphi}(\mathbf{P}) = (0, \ldots, 0, p_{n+1}, \ldots, p_{n+k}) \in \bar{E}^k,$$

and

$$\boldsymbol{\varphi}^{-1}(0, \ldots, 0, p_{n+1}, \ldots, p_{n+k}) = \mathbf{P}.$$

We denote by $\bar{\boldsymbol{\psi}}$ the mapping $(f_1, \ldots, f_n, 0, \ldots, 0)$; $\bar{\boldsymbol{\psi}}$: $E^{n+k} \to E^{n+k}$, and more precisely,

$$\bar{\boldsymbol{\psi}}: E^{n+k} \to \bar{E}^n, \qquad \bar{\boldsymbol{\psi}} = \boldsymbol{\pi}_{\bar{E}^n} \circ \boldsymbol{\varphi},$$

where $\boldsymbol{\pi}_{\bar{E}^n}$ is the projection mapping of $E^{n+k} \to \bar{E}^n$. Define a new mapping $\bar{\boldsymbol{\theta}}$: $E^{n+k} \to \bar{E}^n$ by

$$\bar{\boldsymbol{\theta}} = \boldsymbol{\pi}_{\bar{E}^n} \circ \boldsymbol{\varphi}^{-1}.$$

It is apparent that $\bar{\boldsymbol{\theta}}(0, \ldots, 0, p_{n+1}, \ldots, p_{n+k}) = (p_1, \ldots, p_n, 0, \ldots, 0)$.

Since $\boldsymbol{\varphi}$ is a diffeomorphism on A, $\boldsymbol{\varphi}(A)$ is a neighborhood of $\boldsymbol{\varphi}(\mathbf{P})$. Let $\bar{\mathbf{t}} \in \bar{E}^k$ be an arbitrary point of $\boldsymbol{\varphi}(A) \cap \bar{E}^k$; this set is an $\bar{E}^k$-neighborhood of $\boldsymbol{\varphi}(\mathbf{P})$. For each such $\bar{\mathbf{t}}$, there is a unique $\mathbf{x}_t = \boldsymbol{\varphi}^{-1}(\bar{\mathbf{t}}) \in A$, and we can write

$$(\bar{\boldsymbol{\theta}}\ (\bar{\mathbf{t}}) \oplus \bar{\mathbf{t}}) = \boldsymbol{\pi}_{\bar{E}^n}\boldsymbol{\varphi}^{-1}(\bar{\mathbf{t}}) \oplus \bar{\mathbf{t}} = \boldsymbol{\pi}_{\bar{E}^n}\mathbf{x}_t \oplus \boldsymbol{\varphi}(\mathbf{x}_t) = \boldsymbol{\pi}_{\bar{E}^n}\mathbf{x}_t \oplus \boldsymbol{\pi}_{\bar{E}^k}\mathbf{x}_t = \mathbf{x}_t.$$

Hence

$$\bar{\boldsymbol{\psi}}\ (\bar{\boldsymbol{\theta}}\ (\bar{\mathbf{t}}) \oplus \bar{\mathbf{t}}) = \bar{\boldsymbol{\psi}}\ (\mathbf{x}_t) = \boldsymbol{\pi}_{\bar{E}^n} \circ \boldsymbol{\varphi}(\mathbf{x}_t) = \boldsymbol{\pi}_{\bar{E}^n}\bar{\mathbf{t}} = \mathbf{0}.$$

The mappings $\bar{\boldsymbol{\psi}}, \bar{\boldsymbol{\theta}}$ are uniquely defined; both are C^1-mappings on A, being composites of such mappings. We simply identify $\bar{\boldsymbol{\psi}}$ with $\boldsymbol{\psi}$, restrict $\bar{\boldsymbol{\theta}}$ to $\bar{E}^k$, and then identify $\boldsymbol{\theta}$: $E^k \to E^n$ with $\bar{\boldsymbol{\theta}}|_{\bar{E}^k}$. $\boldsymbol{\theta}$ is therefore unique, a C^1-mapping on an E^k-neighborhood of the point $(p_{n+1}, \ldots, p_{n+k}) \in E^k$, and for all $\mathbf{t}$ in this neighborhood, $\boldsymbol{\psi}(\boldsymbol{\theta}(\mathbf{t}), \mathbf{t}) \equiv \mathbf{0} \in E^n$. □

Consider the set of n equations

$$y_1 = f_1(x_1, \ldots, x_n, x_{n+1}, \ldots, x_{n+k}),$$
$$\vdots$$
$$y_n = f_n(x_1, \ldots, x_n, x_{n+1}, \ldots, x_{n+k}),$$

or otherwise put, $\mathbf{Y} = \boldsymbol{\psi}\mathbf{X}$.

Suppose $\psi\overline{\mathbf{X}}=\mathbf{0}$ and ψ is C^1 in a neighborhood of $\overline{\mathbf{X}}$ and

$$\det\frac{\partial(f_1,\ldots,f_n)}{\partial(x_1,\ldots,x_n)}\overline{\mathbf{X}}\neq 0.$$

The theorem says there is a k-dimensional neighborhood of $(\bar{x}_{n+1},\ldots,\bar{x}_{n+k}) \in E^k$ in which we have

$$x_1=\theta_1(x_{n+1},\ldots,x_{n+k}),$$
$$\vdots$$
$$x_n=\theta_n(x_{n+1},\ldots,x_{n+k})$$

as solutions to the n equations $f_i(\mathbf{X})=0$; furthermore, the n functions θ_i are C^1-functions in this neighborhood.

Evidently, if $k=0$, the implicit function theorem ensures the existence of a unique numerical solution to the system of equations $\{f_i(\mathbf{X})=0\}_{i=1}^n$.

Suppose $\{y_1,\ldots,y_n\}$ are a fixed set of numbers; the system of equations above is equivalent to the system

$$\{g_i(\mathbf{X})=0\}_{i=1}^n,$$

where for each i, $g_i(\mathbf{X})=f_i(\mathbf{X})-y_i$. If $\overline{\mathbf{X}}$ is a solution for the system $\{g_i(\mathbf{X})=0\}$, the theorem ensures that we can find C^1-functions θ_i, such that

$$x_i=\theta_i(x_{n+1},\ldots,x_{n+k}),\qquad i=1,\ldots,n,$$

in a neighborhood of $(\bar{x}_{n+1},\ldots,\bar{x}_{n+k})$. If we substitute these functions back into the g_is, we then end up with solutions for the y_is in terms of $\{x_{n+1},\ldots,x_{n+k}\}$. When $k=0$, the inverse function theorem guarantees that we can find solutions for the x_is in terms of the y_is.

Now consider the system

$$y_1=f_1(x_1,\ldots,x_n),$$
$$\vdots$$
$$y_m=f_m(x_1,\ldots,x_n)\qquad\text{with}\quad m\geqslant n.$$

Assume this mapping $\psi:\mathbf{X}\mapsto\mathbf{Y}$ is C^1 and J_ψ is of rank n. Then the inverse function theorem says we can, for some n of these equations, say the first n, solve to get

$$x_i=g_i(y_i,\ldots,y_n),\qquad i=1,\ldots,n,$$

where the g_i are C^1-functions. Now, substitute these solutions into the expressions for y_j, $j=n+1,\ldots,m$, and we have not only all the xs but the last $m-n$ ys expressed in terms of the first n ys, and uniquely so. At least, we can do all this in a neighborhood of some point $\bar{\mathbf{x}}$ where

$$\det\frac{\partial(f_1,\ldots,f_n)}{\partial(x_1,\ldots,x_n)}\neq 0.$$

Perhaps we have talked too much about this theorem, but it is an important one, and experience shows that most students have trouble visualizing what is going on here. But we won't say more. We have a few applications to discuss instead.

Suppose we have a C^1-function $y = f(x_1, \ldots, x_n)$, and suppose further that $\mathbf{P} = (\bar{x}_1, \ldots, \bar{x}_n)$ is a point in an open set A where $\bar{y} = f(\bar{x}_1, \ldots, \bar{x}_n)$ is an extreme value for f on A. Now suppose that **grad** $f(\mathbf{P}) \neq 0$. This means that at $\overline{\mathbf{P}} = (\bar{x}_1, \ldots, \bar{x}_n, \bar{y}) \in E^{n+1}$ if φ is the function $f(x_1, \ldots, x_n) - y$ mapping $E^{n+1} \to E^1$, the rank of the Jacobian matrix $J_\varphi(\overline{\mathbf{P}})$ is 1, and $\varphi(\overline{\mathbf{P}}) = 0$. Better said, at least one of the partial derivatives in **grad** $\varphi(\overline{\mathbf{P}})$, besides the last one, does not vanish, so we can solve for one of the xs, say x_1, in terms of $x_2, \ldots, x_n$, and y:

$$x_1 = \theta_1(x_2, \ldots, x_n, y),$$

this solution being a C^1-solution in an n-dimensional neighborhood of $(\bar{x}_2, \ldots, \bar{x}_n, \bar{y})$. Hence for some sufficiently small $\varepsilon > 0$ we have

$$x_1' = \theta_1(\bar{x}_2, \ldots, \bar{x}_n, \bar{y} \pm \varepsilon),$$

and

$$\varphi(x_1', \bar{x}_2, \ldots, \bar{x}_n, \bar{y} \pm \varepsilon) = 0,$$

which means that

$$\bar{y} \pm \varepsilon = f(x_1', \bar{x}_2, \ldots, \bar{x}_n), \qquad \text{with} \quad (x_1', \bar{x}_2, \ldots, \bar{x}_n) \in A.$$

This contradicts the fact that $\bar{y}$ is an extreme value for f on A. We can summarize what we have just proved:

Theorem 9.4.3. *Suppose $f(x)$ is a C^1-function on E^n into E^1 and* $\mathbf{P} \in E^n$ *is a point. If f is extreme at* $\mathbf{P}$, *then* **grad** $f(\mathbf{P}) = \mathbf{0}$.

It is well to remark that if $f\colon E^n \to \mathbb{R}$ is a mapping and A is a subset of E^n, f may or may not assume extreme values on A.

Definition 9.4.4. If f is a real-valued function on a topological space S, a point $\bar{\mathbf{x}}$ is called a *local maximum* for f if there exists a neighborhood N of $\bar{\mathbf{x}}$ such that $\forall \mathbf{x} \in N$, $f(\bar{\mathbf{x}}) \geqslant f(\mathbf{x})$. A *local minimum* for f is defined analogously.

If $\bar{\mathbf{x}}$ is such that $f(\bar{\mathbf{x}}) \geqslant f(\mathbf{x})$, $\forall \mathbf{x} \in S$, then $\bar{\mathbf{x}}$ is a *maximum point* for f. A *minimum point* for f is defined analogously. A (local) *extreme point* for f is one which is either a (local) maximum point or a (local) minimum point. The value that f assumes at an extreme point is called an *extreme value*. A point $\mathbf{x} \in \operatorname{dom} f$ is called a *critical point* of f if $\nabla f(\mathbf{x}) = \mathbf{0}$ or if $\nabla f(\mathbf{x})$ does not exist.

If we are looking for extreme values for a real-valued function f, the first step is to find points at which ∇f vanishes and any other critical points. These points are possible candidates for extreme points for f. But, we must add that the converse to the last theorem is not true. Counterexamples abound.

Definition 9.4.5. Let $f: E^n \to E^1$ be a real-valued C^2-map. The *Hessian* of f is the determinant of the $n \times n$ matrix consisting of the second partials of f. This is to say, the Hessian of f is

$$\det H_f = \det \begin{bmatrix} D_{11}f & \cdots & D_{1n}f \\ \vdots & & \vdots \\ D_{n1}f & \cdots & D_{nn}f \end{bmatrix}.$$

H_f is the Hessian matrix of f.

Theorem 9.4.6. *Suppose $f: E^2 \to E^1$ is a C^2-mapping and at* $\mathbf{P} \in E^2$ **grad** $f(\mathbf{P}) = 0$.

1. *If* $\det H_f(\mathbf{P}) < 0$, $\mathbf{P}$ *is not a local extreme point for f, but is a point such that in the E^3 containing the surface $f(x, y) = z$ there is a horizontal tangent plane to this surface at* $(\mathbf{P}, f(\mathbf{P}))$. *The point* $\mathbf{P}$ *is called a saddle point for f.*
2. *If* $\det H_f(\mathbf{P}) > 0$ *and if* $D_{11}f(\mathbf{P}) < 0$ *or* $D_{22}f(\mathbf{P}) < 0$, $\mathbf{P}$ *is a local maximum for f.*
3. *If* $\det H_f(\mathbf{P}) > 0$ *and* $D_{11}f(\mathbf{P}) > 0$ *or* $D_{22}f(\mathbf{P}) > 0$, $\mathbf{P}$ *is a local minimum for f.*
4. *If* $\det H_f(\mathbf{P}) = 0$, *we can draw no conclusions other than* $\mathbf{P}$ *is still a candidate for being an extreme point for f.*

The Hessian is analogous to the second derivative of a real-valued function of a single real variable, just as the Jacobian is to the first derivative. If f were a function from $E^n \to E^1$, we could find some criteria that would allow us to conclude that a zero of the gradient is a local max, or a local min, for f, but the problem is difficult. Leave it to the reader! (See Problem 56.)

PROOF: Using Taylor's formula, and the fact that **grad** $f(\mathbf{P}) = 0$, we can write for $\mathbf{P} = (p_1, p_2)$ and $\mathbf{H} = (h_1, h_2)$ in E^2

$$f(\mathbf{P}+\mathbf{H}) = f(\mathbf{P}) \\ + \left(h_1^2 D_{11} f(\mathbf{P}+t\mathbf{H}) + 2h_1h_2 D_{12} f(\mathbf{P}+t\mathbf{H}) + h_2^2 D_{22} f(\mathbf{P}+t\mathbf{H})\right)/2,$$

where $0 \leqslant t \leqslant 1$. The expression in parentheses can be thought of as a quadratic in h_1 having as discriminant the expression

$$h_2^2\left[\left(D_{12}f(\mathbf{P}+t\mathbf{H})\right)^2 - D_{11}f(\mathbf{P}+t\mathbf{H})D_{22}f(\mathbf{P}+t\mathbf{H})\right]$$
$$= -h_2^2 \det H_f(\mathbf{P}+t\mathbf{H}).$$

If the Hessian is positive at $\mathbf{P}$, then being continuous it is positive at $\mathbf{P}+t\mathbf{H}$ for $\mathbf{H}$ sufficiently close to $\mathbf{0}$, in which case the discriminant of the quadratic in h_1 is negative. This is to say, the quadratic is always positive, or always negative, in a neighborhood of $\mathbf{P}+t\mathbf{H}$, and hence in a neighborhood of $\mathbf{P}$ if $\mathbf{H}$ is sufficiently close to $\mathbf{0}$. For all such $\mathbf{H}$, $f(\mathbf{P}+\mathbf{H})$ is either always greater than or always less than $f(\mathbf{P})$. Equality might occur if h_1 or h_2 is zero.

If both $D_{11}f(\mathbf{P}+t\mathbf{H})$, $D_{22}f(\mathbf{P}+t\mathbf{H})$ are negative, which will be the case if both $D_{11}f(\mathbf{P})$, $D_{22}f(\mathbf{P})$ are negative for small enough $\mathbf{H}$, then the quadratic is nonpositive for all values of h_1 and h_2, and we have $f(\mathbf{P})$ is a local maximum value, or a relative maximum value. If both $D_{11}f(\mathbf{P})$, $D_{22}f(\mathbf{P})$ are positive, $\mathbf{P}$ is a local minimum for f.

If the Hessian is negative, then the quadratic expression of the hs takes on both positive and negative values at $\mathbf{P}+t\mathbf{H}$; hence $f(\mathbf{P})$ is not a local extreme value. However, there is a tangent plane to the surface $f(x,y)=z$ at the point $(\mathbf{P}, f(\mathbf{P}))$ given by the equation $z = f(\mathbf{P})$. To see this, consider the real-valued function $g(x,y,z)=f(x,y)-z$. g vanishes at $(p_1, p_2, f(p_1,p_2))$, and $\mathbf{grad}\, g = (D_1 f, D_2 f, -1)$; at $(\mathbf{P}, f(\mathbf{P}))$, $\mathbf{grad}\, g(\mathbf{P}, f(\mathbf{P})) = (0,0,-1)$. By Theorem 9.1.13 the equation of the tangent plane at $(\mathbf{P}, f(\mathbf{P}))$ is

$$(0,0,-1)\cdot\left((x,y,z)-(p_1,p_2,f(p_1,p_2))\right) = -(z-f(\mathbf{P}))=0$$

or

$$z=f(\mathbf{P}).$$

Finally, suppose the Hessian is zero at $\mathbf{P}$. Then for different choices of $\mathbf{H}$, the Hessian at $\mathbf{P}+t\mathbf{H}$ might be positive or negative; hence we can draw no conclusions about $\mathbf{P}$. We could further expand the formula and perhaps determine that $\mathbf{P}$ is, or is not, a local extreme point, but we do not choose to push further in that direction. □

It could well be that $f\colon E^n \to E^1$ is C^1, that A is a compact set in E^n, and that f does not have an extreme point in int A but does have one on the boundary of A. Or more generally, suppose f is C^1 over a set A and the argument of f is restricted to some certain subset $S \subset A$ (generally, S is defined by a set of restrictive equations called constraints), how can we find points of S that are candidates for extreme points for $f|_S$? We now discuss a method for finding such points, the method of Lagrange multipliers.

9.5 LAGRANGE MULTIPLIERS

Suppose we have the C^1 system

$$\begin{aligned} y &= f(x_1,\ldots,x_n),\\ y_1 &= f_1(x_1,\ldots,x_n),\\ &\vdots\\ y_k &= f_k(x_1,\ldots,x_n), \end{aligned}$$

and suppose $\mathbf{P}=(\bar{x}_1,\ldots,\bar{x}_n)$ is a point such that $y_j(\mathbf{P})=0$, $j=1,\ldots,k$, and $y(\mathbf{P})$ is extreme. If $dy_1\,dy_2\cdots dy_k\neq 0$ at $\mathbf{P}$, which is tantamount to saying that at $\mathbf{P}$ the rank of the matrix

$$\frac{\partial(f_1,\ldots,f_k)}{\partial(x_{i_1},\ldots,x_{i_k})}$$

is k for some k of the variables x_i, then we can solve for, say, $x_1,\ldots,x_k$ in terms of $x_{k+1},\ldots,x_n$. That is, for each $i=1,\ldots,k$, we can find C^1-functions θ_i such that

$$x_i=\theta_i(x_{k+1},\ldots,x_n),$$

these solutions being valid in some neighborhood of $(\bar{x}_{k+1},\ldots,\bar{x}_n)$. Substituting back into f gives

$$y=f(\theta_1(x_{k+1},\ldots,x_n),\ldots,\theta_k(x_{k+1},\ldots,x_n),x_{k+1},\ldots,x_n)=\varphi(x_{k+1},\ldots,x_n).$$

For $y=f(\bar{x}_1,\ldots,\bar{x}_n)=f(\mathbf{P})$ to be extreme, we need **grad** φ to vanish at the point $\bar{\mathbf{P}}=(\bar{x}_{k+1},\ldots,\bar{x}_n)$.

Thus, given the original system of functions $\{f,f_1,\ldots,f_k\}$, to find a candidate for a point $\mathbf{P}\in E^n$ which will make f extreme and make the remaining k functions vanish simultaneously, we solve for k of the x_is to find the θ_is, then obtain φ, and solve for the remaining x_is by equating the $n-k$ partial derivatives of φ to zero. Having obtained these solutions, substitute them into the θ_is and (εὑρήκας!) you have your candidate for $\mathbf{P}$; all the x_is have been found!

This does seem a bit difficult. If we write instead

$$y+\lambda_1 y_1+\cdots+\lambda_k y_k\equiv g(x_1,\ldots,x_n,\lambda_1,\ldots,\lambda_k),$$

and look for a point $\mathbf{P}^*$ where **grad** g will vanish, we have to solve

$$\frac{\partial g}{\partial x_i}=0,\qquad i=1,\ldots,n,$$

$$\frac{\partial g}{\partial \lambda_j}=0,\qquad j=1,\ldots,k,\quad k<n.$$

If we solve these $n+k$ equations in the $n+k$ unknowns to get a point $\mathbf{P}^*$, the point $\mathbf{P}$ will automatically make the $y_j=0$, $j=1,\ldots,k$, since

$(\partial g/\partial\lambda_j)(\mathbf{P}^*) = y_j$. Now if $\mathbf{P}^*$ also satisfies $(\partial g/\partial x_i)(\mathbf{P}^*) = 0$, $i = 1,\ldots,n$, then we have the n equations

$$-\frac{\partial y}{\partial x_i}(\mathbf{P}) = \sum_{j=1}^{k} \bar{\lambda}_j \frac{\partial y_j}{\partial x_i}(\mathbf{P}), \qquad i = 1,\ldots,n,$$

where $\mathbf{P} = (\bar{x}_1,\ldots,\bar{x}_n)$, $\mathbf{P}^* = (\bar{x}_1,\ldots,\bar{x}_n,\bar{\lambda}_1,\ldots,\bar{\lambda}_k)$, and $\overline{\mathbf{P}} = (\bar{x}_{k+1},\ldots,\bar{x}_n)$.

But under the hypothesis that $dy_1 \cdots dy_k \neq 0$ at $\mathbf{P}$, we could express each $\bar{x}_i$, $i = 1,\ldots,k$, in terms of the remaining x_i; if we substitute these expressions $\theta_i(\bar{x}_{k+1},\ldots,\bar{x}_n) = \theta_i(\overline{\mathbf{P}})$ for $\bar{x}_i$ in the n equations above, the right-hand sides must vanish since for each $j = 1,\ldots,k$, $y_j \equiv 0$ in a neighborhood of $\overline{\mathbf{P}}$. The vanishing of the left-hand sides is equivalent to the vanishing of **grad** φ at $\overline{\mathbf{P}}$.

Theorem 9.5.1. *If $f(x_1,\ldots,x_n)$ is a real-valued C^1-function on E^n and the vector $(x_1,\ldots,x_n)$ is subject to the constraints*

$$f_i(x_1,\ldots,x_n) = 0, \qquad i = 1,\ldots,k, \quad k < n,$$

with $df_1\, df_2 \cdots df_k \neq 0$ at a point $\mathbf{P}$, then if $\mathbf{P}$ is to satisfy the k constraints and also be an extreme point for f, there exist constants $\{\bar{\lambda}_1,\ldots,\bar{\lambda}_k\}$, called Lagrange multipliers, such that the gradient of the function

$$g(x_1,\ldots,x_n,\lambda_1,\ldots,\lambda_k) = f(x_1,\ldots,x_n) + \sum_{j=1}^{k} \lambda_j f_j(x_1,\ldots,x_n)$$

will vanish at $\mathbf{P}^* = (\mathbf{P},\bar{\lambda}_1,\ldots,\bar{\lambda}_k)$. Hence, to find a candidate for such a point $\mathbf{P}$, we solve the system **grad** $g(x_1,\ldots,x_n,\lambda_1,\ldots,\lambda_k) = \mathbf{0}$. *If $\mathbf{P}^*$ is a solution to this system, the first n components of $\mathbf{P}^*$ comprise an n-tuple which serves as a candidate for $\mathbf{P}$.*

The method of Lagrange multipliers sometimes makes the problem of finding extreme points simpler; at other times the direct attack works just as well. No matter how you find the candidate, it must be checked to see whether you have a minimum, a maximum, or a saddle point. We will give an example to illustrate the Lagrange multiplier process.

Suppose a particle lies on the line $2x + 3y = 25$ and another particle lies on the ellipse $u^2 + 4v^2 = 4$. If these particles attract each other gravitationally, what will be their equilibrium positions?

The function we want to make extreme is

$$f(x, y, u, v) = (x-u)^2 + (y-v)^2,$$

and the constraints are

$$f_1(x, y, u, v) = 2x + 3y - 25 = 0,$$

$$f_2(x, y, u, v) = u^2 + 4v^2 - 4 = 0.$$

Put

$$g=(x-u)^2+(y-v)^2+\lambda_1(2x+3y-25)+\lambda_2(u^2+4v^2-4),$$

and solve

$$\begin{aligned} 2(x-u)+2\lambda_1&=0, & 2(y-v)+3\lambda_1&=0,\\ -2(x-u)+2u\lambda_2&=0, & -2(y-v)+8v\lambda_2&=0,\\ 2x+3y&=25, & u^2+4v^2&=4. \end{aligned}$$

We note that

$$df_1\,df_2=4u\,dx\,du+16v\,dx\,dv+6u\,dy\,du+24v\,dy\,dv,$$

and this form vanishes only at $u=0=v$, a point not satisfying the second constraint. Hence any points $\mathbf{P}$ that we find will not cause $df_1\,df_2$ to vanish.

We don't need to solve for λ_1 and λ_2 separately; find λ_1/λ_2 first. We get

$$\lambda_1/\lambda_2=-u=-8v/3,$$

from which it follows that $(u,v)=(\pm\frac{8}{5},\pm\frac{3}{5})$. Since $\lambda_1=u-x=\frac{2}{3}(v-y)$, we can find x and y quickly. When $(u,v)=(\frac{8}{5},\frac{3}{5})$, we find that $(x,y)=(\frac{304}{65},\frac{339}{65})$, and when $(u,v)=(-\frac{8}{5},-\frac{3}{5})$, we find that $(x,y)=(\frac{196}{65},\frac{411}{65})$. The stable equilibrium is $(x,y)=(\frac{304}{65},\frac{339}{65})$ and $(u,v)=(\frac{8}{5},\frac{3}{5})$. The other pair of points give an unstable equilibrium; the distance between the particles is the maximal distance between line and ellipse.

Note that we never did bother to solve for λ_1 or λ_2.

9.6 SOME PARTICULAR PARAMETRIC MAPS

Suppose we have in E^m some sort of n-dimensional surface S described by one or several equations. If the surface S is nice enough, we can represent S more effectively, at least as far as our theory is concerned, by finding a C^1-mapping from E^n into E^m that maps the unit cube K, or some relatively simple measurable set A, onto S. In the event that we can find a $\varphi\colon E^n\to E^m$ that maps a box B onto S, we call φ a parametric mapping. Think of the golden bell of Gabriel's trumpet held high, pointing straight to the heavens; we can describe it, not mellifluously perhaps, but parametrically by the map $\boldsymbol{\varphi}$:

$$x=r\cos\theta,\qquad y=r\sin\theta,\qquad z=13\ln r,$$

$(r,\theta)\in[1,8]\times[0,2\pi)$. The symbols r,θ are the parameters. Do not confuse them with polar coordinates; they are Cartesian coordinates in E^2. $\boldsymbol{\varphi}$ maps the indicated box onto the bell of the trumpet.

We can generalize the mapping that transforms the spherical coordinates of E^3 into the Cartesian coordinates of the same space by exhibiting a mapping that sends a box in E^n into a ball in E^n. The mapping φ is

$$x_1 = \rho\cos\theta_1,$$
$$x_2 = \rho\sin\theta_1\cos\theta_2,$$
$$\vdots$$
$$x_k = \rho\left[\prod_{i=1}^{k-1}\sin\theta_i\right]\cos\theta_k, \qquad k = 2,3,\ldots,n-1,$$
$$\vdots$$
$$x_n = \rho\prod_{i=1}^{n-1}\sin\theta_i,$$

$0 \leqslant \rho \leqslant R$, $0 \leqslant \theta_i < \pi$, $i = 1,\ldots,n-2$, $0 \leqslant \theta_{n-1} < 2\pi$. If you calculate $\det J_\varphi$, you will find it to be positive, so this particular representation of the mapping is orientation preserving. For example, if $n=2$, $\det J_\varphi = \rho$; if $n=3$, $\det J_\varphi = \rho^2\sin\theta_1$, and if $n=5$, $\det J_\varphi = \rho^4\sin^3\theta_1\sin^2\theta_2\sin\theta_3$. You might check the case $n=4$ to see if it fits the pattern; it looks very much as if the magnification factor for the general case is

$$\det J_\varphi = \rho^{n-1}\sin^{n-2}\theta_1\sin^{n-3}\theta_2\cdots\sin\theta_{n-2}.$$

If $\alpha > 0$ is a positive constant, the sphere in E^3 of radius α has area $4\pi\alpha^2$. In E^4, the locus of points (x_1, x_2, x_3, x_4) whose distance from the origin is α is a hypersphere, a three-dimensional surface satisfying $\Sigma_{i=1}^4 x_i^2 = \alpha^2$. Now consider the following mapping ψ:

$$\left.\begin{aligned} x_1 &= \alpha\sin\theta_1\sin\theta_2,\\ x_2 &= \alpha\sin\theta_1\cos\theta_2,\\ x_3 &= \alpha\cos\theta_1\sin\theta_2,\\ x_4 &= \alpha\cos\theta_1\cos\theta_2. \end{aligned}\right\} \quad \begin{aligned} &\theta_1 \in [0, \pi/2],\\ &\theta_2 \in [0, \pi/2], \end{aligned}$$

ψ maps the square $[0, \pi/2]^2 \subset E^2$ onto a two-dimensional surface satisfying $\Sigma_{i=1}^4 x_i^2 = \alpha^2$, $\alpha \geqslant x_i \geqslant 0$, $\forall i$. The area of this 2-surface contained in the hypersphere centered at the origin and having radius α can be found by calculating the g factor associated with ψ.

$$g^2/\alpha^4 = (-\sin\theta_1\cos\theta_1)^2 + (\sin\theta_2\cos\theta_2)^2 + \left[\sin^2\theta_1\cos^2\theta_2 - \cos^2\theta_1\sin^2\theta_2\right]^2$$
$$+\left[\cos^2\theta_1\cos^2\theta_2 - \sin^2\theta_1\sin^2\theta_2\right]^2 + (-\sin\theta_2\cos\theta_2)^2 + (\sin\theta_1\cos\theta_1)^2,$$

and, as one might hope, the right-hand side reduces to 1. Hence the total area is $\pi^2\alpha^2/4$. Where does the extra π come from? A circle of radius R has circumference $2\pi R$ whether it lies in E^2 or E^3. As you ponder the question, consider: Do the points of that surface span a three-dimensional or four-dimensional subspace of E^4?

Consider the map $\boldsymbol{\varphi}: E^1 \to E^n$ given by

$$\boldsymbol{\varphi} = (f_1(t),\ldots,f_n(t)), \qquad t \in [a,b].$$

This is a parametric mapping. Assume that the image of $[a, b]$ under $\boldsymbol{\varphi}$ is a "smooth" curve in E^n.

Definition 9.6.1. We call a mapping $\boldsymbol{\varphi}: E^n \to E^m$ *smooth* if it is of class C^k, for some $k \geqslant 1$. If $g_{\boldsymbol{\varphi}} > 0$ at every point, we call $\boldsymbol{\varphi}$ *regular*. The reader should take note of the fact that in some other areas of mathematics a mapping must be of class C^∞ to be smooth.

Suppose $\boldsymbol{\varphi}([a, b])$ is a smooth, regular curve in E^n and $\boldsymbol{\varphi}$ is of class C^{n-1}. If the curve is to be a "plane" curve, i.e., if the points of the curve span an $(n-1)$-dimensional "flat" (affine) subspace, then there must be a nonzero constant vector $\mathbf{A} \subset E^n$ such that

$$\mathbf{A} \cdot \boldsymbol{\varphi}(t) \equiv \mathbf{A} \cdot \boldsymbol{\varphi}(a), \qquad \forall t \in [a,b];$$

i.e., there must be a normal vector to the flat $\boldsymbol{\varphi}(t) - \boldsymbol{\varphi}(a)$. Assume, to make things more convenient, that $\boldsymbol{\varphi}(a) = \mathbf{0}$ (or that for some point $x_0 \in [a, b]$ we have $\boldsymbol{\varphi}(x_0) = \mathbf{0}$). If such is not the case, we can always compose $\boldsymbol{\varphi}$ with a translation map so that the curve does pass through the origin.

If the curve, going through the origin is "planar," then there are constants, not all zero, $\{a_1,\ldots,a_n\}$ such that

$$a_1 f_1(t) + \cdots + a_n f_n(t) \equiv 0, \qquad \forall t \in [a,b].$$

Since this is an identity, we can differentiate $n-1$ times and get $n-1$ additional identities. In matrix form

$$W_{\boldsymbol{\varphi}} \begin{bmatrix} a_1 \\ \vdots \\ a_n \end{bmatrix} = \begin{bmatrix} f_1(t) & \cdots & f_n(t) \\ f_1'(t) & \cdots & f_n'(t) \\ \vdots & & \vdots \\ f_1^{(n-1)}(t) & \cdots & f_n^{(n-1)}(t) \end{bmatrix} \begin{bmatrix} a_1 \\ \vdots \\ a_n \end{bmatrix} \equiv \mathbf{0}.$$

Definition 9.6.2. If $\boldsymbol{\varphi}$ is a vector-valued C^{n-1}-mapping of a real variable, the Wronskian of $\boldsymbol{\varphi}$ is the determinant of the matrix $W_{\boldsymbol{\varphi}}$ displayed immediately above.

We know that the Wronskian must be identically equal to zero if the matrix equation above is to be satisfied. Conversely, if the Wronskian is identically equal to zero on $[a,b]$, we can conclude that the rows of the matrix are linearly dependent, so each of the functions $f_i(t)$ is a solution to a homogeneous linear differential equation with constant coefficients of the form

$$\sum_{k=0}^{n-1} b_k y^{(k)}(t)=0, \qquad t\in[a,b], \quad y(a)=0.$$

Furthermore, if the Wronskian is identically zero, the columns are linearly dependent, so there are constants a_i such that $\Sigma_{i=1}^n a_i f_i(t)\equiv 0$. Thus we have the following result.

Theorem 9.6.3. *Let* $\boldsymbol{\varphi}$: $E^1\to E^n$ *be a* C^{n-1}*-mapping of an interval* $[a,b]$ *to a curve* $\gamma\subset E^n$ *which passes through the origin. A necessary and sufficient condition for* γ *to be a "flat" curve is that the Wronskian of* $\boldsymbol{\varphi}$, $\det W_{\boldsymbol{\varphi}}$, *vanish identically on* $[a,b]$. *If the rank of* $W_{\boldsymbol{\varphi}}$ *is* k, *the points of* γ *span a* k*-dimensional flat.*

Suppose $\boldsymbol{\varphi}$: $E^{n-1}\to E^n$ is a C^1-mapping such that the rank of $J_{\boldsymbol{\varphi}}$ is $n-1$ at every point $\mathbf{P}=(u_1,\dots,u_{n-1})\in E^{n-1}$. There are n $(n-1)\times(n-1)$ submatrices in $J_{\boldsymbol{\varphi}}(\mathbf{P})$, not all of which are singular. Let $\mathbf{N}(\mathbf{P})$ be the vector in E^n whose components are the determinants of these submatrices. $\mathbf{N}(\mathbf{P})=(N_1(\mathbf{P}),\dots,N_n(\mathbf{P}))$, where

$$N_k(\mathbf{P})=\det\frac{\partial(x_1,\dots,x_{k-1},x_{k+1},\dots,x_n)}{\partial(u_1,\dots,u_{n-1})}(\mathbf{P})(-1)^{k-1}.$$

If $\mathbf{N}(\mathbf{P})$ is divided by $g_{\boldsymbol{\varphi}}(\mathbf{P})=(\Sigma_{k=1}^n N_k^2(\mathbf{P}))^{1/2}$, we get a unit vector having the same direction.

By the implicit function theorem, the system of equations for $\boldsymbol{\varphi}$,

$$\begin{aligned} x_1&=x_1(u_1,\dots,u_{n-1}),\\ &\vdots\\ x_n&=x_n(u_1,\dots,u_{n-1}),\end{aligned}$$

defines implicitly a function h such that

$$h(\mathbf{x}(\mathbf{u}))\equiv 0.$$

Hence the equation $h(\mathbf{x})=0$ describes the hypersurface described by $\boldsymbol{\varphi}$, and, as we have already seen,

$$\mathbf{grad}\, h(\boldsymbol{\varphi}(\mathbf{P}))=\left(\frac{\partial h}{\partial x_1}(\boldsymbol{\varphi}(\mathbf{P})),\dots,\frac{\partial h}{\partial x_n}\boldsymbol{\varphi}(\mathbf{P})\right)$$

is a vector normal to this hypersurface at $\boldsymbol{\varphi}(\mathbf{P})$.

Since $h(\mathbf{x}(\mathbf{P})) \equiv 0$, we can write

$$\frac{\partial h}{\partial u_1} \equiv 0 \equiv \frac{\partial h}{\partial x_1}\frac{\partial x_1}{\partial u_1} + \cdots + \frac{\partial h}{\partial x_n}\frac{\partial x_n}{\partial u_1},$$

$$\vdots$$

$$\frac{\partial h}{\partial u_{n-1}} \equiv 0 \equiv \frac{\partial h}{\partial x_1}\frac{\partial x_1}{\partial u_{n-1}} + \cdots + \frac{\partial h}{\partial x_n}\frac{\partial x_n}{\partial u_{n-1}}.$$

We have $n-1$ equations in the n unknowns $\partial h/\partial x_k$. Now suppose that $(\partial h/\partial x_n)(\boldsymbol{\varphi}(\mathbf{P})) \neq 0$, and let

$$y_k = \frac{\partial h}{\partial x_k} \Big/ \frac{\partial h}{\partial x_n}.$$

Solve the system

$$\frac{\partial x_1}{\partial u_1}y_1 + \cdots + \frac{\partial x_{n-1}}{\partial u_1}y_{n-1} = -\frac{\partial x_n}{\partial u_1},$$

$$\vdots$$

$$\frac{\partial x_1}{\partial u_{n-1}}y_1 + \cdots + \frac{\partial x_{n-1}}{\partial u_{n-1}}y_{n-1} = -\frac{\partial x_n}{\partial u_{n-1}}$$

to get

$$y_k = \pm\frac{(-1)^{k-1}\det J_{\boldsymbol{\varphi}}^k}{\det J_{\boldsymbol{\varphi}}^n},$$

where $J_{\boldsymbol{\varphi}}^i$ is the Jacobian matrix of $\boldsymbol{\varphi}$ with the ith row deleted. Note that each y_k, and hence each $\partial h/\partial x_k$, for $k=1,\ldots,n-1$, is proportional to N_k. We can conclude that the vector $\mathbf{N}(\mathbf{P})$ is normal to the surface defined by $\boldsymbol{\varphi}$ at $\boldsymbol{\varphi}(\mathbf{P})$.

Theorem 9.6.4. *Let $\boldsymbol{\varphi}: E^{n-1} \to E^n$ be a C^1-mapping having rank $n-1$ everywhere on a region $A \subset E^{n-1}$. If $\mathbf{P} \in A$, then $\mathbf{N}(\mathbf{P}) = (N_1(\mathbf{P}),\ldots,N_n(\mathbf{P}))$ is a unit normal to the hypersurface $\boldsymbol{\varphi}(A)$ at $\boldsymbol{\varphi}(\mathbf{P})$, where*

$$N_k(\mathbf{P}) = \frac{(-1)^{k-1}\det J_{\boldsymbol{\varphi}}^k(\mathbf{P})}{g_{\boldsymbol{\varphi}}(\mathbf{P})}, \qquad \forall k=1,\ldots,n.$$

Note that the vector $\mathbf{N}(\mathbf{P})$ is given in terms of the us; i.e., its components are evaluated at $\mathbf{P}=(u_1,\ldots,u_{n-1})$.

Corollary 9.6.5. *The tangent hyperplane to $\boldsymbol{\varphi}(A)$ at $\boldsymbol{\varphi}(\mathbf{P})$ is given by the equation $\mathbf{N}(\mathbf{P}) \cdot (\mathbf{X} - \boldsymbol{\varphi}(\mathbf{P})) = 0$. Moreover, up to a constant of proportionality,*

$$\|\mathbf{grad}\, h(\boldsymbol{\varphi}(\mathbf{P}))\| = g_{\boldsymbol{\varphi}}(\mathbf{P}).$$

The angles whose cosines are $N_k(\mathbf{P})$ are those made by the vector $\mathbf{N}(\mathbf{P})$ and the kth coordinate axis; they are called the *direction angles* of $\mathbf{N}(\mathbf{P})$. Hence any nonzero vector $\mathbf{V}$ in E^n has its set of direction angles $\{\alpha_k\}_{k=1}^n$; the cosines of these angles are the *direction cosines* of $\mathbf{V}$, and they are simply the components of the normalized vector $\mathbf{V}/\|\mathbf{V}\|$.

If S is a surface in E^3 given parametrically by the mapping $\boldsymbol{\varphi}$,

$$x = x(u,v), \qquad y = y(u,v), \qquad z = z(u,v),$$

and if $\mathbf{P} = (u_0, v_0)$, we can hold v fixed at $v = v_0$. $\boldsymbol{\varphi}(u, v_0)$ is then a curve on S called a *coordinate curve*. The tangent vector to this curve at $\boldsymbol{\varphi}(u_0, v_0)$ has components

$$\left(\frac{\partial x}{\partial u}, \frac{\partial y}{\partial u}, \frac{\partial z}{\partial u}\right)\bigg|_{(u,v)=(u_0,v_0)}.$$

Similarly, the tangent vector to the coordinate curve $\boldsymbol{\varphi}(u_0, v)$ at $\boldsymbol{\varphi}(\mathbf{P})$ has components

$$\left(\frac{\partial x}{\partial v}, \frac{\partial y}{\partial v}, \frac{\partial z}{\partial v}\right)\bigg|_{(u,v)=\mathbf{P}}.$$

Hence the unit normal vector to S at $\boldsymbol{\varphi}(\mathbf{P})$ can be obtained by taking the cross product (in the order shown) and normalizing:

$$\mathbf{N}(\mathbf{P}) = \left[\left(\frac{\partial x}{\partial u}(\mathbf{P}), \frac{\partial y}{\partial u}(\mathbf{P}), \frac{\partial z}{\partial u}(\mathbf{P})\right) \times \left(\frac{\partial x}{\partial v}(\mathbf{P}), \frac{\partial y}{\partial v}(\mathbf{P}), \frac{\partial z}{\partial v}(\mathbf{P})\right)\right] \Big/ g_{\boldsymbol{\varphi}}(\mathbf{P}).$$

If $\boldsymbol{\varphi}\colon E^1 \to E^n$ is a C^1-mapping given parametrically by

$$x_i = x_i(t), \qquad i = 1, \ldots, n,$$

such that $J_{\boldsymbol{\varphi}}$ has rank 1 at every point, the image of $\boldsymbol{\varphi}$ is a regular smooth curve. The vector

$$\boldsymbol{\varphi}(p+h) - \boldsymbol{\varphi}(p), \qquad p \in E^1, \quad h \neq 0,$$

can be thought of as a free vector initiating at $\boldsymbol{\varphi}(p)$ on the curve and terminating at $\boldsymbol{\varphi}(p+h)$; it is parallel to the vector

$$\frac{1}{h}[\boldsymbol{\varphi}(p+h) - \boldsymbol{\varphi}(p)], \qquad h \in \mathbb{R}.$$

The limit of this vector as $h \to 0$ is of course a tangent vector to the curve at $\boldsymbol{\varphi}(p)$; i.e.,

$$\lim_{h\to 0}\left(\frac{x_1(p+h) - x_1(p)}{h}, \ldots, \frac{x_n(p+h) - x_n(p)}{h}\right) = (x_1'(p), \ldots, x_n'(p))$$

is tangent to the curve at $\boldsymbol{\varphi}(p)$.

Theorem 9.6.6. *If $\boldsymbol{\varphi}\colon A \subset E^1 \to E^n$ is a regular C^1-mapping, given parametrically by*

$$x_i = x_i(t), \qquad i = 1, \ldots, n, \quad t \in A,$$

then for each $p \in A$, the unit tangent vector to $\boldsymbol{\varphi}(A)$ at $\boldsymbol{\varphi}(p)$ is given by

$$\mathbf{T}(p)=\frac{(x_1'(p),\ldots,x_n'(p))}{g_{\varphi}(p)},$$

where $g_{\varphi}(p)=[\sum_{i=1}^{n}(x_i'(p))^2]^{1/2}$.

Definition 9.6.7. Suppose $\boldsymbol{\varphi}(A)$ is the curve described in the previous theorem and $\mathbf{T}(t)$ is the unit tangent vector to $\boldsymbol{\varphi}(A)$ at an arbitrary point $\boldsymbol{\varphi}(t)$. We define the *curvature* of $\boldsymbol{\varphi}(A)$ at $\boldsymbol{\varphi}(t)$ by

$$K(t)=\|\mathbf{T}'(t)\|/\|\boldsymbol{\varphi}'(t)\|=\|\mathbf{T}'(t)\|/g_{\varphi}(t).$$

We define the *principal normal* to $\boldsymbol{\varphi}(A)$ by

$$\mathbf{N}(t)=\mathbf{T}'(t)/\|\mathbf{T}'(t)\|.$$

If $\boldsymbol{\varphi}(A)\subset E^3$, we define the *binormal vector* to $\boldsymbol{\varphi}(A)$ by

$$\mathbf{B}(t)=\mathbf{T}(t)\times\mathbf{N}(t).$$

It is not hard to see that $(d/dt)(\mathbf{T}(t)\cdot\mathbf{T}(t))=2\mathbf{T}(t)\cdot\mathbf{T}'(t)$. But since $\mathbf{T}(t)\cdot\mathbf{T}(t)\equiv 1$, we have $\mathbf{T}(t)\cdot\mathbf{T}'(t)\equiv 0$, so $\mathbf{T}$ and $\mathbf{T}'$ are orthogonal. Hence $\mathbf{T}$ and $\mathbf{N}$ are orthogonal, and $\{\mathbf{T},\mathbf{N},\mathbf{B}\}$ is an orthonormal set in E^3.

Definition 9.6.8. $\mathrm{sp}(\mathbf{T},\mathbf{N})$ is called the *osculating plane* of $\boldsymbol{\varphi}(A)$.

9.7 A FIXED POINT THEOREM

The theorem we are going to present in this section has a number of applications, one in the theory of differential equations, one in the theory of integral equations, and of course a number of applications in other areas of mathematics. The trick is to recognize when the theorem can be applied.

Definition 9.7.1. Let B be a normed linear space, M a subset of B, and $\boldsymbol{\varphi}: M\to B$ a mapping. Suppose there is a number α, $0<\alpha<1$, such that $\|\boldsymbol{\varphi}(\mathbf{x})-\boldsymbol{\varphi}(\mathbf{y})\|\leqslant\alpha\|\mathbf{x}-\mathbf{y}\|$ for each $\mathbf{x},\mathbf{y}\in M$. Then $\boldsymbol{\varphi}$ is called a *contraction* mapping on M. If $\mathbf{x}_0\in M$ is such that $\boldsymbol{\varphi}(\mathbf{x}_0)=\mathbf{x}_0$, we call $\mathbf{x}_0$ a fixed point of $\boldsymbol{\varphi}$.

Theorem 9.7.2. *Suppose M is a closed subset of a Banach space B and $\boldsymbol{\varphi}: M\to M$ is a contraction mapping. Then $\boldsymbol{\varphi}$ has a unique fixed point $\mathbf{x}_0\in M$, and if $\mathbf{x}\in M$ and $\boldsymbol{\varphi}^n$ is the repeated application of $\boldsymbol{\varphi}$ n times, then $\{\boldsymbol{\varphi}^n(\mathbf{x})\}_{n=1}^{\infty}$ is a Cauchy sequence in M which converges to $\mathbf{x}_0$.*

PROOF: Since B is complete and $M\subset B$ is closed, every Cauchy sequence in M must converge to some point in M. We also observe that if $\mathbf{x}_0$ is a fixed point for $\boldsymbol{\varphi}$ in M, it is unique, for if $\mathbf{x}_1$ were another one, then we would have

$$\|\boldsymbol{\varphi}(\mathbf{x}_0)-\boldsymbol{\varphi}(\mathbf{x}_1)\| = \|\mathbf{x}_0-\mathbf{x}_1\| \leqslant \alpha\|\mathbf{x}_0-\mathbf{x}_1\|$$

for some α, $0<\alpha<1$. This is impossible unless $\mathbf{x}_0=\mathbf{x}_1$.

Now let $\mathbf{x}\in M$ be arbitrary. Since $\boldsymbol{\varphi}(\mathbf{x})\in M$, we have, for some α, $0<\alpha<1$,

$$\|\boldsymbol{\varphi}(\boldsymbol{\varphi}(\mathbf{x}))-\boldsymbol{\varphi}(\mathbf{x})\| \leqslant \alpha\|\boldsymbol{\varphi}(\mathbf{x})-\mathbf{x}\|,$$

$$\vdots$$

$$\|\boldsymbol{\varphi}^{n+1}(\mathbf{x})-\boldsymbol{\varphi}^{n}(\mathbf{x})\| \leqslant \alpha\|\boldsymbol{\varphi}^{n}(\mathbf{x})-\boldsymbol{\varphi}^{n-1}(\mathbf{x})\| \leqslant \cdots \leqslant \alpha^{n}\|\boldsymbol{\varphi}(\mathbf{x})-\mathbf{x}\|.$$

Moreover,

$$\|\boldsymbol{\varphi}^{n+1}(\mathbf{x})-\mathbf{x}\| \leqslant (\alpha^n+\alpha^{n-1}+\cdots+1)\|\boldsymbol{\varphi}(\mathbf{x})-\mathbf{x}\|$$
$$=\frac{1-\alpha^{n+1}}{1-\alpha}\|\boldsymbol{\varphi}(\mathbf{x})-\mathbf{x}\| < \frac{1}{1-\alpha}\|\boldsymbol{\varphi}(\mathbf{x})-\mathbf{x}\|.$$

Hence, for any m, $\|\boldsymbol{\varphi}^m(\mathbf{x})-\mathbf{x}\|$ is bounded by $(1-\alpha)^{-1}[\boldsymbol{\varphi}(\mathbf{x})-\mathbf{x}\|$, and this is the case for each $\mathbf{x}\in M$.

It follows that for any m and any k,

$$\|\boldsymbol{\varphi}^{m+k}(\mathbf{x})-\boldsymbol{\varphi}^{k}(\mathbf{x})\| \leqslant \alpha^k(1-\alpha)^{-1}\|\boldsymbol{\varphi}(\mathbf{x})-\mathbf{x}\|,$$

so for any $\varepsilon>0$, we can find an integer n such that for all $m\geqslant 1$, $k\geqslant n$, $\|\boldsymbol{\varphi}^{m+k}(\mathbf{x})-\boldsymbol{\varphi}^{k}(\mathbf{x})\|<\varepsilon$. This proves $\{\boldsymbol{\varphi}^n(\mathbf{x})\}$ is Cauchy, and hence converges to some $\mathbf{x}_0\in M$. Since $\|\mathbf{x}_0-\boldsymbol{\varphi}^n(\mathbf{x})\|<\varepsilon$ for all sufficiently large n, so is $\|\boldsymbol{\varphi}(\mathbf{x}_0)-\boldsymbol{\varphi}^{n+1}(\mathbf{x})\|<\varepsilon$ for all sufficiently large n. Hence $\boldsymbol{\varphi}(\mathbf{x}_0)$ is also a limit of $\{\boldsymbol{\varphi}^n(\mathbf{x})\}$. Since B is a Hausdorff space, we must have $\mathbf{x}_0=\boldsymbol{\varphi}(\mathbf{x}_0)$; $\mathbf{x}_0$ is a fixed point for $\boldsymbol{\varphi}$ on M. □

To illustrate an application of this theorem we present the following problem. Suppose we have the space $\mathcal{C}[a,b]$, the space of continuous real functions on $[a,b]$ normed by

$$\|f\| = \sup\{|f(t)| : t\in[a,b]\}.$$

This space is a Banach space. Let $g\in\mathcal{C}[a,b]$, and suppose we have the integral equation

$$f(x)=g(x)+\lambda\int_a^b f(t)e^{-x^2/2}t\,dt,$$

where λ is some nonzero constant. It is clear that the integral exists and that the right-hand side of the equation is a continuous function of x if $f(t)$ is continuous on $[a,b]$; our question is whether or not there is one and only one continuous function $f_0\in\mathcal{C}[a,b]$ which will satisfy the equation, making it an identity in x for $x\in[a,b]$. We apply the theorem as follows.

Note that on $[a, b]\times[a, b]$ the function $|K(x,t)|=|te^{-x^2/2}|$ is bounded by some positive number M. Suppose the constant λ is such that

$$|\lambda|<1/M(b-a);$$

it will be clear why we put this restriction on λ. Now, regard this integral equation as the defining expression for a mapping $\varphi: \mathcal{C}[a,b]\to\mathcal{C}[a,b]$: $\varphi: f\mapsto h\in\mathcal{C}[a,b]$, where $h(x)\equiv g(x)+\lambda\int_a^b tf(t)e^{-x^2/2}\,dt$. With the restriction on λ, φ is a contraction mapping, for if $f_1, f_2\in\mathcal{C}[a,b]$,

$$\begin{aligned}\|\varphi f_1-\varphi f_2\| &= \left\|\lambda\int_a^b (f_1(t)-f_2(t))te^{-x^2/2}\,dt\right\|\\ &\leqslant |\lambda|M(b-a)\|f_1-f_2\|=\alpha\|f_1-f_2\|,\end{aligned}$$

where $\alpha=|\lambda|M(b-a)<1$. The theorem guarantees the existence of a unique $f_0\in\mathcal{C}[a,b]$ such that $\varphi(f_0)=f_0$, so a solution exists. It is a pity that the theorem does not tell us how to find f_0 explicitly. (It does suggest how to approximate f_0.)

Consider this question. Let f be a continuous real function on the compact rectangle $[a,b]\times[c,d]=R$, which contains the point (x_0, y_0) in its interior. Is there a unique, real function y which is C^1 on its domain with $y(x_0)=y_0$, such that if $(x, y(x))\in R$, then $f(x, y(x))=y'(x)$? First, we define a function $y(x)$ by

$$y(x)=y_0+\int_{x_0}^x f(t, y(t))\,dt, \qquad x\in[a,b],$$

since such a $y(x)$ clearly has the value y_0 at $x=x_0$, and $y'(x)=f(x, y(x))$. This is only a candidate; we have no assurance that $y(x)\in[c,d]$ for all $x\in[a,b]$. If $y(x)\notin[c,d]$, $f(x, y(x))$ is not guaranteed to be continuous or even defined.

Let φ be the mapping which sends a function y which is C^1 on its domain, with dom $y\subset[a,b]$ and im $y\subset[c,d]$, to a function of the same type. Let y_1, y_2 be two such functions; then

$$\|\varphi(y_2)-\varphi(y_1)\|=\left\|\int_{x_0}^x [f(t, y_2(t))-f(t, y_1(t))]\,dt\right\|.$$

Note that $|f|$ is bounded on R; hence for any $(x, y_1),(x, y_2)\in R$, we can find a constant $M>0$ such that $|f(x,y_2)-f(x,y_1)|\leqslant M|y_2-y_1|$. Therefore, we can write

$$\begin{aligned}\|\varphi(y_2)-\varphi(y_1)\| &= \left\|\int_{x_0}^x [f(t, y_2(t))-f(t, y_1(t))]\,dt\right\|\\ &\leqslant M\sup|y_2(t)-y_1(t)|\left|\int_{x_0}^x dt\right|\leqslant M|x-x_0|\|y_2-y_1\|,\end{aligned}$$

provided that for all t between x and x_0 we have $y_1(t), y_2(t)\in[c,d]$.

Restrict x by putting $|x-x_0| \leqslant |a-x_0| \wedge |b-x_0| \wedge \alpha/M = \delta$, where $0 < \alpha < 1$ is fixed, then let K be the subset of the space $\mathcal{C}^1[x_0-\delta, x_0+\delta]$, a Banach space, consisting of those functions whose values lie in $[c, d]$. K is a closed set. On K, φ is a contraction mapping. Hence our theorem says that there is a unique solution for our problem.

Why are we interested in this problem in the first place? Suppose we have a linear differential equation, first order, with a given initial condition, of the type

$$y'(x)+a(x)y(x)=b(x), \qquad y(x_0)=y_0.$$

Our theorem says there is a unique solution $y(x)$, valid in some neighborhood of (x_0, y_0), for this differential equation.

Although the theorem does not explicitly give the solution, it points out that if you apply the mapping φ enough times to a suitably chosen element, an approximate solution for problems of these types will be forthcoming.

We bring this section to a close by taking note of an "application." If you stir a cup of hot coffee (with considerable finesse!) so that after stirring no two points of the liquid are as far apart as when you began, then—when the liquid has cooled enough to drink—exactly one point of coffee will be right where it was originally.

PROBLEMS

1. Expand to several terms the mapping $\boldsymbol{\varphi}$: $E^2 \to E^4$ given by

$$y_1 = \sin x_1 x_2, \qquad y_2 = x_2 e^{x_1},$$
$$y_3 = \sqrt{x_1^2 + x_2^2}, \qquad y_4 = x_1^2 + x_2^2 + x_1 x_2$$

about the point $(1,1)$.

2. The following map $\boldsymbol{\varphi}$ is a standard orientation preserving transformation of spherical coordinates to Cartesian coordinates:

$$\begin{aligned}
x_1 &= r\cos\theta_1,\\
x_2 &= r\sin\theta_1\cos\theta_2,\\
x_3 &= r\sin\theta_1\sin\theta_2\cos\theta_3,\\
x_4 &= r\sin\theta_1\sin\theta_2\sin\theta_3\cos\theta_4,\\
&\ \ \vdots\\
x_{n-1} &= r\left(\prod_{i=1}^{n-2}\sin\theta_i\right)\cos\theta_{n-1},\\
x_n &= r\left(\prod_{i=1}^{n-1}\sin\theta_i\right),
\end{aligned}$$

$$0 \leqslant r < \infty, \qquad 0 \leqslant \theta_i < \pi, \quad i = 1, \ldots, n-2, \qquad 0 \leqslant \theta_{n-1} < 2\pi.$$

Find $J_{\boldsymbol{\varphi}}$, and calculate the volume of the 5-ball having radius R. [*Hint*: Start with $n=2$, 3, and 4, and attempt a generalization.]

3. If A is a region in E^n and $\varphi: E^n \to E^n$ a homeomorphism, then $\varphi(A)$ is measurable. Prove or disprove this statement. What if φ is a diffeomorphism?

4. Let φ be given by

$$x = e^u + \ln(1+v^2), \qquad y = u^2v^2 + 1.$$

Find the derivative of φ^{-1}. Indicate a reasonable region of the uv plane where φ is locally injective. Is φ injective on the coordinate axes?

5. If

$$x_1 = a\sin\theta_1\sin\theta_2, \qquad x_2 = a\sin\theta_1\cos\theta_2,$$
$$x_3 = a\cos\theta_1\sin\theta_2, \qquad x_4 = a\cos\theta_1\cos\theta_2,$$

where $\theta_i \in [0, \pi/2]$ and $a \in [0, R]$, is a mapping of $D = [0, R] \times [0, \pi/2] \times [0, \pi/2] \to E^4$, find the three-dimensional measure of $\varphi(D)$. Solve for a in terms of the xs.

6. Find a function $y(x)$ which will satisfy

$$y'(x) = x^2 - 2y/x, \qquad y(1) = 1.$$

7. Find the minimum distance between the curves

$$x^2 + y^2 = 1 \qquad \text{and} \qquad x^2y = 16.$$

8. Find the maximum value of $\Pi_{i=1}^n x_i^2$ subject to the constraint $\Sigma_{i=1}^n x_i^2 = R^2$.

9. Find the extreme values of

$$Q(x, y, z) = ax^2 + by^2 + cz^2 + 2\alpha yz + 2\beta xz + 2\gamma xy$$

over the ball $x^2 + y^2 + z^2 \leqslant 1$.

10. Solve for x_1, x_2, y_3 in terms of y_1 and y_2:

$$y_1 = x_1 + 2x_2, \qquad y_2 = x_1^2 - 1, \qquad y_3 = x_1 - x_2 + 1.$$

11. Find the area bounded by the curve

$$(x^2 + y^2)^2 - a^2x^2 - b^2y^2 = 0.$$

12. Find the length of the arc $y = \ln(1 - x^2)$, $0 \leqslant x \leqslant \frac{1}{2}$.

13. Let $f(t)$ be a C^3-function on E^1 and γ an arc in E^2 given by

$$\begin{aligned} x &= f''(t)\cos t + f'(t)\sin t, \\ y &= -f''(t)\sin t + f'(t)\cos t, \end{aligned} \qquad t_1 \leqslant t \leqslant t_2.$$

Find the length of γ.

14. Prove that if γ is an ellipse with semiaxes a and b, the area enclosed by γ is equal to πab, and the length of γ satisfies the inequalities

$$\pi(a+b) < \mu(\gamma) < \pi\sqrt{2}\sqrt{a^2 + b^2}.$$

15. Let a circular ring have uniform density λ and radius c; suppose a unit point mass is at a distance p from the center of the circular ring, lying on the axis through the center perpendicular to the plane of the ring. Calculate the magnitude and direction of the force of gravitational attraction acting upon the unit point mass. Let G be the gravitational constant.

Now, suppose a unit point mass is at a distance r from the center of a spherical shell of radius a, and the shell is of uniform surface density σ. Calculate the magnitude and direction of the gravitational attraction acting upon the unit point mass when $r>a$, $r<a$, and $r=a$.

Finally, let the unit point mass be at a distance r from the center of a uniform ball of density σ, and radius a. Calculate the magnitude and direction of the gravitational force when $r>a$, $r<a$, $r=a$.

16. Find the critical points of the function

$$f(x,y,z)=(ax^2+by^2+cz^2)e^{-x^2-y^2-z^2},$$

where $a>b>c>0$. (A critical point is one where either $\nabla f=0$ or ∇f does not exist.) Determine which are maxima, which are minima, and which are neither.

17. Let $F(x)=\int_{h(x)}^{g(x)} f(x,t)\,dt$. Express $dF(x)/dx$.

18. Compute $\int_\gamma (x+y)\,dS$, where γ is the quarter circle in the first octant given by $x^2+y^2+z^2=R^2$, $y=x$, and dS is measure along the arc.

19. Let S be the surface of the solid in the first octant bounded by $z=x^2+y^2$, $x^2+y^2=1$, $z=0$, and ω the 2-form

$$\omega=y^2z\,dx\,dy+xz\,dy\,dz+x^2y\,dx\,dz.$$

Compute $\int_S \omega$.

20. Compute

$$\int_0^1 \frac{\arctan ax}{x\sqrt{1-x^2}}\,dx.$$

21. Find $D_u f(3,1)$ if $f(x,y)=x^3-3x^2y+3xy^2+1$ and $u=(0.6,0.8)$.

22. If $y=f(x,t)$ and $F(x,y,t)=0$, express dy/dx. More generally, if $F(x,y,t)=0$ and $G(x,y,t)=0$, express dy/dx.

23. Show that in general a homeomorphism $\varphi\colon E^n\to E^n$ does not preserve measurability.

[*Hint*: Let $n=1$. Refer to the Cantor function f described immediately after Theorem 6.9.1. Let $\psi(x)=x+f(x)$ on $[0,1]$. Then ψ is continuous on $[0,1]$ and strictly increasing. Hence $\psi\colon[0,1]\to[0,2]$ is a homeomorphism. Prove this. Next, show that $\mu(\psi([0,1]\setminus C))=1$, where $C\subset[0,1]$ is the Cantor set. What then is $\mu(\psi(C))$? Let $D\subset\psi(C)$ be a nonmeasurable set. ψ^{-1} is a continuous function, and $\mu(\psi^{-1}(D))=0$. If $N=\psi^{-1}(D)$, $\psi(N)$ is a nonmeasurable homeomorph of a measurable set.]

24. Prove that

$$f(x,y)=\begin{cases}\dfrac{xy}{x^2+y^2} & \text{if } x^2+y^2\neq 0,\\ 0 & \text{if } x^2+y^2=0\end{cases}$$

is discontinuous as a function of two variables, but is continuous in each variable separately, and that **grad** f exists everywhere.

25. Show that

$$f(x,y)=\begin{cases}x\sin(1/y)+y\sin(1/x), & \text{if } xy\neq 0,\\ 0, & \text{if } xy=0\end{cases}$$

is a continuous function of (x,y) at (0,0), but that neither of the iterated limits exists at (0,0).

26. Let

$$f(x,y)=\begin{cases}xy\dfrac{x^2-y^2}{x^2+y^2} & \text{if } x^2+y^2\neq 0,\\ 0 & \text{otherwise.}\end{cases}$$

Compute $D_{12}f(0,0)$ and $D_{21}f(0,0)$.

27. Refer to Theorem 9.2.2, and reword the hypothesis in such a way as to make it less restrictive but still fit the proof of the text.

28. Suppose $f:\mathbb{R}\to\mathbb{R}$ is a contraction mapping on an interval $[a,b]$. Prove that f is continuous on $[a,b]$. Prove that $g(x)=f(x)-x$ vanishes at some $r\in[a,b]$. Assume now that f is differentiable on (a,b). What can you say about the bounds on the value of f' on (a,b)? Suppose $f'(x)<1$ on (a,b). Can you show that if $\{x_n\}\subset(a,b)$ is a sequence defined recursively by

$$x_1\in(a,b),\qquad x_{n+1}=x_n-g(x_n)/g'(x_n),$$

then $\lim_n\{x_n\}=r$, where $g(r)=0$?

29. B is a box with no top and a fixed lateral area. Using Lagrange multipliers, find the dimensions of B that will maximize the volume of B.

30. The pair of equations

$$x^2+y^2=z^2 \qquad \text{and} \qquad y^2=x,$$

together with the restriction $0\leqslant x\leqslant 1$, $0\leqslant y\leqslant 1$, $0\leqslant z\leqslant\sqrt{2}$, describe an arc $\boldsymbol{\gamma}$ in E^3. Find several sets of parametric equations for this arc.

Attempt to find the length of this arc $\boldsymbol{\gamma}$.

To find the centroid of $\boldsymbol{\gamma}$, you need to find the average values of x, y, and z along $\boldsymbol{\gamma}$. This is to say, if σ represents measure along $\boldsymbol{\gamma}$, the average x, y, and z are given by

$$\bar{x}=\int_0^L x\,d\sigma/L,\qquad \bar{y}=\int_0^L y\,d\sigma/L,\qquad \bar{z}=\int_0^L z\,d\sigma/L,$$

respectively, where L is the length of γ. If you find a parametrization for γ, then

$$\bar{x}=\frac{\int_a^b x(t)\sqrt{x'(t)^2+y'(t)^2+z'(t)^2}\,dt}{\int_a^b \sqrt{x'(t)^2+y'(t)^2+z'(t)^2}\,dt}.$$

If you can find the centroid $(\bar{x},\bar{y},\bar{z})$, send the result to the author.

31. Find the extreme values for
(a) $f(x,y)=y^2+x^2y+x^4$,
(b) $f(x,y)=(x-1)^4+(x-y)^4$,
(c) $f(x,y)=x^2+y^2+x+y+xy+1$.

32. Find, using any method, the shortest distance from (1,2,3) to the surface $2x^2+y^2-4z=0$.

33. Let

$$f(x,y,z,w)=x^2+y^2,$$
$$f_1(x,y,z,w)=x^2+z^2+w^2-4=0,$$
$$f_2(x,y,z,w)=y^2+2z^2+3w^2-9=0.$$

Find the extreme points of f subject to the two given constraints.

34. Find the directional derivative for $f(x,y)=x^2y^2\ln(x^2+y^2)$ in the direction (2,3).
If $f(0,0)=0$, does $\nabla f(0,0)$ exist?

35. Can you solve $\sqrt{x^2+y^2+z^2}-\cos z=0$ for y in a neighborhood of (0,1,0). If so, do so. How about solving for z in terms of x and y in such a neighborhood?

36. Let

$$x^2+y^2-2ux+1=0, \qquad x^2+y^2+2vy-1=0.$$

Solve for x and y in terms of u and v, describing the region where a unique solution is possible.

37. Suppose $\varphi\colon A\subset E^2\to E^2$ given by

$$u=f(x,y), \qquad v=g(x,y)$$

is a C^1 mapping.
If there exists a function $h(u,v)$ which does not vanish identically on any neighborhood in $\varphi(A)$ but $h(f(x,y),g(x,y))$ vanishes identically on the domain of φ, we say f and g are *functionally dependent*.
Show that if $\det \partial(f,g)/\partial(x,y)\equiv 0$ but $\partial f/\partial x$ or $\partial f/\partial y$ are not identically equal to 0, then f and g are functionally dependent.

38. If $f(x,y)=(1-x)(1-y)(x+y-1)$, write a Taylor series expansion for f about the point $(\frac{3}{4},\frac{3}{4})$. Find and identify extreme points and saddle points of f.

39. Suppose $z=f(x,y)=(x^3-y^3)/(x^2+y^2)$ if $x^2+y^2\neq 0$ and $f(0,0)=0$. Give the equation of the tangent plane to the surface defined by this equation at a point (a,b,c) other than the origin. Is there a tangent plane at the origin? Find a mapping $\varphi: E^2\to E^3$ that describes this surface parametrically, and express the normal vector in terms of the parameters.

40. Express the differential of the composite function

$$f(x(u,v),y(u,v),z(u,v))$$

in terms of du and dv.

41. Find extreme points, and test for maxima and minima using the Hessian, for the function

$$f(x,y)=xy(12-3x-4y).$$

42. If $f(x,y,z)=x^2+2y^2+3z^2-2xy-2yz-2$, find the extreme points for f.
Suppose $f(x,y,z)$ is fixed at a value zero, so that now z is defined implicitly as a function of x and y. What values of x and y make z extreme?

43. Suppose $F(x,y,z)=0$ defines z as a function of x and y. Consider the fact that $dF\equiv 0$ and that $dz=(\partial z/\partial x)\,dx+(\partial z/\partial y)\,dy$, and show that

$$\frac{\partial z}{\partial x}=-\frac{\partial F}{\partial x}\Big/\frac{\partial F}{\partial z},\qquad \frac{\partial z}{\partial y}=-\frac{\partial F}{\partial y}\Big/\frac{\partial F}{\partial z}.$$

44. Solve the following by using Cramer's rule. If $F(x,y,z,w)=0$, $G(x,y,z,w)=0$, define z and w implicitly as functions of x and y; show that

$$F_z z_x+F_w w_x=-F_x,\qquad F_z z_y+F_w w_y=-F_y,$$
$$G_z z_x+G_w w_x=-G_x,\qquad G_z z_y+G_w w_y=-G_y,$$

where $A_B=\partial A/\partial B$.
Solve for z_x, w_x, z_y, and w_y in terms of the partial derivatives of F and G.
If $22+3x^3-x^2y+z^2-4w^3=0=x+y^2+zw^3-29$, evaluate $\partial z/\partial x$ at (1,2,3,2).

45. Are the functions $x+y$ and x^2+y^2 functionally dependent in any neighborhood?

46. If the functions $u=x+y+z$, $v=xy+yz+zx$, and $w=x^2+y^2+z^2$ are functionally dependent, find the functional relationship. Do the same for $u=x/(y-z)$, $v=y/(z-x)$, and $w=z/(x-y)$.

47. Let the mapping φ be given by

$$u=x/(x^2+y^2+z^2),\quad v=y/(x^2+y^2+z^2),\quad w=z/(x^2+y^2+z^2).$$

Express φ^{-1}, and find the derivatives of the two maps. Solve for x, y, and z in terms of u, v, and w.

48. Let $\mathbf{F},\mathbf{G}$ be differentiable vector fields in E^3. Show that the divergence of $\mathbf{F}\times\mathbf{G}$ is equal to

$$\mathbf{G}\cdot(\nabla\times\mathbf{F})-\mathbf{F}\cdot(\nabla\times\mathbf{G})=\mathbf{G}\cdot\mathbf{curl}\,\mathbf{F}-\mathbf{F}\cdot\mathbf{curl}\,\mathbf{G}.$$

49. If $\mathbf{F}$ is a C^2-vector in E^3, show that

$$\nabla\cdot(\nabla\times\mathbf{F})=\operatorname{div}(\mathbf{curl}\,\mathbf{F})=0.$$

Find $\mathbf{curl}\,\mathbf{F}$ if $\mathbf{F}=(x^2z, yz, xyz)$. Find $\operatorname{div}\mathbf{F}$ if $\mathbf{F}=(xyz, yz, x^2z)$.

50. Let $A\subset E^2$ be the region $[1,2]\times[1,2]$ and $\boldsymbol{\varphi}$: $E^2\to E^4$ be given by

$$x_1=u+v,\qquad x_2=u^2/2,\qquad x_3=-v,\qquad x_4=u+v.$$

Let f: $E^4\to\mathbb{R}$ be given by

$$f(x_1,x_2,x_3,x_4)=x_3(x_1+2x_3+x_4).$$

Evaluate $\int_{\boldsymbol{\varphi}(A)}f\,d\mu$ and $\int_{\boldsymbol{\varphi}(A)}f\,dx_1\,dx_2$.

51. Suppose $\mathbf{x}=(x_1,x_2,x_3)$ and for $i=1,2,3$, f_i: $E^1\to E^1$ is a C^2-mapping. Let g: $E^3\to\mathbb{R}$ be defined by

$$g(\mathbf{x})=f_1(x_1)f_2(x_2)f_3(x_3).$$

Show that when $g(\mathbf{x})\neq 0$,

$$\frac{\nabla^2 g(\mathbf{x})}{g(\mathbf{x})}=\frac{f_1''(x_1)}{f_1(x_1)}+\frac{f_2''(x_2)}{f_2(x_2)}+\frac{f_3''(x_3)}{f_3(x_3)},$$

where $\nabla^2=\nabla\cdot\nabla$. ∇^2 is called the *Laplacian* operator, the divergence of the gradient.

52. Show that if f is a real function on E^3, then

$$\nabla^2 f=\frac{\partial^2 f}{\partial x^2}+\frac{\partial^2 f}{\partial y^2}+\frac{\partial^2 f}{\partial z^2}.$$

If we define the operator ∇^2 on a vector field (a vector-valued function) $\mathbf{F}=(f_1,\ldots,f_n)$ by $\nabla^2\mathbf{F}=(\nabla^2 f_1,\ldots,\nabla^2 f_n)$, show that if $\mathbf{F}$ is a vector field in E^3, then

$$\mathbf{curl}(\mathbf{curl}\,\mathbf{F})=\mathbf{grad}(\operatorname{div}\mathbf{F})-\nabla^2\mathbf{F},\qquad\text{or}\qquad\nabla\times(\nabla\times\mathbf{F})=\nabla(\nabla\cdot\mathbf{F})-\nabla^2\mathbf{F}.$$

53. Refer to Theorem 9.1.6, but assume only that f is defined in a neighborhood of a point $\mathbf{p}\in E^n$, that $\mathbf{grad}\,f$ exists at $\mathbf{p}$, and that $\mathbf{grad}\,f$ is continuous at $\mathbf{p}$. Can you prove that f is differentiable at $\mathbf{p}$?

54. Prove that the span of the surface in E^4 in the example preceding Definition 9.6.1 is four dimensional.

55. Let an arc $\boldsymbol{\gamma}\subset E^3$ be given by

$$x=2t^2,\qquad y=\cos t,\qquad z=\sin t,\qquad 0\leqslant t\leqslant 2\pi.$$

Find the unit tangent, unit normal, and unit binormal vectors to $\boldsymbol{\gamma}$ at the point $(\pi^2/18, \sqrt{3}/2, 1/2)$. What is the curvature at this point? Express the osculating plane to $\boldsymbol{\gamma}$ at this point in terms of x, y, z, and in terms of parameters u, v. Compute the length of $\boldsymbol{\gamma}$. If the density at a point on $\boldsymbol{\gamma}$ is given by $\delta(x,y,z) = \sqrt{x} + y^2 + z^2 - 1$, find the mass of $\boldsymbol{\gamma}$ and, finally, obtain the moment of inertia of this mass about the origin. (If m is the mass of $\boldsymbol{\gamma}$, you must evaluate $\int_{\boldsymbol{\gamma}}(x^2+y^2+z^2)\,dm$.)

56. Refer to Theorem 9.4.6 and Definition 9.4.5. Suppose f is a real-valued C^2-function on $A \subset E^n$, $\mathbf{p} \in A$ is an interior point at which **grad** f vanishes, and $\det H_f(\mathbf{p})$ is the Hessian evaluated at $\mathbf{p}$. Assume $\det H_f(\mathbf{p}) \neq 0$. Denote by H_k the value of the subdeterminant of $H_f(\mathbf{p})$ consisting of the first k rows and columns, with $H_0 = 1$. If the numbers in the sequence $H_0, H_1, \ldots, H_n$ are all positive, then f has a local minimum at $\mathbf{p}$. If these numbers are alternately positive and negative, then f assumes a local maximum at $\mathbf{p}$.

This theorem provides a sufficient condition for f to have an extremum at $\mathbf{p}$. See if you can prove this using Taylor's formula.

57. Let $f: S \subset E^n \to \mathbb{R}$ be such that for every real number λ and every $\mathbf{x} \in S$, $f(\lambda\mathbf{x}) = \lambda^p f(\mathbf{x})$. Then f is said to be *homogeneous of degree p*. If f is C^1 on S, show that

$$\mathbf{x} \cdot \nabla f(\mathbf{x}) = pf(\mathbf{x}).$$

This is Euler's theorem for homogeneous functions. To prove it, put $g(\lambda) = f(\lambda\mathbf{x})$, and compute $g'(1)$. Prove that if this equation holds for all $\mathbf{x} \in S$, then f is homogeneous of degree p on S.

58. Suppose f is C^2 on a closed region $D \subset E^2$. If f is such that $\forall \mathbf{x} \in \mathring{D}$, $\nabla^2 f(\mathbf{x}) = \nabla \cdot \nabla f(\mathbf{x}) = D_{11}f(\mathbf{x}) + D_{22}f(\mathbf{x}) = 0$, f is said to be *harmonic* on $\mathring{D}$. (The expression $\nabla^2 f$, sometimes denoted Δf, is called the *Laplacian* of f.) Suppose that f is harmonic on the interior of D. Prove that f has no extreme points in $\mathring{D}$, and hence any extreme values for f must be assumed on ∂D.

Suppose $f(x,y) = x^2 - y^2 + 2xy - 4y$, and let D be the region bounded by the curve

$$4x^2 + 9y^2 - 8x + 18y - 23 = 0.$$

Find the maximum and minimum values of f on $\overline{D}$.
[*Hint*: Parametrize the boundary.]

59. Refer to Theorem 9.3.3 and replace the scalar-valued function f on $\boldsymbol{\varphi}A$ by a vector-valued function $\boldsymbol{\psi}$ on $\boldsymbol{\varphi}A$, where $\boldsymbol{\psi} = (f_1, \ldots, f_K)$ and each f_i is a measurable real function on $\boldsymbol{\varphi}A$. Define

$$\int_{\boldsymbol{\varphi}A} \boldsymbol{\psi}\,d\mu = \left(\int_{\boldsymbol{\varphi}A} f_1\,d\mu, \ldots, \int_{\boldsymbol{\varphi}A} f_K\,d\mu\right).$$

Use the theorem to evaluate $\int_{\boldsymbol{\varphi}A}\boldsymbol{\psi}\,d\mu$, where $\boldsymbol{\varphi}A$ is the closed unit disk in E^2 and $\boldsymbol{\psi}(x,y) = (x, y^2, x^2+y^2)$. Be sure to find a suitable $A \subset E^2$ and $\boldsymbol{\varphi}: E^2 \to E^2$.

60. If f is C^2, show that curl(grad f) $= \nabla \times \nabla f \equiv 0$.

CHAPTER

10

SEQUENCES AND SERIES

We want to expand some of the notions about sequences we have already introduced; earlier, we talked about sequences of points and numbers, and upon occasion, sequences of sets, and we defined what we meant by the convergence of such sequences. But it is not quite so clear what we mean by the convergence of a sequence of functions, as you shall see.

All of you have been introduced to infinite series, and perhaps infinite products as well. A number of things we shall say in this chapter will be repetitious; you will already know what we're telling you, or you will recognize that we have said it before. But don't be lulled into a false sense of security. The concepts in this chapter are important, and a number of things in the following chapter depend upon results obtained in this one.

10.1 CONVERGENCE OF SEQUENCES OF FUNCTIONS

Convergence is a topological property, and the classes of functions we are interested in are generally normed linear spaces. Quite naturally, then, the "basic" kind of convergence that a sequence of functions might enjoy is what we call norm convergence. Now we have to qualify that statement a bit. If $\mathfrak{F}$ is a normed linear space consisting of a class of functions $\mathbf{f}: B \to \mathbb{R}$, there are many times when we really don't care about the structure $\mathfrak{F}$ at all; we are only interested in a given sequence of functions $\{\mathbf{f}_n(\mathbf{x})\}_{n=1}^{\infty} \subset \mathfrak{F}$, and in what sense $\mathbf{f}_n(\mathbf{x})$ approximates a functions $\mathbf{f}(\mathbf{x})$ for sufficiently large n. Hence the norm convergence of a sequence $\{\mathbf{f}_n\}$ might be totally unimportant to us, whereas the pointwise limit of $\{\mathbf{f}_n(\mathbf{x})\}$ might be crucial.

Definition 10.1.1. Let $\mathfrak{F}$ be a normed linear space of functions on a Banach space B into another Banach space, and let $\{\mathbf{f}_n\}_{n=1}^{\infty} \subset \mathfrak{F}$. We say

1. $\mathbf{f}_n(\mathbf{x}) \to \mathbf{f}(\mathbf{x})$ pointwise iff for each $\mathbf{x}$, given $\varepsilon > 0$, $\exists n_0$ s.t. $\forall n \geqslant n_0$, $\|\mathbf{f}_n(\mathbf{x}) - \mathbf{f}(\mathbf{x})\| < \varepsilon$;
2. $\mathbf{f}_n(\mathbf{x}) \to \mathbf{f}(\mathbf{x})$ uniformly iff given $\varepsilon > 0$, $\exists n_0$ s.t. $\forall n \geqslant n_0$, $\forall \mathbf{x}$, $\|\mathbf{f}_n(\mathbf{x}) - \mathbf{f}(\mathbf{x})\| < \varepsilon$;
3. $\mathbf{f}_n(\mathbf{x}) \to \mathbf{f}(\mathbf{x})$ pointwise a.e. iff $\mathbf{f}_n(\mathbf{x}) \to \mathbf{f}(\mathbf{x})$ pointwise except on a set of measure zero (a.e. means almost everywhere);
4. $\mathbf{f}_n(\mathbf{x}) \to \mathbf{f}(\mathbf{x})$ uniformly a.e. iff the convergence is uniform almost everywhere, i.e., except on a set of measure zero;
5. $\mathbf{f}_n(\mathbf{x}) \to \mathbf{f}(\mathbf{x})$ almost uniformly iff for each $\varepsilon > 0$, there is a measurable set $F \subset B$ with $\mu(F) \leqslant \varepsilon$ such that $\mathbf{f}_n \to \mathbf{f}$ uniformly on $B \setminus F$;
6. $f_n(\mathbf{x}) \to f(\mathbf{x})$ in the p-mean iff $\int_B |f_n - f|^p \, d\mu \to 0$, where $p \geqslant 1$ is a fixed real number and the f_n are scalar valued;
7. $\mathbf{f}_n(\mathbf{x}) \to \mathbf{f}(\mathbf{x})$ in measure iff $\forall \varepsilon > 0$, $\forall \delta > 0$, $\exists n_0$ s.t. $\forall n \geqslant n_0$, $\mu(\{\mathbf{x} \in B : \|\mathbf{f}_n(\mathbf{x}) - \mathbf{f}(\mathbf{x})\| \geqslant \varepsilon\} < \delta$;
8. $\mathbf{f}_n \to \mathbf{f}$ in norm iff $\|\mathbf{f}_n - \mathbf{f}\| \to 0$.

In certain special cases, such as if B is the space of positive integers, and if each $\mathbf{f}_n$ is in fact a sequence of real numbers $\{f_{nk}\}_{k=1}^{\infty}$, or if each $\mathbf{f}_n$ is a vector-valued mapping $B \to H$ where the dimension of H is a number greater than 1 (perhaps even infinite), we say

9. $\mathbf{f}_n \to \mathbf{f}$ coordinatewise iff for each k, $f_{nk} \to f_k$ pointwise, where f_{nk}, f_k are the kth coordinates respectively of $\mathbf{f}_n$, $\mathbf{f}$.

This is a rather lengthy definition; our purpose is simply to introduce you to different notions of convergence. It is not hard to see that if $\mathfrak{F}$ is normed with the sup norm, i.e.,

$$\|\mathbf{f}\| = \sup\{\|\mathbf{f}(\mathbf{x})\| : \mathbf{x} \in B\} \qquad \forall f \in \mathfrak{F},$$

then norm convergence implies uniform convergence, which in turn obviously implies pointwise convergence. On the other hand, if $\mathfrak{F} = L^2(\mathbb{R})$, the space of (real-valued) measurable functions on $\mathbb{R}$, normed by $\forall f \in L^2(\mathbb{R})$,

$$\|f\|^2 = \int_{\mathbb{R}} |f|^2 \, d\mu,$$

for which $\|f\| < \infty$, then norm convergence does not even imply pointwise convergence. It does imply, in this case, convergence in the 2–mean since they are both the same thing here.

When $p = 1$, mean convergence implies convergence in measure, the weakest type of convergence we have described, and almost uniform convergence implies convergence in measure as well as pointwise a.e. convergence. Clearly, it would be a substantial problem to establish the hierarchy of convergence, a problem we believe belongs in a subsequent course.

In E^m, all topologies are equivalent, so convergence does imply coordinatewise convergence, for suppose $\{\boldsymbol{\varphi}_k\}$ is a sequence of mappings from E^n into E^m, and suppose $\boldsymbol{\varphi}_k \to \boldsymbol{\varphi}$. (Unless a particular type of convergence is noted, the convergence is assumed to be pointwise.) Fix $\mathbf{x}$. Then we have $\|\boldsymbol{\varphi}_k(\mathbf{x})-\boldsymbol{\varphi}(\mathbf{x})\| \to 0$. This implies that each coordinate function f_{kj} of $\boldsymbol{\varphi}_k$ must satisfy $|f_{kj}(\mathbf{x})-f_j(\mathbf{x})| \to 0$. Thus we have (pointwise) coordinatewise convergence. The converse is also true: The coordinatewise convergence of $\{\boldsymbol{\varphi}_k\}$ implies the pointwise convergence of $\{\boldsymbol{\varphi}_k\}$.

Now consider the Hilbert space we considered in Chapter 7, where a basis vector $\mathbf{e}_i$ was the $\aleph_0$-tuple with all zeros, except for a 1 in the ith position. The sequence $\{\mathbf{e}_i\}_{i=1}^{\infty}$ converges coordinatewise to the zero vector, but it certainly is not norm convergent to $\mathbf{0}$.

Next, consider the space $L^1[0,1]$, and let $\{f_n\}$ be the sequence of functions $\{x^n\}_{n=1}^{\infty}$. The norm limit of this sequence is the zero function (any $f=0$ a.e. is considered equivalent to the zero function in $L^1[0,1]$). The pointwise limit of this sequence is the function

$$f(x)=\begin{cases} 0 & \text{if} \quad 0\leqslant x<1, \\ 1 & \text{if} \quad x=1, \end{cases}$$

which is equivalent to the zero function. There is no uniform limit.

Finally, consider the Hilbert space $L^2[0,2\pi]$, and let $\{f_n\}$ be a sequence of functions converging in norm to f. This is to say, $\|f_n-f\| \to 0$. If $(\alpha_{n1}, \alpha_{n2}, \ldots)$ are the Fourier coefficients for each f_n relative to some particular complete orthonormal set, and $(\alpha_1, \alpha_2, \ldots)$ are the coefficients for f, then

$$\|f_n-f\|^2=\sum_{i=1}^{\infty}|\alpha_{ni}-\alpha_i|^2 \to 0$$

implies that, for each i, $\alpha_{ni} \to \alpha_i$. Thus the norm convergence in this Hilbert space implies coordinatewise convergence, and conversely.

The following is a theorem of considerable importance.

Theorem 10.1.2

1. *Let $\{\mathbf{f}_n\}$ be a sequence of functions belonging to a normed linear space of functions. Then $\mathbf{f}_n \xrightarrow{\text{norm}} \mathbf{f} \Rightarrow \|\mathbf{f}_n\| \to \|\mathbf{f}\|$, but not conversely, and $\mathbf{f}_n \xrightarrow{\text{norm}} \mathbf{0} \Leftrightarrow \|\mathbf{f}_n\| \to 0$.*
2. *Let $\{\mathbf{f}_n\}$ be a sequence of continuous mappings from one Banach space to another. The pointwise limit, if it exists, need not be continuous, but the uniform limit, if it exists, is continuous.*
3. *If $\{\mathbf{f}_n\}$ is a sequence of C^1-mappings $E^m \to E^k$ converging uniformly to $\mathbf{f}$ and if $\{\mathbf{f}'_n\}$ converges uniformly to $\mathbf{g}$, then $\mathbf{f}$ is C^1, and $\mathbf{f}'=\mathbf{g}$.*
4. *If $\{f_n\}$ is a sequence of real-valued C^0-functions on a compact set $D \subset E^m$, converging uniformly to f on D, then f is integrable over D, and*

$$\int_D f_n \, d\mu \to \int_D f \, d\mu.$$

PROOF: The proof of 1 is trivial. For part 2, we have already seen that the pointwise limit of $\{x^n\}$ on $[0,1]$ is a discontinuous function, so if we define

$f_n(x)$ by

$$f_n(x)=|x|^n \wedge 1 \qquad \forall x \in E^1,$$

the pointwise limit is discontinuous. Now, suppose $\mathbf{f}_n \to \mathbf{f}$ uniformly. Let $\varepsilon>0$, and let $\mathbf{x}$ be an arbitrary fixed point of the common domain. For all n sufficiently large, say $n \geqslant n_0$, and any $\mathbf{x}$, $\|\mathbf{f}_n(\mathbf{x})-\mathbf{f}(\mathbf{x})\|<\varepsilon/3$. Since each $\mathbf{f}_n$ is continuous at $\mathbf{x}$, there is a $\delta>0$ such that if $\|\mathbf{x}-\mathbf{y}\|<\delta$, $\|\mathbf{f}_{n_0}(\mathbf{x})-\mathbf{f}_{n_0}(\mathbf{y})\|<\varepsilon/3$. For any such $\mathbf{y}$ and all $n \geqslant n_0$, we have $\|\mathbf{f}_n(\mathbf{y})-\mathbf{f}(\mathbf{y})\|<\varepsilon/3$. Then, if $\|\mathbf{x}-\mathbf{y}\|<\delta$,

$$\|\mathbf{f}(\mathbf{x})-\mathbf{f}(\mathbf{y})\| \leqslant \|\mathbf{f}(\mathbf{x})-\mathbf{f}_{n_0}(\mathbf{x})\|+\|\mathbf{f}_{n_0}(\mathbf{x})-\mathbf{f}_{n_0}(\mathbf{y})\|+\|\mathbf{f}_{n_0}(\mathbf{y})-\mathbf{f}(\mathbf{y})\|<\varepsilon.$$

This is to say, $\mathbf{f}$ is a continuous function since $\mathbf{x}$ was an arbitrary fixed point. Note that the choice of δ depended upon $\mathbf{x}$ as well as upon ε, so $\mathbf{f}$ need not be uniformly continuous. It will be, certainly, if all the $\mathbf{f}_n$ are uniformly continuous. This proves the second part of the theorem.

Part 3 is a bit more difficult. We have from 2 that $\mathbf{f}$ is a continuous function, as is $\mathbf{g}$. It is well to remark that just because the mappings are C^1 and converge uniformly, we can't assume that the sequence of derivatives will converge. The partial derivatives may be very ill behaved. Hence we need the additional hypothesis that the sequence of derivatives is also uniformly convergent.

For an arbitrary fixed $\mathbf{P} \in E^m$ and each n, we have

$$\mathbf{f}_n(\mathbf{P}+\mathbf{H})=\mathbf{f}_n(\mathbf{P})+\mathbf{f}'_n(\mathbf{P}) \cdot \mathbf{H}+o_n(\mathbf{H}).$$

Letting $n \to \infty$, we have

$$\mathbf{f}(\mathbf{P}+\mathbf{H})=\mathbf{f}(\mathbf{P})+\mathbf{g}(\mathbf{P}) \cdot \mathbf{H}+o(\mathbf{H}).$$

Since the convergences are all uniform, this last equation holds for all $\mathbf{P} \in E^m$. Hence $\mathbf{g}$ is the derivative of $\mathbf{f}$. We should remark that $\mathbf{f}'_n(\mathbf{P})$ is a matrix, the Jacobian matrix of $\mathbf{f}_n$ evaluated at $\mathbf{P}$. By the uniform limit of this sequence of matrices we mean of course the matrix whose ijth element is the uniform limit of the sequence of the ijth elements of the matrices $\mathbf{f}'_n(\mathbf{P})$. Strictly speaking, the sequence $\{o_n(\mathbf{H})\}$ does not converge to $o(\mathbf{H})$, but it is such a sequence that $\{o_n(\mathbf{H})/\|\mathbf{H}\|\}$ does tend to $\mathbf{0}$ as $\mathbf{H} \to \mathbf{0}$. Hence what we have written is not incorrect.

To prove part 4 we note that f is continuous on D, and hence integrable on D. Also, each integral $\int_D f_n \, d\mu$ exists. The uniform convergence of f_n to f on D means that $\sup\{|f_n(\mathbf{x})-f(\mathbf{x})|:\mathbf{x} \in D\} \to 0$, and the fact that $\mu(D)<\infty$ guarantees that:

$$\left|\int_D f_n \, d\mu-\int_D f \, d\mu\right| \leqslant \int_D |f_n-f| \, d\mu \leqslant \sup\{|f_n(\mathbf{x})-f(\mathbf{x})|:\mathbf{x} \in D\} \cdot \mu(D) \to 0.$$

This proves 4. □

We should remark that if $\{f_n\}$ is a sequence of integrable functions converging uniformly to an integrable function f and D is a measurable set over which f and each f_n are integrable, then

$$\int_D f_n \, d\mu \to \int_D f \, d\mu.$$

Theorem 10.1.3. *Let $\{f_n\}$ be a sequence of measurable functions. The functions*

$$\bigvee_{n=1}^{\infty} f_n, \qquad \bigwedge_{n=1}^{\infty} f_n, \qquad \bigvee_{k=1}^{\infty} \bigwedge_{n=k}^{\infty} f_n, \qquad \bigwedge_{k=1}^{\infty} \bigvee_{n=k}^{\infty} f_n$$

are all measurable. If $f_n \to f$, then f is measurable.

PROOF: $\{\mathbf{x} : \wedge_{n=1}^{\infty} f_n(\mathbf{x}) \leqslant b\} = \bigcup_{n=1}^{\infty} \{\mathbf{x} : f_n(\mathbf{x}) \leqslant b\}$, and $\{\mathbf{x} : \wedge_{n=1}^{\infty} f_n(\mathbf{x}) \geqslant a\} = \bigcap_{n=1}^{\infty} \{\mathbf{x} : f_n(\mathbf{x}) \geqslant a\}$ for extended real numbers a, b. It follows that $\wedge_n f_n$ is measurable. Analogously,

$$\left\{\mathbf{x} : \bigvee_n f_n(\mathbf{x}) \leqslant b\right\} = \bigcap_n \{\mathbf{x} : f_n(\mathbf{x}) \leqslant b\},$$

and

$$\left\{\mathbf{x} : \bigvee_n f_n(\mathbf{x}) \geqslant a\right\} = \bigcup_n \{\mathbf{x} : f_n(\mathbf{x}) \geqslant a\}.$$

Thus it follows that $\vee f_n$ is measurable.

The measurability of the remaining two functions is an immediate consequence of what we have just said. Finally, if $f_n \to f$ (pointwise), then

$$f = \bigvee_{k=1}^{\infty} \bigwedge_{n=k}^{\infty} f_n = \bigwedge_{k=1}^{\infty} \bigvee_{n=k}^{\infty} f_n,$$

so f is measurable. □

We must point out to the reader that this theorem shows that the class of measurable functions is a σ-complete lattice, as well as a ring. In fact, if $\{f_n\}$ is a sequence of positive measurable functions, then each partial sum $S_k = \Sigma_{n=1}^{k} f_n$ is measurable, and $\vee_{k=1}^{\infty} S_k = \Sigma_{n=1}^{\infty} f_n$, so the class of positive measurable functions is a σ-ring of functions as well as a σ-complete lattice. Such is not the case for the class of positive continuous functions. It has been pointed out to the author by Professor J. Diestel that the whole notion of the Lebesgue integral with its sophisticated complexities was developed just to treat the question of integrating sequences of continuous functions. It did what it was asked to do, and much, much more, but this was perhaps its crowning glory.

10.2 SERIES OF FUNCTIONS AND CONVERGENCE

An infinite series or, more commonly, a series is a sum of a countable number of terms. If the sum exists in a certain sense, we say the series is convergent in that sense; otherwise, it is divergent. A series $\Sigma_{n=1}^{\infty} a_n$ is called absolutely convergent if $\Sigma_{n=1}^{\infty}|a_n|$ converges and conditionally convergent if $\Sigma_{n=1}^{\infty} a_n$ converges but $\Sigma_{n=1}^{\infty}|a_n|$ diverges to ∞. Series can be divergent in basically two ways; either they don't settle down to one specific sum, like the series

$$1-\tfrac{1}{2}+\tfrac{2}{3}-\tfrac{3}{4}+\tfrac{5}{6}-\tfrac{7}{8}+\tfrac{9}{10}-\tfrac{11}{12}+\cdots,$$

or they tend to an infinite sum, like the series $\Sigma_{n=1}^{\infty}(1/n)$.

If $\Sigma_{n=1}^{\infty} a_n$ is a series, there are two sequences intimately related to this series. One is the sequence of terms $\{a_n\}_{n=1}^{\infty}$ and the other is the sequence of partial sums $\{S_n\}_{n=1}^{\infty}$, where $S_n = \Sigma_{i=1}^{n} a_i$.

Definition 10.2.1. If $\Sigma_{n=1}^{\infty} \mathbf{a}_n$ is a series of terms from a Banach space B, we say the series *converges to* $\mathbf{S} \in B$ iff $\lim_{n\to\infty}\|\Sigma_{i=1}^{n}\mathbf{a}_i - \mathbf{S}\| \to 0$.

As was the case with sequences of functions from a Banach space, we are not always interested in the convergence of a series within the norm structure of the Banach space. Often, we simply want to know in what sense does $\Sigma_{i=1}^{n}\mathbf{g}_i(\mathbf{x})$ approximate a function $\mathbf{g}(\mathbf{x})$ if $\{\mathbf{g}_i(\mathbf{x})\}$ is a sequence of functions? Hence we could simply duplicate Definition 10.1.1 and then add: If $\mathbf{f}_n(\mathbf{x})$ is the nth partial sum of a series $\Sigma_{i=1}^{\infty}\mathbf{g}_i(\mathbf{x})$, we say that $\Sigma_{i=1}^{\infty}\mathbf{g}_i(\mathbf{x}) \to \mathbf{f}(\mathbf{x})$ in the same manner that $\mathbf{f}_n(\mathbf{x}) \to \mathbf{f}(\mathbf{x})$. When we say $\mathbf{f}(\mathbf{x})$ is the limit of the series $\Sigma\mathbf{g}_i(\mathbf{x})$, we mean the pointwise limit, unless another specific type of convergence is indicated.

If $\{a_n\}$ is a sequence of real (complex) numbers, a series of the form

$$\sum_{n=0}^{\infty} a_n(x-x_0)^n$$

is called a real (complex) *power series*. The point x_0 is often called the center of the expansion since when $x = x_0$, we agree that the series has the value a_0 and if the series of numbers $\Sigma_{n=1}^{\infty} a_n|x_1 - x_0|^n$ converges, then for every x such that $|x-x_0| < |x_1 - x_0|$, $\Sigma_1^{\infty} a_n(x-x_0)^n$ converges.

We now give a result analogous to Theorem 10.1.2.

Theorem 10.2.2. *Let* $\{\mathbf{g}_i(\mathbf{x})\}$ *be a sequence of (uniformly) continuous mappings from one Banach space to another such that* $\Sigma\mathbf{g}_i(\mathbf{x})$ *converges to* $\mathbf{g}(\mathbf{x})$. *If the convergence is uniform,* $\mathbf{g}(\mathbf{x})$ *is (uniformly) continuous.*

Let $\{\mathbf{g}_i(\mathbf{x})\}$ *be a sequence of* C^1*-mappings* $E^m \to E^k$ *such that* $\Sigma\mathbf{g}_i(\mathbf{x})$ *converges uniformly to* $\mathbf{g}(\mathbf{x})$ *and* $\Sigma\mathbf{g}_i'(\mathbf{x})$ *converges uniformly to* $\mathbf{h}(\mathbf{x})$. *Then* $\mathbf{g}(\mathbf{x})$ *is* C^1, *and* $\mathbf{g}'(\mathbf{x}) = \mathbf{h}(\mathbf{x}) = \Sigma\mathbf{g}_i'(\mathbf{x})$.

If $\{g_i(\mathbf{x})\}$ is a sequence of real-valued continuous functions on a compact set $D \subset E^m$ such that $\Sigma g_i(\mathbf{x})$ converges uniformly to $f(\mathbf{x})$ on the compact set D, then f is continuous, and

$$\int_D f\,d\mu = \int_D \sum_{i=1}^{\infty} g_i(\mathbf{x})\,d\mu = \int_D \lim_n \sum_{i=1}^{n} g_i(\mathbf{x})\,d\mu = \lim_n \int_D \sum_{i=1}^{n} g_i(\mathbf{x})\,d\mu$$
$$= \lim_n \sum_{i=1}^{n} \int_D g_i(\mathbf{x})\,d\mu = \sum_{i=1}^{\infty} \int_D g_i(\mathbf{x})\,d\mu.$$

The last two statements assert that if the series $\Sigma g_i(\mathbf{x})$ and $\Sigma g_i'(\mathbf{x})$ are uniformly convergent series, then we can integrate or differentiate term by term to get the respective integral or derivative of the series $\Sigma g_i(\mathbf{x})$.

PROOF: Simply use the definition of uniform convergence of a series and apply Theorem 10.1.2. □

As a very simple application of this theorem, consider that $\Sigma_{n=0}^{\infty} x^n \to (1-x)^{-1}$ uniformly on $[-\delta, \delta]$, where $0<\delta<1$. To prove this, let $\varepsilon>0$ be given and be arbitrarily small. Let n be large enough so that

$$\left|\frac{1}{1-x} - \sum_{i=0}^{n} x^i\right| = \left|\frac{1}{1-x} - \frac{1-x^{n+1}}{1-x}\right|$$
$$\leqslant \frac{\delta^{n+1}}{1-\delta} < \varepsilon.$$

Then for all $x \in [-\delta, \delta]$ and $n > (\ln \varepsilon + \ln(1-\delta))/\ln \delta$, we have

$$\left|\frac{1}{1-x} - \sum_{i=0}^{n} x^i\right| < \varepsilon.$$

Hence

$$\int \frac{1}{1-x}\,dx = -\ln(1-x) = \sum_{n=1}^{\infty} \frac{x^n}{n} = \ln\left(\frac{1}{1-x}\right),$$

valid within $[-\delta, \delta]$.

Theorem 10.2.3. *Suppose that $\Sigma_{i=1}^{\infty} \mathbf{g}_i$ is a series of mappings from E^n to E^m and*

$$\lim_{p,q\to\infty} \left\| \sum_{i=p}^{q} \mathbf{g}_i \right\| = 0,$$

where the norm here is given by $\forall \mathbf{f}$: $E^n \to E^m$,

$$\|\mathbf{f}\| = \max_{i=1,\dots,m} \sup\{|f_i(\mathbf{x})| : \mathbf{x} \in E^n\},$$

then $\Sigma_{i=1}^{\infty} \mathbf{g}_i(\mathbf{x})$ converges uniformly on E^n, and conversely.

PROOF: The hypothesis of the theorem says that the sequence of partial sums is a Cauchy sequence. Let $\mathbf{f}_i$, $\mathbf{f}_j$, be the ith, jth partial sums, respectively, of $\Sigma \mathbf{g}_i$. Let $\varepsilon > 0$. Then for any i, j sufficiently large, we have

$$|f_{ik}(\mathbf{x}) - f_{jk}(\mathbf{x})| < \varepsilon, \qquad \forall \mathbf{x} \in E^n, \quad k = 1, \ldots, m.$$

This ensures that the sequence of kth coordinates of the partial sums is also a Cauchy sequence, which is to say, for each $k = 1, \ldots, m$, if $\mathbf{f}_n = (f_{n1}, \ldots, f_{nm})$ is the nth partial sum, the sequence $\{f_{nk}\}_{n=1}^{\infty}$ is a Cauchy sequence. Now the space of real-valued mappings $E^n \to E^1$ with the supnorm is a Banach space. This is to say, for each $\mathbf{x} \in E^n$, $\{f_{nk}(\mathbf{x})\}_{n=1}^{\infty}$ is a Cauchy sequence of real numbers. Hence there is a pointwise limit for the sequence of functions $\{f_{nk}(\mathbf{x})\}$; call it $g_k(\mathbf{x})$. But $g_k(\mathbf{x})$ is also the uniform limit of $\{f_{nk}(\mathbf{x})\}$ since we have $\sup_{x \in E^n} |f_{nk}(\mathbf{x}) - g_k(\mathbf{x})| < \varepsilon$ for all sufficiently large n. Thus $\mathbf{g} = (g_1, \ldots, g_m)$ is our clear candidate for the limit of $\{\mathbf{f}_n\}$. Since for all sufficiently large n

$$\|\mathbf{f}_n - \mathbf{g}\| = \max_i \sup_{\mathbf{x} \in E^n} |f_{ni}(\mathbf{x}) - g_i(\mathbf{x})| \leqslant \sum_{i=1}^{m} \| f_{ni} - g_i \| \leqslant m \max_i \| f_{ni} - g_i \|$$

if $\varepsilon > 0$ is given, we can choose n sufficiently large so that the largest of $\| f_{ni} - g_i \|$ is less than ε / m. Thus the convergence of $\mathbf{f}_n$ to $\mathbf{g}$ is uniform. The converse is clear. □

Corollary 10.2.4 (Weierstrass M-Test). *If $\Sigma_{i=1}^{\infty} g_i(\mathbf{x})$ is a series of scalar-valued functions on E^n, and, if for each i, $|g_i(\mathbf{x})|$ is uniformly bounded by a real constant $M_i \geqslant 0$ (i.e., $\sup\{|g_i(\mathbf{x})| : \mathbf{x} \in E^n\} \leqslant M_i$) and if $\Sigma_{i=1}^{\infty} M_i < \infty$, then $\Sigma_{i=1}^{n} g_i(\mathbf{x})$ converges uniformly on E^n.*

PROOF: Let $\varepsilon > 0$. For any $\mathbf{p} \in E^n$ we have for all n sufficiently large

$$\left| \sum_{i=n}^{\infty} g_i(\mathbf{p}) \right| \leqslant \sum_{i=n}^{\infty} M_i < \varepsilon.$$

But this means $\|\Sigma_{i=n}^{\infty} g_i\| \to 0$, so $\Sigma_{i=1}^{\infty} g_i(\mathbf{x})$ converges uniformly. □

We must assume that the reader has had experience with series of scalars and is familiar with the standard tests for the absolute convergence of such series. We list a number of these tests for reference. Suppose Σu_n is a series of scalars.

Theorem 10.2.5 (d'Alembert Ratio Test). *If there is a positive constant $\lambda < 1$ such that for all n sufficiently large $|u_{n+1}/u_n| < \lambda$, then Σu_n converges absolutely. This is to say, if*

$$\limsup_n |u_{n+1}/u_n| < 1,$$

Σu_n converges absolutely, and if

$$\liminf_n |u_{n+1}/u_n| > 1,$$

Σu_n diverges. If

$$\liminf_n |u_{n+1}/u_n| \leqslant 1 \leqslant \limsup_n |u_{n+1}/u_n|,$$

the test is inconclusive.

Theorem 10.2.6 (Cauchy Root Test). *If there is a positive number $\lambda < 1$ such that for all sufficiently large n*

$$\sqrt[n]{|u_n|} < \lambda,$$

then Σu_n converges absolutely. If, however, $\sqrt[n]{|u_n|} \geqslant 1$ for infinitely many n, the series is divergent. If

$$\limsup_n \sqrt[n]{|u_n|} = 1$$

the test is inconclusive.

Theorem 10.2.7 (Raabe's Test). *If for each n we write*

$$\left|\frac{u_{n+1}}{u_n}\right| = 1 - \frac{\lambda_n}{n} \qquad \text{or} \qquad \lambda_n = n\left(1 - \left|\frac{u_{n+1}}{u_n}\right|\right),$$

and if λ_n has a limit λ as $n \to \infty$, then the series Σu_n is absolutely convergent if $\lambda > 1$ and not absolutely convergent (i.e., divergent or conditionally convergent) if $\lambda < 1$. If $\lambda = 1$, the test is inconclusive.

Theorem 10.2.8 (Gauss's Test). *If for each n, $|u_{n+1}/u_n|$ can be expressed in the form*

$$\left|\frac{u_{n+1}}{u_n}\right| = 1 - \frac{p}{n} + \frac{A_n}{n^q},$$

where p is fixed and $q > 1$ is fixed, and if the sequence $\{A_n\}$ is bounded, then the series Σu_n is absolutely convergent if $p > 1$ and not absolutely convergent (i.e., divergent or conditionally convergent) if $p \leqslant 1$.

Theorem 10.2.9 (The Integral Test). *If there is a function $f(x)$, positive and decreasing on $[1, \infty)$, such that for each n, $f(n) = |u_n|$, then Σu_n is absolutely convergent iff the sequence*

$$\left\{\int_1^n f(x)\,dx\right\}_{n=1}^{\infty}$$

is bounded.

Theorem 10.2.10 (Limit Comparison Test). *If Σv_n is a series whose behavior is known and if*

$$\lim_n |u_n/v_n| = c \neq 0,$$

then Σu_n converges absolutely iff Σv_n converges absolutely. If $c=0$ and Σv_n converges absolutely, then so does Σu_n converge absolutely.

Theorem 10.2.11. *Σu_n converges iff $\forall \varepsilon > 0$, $\exists n_0$ s.t. $\forall p, q$, $n_0 \leqslant p \leqslant q$, $|\Sigma_{n=p}^{q} u_n| < \varepsilon$. In particular, if $u_n \nrightarrow 0$, the series is divergent.*

Compare this with Theorem 10.2.3.

The proofs of the past seven theorems will be omitted; they are left for the reader to struggle with. The next theorem goes to the other extreme; it generally appears in more advanced texts.

Theorem 10.2.12 (Lebesgue Dominated Convergence Theorem). *Suppose $\{f_n(\mathbf{x})\}$ is a sequence of measurable real-valued functions converging pointwise a.e. to a function $f(\mathbf{x})$ on a measurable set T. Suppose further that there exists a positive function $F(\mathbf{x})$ such that for each n, $|f_n(\mathbf{x})| \leqslant F(\mathbf{x})$ a.e. on T, and that $\int_T F\,d\mu$ exists. Then $\int_T f\,d\mu$ exists, $\int_T f_n\,d\mu$ exists for each n, and $\lim_n \int_T |f_n - f|\,d\mu = 0$.*

PROOF: We assume without loss of generality that $\mu(T) > 0$ and that $\int_T F\,d\mu = L > 0$. It is clear that f is measurable (Theorem 10.1.3) and that the integrals $\int_T f_n\,d\mu$, $\int_T f\,d\mu$ all exist and have absolute values less than or equal to L (see Problem 55). Since for each n, $|f_n(\mathbf{x})| \leqslant F(\mathbf{x})$ a.e., we can be sure that $|f_n(\mathbf{x}) - f(\mathbf{x})| \leqslant 2F(\mathbf{x})$ a.e. for each n, so it follows that $|f_n - f|$ is integrable, and

$$0 \leqslant \int_T |f_n - f|\,d\mu \leqslant 2L < \infty.$$

Let $\varepsilon > 0$ be arbitrary. Let $\{T_k\}$ be an increasing sequence of measurable sets, each having positive finite measure, such that $T = \bigcup_{k=1}^{\infty} T_k$. Take a T_k. Let $E_n = \bigcup_{i=n}^{\infty} \{\mathbf{x} \in T : |f_i(\mathbf{x}) - f(\mathbf{x})| \geqslant \delta_k\}$, where $\delta_k > 0$ is a fixed number, such that $\delta_k < \varepsilon / 2\mu(T_k)$. Clearly, $\{E_n\}$ is a decreasing sequence of measurable sets for which

$$\mu\left(\bigcap_{n=1}^{\infty} E_n\right) = 0.$$

Moreover, $E_n \subset \{\mathbf{x} : F(\mathbf{x}) \geqslant \delta_k/2\}$ for each n, and this means $\mu(E_n) < \infty$. Applying Theorem 8.1.3 (continuity of measure from above and below), we can conclude that $\lim_n \mu(E_n) = \mu(\bigcap_{n=1}^{\infty} E_n) = 0$.

Since $\mu(E_n) \to 0$, for all n sufficiently large, $\mu(E_n)$ is necessarily small. We can write for all n

$$\int_{E_n} |f_n - f|\,d\mu \leqslant \int_{E_n} 2F\,d\mu.$$

Since $\int_T F\,d\mu = L < \infty$ and $F > 0$, then the set function $\nu(S)$, defined on all the measurable subsets of T by

$$\nu(S) = \int_S F\,d\mu,$$

is a finite positive measure on the σ-ring of measurable subsets of T. As a measure, ν is continuous from above; hence we must have $\lim_n \nu(E_n) = \nu(\bigcap_{n=1}^{\infty} E_n) = 0$. Therefore, for all n sufficiently large, we can write

$$0 \leqslant \int_{E_n} |f_n - f|\,d\mu \leqslant \int_{E_n} 2F\,d\mu < \frac{\varepsilon}{2}.$$

Now, write for all n sufficiently large

$$\int_{T_k} |f_n - f|\,d\mu \leqslant \int_{T_k \setminus E_n} |f_n - f|\,d\mu + \int_{T_k \cap E_n} |f_n - f|\,d\mu \leqslant \delta_k \mu(T_k) + \frac{\varepsilon}{2}.$$

Since δ_k was fixed small enough so that $\delta_k \mu(T_k) < \varepsilon/2$, we have for all sufficiently large n, $\int_{T_k} |f_n - f|\,d\mu < \varepsilon$. For each k, it is true that for all n sufficiently large, we have

$$\int_{T_k} |f_n - f|\,d\mu < \varepsilon.$$

Hence for each k, it is also true that

$$\lim_{n\to\infty} \int_{T_k} |f_n - f|\,d\mu < \varepsilon.$$

It follows that

$$\lim_{n\to\infty} \int_T |f_n - f|\,d\mu \leqslant \varepsilon$$

since $\{T_k\} \nearrow T$, and since ε was arbitrary, we conclude that

$$\lim_{n\to\infty} \int_T |f_n - f|\,d\mu = 0.$$

In the event that $\mu(T) = 0$ or $L = 0$, the necessary modifications in the above proof are simple enough to make. □

The following test usually has to do with the convergence of a series of numbers rather than a series of functions. However, its hypothesis can be modified a bit so that it will also yield a sufficient condition for the convergence of an improper integral.

Theorem 10.2.13 (Dirichlet Test). *We present the hypothesis in two forms. First, let* $\{a_n\}$, $\{b_n\}$ *be two sequences of scalars which satisfy*

1. $\lim_n a_n = 0$,
2. $\sum_{n=1}^{\infty} |a_{n+1} - a_n|$ *converges, and*
3. *the partial sums* $|\sum_{n=1}^{m} b_n|$ *are bounded uniformly with respect to* m; *i.e., there is a real number* M *such that for each* m, $|\sum_{n=1}^{m} b_n| \leqslant M$.

Second, let $a(x)$, $b(x)$, $a'(x)$ be continuous scalar-valued functions on the interval $[c,\infty)$ which satisfy

1. $\lim_{x\to\infty} a(x)=0$,
2. $\int_c^\infty |a'(x)|\,dx<\infty$, *and*
3. $|\int_c^m b(x)\,dx|$ *is bounded uniformly with respect to* m *by* M.

Then both the series $\Sigma a_n b_n$ *and the integral* $\int_c^\infty a(x)b(x)\,dx$ *converge.*

PROOF: Let $B_k=b_1+\cdots+b_k$ for each k. Then

$$\begin{aligned}\sum_{k=1}^{n} a_k b_k &= a_1b_1+a_2b_2+\cdots+a_nb_n\\ &= a_1B_1+a_2(B_2-B_1)+\cdots+a_n(B_n-B_{n-1})\\ &= (a_1-a_2)B_1+(a_2-a_3)B_2+\cdots+(a_{n-1}-a_n)B_{n-1}+a_nB_n\\ &= a_nB_n-\sum_{k=1}^{n-1}(a_{k+1}-a_k)B_k.\end{aligned}$$

What we have done looks like integration by parts; indeed, the identity we have set down is a "summation by parts" formula.

Since $a_nB_n\to 0$ and $|a_{k+1}-a_k||B_k|\leqslant M|a_{k+1}-a_k|$, we have that $\Sigma_{k=1}^\infty(a_{k+1}-a_k)B_k$ converges, and so $\Sigma_{k=1}^\infty a_kb_k$ converges.

Corollary 10.2.14. *If* $\{a_n\}\searrow 0$ *or* $\{a_n\}\nearrow 0$, *then conditions* (1) *and* (2) *are both satisfied, so if* $|\Sigma_{n=1}^m b_n|$ *is uniformly bounded as well, then* $\Sigma a_n b_n$ *converges.*

If $\{a_n\}$ *monotonically converges to* 0 *and* $b_n=(-1)^n$, *all three conditions are satisfied, so* $\Sigma(-1)^n a_n$ *converges.*

Definition 10.2.15 (Abel's Summation Formula). If $\{a_k\}$, $\{b_k\}$ are sequences of scalars and $B_k=\Sigma_{j=1}^k b_j$, then for each n we have the identity

$$\sum_{k=1}^{n} a_kb_k=a_nB_n-\sum_{k=1}^{n-1}(a_{k+1}-a_k)B_k.$$

We return to the proof of the theorem. Since $\int_c^R a(x)b(x)dx$ exists for any real R, if we put $B(x)=\int_c^x b(t)dt$, we can write

$$\int_c^R a(x)\,dB(x)=a(x)B(x)|_c^R-\int_c^R B(x)a'(x)\,dx.$$

Now

$$\begin{aligned}\left|\int_c^R B(x)\alpha'(x)\,dx\right| &\leqslant \int_c^R |B(x)||\alpha'(x)|\,dx\\ &\leqslant M\int_c^R |a'(x)|\,dx\leqslant M\int_c^\infty |a'(x)|\,dx<\infty,\end{aligned}$$

and

$$a(x)B(x)|_c^R = a(R)B(R) - a(c)B(c) \leqslant |a(R)|M + |a(c)||B(c)|.$$

As $R \to \infty$, $|a(R)| \to 0$, and $|B(c)| = 0$, so $\lim_{R\to\infty} a(x)B(x)|_c^R = 0$. We can conclude that for any c

$$\left|\int_c^\infty a(x)b(x)\,dx\right| \leqslant M\int_c^\infty |a'(x)|\,dx < \infty.$$

Now, for a sufficiently large $d > c$ we have that $M\int_d^\infty |a'(x)|\,dx$ is necessarily small. Hence the integral $\int_c^\infty a(x)b(x)\,dx$ converges since $|\int_d^\infty a(x)b(x)\,dx|$ can be made as small as we like by choosing d sufficiently large. □

Corollary 10.2.16. *If $a(x)$ approaches zero monotonically as $x \to \infty$, then conditions (1) and (2) both hold since $\int_c^\infty |a'(x)|\,dx = |\int_c^\infty a'(x)\,dx| = |a(\infty) - a(c)| = |a(c)|$, so if (3) also holds, then $\int_c^\infty a(x)b(x)\,dx$ exists. In particular, if $a(x)$ is a C^1-function which monotonically approaches zero as $x \to \infty$, then the integrals*

$$\int_c^\infty a(x)\sin kx\,dx \qquad \textit{and} \qquad \int_c^\infty a(x)\cos kx\,dx$$

both exist.

Theorem 10.2.17 (Ermakoff). *Suppose $\Sigma_{n=1}^\infty u_n$ is a series of scalars and there exists a positive decreasing function $f(x)$ on $[1,\infty)$ such that for each n, $f(n) = |u_n|$. Suppose further that $\varphi(x)$ is any increasing positive differentiable function on $[1,\infty)$ such that $\varphi(x) > x$ for all $x > 1$. Then Σu_n converges absolutely if for all sufficiently large x*

$$\varphi'(x)f(\varphi(x))/f(x) \leqslant \lambda < 1,$$

and $\Sigma|u_n|$ diverges if for all sufficiently large x

$$\varphi'(x)f(\varphi(x))/f(x) \geqslant 1.$$

Some obvious possible choices for the function $\varphi(x)$ would be e^x, x^2, $x+1$, etc.

PROOF: Suppose for all $x \geqslant x_0 > 1$, $\varphi'(x)f(\varphi(x))/f(x) \leqslant \lambda < 1$. Then

$$\int_{\varphi(x_0)}^{\varphi(x)} f(t)\,dt = \int_{x_0}^{x} \varphi'(t)f(\varphi(t))\,dt \leqslant \lambda\int_{x_0}^{x} f(t)\,dt.$$

Therefore

$$(1-\lambda)\int_{\varphi(x_0)}^{\varphi(x)} f(t)\,dt \leqslant \lambda\left[\int_{x_0}^{x} f(t)\,dt - \int_{\varphi(x_0)}^{\varphi(x)} f(t)\,dt\right]$$

$$= \lambda\left[\int_{x_0}^{x} f(t)\,dt + \int_{x}^{\varphi(x_0)} f(t)\,dt - \int_{x}^{\varphi(x_0)} f(t)\,dt \right.$$

$$\left. - \int_{\varphi(x_0)}^{\varphi(x)} f(t)\,dt\right]$$

$$= \lambda\left[\int_{x_0}^{\varphi(x_0)} f(t)\,dt - \int_{x}^{\varphi(x)} f(t)\,dt\right]$$

$$\leqslant \lambda\int_{x_0}^{\varphi(x_0)} f(t)\,dt = \lambda M,$$

where $M = \int_{x_0}^{\varphi(x_0)} f(t)\,dt > 0$. Hence we have for any $x \geqslant x_0$,

$$\int_{\varphi(x_0)}^{\varphi(x)} f(t)\,dt \leqslant \lambda M/(1-\lambda);$$

furthermore,

$$\int_{x_0}^{x} f(t)\,dt \leqslant M + \int_{\varphi(x_0)}^{\varphi(x)} f(t)\,dt \leqslant M + \lambda M/(1-\lambda) = M/(1-\lambda).$$

Thus, by the integral test, $\Sigma|u_n|$ converges.

Now, suppose that for all $x \geqslant x_0 > 1$, $\varphi'(x)f(\varphi(x))/f(x) \geqslant 1$. For any such x we have

$$\int_{\varphi(x_0)}^{\varphi(x)} f(t)\,dt = \int_{x_0}^{x} \varphi'(t)f(\varphi(t))\,dt \geqslant \int_{x_0}^{x} f(t)\,dt.$$

It follows that for any $x > x_0$

$$\int_{x}^{\varphi(x)} f(t)\,dt = \int_{x_0}^{\varphi(x_0)} f(t)\,dt + \int_{\varphi(x_0)}^{\varphi(x)} f(t)\,dt - \int_{x_0}^{x} f(t)\,dt \geqslant M > 0.$$

If x is a large integer, this inequality shows that the integral

$$\int_{n}^{[\![\varphi(n)]\!]+1} f(t)\,dt$$

is bounded away from zero, so the sequence of remainders of the series $\Sigma|u_n|$ is likewise bounded away from zero. Thus $\Sigma|u_n|$ cannot converge. □

We bring this section to a close by looking at a few examples. Let

$$f(x) = \sum_{n=1}^{\infty} \frac{x}{n(x+n)}, \qquad x \in [0,1],$$

where by $f(x)$ we mean the pointwise limit of the series. Apply the Weierstrass M-test; each term of the series is uniformly bounded on $[0,1]$ by

n^{-2}, and $\Sigma n^{-2}<\infty$. Hence the convergence of the series is uniform, and the (uniform) continuity implies the (uniform) continuity of $f(x)$, which in turn ensures that $\int_0^1 f(x)\,dx$ exists. Call this number γ, and we have

$$\begin{aligned}\gamma=\int_0^1 f(x)\,dx&=\int_0^1\left(\Sigma\frac{x}{n(x+n)}\right)dx=\sum_1^\infty\int_0^1\frac{x\,dx}{n(x+n)}\\&=\sum_{n=1}^\infty\int_0^1\left[\frac{1}{n}-\frac{1}{x+n}\right]dx=\sum_{n=1}^\infty\left[\frac{1}{n}-\ln(n+1)+\ln n\right]\\&=\lim_{N\to\infty}\left\{\sum_{n=1}^N\frac{1}{n}-\ln(N+1)\right\}.\end{aligned}$$

To estimate γ, let $N=10^{10}$, compute, and obtain

$$\gamma=.5772157\ldots.$$

This number is called Euler's constant, and we see from our work that we can approximate $\Sigma_{n=1}^N 1/n$ by $\ln(N+1)+.5772157$.

Now consider $f(x)$, the pointwise limit of the series

$$\sum_{n=1}^\infty\frac{x}{1+n^2x^2},\qquad x\in[0,\infty).$$

It is clear that $f(0)=0$, and that for any $\delta>0$, if $x\geqslant\delta$,

$$\frac{x}{1+n^2x^2}\leqslant\frac{1}{x^{-1}+n^2\delta}<\frac{1}{\delta n^2},$$

so the series converges uniformly on $[\delta,\infty)$, for each positive δ. Hence we can write for each n and $0<\delta\leqslant x$

$$\frac{x}{1+(n+1)^2x^2}\leqslant\int_n^{n+1}\frac{x}{1+t^2x^2}\,dt\leqslant\frac{x}{1+n^2x^2},$$

and

$$\sum_{n=0}^\infty\frac{x}{1+(n+1)^2x^2}\leqslant\int_0^\infty\frac{x}{1+t^2x^2}\,dt\leqslant\sum_{n=0}^\infty\frac{x}{1+n^2x^2}.$$

This is to say

$$f(x)\leqslant\arctan tx\Big|_{t=0}^{t=\infty}\leqslant x+f(x),$$

or $f(x)\leqslant\pi/2\leqslant x+f(x)$.

Now, let $x\searrow 0$ and $\lim_{x\searrow 0}f(x)=\pi/2$. But $f(0)=0$, so we conclude that the convergence of the series is not uniform on $[0,\infty)$; if it were, $f(x)$ would be continuous since each term of the series is continuous.

Let $f(x)$ be the pointwise limit of the series

$$\sum_{n=1}^\infty\frac{\sin n^2x}{n^2},\qquad x\in\mathbb{R}.$$

The Weierstrass M-test ensures that $f(x)$ is the uniform limit of the series, so $f(x)$ is certainly continuous. If we differentiate the series term by term, we get the series

$$\sum_{n=1}^{\infty} \cos n^2 x.$$

If x is any rational multiple of π, then infinitely many terms of this series have an absolute value of 1, so this series cannot converge on any interval. If $f'(x)$ exists, it is not the pointwise limit of this series, $\Sigma \cos n^2 x$. We must confess that we do not know whether the continuous function $f(x)$ is differentiable, even at $x = 0$. $f(x)$ is known as Riemann's function.

10.3 POWER SERIES

Suppose $f(x)$ is the pointwise limit of the series

$$\sum_{i=0}^{\infty} a_i (x-c)^i.$$

It is to be understood that even if $x = c$, we consider the first term to be equal to a_0. Is the convergence to $f(x)$ uniform? If we apply the M-test, we have each term of the series uniformly bounded, provided we stay within some closed region which lies within the region of absolute convergence. This is to say, if the series is absolutely convergent for any x in the region $|x-c| < R$, then if $0 < B < R$, the series will converge uniformly in the region $|x-c| \leqslant B$ since the nth term of the series is uniformly bounded by $|a_n| B^n$ and $\Sigma_{i=0}^{\infty} |a_i| B^i$ converges.

As we have mentioned, such series $\Sigma_{i=0}^{\infty} a_i (x-c)^i$ are called power series. The coefficients may be real or complex, and the indeterminant x may be a real or complex variable. By the radius of convergence of a series centered at c we mean a real number $R \geqslant 0$ such that the series is absolutely convergent for all x such that $|x-c| < R$ and the series is divergent if $|x-c| > R$.

Theorem 10.3.1. *If $\Sigma_{n=0}^{\infty} a_n (x-c)^n$ is a power series, the radius of convergence R is given by*

$$\frac{1}{R} = \limsup_n \sqrt[n]{|a_n|}.$$

If $\limsup_n |a_{n+1}/a_n|$ *exists, this must also have the value* $1/R$; $0 \leqslant R \leqslant \infty$.

PROOF: The proof is a direct consequence of the definition of radius of convergence. □

If $|x-c| = R$, the series may converge conditionally, absolutely, or it may diverge.

Theorem 10.3.2. *Let $R>0$ be the radius of convergence of the real series $\Sigma_{n=0}^{\infty}a_n(x-c)^n$, where x is a real variable. Let $0<B<R$. If $f(x)$ is the uniform limit of the series on $|x-c|\leqslant B$, then f is C^1, and $f'(x)$ is the uniform limit of the series*

$$\sum_{n=1}^{\infty} na_n(x-c)^{n-1} \qquad \text{on} \quad |x-c|\leqslant B.$$

Furthermore, $F(x)=\int_c^x f(t)\,dt$ is the uniform limit of

$$\sum_{n=0}^{\infty} \frac{a_n(x-c)^{n+1}}{n+1} \qquad \text{on} \quad |x-c|\leqslant B.$$

Moreover, R is the radius of convergence of each of the two series obtained from $\Sigma a_n(x-c)^n$, one by differentiation, the other by antidifferentiation. It follows that on $|x-c|\leqslant B$, the function $f(x)$ is uniformly C^∞. Finally, for each n, the number a_n is necessarily equal to the value

$$\frac{f^{(n)}(c)}{n!},$$

so there is only one power series of the form $\Sigma_{n=0}^{\infty}a_n(x-c)^n$ which represents $f(x)$ on the interval $|x-c|\leqslant B$.

We leave the proof of this theorem as a project for the reader.

Definition 10.3.3. A function $f(x)$ is said to be *analytic* at c if it can be represented on a neighborhood of c in terms of a power series of the form $\Sigma_{n=0}^{\infty}a_n(x-c)^n$ having a positive radius of convergence.

The results of the previous theorem imply that if f is analytic at c, then f is of class C^∞ in a neighborhood of c. Note that e^{-1/x^2} is C^∞ on $\mathbb{R}$ but *not* analytic at 0.

The next theorem, due to Abel, is not so obvious as it may seem, and the proof is cumbersome.

Theorem 10.3.4 (Abel). *Let $R>0$ be the radius of convergence of the complex series $\Sigma_{n=0}^{\infty}a_nz^n$, and suppose that the series converges conditionally for all z for which $|z|=R$. If $f(z)$ is the pointwise limit of this series on the disk $|z|\leqslant R$, then the series Σa_nx^n converges uniformly to $f(x)$ on the real interval $-R\leqslant x\leqslant R$, and $f(x)$ is continuous on this closed interval.*

PROOF: The a_n are generally complex numbers, and z is a complex variable. Now, assume without loss of generality that $R=1$. We need to show that, given $\varepsilon>0$, there is an index n_0 such that for all $n\geqslant n_0$ and any real x, $|x|\leqslant 1$, we have $|\Sigma_{k=n}^{\infty}a_kx^k|<\varepsilon$.

Since Σa_n converges, if we put $B_n = \Sigma_{k=n}^{\infty} a_k$, we know that $B_n \to 0$ as $n \to \infty$. Assume now that $0 \leqslant x < 1$. Then

$$\left|\sum_{k=n}^{\infty} a_k x^k\right| = |(B_n - B_{n+1})x^n + (B_{n+1} - B_{n+2})x^{n+1} + \cdots|$$

$$= |B_n x^n + B_{n+1}(x^{n+1} - x^n) + B_{n+2}(x^{n+2} - x^{n+1}) + \cdots|$$

$$= \left|B_n x^n + x^n(x-1)\left[B_{n+1} + B_{n+2}x + B_{n+3}x^2 + \cdots\right]\right|.$$

Let $\varepsilon > 0$. Choose N so large that $n \geqslant N \Rightarrow |B_n| < \varepsilon/2$. Then for all $n \geqslant N$ we have

$$\left|\sum_{k=n}^{\infty} a_k x^k\right| \leqslant |B_n||x|^n + |x|^n(1-x)[|B_{n+1}| + |B_{n+2}x| + \cdots]$$

$$< \frac{\varepsilon}{2}x^n + \frac{\varepsilon}{2}x^n(1-x)\left[1 + x + x^2 + \cdots\right] < \varepsilon,$$

and this inequality holds for all x, $0 \leqslant x < 1$. If $x = 1$, $|\Sigma_{k=n}^{\infty} a_k x^k| = |B_n| < \varepsilon/2$. Now, assume $-1 < x \leqslant 0$, and put $c_n = \Sigma_{k=n}^{\infty} a_k(-1)^k$. By hypothesis, $\Sigma_{k=0}^{\infty} a_k(-1)^k$ converges, so $c_n \to 0$. Write

$$\left|\sum_{k=n}^{\infty} a_k x^k\right| = \left|\sum_{k=n}^{\infty} a_k(-1)^k |x|^k\right|$$

$$= |(c_n - c_{n+1})|x|^n + (c_{n+1} - c_{n+2})|x|^{n+1} + \cdots|$$

$$= \left|c_n|x|^n + |x|^n(|x|-1)[c_{n+1} + c_{n+2}|x| + \cdots]\right|.$$

Choose M so large that for all $n \geqslant M$, $|c_n| < \varepsilon/2$. Then, for any x, $-1 < x \leqslant 0$, and any $n \geqslant M$, we have

$$\left|\sum_{k=n}^{\infty} a_k x^k\right| < \frac{\varepsilon}{2}|x|^n + |x|^n \frac{\varepsilon}{2}(1-|x|)\left(1 + |x| + |x|^2 + \cdots\right) < \varepsilon.$$

If $x = -1$, we have $|\Sigma_n^{\infty} a_k(-1)^k| = |c_n| \leqslant \varepsilon/2$. If $n_0 = N \vee M$, we have $n \geqslant n_0 \Rightarrow |\Sigma_{k=n}^{\infty} a_k x^k| < \varepsilon$ for any $x \in [-1,1]$. This is to say, $\Sigma_0^{\infty} a_k x^k$ converges uniformly on $[-1,1]$ to its pointwise limit $f(x)$ and $f(x)$ is continuous, in fact uniformly continuous, on this interval since each term of the series is uniformly continuous on $[-1,1]$. □

Corollary 10.3.5. *If $\Sigma_{n=0}^{\infty} a_n x^n$ is a real series with radius of convergence $R > 0$ and if this series converges conditionally for $x = R$ (or $x = -R$), then the convergence of this series to its pointwise limit $f(x)$ is uniform on* $[0, R]$ *(or* $[-R, 0]$*), and $f(x)$ is thus continuous on* $[0, R]$ *(or* $[-R, 0]$*). Hence, if $x_0 \in [0, R]$ (or* $[-R, 0]$*), then*

$$\lim_{x \to x_0} \sum a_n x^n = \lim_{x_0} f(x) = f(x_0) = \sum a_n x_0^n = \sum a_n \left(\lim_{x_0} x\right)^n = \sum a_n \lim_{x_0} x^n.$$

Note that this last statement is a specialization of a result of Theorem 10.2.3. If $\Sigma g_n(x)$ is a series of continuous mappings converging uniformly to $g(x)$, then $g(x)$ is continuous and we have

$$\lim_{x_0} \sum g_n(x) = \lim_{x_0} g(x) = \sum g_n(x_0) = \sum \lim_{x_0} g_n(x).$$

Thus, in this case, the limit and summation operations commute.

We now present a theorem which is an extension of Abel's theorem.

Theorem 10.3.6. *Suppose $R>0$ is the radius of convergence of the complex power series $\Sigma_{n=0}^{\infty} a_n z^n$ and $f(z)$ is the pointwise limit of this power series in the disk $D=\{z:|z|<R\}$. Suppose further that this series converges, perhaps conditionally, to g at $z_0 \in \partial D$. Then there exists a closed triangle T contained in $D \cup \{z_0\}$ having a vertex at z_0 such that, as $z \to z_0$ within this triangle, $f(z) \to g$.*

If $\Sigma_{n=0}^{\infty} a_n z_0^n$ converges absolutely, and we define $f(z_0)=g$, then the convergence of $\Sigma_{n=0}^{\infty} a_n z^n$ to $f(z)$ is uniform on $D \cup \{z_0\}$, and f may be continuously extended to $\overline{D}$.

PROOF: We show first how we construct the (by no means unique) triangle T and then show that there is a constant K such that, for all $z \in T \cap D$,

$$|z_0 - z|/(|z_0| - |z|) \leqslant K.$$

Let z_0 be an arbitrary point on the circle centered at 0 with radius $|z_0|$, and from z_0 construct the two equal chords $\overline{z_0 z_3}$ and $\overline{z_0 z_4}$, each making an angle θ_0 with the radius $\overline{0 z_0}$. Let z_1 and z_2 be the midpoints of the chords $\overline{z_0 z_3}$and $\overline{z_0 z_4}$, respectively. The triangle T is $\triangle z_1 z_0 z_2$. (See Figure 10.1.)

FIGURE 10.1 The triangle T. The angle $z_3 z_0 z_4$ is often called the Stolz angle.

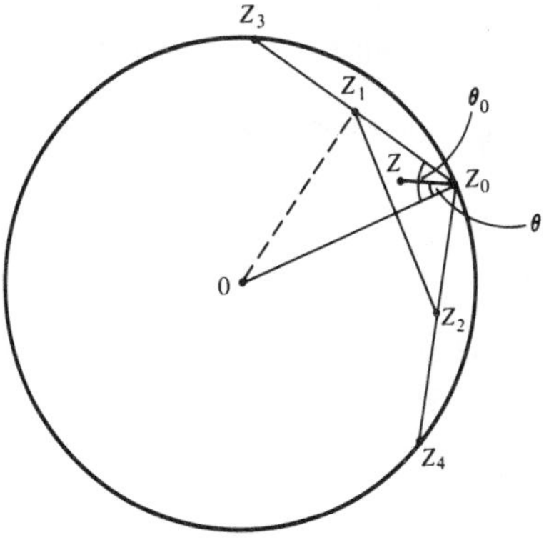

If $z \in T \cap D$, we denote by θ the angle $zz_0 0$. It is apparent that $0 < \cos\theta_0 \leqslant \cos\theta \leqslant 1$ and that $|z_0 - z| \leqslant |z_0|\cos\theta_0$. Let K be the constant $2/\cos\theta_0$, and write using the law of cosines

$$|z_0| - |z| = |z_0| - \left(|z_0|^2 + |z_0 - z|^2 - 2|z_0||z_0 - z|\cos\theta\right)^{1/2}$$

Then

$$\frac{|z_0 - z|}{|z_0| - |z|} \leqslant \frac{2}{\cos\theta_0}$$

if and only if

$$|z_0 - z|^2 - 2|z_0||z_0 - z|\cos\theta \leqslant -|z_0||z_0 - z|\cos\theta_0 + \tfrac{1}{4}|z_0 - z|^2\cos^2\theta_0.$$

We increase the left-hand side of this inequality by replacing $|z_0 - z|$ by $|z_0|\cos\theta_0$ and $\cos\theta$ by $\cos\theta_0$ to obtain $-|z_0|^2\cos^2\theta_0$. Since

$$-|z_0|^2\cos^2\theta_0 + |z_0||z_0 - z|\cos\theta_0 \leqslant 0 \leqslant \tfrac{1}{4}|z_0 - z|^2\cos^2\theta_0,$$

we have proved that for all $z \in T \cap D$,

$$\frac{|z_0 - z|}{|z_0| - |z|} \leqslant \frac{2}{\cos\theta_0} = K.$$

We can simplify the proof that $\sum_{n=0}^{\infty} a_n(z^n - z_0^n) \to 0$ as $z \to z_0$ in T by transforming the series $\sum_{n=0}^{\infty} a_n z^n$ into one that has a radius of convergence equal to 1 and converges, perhaps conditionally, at $z = 1$. We do this in the following way: If $Re^{i\alpha_0}$ is the polar representation for z_0, then the series $\sum_{n=0}^{\infty} c_n z^n$ converges absolutely for all z in the unit disk $D' = \{z : |z| < 1\}$ and converges, perhaps conditionally, at $z = 1$ if, for each n, $c_n = a_n R^n e^{in\alpha_0}$. In fact, $\sum_{n=0}^{\infty} c_n = g$. Next, we put $d_0 = c_0 - g$ and $d_n = c_n$ for all $n \geqslant 1$; the series $\sum_{n=0}^{\infty} d_n z^n$ is the transformed series we consider.

Let T' be the image of T under the similarity mapping $z \mapsto z/(Re^{i\alpha_0})$; if we can show that $\sum_{n=0}^{\infty} d_n z^n \to 0$ as $z \to 1$ in T', we shall have shown that $f(z) \to g$ as $z \to z_0$ in T. Note that for all $z \in T'$, we have $|1 - z| \leqslant K(1 - |z|)$, and that $\sum_{n=0}^{\infty} d_n = 0$. Denote by $g(z)$ the pointwise limit of $\sum_{n=0}^{\infty} d_n z^n$ in D'.

Let $D_n = \sum_{k=0}^{n} d_k$, and use Abel's summation formula (Definition 10.2.15) to obtain

$$\sum_{k=0}^{n} d_k z^k = D_n z^n - \sum_{k=0}^{n-1} (z^{k+1} - z^k) D_k = (1 - z)\sum_{k=0}^{n-1} D_k z^k + D_n z^n.$$

Since $|z| < 1$ and $D_n \to 0$, we have that $\lim_{n\to\infty} D_n z^n = 0$, so we can write

$$g(z) = \sum_{n=0}^{\infty} d_n z^n = (1 - z)\sum_{k=0}^{\infty} D_k z^k.$$

Let $\varepsilon > 0$, and let $\delta = \varepsilon/2K$. Choose N sufficiently large that, for all $n \geqslant N$,

$|D_n|<\delta$. Then the Nth remainder of the series $\Sigma_{n=0}^{\infty} D_n z^n$ will be dominated by

$$\delta \sum_{n=N}^{\infty} |z|^n = \frac{\delta|z|^N}{1-|z|} < \frac{\delta}{1-|z|}.$$

It follows that for $z \in T' \cap D'$

$$|g(z)| \leqslant |1-z|\left(\left|\sum_{n=0}^{N-1} D_n z^n\right| + \frac{\delta}{1-|z|}\right) \leqslant |1-z|\left|\sum_{n=0}^{N-1} D_n z^n\right| + K\delta$$

$$\leqslant |1-z| \sum_{n=0}^{N-1} |D_n| + \frac{\varepsilon}{2}.$$

If $z \in T' \cap D'$ is chosen sufficiently close to 1, it is clear that $|g(z)|<\varepsilon$. Since $\varepsilon>0$ was arbitrary, we conclude that $g(z) \to 0$ as $z \to 1$ in T'. Hence, $f(z) \to g$ as $z \to z_0$ in T.

Finally, suppose $\Sigma_0^{\infty} a_n z_0^n$ converges absolutely. For each $n=0,1,2,\ldots,$ we have

$$\left|\sum_{k=n}^{\infty} a_k z^k\right| \leqslant \sum_{k=n}^{\infty} |a_k||z_0|^k < \infty.$$

Let $\varepsilon>0$; choose N sufficiently large that

$$\sum_{n=N}^{\infty} |a_n||z_0|^n = \sum_{n=N}^{\infty} |a_n| R^n < \varepsilon.$$

Then for all $z \in D \cup \{z_0\}$ we have $|\Sigma_{n=N}^{\infty} a_n z^n| < \varepsilon$. Thus the convergence is uniform if we define $f(z_0)=g$. Since each term of the power series is uniformly continuous on $D \cup \{z_0\}$, $f(z)$ must be uniformly continuous there, and by Theorem 6.7.5 f can be continuously extended to all of $\overline{D}$. □

We close this section with a question. If the series $\Sigma_{n=0}^{\infty} a_n z^n$ of Theorem 10.3.6 converges conditionally for all z_0 on the boundary of D, then for each such z_0 is the convergence of $\Sigma_{n=0}^{\infty} a_n z^n$ to $f(z)$ uniform on the whole closed segment $\overline{0z_0}$?

10.4 ARITHMETIC WITH SERIES

Multiplication and division of series are processes that frequently help a great deal in many applications, particularly when the series are power series. For a general series of functions we don't always have a nice simple region about a point where the series converges absolutely, but we do for convergent power series, and two power series often have a common region of absolute convergence.

There are several things the reader should keep in mind. If Σu_n, Σv_n are two series converging respectively to u and v, then an arithmetic product of these two series had better be a series which converges to uv. However, we are all well aware that a series which is not absolutely convergent can be altered considerably by rearranging (infinitely many of) its terms. That is, a conditionally convergent series can by such rearranging be made to converge to any given number, or to diverge to infinity. Some divergent series can be made to converge conditionally. Hence multiplication of two conditionally convergent series might upset the delicate balances involved.

On the other hand, an absolutely convergent series can be twisted, trampled, and permuted beyond recognition, but it will still converge to the same limit as before, as long as none of the terms themselves are changed in value. Hence it seems reasonable to restrict our attention to the set of absolutely convergent series of scalars and attempt to make a division ring out of this set.

Definition 10.4.1. Let $\Sigma_{n=0}^{\infty} u_n$, $\Sigma_{n=0}^{\infty} v_n$ be two absolutely convergent series of scalars. Then

$$\sum_{n=0}^{\infty} u_n + \sum_{n=0}^{\infty} v_n = \sum_{n=0}^{\infty} (u_n + v_n)$$

and

$$\left(\sum_{n=0}^{\infty} u_n\right)\left(\sum_{n=0}^{\infty} v_n\right) = \sum_{n=0}^{\infty} w_n,$$

where for each n

$$w_n = u_0 v_n + u_1 v_{n-1} + u_2 v_{n-2} + \cdots + u_{n-1} v_1 + u_n v_0.$$

Evidently, these two operations are commutative and associative.

Theorem 10.4.2. *The sum and product of two absolutely convergent series are absolutely convergent. Moreover, the limits of the sum and product equal, respectively, the sum and product of the limits.*

PROOF: Let $\Sigma u_n = u$, $\Sigma v_n = v$, $\Sigma|u_n| = A$, and $\Sigma|v_n| = B$. $\Sigma(u_n + v_n)$ is absolutely convergent since every partial sum of $\Sigma|u_n + v_n|$ is bounded by $A + B$. For each k, $\Sigma_{n=0}^{k}(u_n + v_n) = \Sigma_{n=0}^{k} u_n + \Sigma_{n=0}^{k} v_n$, so the partial sums of the series $\Sigma_{n=0}^{\infty}(u_n + v_n)$ converge to $u + v$.

Consider any way one might multiply together the two series $\Sigma|u_n|$ and $\Sigma|v_n|$; the product, however obtained, must contain once and only once a term of the form $|u_i||v_j|$, where i and j are a certain pair of nonnegative integers. Any partial sum of the product series, however obtained, will be contained in a sum which is the product $(\Sigma_{k=0}^{m}|u_k|)$ $(\Sigma_{k=0}^{m}|v_k|)$ for m sufficiently large, and this sum is bounded by AB. Hence such an arbitrarily formed series is absolutely convergent; being so, it can be rearranged in any way, and it will still converge to the same limit. Our series $\Sigma_{n=0}^{\infty} w_n$ is just one particular arrangement. That Σw_n converges to uv can most easily be seen if we rearrange the series and then consider a sequence of partial sums of the form $S_m = (\Sigma_{k=0}^m u_k)(\Sigma_{k=0}^m v_k)$. It is clear that $S_m \to uv$ as $m \to \infty$. □

This theorem shows that we can form a product of two absolutely convergent series in infinitely many ways. The method given, often referred to as the *Cauchy product*, is a very natural one since the Cauchy product of two power series in x turns out to be a power series with the terms arranged in exactly the right order.

Definition 10.4.3. If $\Sigma_0^\infty w_n$, $\Sigma_0^\infty u_n$ are two absolutely convergent series with $u_0 \neq 0$, the quotient series

$$\sum_{n=0}^{\infty} v_n = \sum_{n=0}^{\infty} w_n \Big/ \sum_{n=0}^{\infty} u_n$$

is the series obtained by long division. That is, we form $\Sigma_{n=0}^\infty v_n$ by solving successively

$$\begin{aligned} u_0 v_0 &= w_0, \\ u_0 v_1 + u_1 v_0 &= w_1, \\ &\vdots \\ u_0 v_n + \cdots + u_n v_0 &= w_n. \end{aligned}$$

It is certainly true that $(\Sigma u_n)(\Sigma v_n) = \Sigma w_n$ as a consequence of the way we have defined the quotient series. But suppose

$$\sum w_n = 1 + 0 + 0 + \cdots \qquad \text{and} \qquad \sum u_n = 1 - 4 + 0 + 0 + \cdots.$$

Then $\Sigma_{n=0}^\infty v_n = \Sigma_{n=0}^\infty 4^n$, a very divergent series! Hence quotient series must be carefully checked to see whether they are absolutely convergent, conditionally convergent, or divergent. It is true that if we have

$$\sum_{n=0}^{\infty} v_n x^n = \sum_{n=0}^{\infty} w_n x^n \Big/ \sum_{n=0}^{\infty} u_n x^n,$$

the coefficients v_n are found in exactly the same way that we found them above, but for all sufficiently small $|x|$, the quotient series will converge absolutely.

The application of this idea that one naturally thinks of first is in finding a Taylor series expansion about a point for a function which is a quotient of two functions for which the Taylor series are known. It might be easier, for example, to do the long division to find the Maclaurin series for $(1+x)^{1/2} \cdot (1+x^2)^{-1/2}$ than to do a lot of differentiation.

Since division of absolutely convergent series of scalars is not always a closed operation, we see that these things comprise an algebraic structure something less than a division ring; indeed, the structure is that of a commutative ring with unity.

Theorem 10.4.4. *If $\Sigma a_n x^n$, $\Sigma b_n x^n$ are power series with radii of convergence A and B, respectively, then the sum, difference, and product of the two series will converge absolutely if $|x| < A \wedge B$.*

PROOF: For the sum and difference the statement is obviously true. The previous theorem ensures that $\Sigma c_n x^n = (\Sigma a_n x^n)(\Sigma b_n x^n)$ is absolutely convergent if $|x| < A \wedge B$. However, $(\Sigma x^n)(\Sigma a_n x^n)$, where $a_0 = 2$, $a_1 = -2$, $a_i = 0$, $i \geqslant 2$, is equal to $\Sigma c_n x^n$, where $c_0 = 2$, $c_i = 0$, $i > 0$, and the product series converges for all x, yet the first factor is absolutely convergent only when $|x| < 1$. □

10.5 INFINITE PRODUCTS

We shall not have much occasion to use the notion of infinite product, but the idea should be introduced. If $\{u_n\}_{n=1}^{\infty}$ is a sequence of scalars, we can form the product

$$\prod_{n=1}^{\infty} u_n = u_1 u_2 u_3 \cdots .$$

To define what we mean by this, we must, as we did with series, introduce a notion of convergence. Let $P_k = \Pi_{n=1}^{k} u_n$. P_k is a partial product.

Definition 10.5.1. If $\Pi_{n=1}^{\infty} u_n$ is an infinite product, we say the following.

1. If none of the factors u_n is zero and the sequence of partial products converges to a number $P \neq 0$, $\Pi_{n=1}^{\infty} u_n$ converges to P.
2. If at most finitely many of the factors are zero, and at least one factor is zero, and the infinite product $\Pi_{n=N}^{\infty} u_n$, for some $N > 1$, converges in the sense of 1, $\Pi_{n=1}^{\infty} u_n$ converges to zero.
3. If infinitely many of the u_n are zero or if the sequence of partial products converges to zero, $\Pi_{n=1}^{\infty} u_n$ diverges to zero.
4. If the sequence of partial products diverges to infinity or does not converge, $\Pi_{n=1}^{\infty} u_n$ is divergent.

It is easy to see that if $u_n \nrightarrow 1$, an infinite product cannot converge. This fact suggests that we might find it convenient to write

$$\prod_{n=1}^{\infty} u_n = \prod_{n=1}^{\infty} (1 + a_n).$$

Theorem 10.5.2. *Assume $a_n > 0$ for each n; then $\Pi_{n=1}^{\infty}(1 + a_n)$ converges iff $\Sigma_{n=1}^{\infty} a_n$ converges.*

PROOF: Put $S_n = \Sigma_{i=1}^n a_i$ and $P_n = \Pi_{i=1}^n (1+a_i)$. Both sequences $\{S_n\}$ and $\{P_n\}$ are increasing. Hence, if $\{P_n\}$ is bounded, it converges to $P = \Pi_{n=1}^\infty (1+a_n)$. Assume $\{P_n\}$ is bounded. For each n, $P_n > S_n$, so $\{S_n\}$ is bounded and thus converges. Now assume $\{S_n\}$ is bounded, so that Σa_n converges. Note that $1+a_n \leqslant e^{a_n}$ for each n, so $P_n \leqslant e^{S_n}$. Hence $\{P_n\}$ is bounded, and $\Pi_{n=1}^\infty (1+a_n)$ converges. □

Definition 10.5.3. We say that $\Pi_{n=1}^\infty u_n = \Pi_{n=1}^\infty (1+a_n)$ converges absolutely if $\Pi_{n=1}^\infty (1+|a_n|)$ converges.

Corollary 10.5.4. $\Pi_{n=1}^\infty (1+a_n)$ *converges absolutely iff* $\Sigma_{n=1}^\infty a_n$ *converges absolutely.*

Theorem 10.5.5. *Absolute convergence of a product implies convergence.*

PROOF: We always have

$$0 \leqslant |(1+a_{n+1})\cdots(1+a_{n+k})-1| \leqslant |(1+|a_{n+1}|)\cdots(1+|a_{n+k}|)-1|.$$

It is not hard to see that an infinite product converges iff, given $\varepsilon > 0$, there exists an index n_ε such that for all n, m with $n_\varepsilon \leqslant n < m$,

$$|u_n u_{n+1} \cdots u_m - 1| < \varepsilon.$$

Hence we can conclude that absolute convergence implies convergence. □

Consider the products

$$\prod_{n=2}^\infty \left(1+\frac{(-1)^n}{\sqrt{n}}\right) \quad \text{and} \quad \prod_{n=2}^\infty \left(1+\frac{(-1)^n(2\sqrt{n}+1)+1}{2n}\right).$$

If either of these converge, the convergence is not absolute. Since $\Sigma_{n=2}^\infty (-1)^n n^{-1/2}$ converges conditionally, perhaps the first product does converge. But note that

$$(1-\tfrac{1}{3})(1-\tfrac{1}{5})(1-\tfrac{1}{7})\cdots > \prod_{n=2}^\infty \left(1+(-1)^n/\sqrt{n}\right) > (1-\tfrac{1}{2})(1-\tfrac{1}{4})(1-\tfrac{1}{6})\cdots.$$

We need only to observe that the partial product $\Pi_{k=1}^n (1-(2k+1)^{-1}) = a_n$ multiplied by the partial product $\Pi_{k=1}^n (1+(2k+1)^{-1}) = b_n$ yields a partial product $c_n = a_n b_n = \Pi_{k=1}^n (1-(2k+1)^{-2}) \leqslant 1$. Since $\Pi_{k=1}^\infty (1+(2k+1)^{-1}) = \lim_n b_n = \infty$, we conclude that $a_n \to 0$, so the first of the infinite products diverges to zero.

Theorem 10.5.6. *If each* $a_n \geqslant 0$, $\Pi_{n=1}^\infty (1-a_n)$ *converges iff* Σa_n *converges.*

PROOF: If Σa_n converges, it does so absolutely, which implies that $\Pi_{n=1}^\infty (1-a_n)$ converges. If $\Sigma_{n=1}^\infty a_n$ diverges, and $a_n \nrightarrow 0$, clearly, $\Pi_{n=1}^\infty (1-a_n)$ diverges.

If $\Sigma_{n=1}^{\infty} a_n$ diverges and $a_n \to 0$, then, discarding terms where $a_n = 0$ or $a_n \geqslant 1$, we use the example above as a model for the remainder of the proof and conclude that $\Pi_{n=1}^{\infty}(1-a_n)$ diverges to zero. □

In the second product of the preceding example,

$$\sum_{n=2}^{\infty} \frac{(-1)^n\left(2\sqrt{n}+1\right)+1}{2n} = \infty,$$

but note that the product is of the form

$$\left[\left(1+\frac{1}{\sqrt{2}}+\frac{1}{2}\right)\left(1-\frac{1}{\sqrt{3}}\right)\right]\cdots\left[\left(1+\frac{1}{\sqrt{2n}}+\frac{1}{2n}\right)\left(1-\frac{1}{\sqrt{2n+1}}\right)\right]\cdots.$$

A partial product is of the form

$$\prod_{n=1}^{k}\left(1+\frac{1}{\sqrt{2n}}+\frac{1}{2n}-\frac{1}{\sqrt{2n+1}}-\frac{1}{\sqrt{2n}\sqrt{2n+1}}-\frac{1}{2n\sqrt{2n+1}}\right)$$

$$=\prod_{n=1}^{k}\left[1-\frac{2n-\left(\sqrt{2n}+1\right)\left(\sqrt{2n+1}-1\right)}{2n\sqrt{2n+1}}\right]=\prod_{n=1}^{k}[1-a_n].$$

Now, a_n here lies between 0 and $(2n\sqrt{2n+1})^{-1}$, so an application of Theorem 10.5.6 allows us to conclude that the sequence of partial products converges. Hence the second of the exhibited infinite products converges.

PROBLEMS

1. Let S be a set of real numbers. Prove that $\Sigma_{x \in S} x$ diverges if more than countably many numbers in S are nonzero.

2. Prove Theorems 10.2.5–10.2.11.

3. Refer to Corollary 10.2.16, and prove that
$$\int_0^{\infty} \frac{\cos x}{x}\,dx = \infty.$$

4. Prove Theorem 10.3.2.

5. Let A be a $d \times d$ matrix. If for some $k > 1$, A^k is the zero matrix, we call A a *nilpotent* matrix. Let $f(x)$ be analytic at $x=0$. Then $f(x)=\Sigma_0^{\infty} a_n x^n$, where each $a_n = f^{(n)}(0)/n!$. We can define $f(A)$ to be the matrix
$$f(A)=\sum_{n=0}^{\infty}\frac{f^{(n)}(0)}{n!}A^n$$

if each of the d^2 series converge. For example, $e^A = I + A + \frac{1}{2}A^2 + \frac{1}{6}A^3 + \cdots$. If A is nilpotent, there is no question of convergence. Let A be the matrix

$$A = \begin{bmatrix} 2 & 0 & -4 \\ 3 & 0 & 0 \\ 1 & 0 & -2 \end{bmatrix}.$$

What is $\sin A$? $\cos A$? e^A? Is $(I-A)^{-1} = I + A + A^2 + A^3 + \cdots$?

6. Test the following for convergence:

(a) $\Sigma(\ln n)^{1-n}$,

(b) $\Sigma(n \ln n \ln\ln n)^{-1}$,

(c) $\Sigma \sin\left(\dfrac{\pi}{2^n}\right)$,

(d) $\Sigma \tan\left(\dfrac{\pi}{4n}\right)$,

(e) $\Sigma \dfrac{1}{n}\left[\sqrt{n^2+n+1} - \sqrt{n^2-n+1}\right]$,

(f) $\Sigma(\ln n)(n^{-5/4})$.

7. Find the radii of convergence for the following:

(a) $\Sigma_{n=0}^{\infty} x^{n^2}$,

(b) $\Sigma_{n=1}^{\infty} \dfrac{x^n}{n+\sqrt{n}}$,

(c) $\Sigma_{n=0}^{\infty} \dfrac{nx^n}{e^{nx}}$,

(d) $\Sigma_{n=1}^{\infty} x^n n^{-x}$,

(e) $\Sigma_{n=0}^{\infty} x^n \tan\left(\dfrac{x}{2^n}\right)$,

(f) $\Sigma_{n=0}^{\infty} \dfrac{x^n}{1+x^{2n}}$.

8. Prove the convergence is uniform:

(a) $\Sigma_{n=0}^{\infty} 2^{1-n}(1+nx)^{-1/2}$, $x \in (0,\infty)$;

(b) $\Sigma_{n=1}^{\infty} \ln(1+nx)/nx^n$, $x \in [1+\delta, \infty)$, $\delta > 0$.

9. If $f(x) = \Sigma_{n=1}^{\infty} ne^{-nx}$, show $f(x)$ is continuous on $(0,\infty)$. Evaluate $\int_{\ln 2}^{\ln 3} f(x)\,dx$. Does $\int_0^1 f(x)\,dx$ exist?

10. Show that $\Sigma_{n=0}^{\infty} 2^{-n} \sin 2^n \pi x$ converges uniformly on $\mathbb{R}$. Why can't this series be differentiated term by term?

11. Solve for y: $y = \ln(1+x) - xy$. Now, suppose $y = \Sigma_{n=0}^{\infty} a_n x^n$. Substitute this series for y in the equation, equate coefficients, and determine the actual series for y.

12. Let $f_n(x) = (2/\pi)x \tan^{-1} nx$. Find $f_n'(x)$. Find $f(x) = \lim_n f_n(x)$. Compare $f'(x)$ with $g(x) = \lim_n f_n'(x)$. Comment.

13. Try to justify the formula

$$\sin x = x \prod_{n=1}^{\infty} \left(1 - \frac{x^2}{n^2\pi^2}\right).$$

Note that the right-hand side vanishes for all integral multiples of π. For any other value of x, does the product converge absolutely?

14. Prove that if $\Sigma w_n x^n$, $\Sigma u_n x^n$ both converge absolutely on $(-1,1)$, then there is a $\delta > 0$ such that

$$\sum v_n x^n = \sum w_n x^n / \sum u_n x^n$$

converges absolutely on $(-\delta, \delta)$.

15. Test the series $\Sigma_{n=1}^{\infty} n^{-\alpha} \sin nx$, where $0 < \alpha \leqslant \frac{1}{2}$, for convergence. Put $f(x) = \Sigma_{n=1}^{\infty} n^{-\alpha} \sin nx$, $-\pi \leqslant x \leqslant \pi$. Prove that $f(x)$ is not Riemann integrable by using Bessel's inequality.
[*Hint*: Assume $\int_{-\pi}^{\pi} f(x)\,dx$ exists. Then so does $\int_{-\pi}^{\pi} [f(x)]^2\,dx$.]

16. Show that $\Sigma_{n=2}^{\infty}(\sin nx/\ln n)$ converges for all x. Let $f(x) = \Sigma_{n=2}^{\infty}(\sin nx/\ln n)$. Is $f(x)$ continuous a.e.? Put $F(x) = \int_0^x f(t)\,dt$. Is F absolutely continuous and periodic? (F is *periodic* iff $\exists p > 0$ s.t. $\forall n \in \mathbb{Z}$, $\forall x$, $F(x) = F(x + np)$.) Show that $F(x)$ is an *even* function, i.e., that $F(x) \equiv F(-x)$. Now, show that $F(x)$ has a Fourier series representation of the form $\Sigma_{n=0}^{\infty} a_n \cos nx$. Compute a_n, for $n \geqslant 2$, to get $a_n = -1/n \ln n$. Finally, $F(0) = \Sigma_{n=1}^{\infty} a_n$. But $F(0) = 0$. This proves $f(x)$ is not integrable. Why?

17. Test by the ratio test and by the root test the two series

$$\sum_{n=1}^{\infty} 2^{(-1)^n - n} \quad \text{and} \quad \sum_{n=1}^{\infty} \left[\frac{5 + (-1)^n}{2}\right]^{-n}.$$

18. Prove that for any series Σa_n of positive terms:

$$\underline{\lim}\, \frac{a_{n+1}}{a_n} \leqslant \underline{\lim}\, \sqrt[n]{a_n} \leqslant \overline{\lim}\, \sqrt[n]{a_n} \leqslant \overline{\lim}\, \frac{a_{n+1}}{a_n}.$$

19. Let

$$f_n(x) = \begin{cases} 2n^2 x, & 0 \leqslant x \leqslant 1/2n, \\ n - 2n^2(x - 1/2n), & 1/2n \leqslant x \leqslant 1/n, \\ 0, & 1/n \leqslant x \leqslant 1. \end{cases}$$

Compute $\lim_{n\to\infty} \int_0^1 f_n(x)\,dx$ and $\int_0^1 \lim_{n\to\infty} f_n(x)\,dx$. Why are these not equal?

20. Evaluate $\int_0^{\infty}(\Sigma_{n=1}^{\infty}(n^4 + x^2)^{-1})\,dx$. Is the convergence uniform on $[0, \infty)$?

21. Prove that every rearrangement of an absolutely convergent series converges to the same limit and that a conditionally convergent series may be rearranged to converge to any limit, finite or infinite.

22. Does the series $\Sigma_{n=2}^{\infty} 1/[\ln n]^p$ converge for $p>1$?
[*Hint*: Use Theorem 10.2.17, with $\varphi(x)=x^2$.]

23. Let $\{a_n\}$, $\{b_n\}$ be two sequences of real numbers. Prove that $\overline{\lim}\{a_n+b_n\} \leqslant \overline{\lim}\{a_n\}+\overline{\lim}\{b_n\}$ and $\overline{\lim}\{a_n b_n\} \leqslant \overline{\lim}\{a_n\}\overline{\lim}\{b_n\}$ if all the a_n, b_n are positive.

24. Find $\overline{\lim}$ and $\underline{\lim}$ for the following sequences.

(a) $\{[1+1/n]\cos n\pi\}$,
(b) $\{n\sin n\pi/3\}$,
(c) $\{(-1)^n n(1+n)^{-n}\}$,
(d) $\{S_{n+1}/S_n\}$, where $S_1=S_2=1$, $S_{n+2}=S_{n+1}+S_n$.
[*Hint*: Evaluate $S_n S_{n+2}-S_{n+1}^2$.]
(e) $\{n/5-[\![n/5]\!]\}$.

25. Let $\{n_k\}$ be the increasing sequence of positive integers that do not have a 0 in their decimal representations. Show that $\Sigma_{k=1}^{\infty} 1/n_k$ converges to a limit $L<90$.

26. Suppose Σa_n is a convergent series of a positive terms. Prove that $\Sigma(a_n a_{n+1})^{1/2}$ converges.

27. $\Sigma_{n=0}^{\infty}(-1)^{n+1}(n+1)^{-1/2}$ is a convergent series; form the Cauchy square. Does this latter series converge? Do the same for the series $\Sigma_{n=0}^{\infty}(-1)^{n+1}(n+1)^{-1}$. You should get $2\,\Sigma_{n=1}^{\infty}(-1)^{n+1}(n+1)^{-1}(1+1/2+\cdots+1/n)$. Why does this series converge or not converge?

28. Define $\zeta(S)=\Sigma_{n=1}^{\infty} 1/n^S$, where $S>1$. Show that $\zeta^2(S)=\Sigma_{n=1}^{\infty} d(n)/n^S$, where $d(n)$ is the number of divisors of n, including 1 and n.

29. For what values of x does the series $\Sigma_{n=1}^{\infty}(1/n)(1+\frac{1}{2}+\cdots+1/n)\sin nx$ converge?

30. If $0<r<1$ and p is any positive integer, show that the series $\Sigma_{n=p+1}^{\infty} n(n-1)(n-2)\cdots(n-p)r^n$ converges.

31. If

$$u_n=\frac{1}{n}\ln\frac{(n+1)^2}{n(n+2)},$$

show that Σu_n converges.
[*Hint*: Can you show that

$$\sum_{i=m+1}^{n} u_i<\frac{1}{m+1}\ln\frac{(m+2)(n+1)}{(m+1)(n+2)}<\frac{1}{m+1}\ln 2?]$$

32. $\Sigma_{n=1}^{\infty}(-1)^{n-1}n^{-1/2}$ is convergent. Show that $1+1/\sqrt{3}-1/\sqrt{2}+1/\sqrt{5}+1/\sqrt{7}-1/\sqrt{4}++-\cdots$ diverges to infinity.

33. Obtain the series

$$1+(1-\tfrac{1}{3})x+(1-\tfrac{1}{3}+\tfrac{1}{5})x^2+(1-\tfrac{1}{3}+\tfrac{1}{5}-\tfrac{1}{7})x^3+\cdots$$

for

$$\frac{1}{1-x}\frac{\tan^{-1}\sqrt{x}}{\sqrt{x}}.$$

34. Prove that for all x, $\Sigma_{n=1}^{\infty}(2n-1)^{-1/2}\sin(2n-1)x$ converges.

35. Prove that if $\{f_n\}$ is a sequence of real-valued continuous functions converging pointwise to a continuous function f on a compact set K and if for each n, $f_n \geqslant f_{n+1}$ on K, then the convergence is uniform. This is a theorem of Dini. Show that compactness of K is essential by showing that the sequence $\{1/(nx+1)\}$ converges pointwise but not uniformly to zero on $(0,1]$.

36. Show that $\Sigma_{n=1}^{\infty}(-1)^n n^{-1/2}\sin(1+x/n)$ converges uniformly on $\mathbb{R}$.

37. Show that $\Sigma_{n=0}^{\infty}(x^{2n+1}/(2n+1)-x^{n+1}/(2n+2))$ converges pointwise but not uniformly on $[0,1]$.

38. Let $f_n(x)=\cos^n x$ on $[0,\pi]$. Show that $\{f_n\}$ does not converge uniformly, does not converge pointwise, but does converge in measure on $[0,\pi]$.

39. Let

$$f_n(x)=\begin{cases}0, & 0\leqslant x\leqslant 1/n \quad \text{or} \quad 2/n\leqslant x\leqslant 1,\\ n, & 1/n<x<2/n.\end{cases}$$

Show that $\{f_n\}\to 0$ pointwise and in measure, but that $\{f_n\}\not\to 0$ in the mean on $[0,1]$.

40. Consider the binomial series $\Sigma_{n=0}^{\infty}\binom{\alpha}{n}x^n$, and try to show that

- **(a)** if $x=-1$, the series converges for $\alpha\geqslant 0$ and diverges for $\alpha<0$;
- **(b)** if $x=1$, the series diverges for $\alpha\leqslant -1$, converges conditionally for $-1<\alpha<0$, and converges absolutely for $\alpha\geqslant 0$.

41. Suppose each $a_n\geqslant 0$ and Σa_n diverges. Show that $\Sigma a_n x^n\to\infty$ as $x\to 1^-$, assuming that 1 is the radius of convergence.

42. Let $f_n(x)=x(1+1/n)$, $\forall x\in\mathbb{R}$, and

$$g_n(x)=\begin{cases}1/n, & \text{if } x \text{ is irrational or } 0,\\ q+1/n & \text{if } x=p/q,\end{cases}$$

where p and q are integers with no common factor and $p\neq 0$, $q>0$. Show that $\{f_n\}$ and $\{g_n\}$ converge uniformly on any finite interval, but that $\{f_n g_n\}$ does not converge uniformly on any finite interval.

43. Show that $\{nx(1-x)^n\}$ converges pointwise, but not uniformly, on $[0,1]$, but that term by term integration leads to the correct result in this case.

44. Let A be 3×3 matrix of real numbers. Define e^A to be the matrix $E=\Sigma_{n=0}^{\infty}A^n/n!$, where A^0 is understood to be the 3×3 identity matrix. Can you show that E exists even though A may not be nilpotent?

45. Suppose $f(y)=\int_0^y x^k(y-x)^m\,dx$, m, n positive integers. What are the terms of the sequence $\{f^{(n)}(y)\}_{n=1}^{\infty}$, where $f^{(n)}(y)$ is the nth derivative of $f(y)$?

If you obtained

$$f^{(m)}(y)=\frac{m!}{k+1}y^{k+1}$$

for the mth derivative, note that $f^{(n)}(0)=0$, $\forall n$, so integrate back, and obtain

$$f(y)=\frac{m!n!}{(m+n+1)!}y^{m+n+1}.$$

46. Exhibit a series (nontrivial) whose convergence may be tested by Raabe's test. Next, find one for which one would use Gauss's test.

47. Prove that

$$\int_1^{\infty}\frac{\ln x}{x}\cos\pi x\,dx \qquad \text{and} \qquad \int_1^{\infty}\frac{\ln x}{x^2}dx$$

converge, and that

$$\sum_{n=1}^{\infty}\left[\frac{5+(-1)^n}{2\ln(x^2+1)}\right]^{-n}$$

converges for $|x|<\sqrt{e^2-1}$.

48. Show that

$$\prod_{k=1}^{\infty}\cos\frac{x}{2^k}=\frac{\sin x}{x}.$$

Integrate

$$\int_{\pi/6}^{\pi/2}\left[\sum_{k=1}^{\infty}\frac{1}{2^k}\tan\frac{x}{2^k}\right]dx.$$

49. Sum the series $\Sigma_{n=1}^{\infty}x^{4n-3}/(4n-3)$.

50. Show that for $|x|<1$

$$\frac{1}{1-x}=\sum_{n=1}^{\infty}\frac{2^{n-1}x^{2^{n-1}-1}}{1+x^{2^{n-1}}}.$$

51. Express $\int_0^1 x^x\,dx$ as a series.

52. Find a series representation for a solution $y(x)$ to the equation $y^3+xy=1$ (three terms are enough).
[*Hint*: Let $y=\Sigma_{n=0}^{\infty}a_n x^n$, and equate coefficients.]

53. Find a series for $y(x)$ if $y(x)$ satisfies $xy'=y(x+1)$. Do the same for

$$y'=y^3-x, \quad \text{where} \quad y(0)=1,$$

$$y'=x^2-y^2, \quad \text{with} \quad y(0)=0.$$

54. See if you can resolve the question posed at the end of Section 10.3.

55. Note that the proof of Theorem 10.2.12 is not complete. We still need to show that if $F>0$ is integrable (in the Lebesgue sense) over a measurable T, and that if f is a measurable function such that $|f|\leqslant F$, then f is integrable over T. Do this.
[*Hint*: Assume that $f>0$ and that $\pi=\{0=a_0, a_1,\ldots,a_n, a_{n+1}=\infty\}$ is an arbitrary partition of $[0,\infty]$. Denote $T\cap F^{-1}((a_i, a_{i+1}])$, $T\cap f^{-1}((a_i, a_{i+1}])$ by A_i, B_i, respectively. Using the fact that $\underline{S}(F,\pi,T)\leqslant\int_T F\,d\mu<\infty$, show that $\mu(\cup_{i=1}^{n}A_i)<\infty$.

Show that $\cup_{i=1}^{n}B_i\subset\cup_{i=1}^{n}A_i$ and that if $\varepsilon>0$, then for all sufficiently large a_n, $a_n\mu(B_n)\leqslant a_n\mu(A_n)<\varepsilon$. Can you show that F, and hence f, must be bounded on $\cup_{i=1}^{n}B_i$ except for a set of measure zero? Now deduce that $\int_{T\setminus B_0}f\,d\mu$ exists for the partition π.

Now let π' be a measurable countable partition of B_0. f is bounded on B_0, so consider the lower Darboux sum $\underline{S}(f,\pi')$. Must the net of sums converge? Conclude that $\int_T f\,d\mu$ exists and that $\int_T f\,d\mu\leqslant\int_T F\,d\mu$.]

56. A real valued function f with domain $D\subset R$ is said to be *piecewise continuous* on D if D can be partitioned into a finite number of parts $\{D_i\}$ such that for each D_i having a non-empty interior, f restricted to the interior of that set is continuous, and for every compact set K which meets a D_i the restriction of f to $K\cap D_i$ can be extended to a uniformly continuous function on the closure of this intersection. Show that if f and g are two piecewise continuous real-valued functions on $\mathbf{R}$ then $f+g$ and fg are also piecewise continuous, and if $h:[a, b]\to\mathbf{R}$ is piecewise continuous, then h can be uniformly approximated by a sequence of step-functions.

CHAPTER

11

APPLICATIONS OF IMPROPER INTEGRALS

We have talked about three basic integrals: the Stieltjes integral with respect to Jordan content, the Darboux integral with respect to a measure, and the Lebesgue integral. (The Riemann integral is simply the special Stieltjes integral having integrator $\alpha(x) \equiv x$.) The first two integrals were partially characterized by the restrictions that the integrands be essentially bounded and that the measures of the sets over which the integrations occur be finite. The fact that the sequence of numbers $\{S_n\}_{n=1}^{\infty}$ is bounded, where $S_n = \int_1^n x^{-2}\,dx$ and the integral is the Riemann or Darboux integral, suggests that we may be able to extend the theory of Darboux integration a bit.

For a function f to be Lebesgue integrable, it must be measurable. We know that if f is a measurable function, then f is Lebesgue integrable iff $|f|$ is Lebesgue integrable. However, a nonmeasurable function f might be such that $|f|$ is Darboux integrable, even if f is not. For example, let $f \equiv 1$ on a nonmeasurable subset of $[0,1]$ and $f \equiv -1$ on the complement of that set in $[0,1]$. Then $\int_0^1 |f|\,dx$ certainly exists, but $\int_0^1 f\,dx$ does not, nor do $\int f^+\,d\mu$, $\int_0^1 f^-\,d\mu$. On the other hand, if a function f is Darboux integrable, so are f^+, f^-, and $|f|$.

It is not hard to see that if f is a nonmeasurable function, then the Darboux integral of f will not exist since every measurable partition of the domain of integration will give rise to upper and lower Darboux sums which will differ by amounts bounded away from zero, no matter how fine the partition. Hence it seems fruitless to attempt to improve our Darboux theory so that the class of integrable functions will include nonmeasurable ones. However, we can try this: Let us see if we can extend the class to include some functions which are not essentially bounded on measurable sets, and, also, let us try to extend the class of sets over which we integrate to include those of infinite measure.

Recall that f is essentially bounded on a measurable set A if f is bounded on a subset $B \subset A$, with $\mu(A \setminus B) = 0$. Moreover, we have postulated that if f is unbounded on A and $\mu(A) = 0$, then $\int_A f\,d\mu = 0$.

11.1 IMPROPER INTEGRALS

As far as Lebesgue integration is concerned, there are no improper integrals in the sense that we use the term. The following definition applies to the Darboux integral.

Definition 11.1.1. The integral $\int_A f\,d\mu$ is said to be *improper* if $|\mu(A \setminus Z_f)| = \infty$, or if f is not essentially bounded on A.

Before we proceed, we want to make a few remarks. As we have said, if f is Darboux integrable, so are f^+, f^-, and $|f|$. Now, if $\int_A f\,d\mu$ is improper, we hope to get a value for this improper integral by "approaching" it in some sense. For example, we could take a sequence of compact regions $\{D_n\}$ such that $\{D_n \cap A\}$ converges in measure to A and see if $\int_{D_n \cap A} f\,d\mu$ converges to some real number. Or, we might take a sequence of functions $\{f_n\}$ converging uniformly to f on A and see if $\int_A f_n\,d\mu$ converges. Whatever method we might choose, it seems clear that we can simplify matters by considering only positive functions. This is what we shall do, and indeed it will turn out that this simplification is necessary. Then if we can establish values for the improper integrals for the positive and negative parts of a function f, we shall have $\int_A f\,d\mu = \int_A f^+\,d\mu - \int_A f^-\,d\mu$. However, by considering just positive functions, we shall be developing a theory that will have the same defect that Lebesgue theory has, namely, an improper integral $\int_A f\,d\mu$ can be assigned a value iff $\int_A |f|\,d\mu$ can be assigned a value. Corollary 10.2.16 ensures that we can assign a value to $\int_1^\infty x^{-1} \sin x\,dx$ but the sequence $\{\int_1^n x^{-1}|\sin x|\,dx\}_{n=1}^\infty \to \infty$.

If we can assign a value to an improper integral in a reasonable way, we say the improper integral converges to that value. The example just given points up the fact that we are setting out to develop a theory for the absolute convergence of improper integrals. As you see, some improper integrals will converge, not absolutely, but conditionally, and to identify them requires ingenuity.

Our next remark is this: Suppose f is a complex-valued function of a complex variable $z = x + iy$. Then $f(z)$ has a representation of the form $u(x, y) + iv(x, y)$ where u and v are real-valued functions. If γ is an arc, or a curve, in the z plane, the integral $\int_\gamma f(z)\,dz$ is defined in such a way as to be the same thing as

$$\int_\gamma [u(x, y) + iv(x, y)][dx + i\,dy],$$

so that

$$\int_\gamma f(z)\,dz=\left[\int_\gamma u(x,y)\,dx-v(x,y)\,dy\right]+i\left[\int_\gamma v(x,y)\,dx+u(x,y)\,dy\right],$$

and the integrals on the right-hand side are simply integrals of real 1-forms in E^2 over γ, with which we are familiar. Hence this complex integration reduces to real integrations of 1-forms in E^2.

Definition 11.1.2. Suppose $\int_A f\,d\mu$ is an improper integral, and $f|_A>0$, which is to say, $\forall x\in A$, $f(x)\geqslant 0$ and for at least one $x\in A$, $f(x)>0$. If there exists an increasing sequence of compact regions $\{D_n\}$ such that

1. $\forall n$, $\int_{A\cap D_n} f\,d\mu$ is a proper integral with a finite value,
2. $A\subset\bigcup_{n=1}^\infty D_n\cup N$, where N is a set of measure zero,
3. $\lim_{n\to\infty}\int_{A\cap D_n} f\,d\mu$ exists and is finite,

we say that the improper integral converges, and its value is that limit.

Theorem 11.1.3. *Suppose f is a positive function on a measurable set $A\subset E^n$, and the integral $\int_A f\,d\mu$ is improper. Suppose further that $\{D_n\}$ is an increasing sequence of compact regions with respect to which $\int_A f\,d\mu$ converges to a limit L. If $\{D_n'\}$ is any other increasing sequence of compact regions satisfying conditions* 1 *and* 2 *of Definition 11.1.2, then $\int_A f\,d\mu$ will converge with respect to this latter sequence to the same limit L. Thus Definition 11.1.2 is "well given," which is to say, the convergence of an improper integral is independent of the choice of a "suitable" sequence $\{D_n\}$.*

PROOF: By a suitable sequence we mean an increasing sequence of compact regions $\{D_n\}$ such that conditions 1 and 2 of Definition 11.1.2 are satisfied. Since each D_n of such a sequence is a compact region, $0<\mu(D_n)<\infty$. We observe that for any suitable sequence $\lim_{n\to\infty}\mu(D_n\cap A)=\mu(A)$, for we have

$$\begin{aligned}\mu(A)&=\mu\left(A\cap\left(\bigcup_{n=1}^{\infty}D_n\cup N\right)\right)\\&=\mu(A\cap N)+\mu\left(A\cap\bigcup_{n=1}^{\infty}D_n\right)\\&=\mu\left(A\cap\lim_n D_n\right)\\&=\lim_n\mu(D_n\cap A)\leqslant\mu(A).\end{aligned}$$

This proves $\mu(A\cap\bigcup D_n)=\lim_n\mu(D_n\cap A)=\mu(A)$ for any suitable sequence $\{D_n\}$.

It follows that if $\{D_n\},\{D_n'\}$ are two suitable sequences,

$$\lim_{m,n\to\infty} \mu([A\cap D_n]\triangle[A\cap D_m'])=0.$$

Suppose now that $\lim_n \int_{A\cap D_n} f\,d\mu = L$. For each n, m we have

$$\left|\int_{A\cap D_n} f\,d\mu - \int_{A\cap D_m'} f\,d\mu\right| = \left|\int_{A\cap(D_n\setminus D_m')} f\,d\mu - \int_{A\cap(D_m'\setminus D_n)} f\,d\mu\right|. \quad (11.1)$$

Since the expression on the right-hand side of (11.1) is bounded by $\int_{A\cap(D_n\setminus D_m')} f\,d\mu$, which in turn is bounded by L, and since $\mu(A\cap(D_n\setminus D_m'))\leqslant \mu([A\cap D_n]\triangle[A\cap D_m'])$, it follows that as $n, m\to\infty$, the right-hand side of (11.1) tends to zero. Hence so does the term on the left, and we have $\lim_m \int_{A\cap D_m'} f\,d\mu = L$. □

It is vitally important to note that in the above inequalities we have made use of the fact that f is positive. If f were positive on $A\cap(D_n\setminus D_m')$ and negative on $A\cap(D_m'\setminus D_n)$, the inequality

$$\left|\int_{A\cap(D_n\setminus D_m')} f\,d\mu - \int_{A\cap(D_m'\setminus D_n)} f\,d\mu\right| \leqslant \int_{A\cap(D_n\setminus D_m')} f\,d\mu$$

would not hold.

Consider the following integral, where Q is the first quadrant of the plane $\int_Q \sin(x^2+y^2)\,d\mu$. We cannot apply our criterion for convergence since $\sin(x^2+y^2)$ is not positive. However, $\int_Q|\sin(x^2+y^2)|\,d\mu$ does diverge to ∞, so if the original integral converges, it does so conditionally. A familiar transformation leads to

$$\int_Q \sin(x^2+y^2)\,d\mu = \lim_n \int_0^{\pi/2}\int_0^n r\sin r^2\,dr\,d\theta$$

$$= \lim_n \frac{\pi}{2}\left(-\frac{\cos r^2}{2}\right)\Bigg|_0^n = \lim_n \frac{\pi}{4}(1-\cos n^2),$$

and this limit does not exist, so we conclude that the original integral does not even converge conditionally.

Note that our sequence of compact sets $\{D_n\}$ in this example was the quarter-disks of radius n. Suppose we had taken $\{D_n\}$ to be the squares $[0,n]\times[0,n]$. Then

$$\iint_{D_n} \sin(x^2+y^2)\,dy\,dx = \iint_{D_n} \sin x^2\cos y^2\,dy\,dx + \iint_{D_n}\cos x^2\sin y^2\,dy\,dx$$

$$= 2\left(\int_0^n \sin x^2\,dx\right)\left(\int_0^n \cos x^2\,dx\right),$$

and as $n\to\infty$, this last product converges to $\pi/4$, a result one gets by using some tricky analysis. In any event, the Dirichlet test ensures that $\int_0^\infty \sin x^2\,dx = \int_0^\infty \frac{1}{2}u^{-1/2}\sin u\,du$ converges, as does $\int_0^\infty \cos x^2\,dx$ (see Problem 1). This example shows that our theory really is a theory of absolute convergence.

We must point out as we close this section that we can extend our theory for improper integrals to include functions which are not always positive if we modify our definition for convergence to read: an improper integral $\int_A f\,d\mu$ converges iff for every possible suitable sequence $\{D_n\}$, $\lim_n \int_{A\cap D_n} f\,d\mu$ exists and is the same finite value, but this criterion is extremely difficult to apply. We shall reserve it for identifying those special cases of conditionally but not absolutely convergent integrals.

11.2 SOME FURTHER CONVERGENCE THEOREMS

Suppose f is measurable and $\int_A f\,d\mu$ is an improper integral. Suppose further that g is a positive integrable function such that on the measurable set A, $|f|\leqslant g$, and the improper integral $\int_A g\,d\mu$ converges. Theorem 10.2.12 guarantees that in the Lebesgue sense the integrals $\int_A f\,d\mu$, $\int_A f^+\,d\mu$, and $\int_A |f|\,d\mu$ exist. As improper integrals, the convergence of $\int_A g\,d\mu$ guarantees the convergence of $\int_A |f|\,d\mu$, $\int_A f^+\,d\mu$, and $\int_A f^-\,d\mu$. Hence $\int_A f\,d\mu$ converges.

Theorem 11.2.1. *If g is a positive function on a measurable set A, and $\int_A g\,d\mu$ is an improper integral, then the convergence of $\int_A g\,d\mu$ ensures the convergence or existence of the integral $\int_A f\,d\mu$ for any measurable function f for which $|f|\leqslant g$ on A.*

PROOF: Let $\{D_n\}$ be a suitable sequence. Since $\int_A g\,d\mu$ converges, $\sum_{n=N}^\infty \int_{A\cap(D_{n+1}\setminus D_n)} g\,d\mu$ is finite and approaches zero as $N\to\infty$. The same must be true for $\sum_{n=N}^\infty \int_{A\cap(D_{n+1}\setminus D_n)} f^+\,d\mu$ and $\sum_{n=N}^\infty \int_{A\cap(D_{n+1}\setminus D_n)} f^-\,d\mu$. $\square$

This little theorem ensures that, just as in the case of series and products, absolute convergence implies convergence.

If $f\colon E^2\to\mathbb{R}$ is a real-valued function, the integral $\int_a^b f(x,y)\,dx$, if it exists, defines a function $g(y)$. When $f(x,y)$ is an integrable function of x, the improper integral $\int_a^\infty f(x,y)\,dx$, if it converges, also defines a function $g(y)$. Suppose this improper integral converges for each $y\in B\subset E^1$ to a function $g(y)$. We say that the convergence is *uniform* if for each $\varepsilon>0$, there exists an n_ε such that for all $n\geqslant n_\varepsilon$ and every $y\in B$ $|g(y)-\int_a^n f(x,y)\,dx|<\varepsilon$.

Theorem 11.2.2. *Suppose $\int_a^\infty f(t,x)\,dt$ converges uniformly for all x in some compact set $[c,d]$ and $f(t,x)$ is continuous on $[a,\infty)\times[c,d]$. Then $g(x)=\int_a^\infty f(t,x)\,dt$ is continuous on $[c,d]$.*

Now assume further than $D_2 f$ exists and is also continuous on $[a,\infty)\times[c,d]$ and, moreover, both the integrals:

$$\int_a^{\infty} |f(t,x)|\,dt \qquad \text{and} \qquad \int_a^{\infty} |D_2 f(t,x)|\,dt$$

converge uniformly on $[c,d]$. Then g is differentiable, and

$$g'(x)=\int_a^{\infty} D_2 f(t,x)\,dt.$$

PROOF: Note the similarity with the case of infinite series. The continuity of the integrand and the uniform convergence of the integral yield continuity of the limit $g(x)$. For any $x\in[c,d]$ we have

$$\begin{aligned}|g(x+h)-g(x)| &= \left|\int_a^{\infty}[f(t,x+h)-f(t,x)]\,dt\right| \\ &\leqslant \left|\int_a^{n}[f(t,x+h)-f(t,x)]\,dt\right| + \left|\int_n^{\infty} f(t,x+h)\,dt\right| \\ &\quad + \left|\int_n^{\infty} f(t,x)\,dt\right|.\end{aligned}$$

Let $\varepsilon>0$. Choose n so large that for any $y\in[c,d]$, $|\int_n^{\infty} f(t,y)\,dt|<\varepsilon/3$. Since $f(t,x)$ is uniformly continuous on $[a,n]\times[c,d]$, we have for all h sufficiently small, $|f(t,x+h)-f(t,x)|<\varepsilon/3(n-a)$. For all h sufficiently small, $|g(x+h)-g(x)|<\varepsilon$, so g is continuous on $[c,d]$, in fact uniformly so. Now we take on the further assumptions. We have that

$$\begin{aligned}&\left|\frac{g(x+h)-g(x)}{h}-\int_a^{\infty} D_2 f(t,x)\,dt\right| \\ &\quad\leqslant \int_a^{\infty}\left|\frac{f(t,x+h)-f(t,x)}{h}-D_2 f(t,x)\right|dt \\ &\quad= \int_a^{\infty}|D_2 f(t,c_t)-D_2 f(t,x)|\,dt,\end{aligned}$$

where c_t is between x and $x+h$, and this last expression is less than or equal to

$$\int_a^{n}|D_2 f(t,c_t)-D_2 f(t,x)|\,dt+\int_n^{\infty}|D_2 f(t,c_t)|\,dt+\int_n^{\infty}|D_2 f(t,x)|\,dt.$$

Because of the uniform convergence of $\int_a^{\infty}|D_2 f(t,x)|\,dt$, by taking n sufficiently large we can make each of the last two terms less than $\varepsilon/3$, and by the uniform continuity of $D_2 f(t,x)$ on $[a,n]\times[c,d]$, we can make the first term less than $\varepsilon/3$ for all h sufficiently small. We conclude that $g'(x)=\int_a^{\infty} D_2 f(t,x)\,dt$. □

Theorem 11.2.3. *Suppose $f(t,x)$ is defined and continuous on $[a,\infty)\times[c,d]$ and that $\int_a^\infty f(t,x)\,dt$ converges uniformly on $[c,d]$. Then*

$$\int_c^d\left[\int_a^\infty f(t,x)\,dt\right]dx=\int_a^\infty\left[\int_c^d f(t,x)\,dx\right]dt.$$

PROOF: Let $\varepsilon>0$. The uniform convergence of the integral on $[c,d]$ means that there is an n_0 such that for all $n\geqslant n_0$ and each $x\in[c,d]$, $|\int_n^\infty f(t,x)\cdot dt|<\varepsilon/(d-c)$. From the previous theorem, $g(x)=\int_a^\infty f(t,x)\,dt$ is continuous on $[c,d]$, so the integral $\int_c^d g(x)\,dx$ exists. Hence

$$\left|\int_c^d\int_a^n f(t,x)\,dt\,dx-\int_c^d\int_a^\infty f(t,x)\,dt\,dx\right|<\varepsilon.$$

But for each n,

$$\int_c^d\int_a^n f(t,x)\,dt\,dx=\int_a^n\int_c^d f(t,x)\,dx\,dt,$$

so

$$\lim_n\int_a^n\int_c^d f(t,x)\,dx\,dt=\int_a^\infty\int_c^d f(t,x)\,dx\,dt=\int_c^d\int_a^\infty f(t,x)\,dt\,dx.\quad\square$$

This theorem allows us to improve Theorem 11.2.2 slightly.

Theorem 11.2.4. *Let $f(t,x)$ be continuous on $[a,\infty)\times[c,d]$, and assume that $D_2 f$ is also continuous on this region. Assume further that $\int_a^\infty D_2 f(t,x)\,dt$ converges uniformly on $[c,d]$ and that $g(x)=\int_a^\infty f(t,x)\,dt$ converges for each $x\in[c,d]$. Then g is differentiable, and $g'(x)=\int_a^\infty D_2 f(t,x)\,dt$.*

PROOF: Note that we have assumed neither the absolute convergence of $g(x)$ or $\int_a^\infty D_2 f(t,x)\,dt$, nor the uniform convergence of $g(x)$ on $[c,d]$. By the previous theorem we can write

$$\int_c^x\int_a^\infty D_2 f(t,v)\,dt\,dv=\int_a^\infty\int_c^x D_2 f(t,v)\,dv\,dt$$

$$=\int_a^\infty[f(t,x)-f(t,c)]\,dt=g(x)-g(c).$$

This means that $g'(x)=\int_a^\infty D_2 f(t,x)\,dt$. $\square$

We are now ready to give the analog of Fubini's theorem, the improper version of the simple form.

Theorem 11.2.5. *Let $f(t,x)$ be continuous on $[a,\infty)\times[c,\infty)$, and assume that the integrals $\int_a^\infty|f(t,x)|\,dt$ and $\int_c^\infty|f(t,x)|\,dx$ converge uniformly in every* *compact interval* $[c, d]$ *and* $[a, b]$, *respectively. Then*

$$\int_c^\infty \int_a^\infty f(t, x)\, dt\, dx = \int_a^\infty \int_c^\infty f(t, x)\, dx\, dt = \iint_{[a,\infty)\times[c,\infty)} f\, d\mu.$$

PROOF: If any one of these three integrals is infinite, then they all are. Assume that f is positive and that $\int_c^\infty \int_a^\infty f(t, x)\, dt\, dx$ converges. Then for any $b \geqslant a$

$$\int_a^b \int_c^\infty f(t, x)\, dx\, dt = \int_c^\infty \int_a^b f(t, x)\, dt\, dx \leqslant \int_c^\infty \int_a^\infty f(t, x)\, dt\, dx < \infty.$$

Since each integral is positive, $\lim_{b\to\infty} \int_a^b \int_c^\infty f(t, x)\, dx\, dt$ exists and we have

$$\int_a^\infty \int_c^\infty f(t, x)\, dx\, dt \leqslant \int_c^\infty \int_a^\infty f(t, x)\, dt\, dx.$$

Having established the convergence of $\int_a^\infty \int_c^\infty f(t, x)\, dx\, dt$ we can now write, for any $d \geqslant c$,

$$\int_c^d \int_a^\infty f(t, x)\, dt\, dx = \int_a^\infty \int_c^d f(t, c)\, dx\, dt \leqslant \int_a^\infty \int_c^\infty f(t, x)\, dx\, dt < \infty,$$

and thus

$$\lim_{d\to\infty} \int_c^d \int_a^\infty f(t, x)\, dt\, dx \leqslant \int_a^\infty \int_c^\infty f(t, x)\, dx\, dt.$$

The two inequalities establish the equality.

It follows almost immediately that under the hypothesis of the theorem

$$\int_c^\infty \int_a^\infty f(t, x)\, dt\, dx = \int_a^\infty \int_c^\infty f(t, x)\, dx\, dt$$

for a continuous function $f(t, x)$, not necessarily positive, since the equality holds for the integrands f^+ and f^-.

Since for each b, d, we have

$$\iint_{[a,b]\times[c,d]} f(t, x)\, d\mu = \int_a^b \int_c^d f(t, x)\, dx\, dt,$$

it follows from the more general form of Fubini's theorem that

$$\iint_{[a,\infty)\times[c,\infty)} f\, d\mu = \int_a^\infty \int_c^\infty f(t, x)\, dx\, dt. \quad \square$$

Consider the improper integral $\int_0^\infty t^{-1} \sin t\, e^{-tx}\, dt$. The discontinuity at zero is removable since we can define the integrand to be equal to 1 at this point. Let

$$g(x) = \int_0^\infty \frac{\sin t}{t} e^{-tx}\, dt,$$

and let $[0, d]$ be a compact interval over which x ranges. We investigate the convergence of $g(x)$ on this interval. If $x=0$, the integral converges; this we know by the Dirichlet test. However, $g(0)$ does not converge absolutely, a fact we ask the reader to prove. If $x \geqslant c > 0$,

$$\int_0^\infty \left|\frac{\sin t}{t}\right| e^{-tx}\,dt \leqslant \int_0^\infty e^{-tc}\,dt = \frac{1}{c} < \infty,$$

so the integral $g(x)$ does converge absolutely on $[c, d]$ for each $c>0$. Moreover, the convergence is uniform, for let $\varepsilon>0$. Choose n_0 so large that $(ce^{n_0 c})^{-1} < \varepsilon$. Then for all $n \geqslant n_0$

$$\left|\int_n^\infty \frac{\sin t}{t} e^{-tx}\,dt\right| \leqslant \int_n^\infty e^{-tx}\,dt \leqslant (ce^{nc})^{-1} < \varepsilon.$$

Now

$$D_2\left(\frac{\sin t}{t} e^{-tx}\right) = (-\sin t)e^{-tx},$$

and this function is continuous on $[0, \infty)\times[0, d]$. Moreover, $\int_0^\infty(-\sin t)\cdot e^{-tx}\,dt$ also converges uniformly on $[c, d]$, so $g(x)$ is differentiable on $[c, d]$, and

$$g'(x) = -\int_0^\infty (\sin t)e^{-tx}\,dt = -\frac{1}{1+x^2}.$$

Therefore, $g(x) = -\arctan x + \text{const}$ on the interval $[c, d]$. Evidently, $g(x)$ is continuous on $[c, d]$ for any $c>0$, so it appears that as $x \to 0^+$, $g(x)$ approaches that constant. If we let $x \to \infty$, we have

$$|g(x)| = \left|\int_0^\infty \frac{\sin t}{t} e^{-tx}\,dt\right| \leqslant \int_0^\infty e^{-tx}\,dt = \frac{1}{x},$$

so $\lim_{x\to\infty} g(x) = 0$. This leads us to conclude that the constant must be equal to $\pi/2$. Hence for all $x>0$ we have $g(x) = \pi/2 - \arctan x$, and $\lim_{x\to 0^+} g(x) = \pi/2$. This means that the value to which $\int_0^\infty t^{-1}\sin t\,dt$ converges must be $\pi/2$. Furthermore, for any $b>0$ we have

$$\int_0^\infty \frac{\sin bt}{t}\,dt = \int_0^\infty \frac{\sin bt}{bt}\,d(bt) = \int_0^\infty \frac{\sin \tau}{\tau}\,d\tau = \frac{\pi}{2}.$$

11.3 SOME SPECIAL FUNCTIONS

Definition 11.3.1. The *gamma function* is defined by

$$\Gamma(x) = \int_0^\infty u^{x-1}e^{-u}\,du.$$

Since this improper integral converges (absolutely) for all $x>0$ and the convergence is uniform on any compact x interval $[\delta, L]$, $\delta>0$, $L \geqslant \delta$, $\Gamma(x)$

is continuous on $[\delta, \infty)$. Repeated integration by parts yields

$$\Gamma(n+1)=\int_0^\infty u^n e^{-u}\,du=n!$$

for each positive integer n, so the gamma function gives the values of the factorials of positive integers. In fact, the formula suggests that 0! should be defined to be 1. In general,

$$\Gamma(x+1)=\int_0^\infty u^x e^{-u}\,du=-u^x e^{-u}\Big|_0^\infty+\int_0^\infty e^{-u}\,d(u^x)$$
$$=0+x\int_0^\infty u^{x-1}e^{-u}\,du=x\Gamma(x).$$

The functional equation $\Gamma(x+1)=x\Gamma(x)$ allows us to compute $\Gamma(x)$ on the whole ray $(0,\infty)$, provided we know the values on one unit interval $[k, k+1)$. $\Gamma(1)=1$. Since $\Gamma(x)=x^{-1}\Gamma(x+1)$, we see that as $x\to 0^+$, $\Gamma(x)$ must approach $+\infty$, and as $x\to 0^-$, $\Gamma(x)$ must approach $-\infty$. Thus

$$\Gamma(0^+)=+\infty,$$
$$\Gamma(0^-)=-\infty,$$
$$\Gamma(-1^+)=-\Gamma(0^+)=-\infty,$$
$$\Gamma(-1^-)=-\Gamma(0^-)=+\infty,$$
$$\Gamma(-2^+)=-\tfrac{1}{2}\Gamma(-1^+)=+\infty,$$
$$\Gamma(-2^-)=-\tfrac{1}{2}\Gamma(-1^-)=-\infty,$$

and so forth.

The formula $\Gamma(x)=\int_0^\infty e^{-u}u^{x-1}\,du$ can be altered by making all sorts of changes of variables; for example, set

$u=t^2$: $\quad \Gamma(x)=\int_0^\infty t^{2x-2}e^{-t^2}\,d(t^2)=2\int_0^\infty t^{2x-1}e^{-t^2}\,dt;$

$u=t^\alpha$: $\quad \Gamma(x)=\alpha\int_0^\infty t^{\alpha x-1}e^{-t^\alpha}\,dt;$

$u=\alpha t$: $\quad \Gamma(x)=\alpha^x\int_0^\infty t^{x-1}e^{-\alpha t}\,dt;$

$u=-\ln t$: $\quad \Gamma(x)=\int_1^0 \ln\left(\frac{1}{t}\right)^{x-1} t\,d\left(\ln\frac{1}{t}\right)=\int_0^1\left(\ln\frac{1}{t}\right)^{x-1}dt;$

and so forth. Using the first of these, we obtain $\Gamma(\frac{1}{2})=2\int_0^\infty e^{-t^2}\,dt$. Since

$$\int_0^\infty\int_0^\infty e^{-(x^2+y^2)}\,dx\,dy=\int_0^{\pi/2}\int_0^\infty e^{-r^2}r\,dr\,d\theta=\left[-\tfrac{1}{2}e^{-r^2}\right]\Big|_0^\infty\frac{\pi}{2}$$
$$=\frac{\pi}{4}=\left(\int_0^\infty e^{-x^2}\,dx\right)\left(\int_0^\infty e^{-y^2}\,dy\right),$$

we conclude that $\Gamma(\frac{1}{2})=2\sqrt{\pi}/2=\sqrt{\pi}$. It follows that $\Gamma(\frac{3}{2})=\frac{1}{2}\Gamma(\frac{1}{2})=\sqrt{\pi}/2$, and in general,

$$\Gamma\left(n+\frac{1}{2}\right)=\frac{(2n)!\sqrt{\pi}}{4^n n!}, \qquad n\geqslant 0.$$

Working backwards, $\Gamma(-\frac{1}{2})=(-\frac{1}{2})^{-1}\Gamma(\frac{1}{2})=-2\sqrt{\pi}$ and $\Gamma(-\frac{3}{2})=(-\frac{3}{2})^{-1}\cdot\Gamma(-\frac{1}{2})=\frac{4}{3}\sqrt{\pi}$, and so forth.

We leave it to the reader to verify that $\Gamma(x)$ is continuous except at the nonpositive integers; in fact, it is a C^∞-function except at these points. A table of values for $\Gamma(x)$, $1\leqslant x\leqslant 2$, is usually available.

Definition 11.3.2. Let p, q be positive numbers. The function

$$B(p,q)=\int_0^1 x^{p-1}(1-x)^{q-1}\,dx$$

of two positive real variables is called the *beta function*.

Its importance lies partially in the fact that some integrals which are apparently intractable can be manipulated into this form. For example,

$$\int_0^1\frac{dx}{\sqrt{1-x^4}}=\int_0^1\frac{d(u^{1/4})}{(1-u)^{1/2}}=\tfrac{1}{4}\int_0^1 u^{-3/4}(1-u)^{-1/2}\,du$$

$$=\tfrac{1}{4}\int_0^1 u^{1/4-1}(1-u)^{1/2-1}\,du=\tfrac{1}{4}B(\tfrac{1}{4},\tfrac{1}{2}),$$

and

$$\int_0^{\pi/2}\sqrt{\sin\theta}\,d\theta=\int_0^1\sqrt{\sqrt{v}}\;d\left(\arcsin\sqrt{v}\right)=\tfrac{1}{2}\int_0^1\frac{v^{1/4}}{v^{1/2}(1-v)^{1/2}}\,dv$$

$$=\tfrac{1}{2}\int_0^1 v^{(3/4)-1}(1-v)^{(1/2)-1}\,dv=\tfrac{1}{2}B(\tfrac{3}{4},\tfrac{1}{2}).$$

To evaluate $B(p,q)$ is a tiresome job, but note $\Gamma(p)=2\int_0^\infty y^{2p-1}e^{-y^2}\,dy$ implies

$$\Gamma(p)\Gamma(q)=4\left(\int_0^\infty y^{2p-1}e^{-y^2}\,dy\right)\left(\int_0^\infty x^{2q-1}e^{-x^2}\,dx\right)$$

$$=4\int_0^\infty\int_0^\infty y^{2p-1}x^{2q-1}e^{-(x^2+y^2)}\,dx\,dy.$$

The integrand is positive; a suitable sequence of increasing compact sets are the quarter-disks, and we find that the integral does converge.

$$\Gamma(p)\Gamma(q)=4\int_0^{\pi/2}\int_0^\infty (r\cos\theta)^{2q-1}(r\sin\theta)^{2p-1}e^{-r^2}r\,dr\,d\theta$$

$$=4\left(\int_0^\infty r^{2p+2q-1}e^{-r^2}\,dr\right)\left(\int_0^{\pi/2}(\cos\theta)^{2q-1}(\sin\theta)^{2p-1}\,d\theta\right).$$

Transform the first integral by changing the variable r to $\sqrt{u}$ and the second by putting $v = \sin^2\theta$ to get

$$\Gamma(p)\Gamma(q) = 4\left(\int_0^\infty \frac{u^{p+q-1/2}e^{-u}}{2u^{1/2}}\,du\right)\left(\int_0^1 \frac{(1-v)^{q-1/2}v^{p-1/2}}{2v^{1/2}(1-v)^{1/2}}\,dv\right)$$

$$= \left(\int_0^\infty u^{p+q-1}e^{-u}\,du\right)\left(\int_0^1 v^{p-1}(1-v)^{q-1}\,dv\right)$$

$$= \Gamma(p+q)B(p,q).$$

Thus we have

$$B(p,q) \equiv \frac{\Gamma(p)\Gamma(q)}{\Gamma(p+q)}.$$

To compute $\int_0^1 x^3(1-x^3)^{-1/2}\,dx$, put $u = x^3$, and get

$$\int_0^1 x^3(1-x^3)^{-1/2}\,dx = \tfrac{1}{3}\int_0^1 u^{(4/3)-1}(1-u)^{(1/2)-1}\,du = \tfrac{1}{3}B\left(\tfrac{4}{3},\tfrac{1}{2}\right)$$

$$= \tfrac{1}{3}\Gamma\left(\tfrac{4}{3}\right)\Gamma\left(\tfrac{1}{2}\right)/\Gamma\left(\tfrac{11}{6}\right) = \left(\sqrt{\pi}/3\right)\Gamma\left(\tfrac{4}{3}\right)/\Gamma\left(\tfrac{11}{6}\right).$$

A deeper study of the gamma function would yield the interesting result that for nonintegral values of x

$$\Gamma(x)\Gamma(1-x) \equiv \pi/\sin\pi x.$$

Hence we can evaluate the following integral by substituting $t(1-t)^{-1}$ for u:

$$\int_0^\infty \frac{u^{x-1}}{1+u}\,du = \int_0^1 t^{x-1}(1-t)^{(1-x)-1}\,dt = B(x,1-x)$$

$$= \frac{\Gamma(x)\Gamma(1-x)}{\Gamma(1)} = \frac{\pi}{\sin\pi x}.$$

Definition 11.3.3. For integers n, k, with $0 \leqslant k \leqslant n$, we define the *binomial coefficient*

$$\binom{n}{k} = \frac{n!}{k!(n-k)!} = \frac{n(n-1)\cdots(n-k+1)}{k!}.$$

This expression suggests the function

$$b(x,y) = \binom{x}{y} = \frac{\Gamma(x+1)}{\Gamma(y+1)\Gamma(x-y+1)},$$

defined for all x, y such that the arguments of the gamma functions are not nonpositive integers.

Analogously, we can generalize to the *multinomial coefficient*

$$\binom{n}{n_1, n_2, \ldots, n_k} = \frac{n!}{(n_1!)(n_2!)\cdots(n_k!)}, \qquad \sum_{i=1}^{k} n_i = n,$$

obtaining the function

$$m(x_1, \ldots, x_k) = \frac{\Gamma\left(1+\sum_{i=1}^{k} x_i\right)}{\Gamma(x_1+1)\Gamma(x_2+1)\cdots\Gamma(x_k+1)},$$

defined when all the factors are defined.

11.4 DIRAC SEQUENCES AND CONVOLUTIONS

Definition 11.4.1. Suppose we have a sequence of real functions $\{K_n\}_{n=1}^{\infty}$ defined over $\mathbb{R}$ which satisfy the following:

1. For each n, $K_n(x) \geqslant 0$.
2. Each K_n is piecewise continuous on every compact interval, and

$$\int_{-\infty}^{\infty} K_n(x)\,dx = 1.$$

3. Given any $\varepsilon > 0$, $\delta > 0$, there is an index N, such that, for all $n \geqslant N$, $\int_{-\infty}^{-\delta} K_n(x)\,dx + \int_{\delta}^{\infty} K_n(x)\,dx < \varepsilon$.

Such a sequence $\{K_n\}$ is called a *Dirac sequence*. For convenience, we usually arrange matters whenever possible so that each K_n is an even function.

Definition 11.4.2. Let f, g be piecewise continuous, bounded functions (at most a finite number of discontinuities). The *convolution* of f and g, denoted $f*g$, is the function defined by the improper integral

$$(f*g)(x) = \int_{-\infty}^{\infty} f(t)g(x-t)\,dt$$

if this integral converges.

Note that if t is replaced by $x-t$, we have

$$\int_{\infty}^{-\infty} f(x-t)g(t)\,d(x-t) = \int_{-\infty}^{\infty} g(t)f(x-t)\,dt = (g*f)(x),$$

so the operation of convolution is commutative. That it is also associative can be shown rather easily:

$$\begin{aligned}[(f*g)*h](y) &= \int_{-\infty}^{\infty}\left[\int_{-\infty}^{\infty} f(t)g(x-t)\,dt\right]h(y-x)\,dx \\ &= \int_{-\infty}^{\infty}\int_{\infty}^{-\infty} f(y-t)g(x-[y-t])h(y-x)\,d(y-t)\,dx \\ &= \int_{\infty}^{-\infty}\int_{\infty}^{-\infty} f(y-t)g(t-[y-x]) \\ &\quad \times h(y-x)\,d(y-x)\,d(y-t) \\ &= \int_{-\infty}^{\infty} f(y-t)(g*h)(t)\,d(y-t) = [f*(g*h)](y).\end{aligned}$$

Furthermore, $(f+g)*h=(f*h)+(g*h)$.

If we have a Dirac sequence $\{K_n\}_{n=1}^{\infty}$ and f is a bounded piecewise continuous function, we can form the convolutions

$$(f*K_n)(x)=\int_{-\infty}^{\infty} f(t)K_n(x-t)\,dt = f_n(x)$$

to get a sequence $\{f_n\}_{n=1}^{\infty}$.

Theorem 11.4.3. *Let f be a bounded continuous real function on $\mathbb{R}$ and $\{K_n\}$ a Dirac sequence, and let $f_n = f*K_n$. If S is a compact set, then $f_n \to f$ uniformly on S.*

PROOF: Since convolution is commutative, we can write

$$f_n(x)=\int_{-\infty}^{\infty} f(x-t)K_n(t)\,dt = (K_n*f)(x).$$

Now $f(x)=f(x)\int_{-\infty}^{\infty}K_n(t)\,dt=\int_{-\infty}^{\infty}f(x)K_n(t)\,dt$, so

$$f_n(x)-f(x)=\int_{-\infty}^{\infty}[f(x-t)-f(x)]K_n(t)\,dt.$$

Let $\varepsilon>0$. Choose $\delta>0$ so that for any $x\in S$, if $|t|<\delta$, then $|f(x-t)-f(x)|<\varepsilon/2$. This we can do since f is uniformly continuous on any closed interval containing S. Let M be an upper bound for $|f(x)|$ on $\mathbb{R}$. Choose N large enough so that for all $n\geqslant N$, $\int_{-\infty}^{-\delta}K_n(t)\,dt+\int_{\delta}^{\infty}K_n(t)\,dt<\varepsilon/8M$. Then

$$|f_n(x)-f(x)|\leqslant\int_{-\infty}^{-\delta}2MK_n(t)\,dt+\int_{\delta}^{\infty}2MK_n(t)\,dt+\int_{-\delta}^{\delta}\frac{\varepsilon}{2}K_n(t)\,dt<\varepsilon.$$

Since we have shown this inequality to be independent of x, and dependent only on n, the convergence $f_n\to f$ is uniform on S. □

Do such functions K_n exist? This is to ask, can we really find any Dirac sequences by which we can transform a function f into functions f_n which will approximate f uniformly? Well, let's see. Let $K(t)=[\pi(1+t^2)]^{-1}$. Then $\int_{-\infty}^{\infty}K(t)\,dt=(1/\pi)\arctan t|_{-\infty}^{\infty}=1$. Define $K_n(t)=nK(nt)$ for each $n=$

1,2,3,.... Certainly, $K_n(t)$ is positive, continuous, and

$$\frac{n}{\pi}\int_{-\infty}^{\infty}\frac{1}{1+n^2t^2}\,dt=\frac{1}{\pi}\arctan(nt)\Big|_{-\infty}^{\infty}=1.$$

To see if condition 3 of Definition 11.4.1 is satisfied, let $\varepsilon>0$, $\delta>0$ be given. Then

$$\frac{n}{\pi}\int_{\delta}^{\infty}\frac{1}{1+n^2t^2}\,dt=\frac{1}{2}-\frac{1}{\pi}\arctan(n\delta).$$

Certainly for all n sufficiently large, $(1/\pi)\arctan n\delta$ will be within $\varepsilon/2$ of $\frac{1}{2}$, so $\{K_n\}$ is a Dirac sequence.

In fact, if $K(t)$ is any positive continuous function which has compact support containing the origin and if $\int_{-\infty}^{\infty}K(t)\,dt=1$, then $\{K_n(t)\}=\{nK(nt)\}$ is a Dirac sequence.

As an application of all this we present a very famous theorem known as the Weierstrass approximation theorem.

Theorem 11.4.4 (Weierstrass). *Let $[a,b]$ be a compact interval of $\mathbb{R}$, and f a continuous real function on $[a,b]$. Then f can be approximated uniformly by polynomials on $[a,b]$.*

We already know how to expand f as a Taylor series if f is suitably differentiable, and hence a Taylor polynomial expanded about $(a+b)/2$ is one possible polynomial approximation. Another we have seen comes from the Lagrange interpolation formula, and a third from the Legendre polynomials.

PROOF: If we set $u=(x-a)/(b-a)$, then $x=(b-a)u+a$, and $0\leqslant u\leqslant 1$. Thus the function $f(x)$ on $[a,b]$ can be transformed to a continuous function $g(u)$ on $[0,1]$. If we can find a polynomial $P(u)$ approximating $g(u)$ uniformly on $[0,1]$, then $P((x-a)/(b-a))$ will be a polynomial in x approximating $f(x)$ uniformly on $[a,b]$. Moreover, we shall assume $f(x)\equiv 0$ outside $[a,b]$, and likewise the transform $g(u)\equiv 0$ outside $[0,1]$. We construct a Dirac sequence by the formula

$$K_n(t)=\begin{cases}(1-t^2)^n/C_n, & t\in[-1,1],\\ 0, & t\notin[-1,1],\end{cases}$$

where the constants $C_n=\int_{-1}^{1}(1-t^2)^n\,dt$. Each $K_n(t)$ is a positive even function, continuous on $\mathbb{R}$, with $\int_{-\infty}^{\infty}K_n(t)\,dt=1$. We should verify that condition 3 of Definition 11.4.1 is satisfied. Note first that

$$\frac{C_n}{2}=\int_0^1(1-t^2)^n\,dt=\int_0^1(1+t)^n(1-t)^n\,dt\geqslant\int_0^1(1-t)^n\,dt=\frac{1}{n+1},$$

so $C_n \geqslant 2/(n+1)$. Given $\delta > 0$ we have

$$\int_\delta^1 K_n(t)\,dt = \int_\delta^1 C_n^{-1}(1-t^2)^n\,dt \leqslant \frac{n+1}{2}\int_\delta^1 (1-\delta^2)^n\,dt$$
$$\leqslant \frac{n+1}{2}(1-\delta^2)^n(1-\delta).$$

Since $1-\delta^2<1$, this last expression approaches zero as $n\to\infty$, so condition 3 is indeed satisfied, and $\{K_n\}$ is a Dirac sequence.

Now put

$$g_n(u) = \int_{-\infty}^{\infty} g(t)K_n(u-t)\,dt = \int_0^1 g(t)K_n(u-t)\,dt$$

since g vanishes outside $[0,1]$ and $|u-t|\leqslant 1$. It remains only to show that $g_n(u)$ is a polynomial in u. Now K_n is a polynomial in its argument, so $K_n(u-t)$ can be written as a polynomial in t and u like this: $K_n(u-t)= P_0(t)+P_1(t)u+\cdots+P_{2n}(t)u^{2n}$, where the P_i are polynomials in t. Then

$$g_n(u) = \sum_{i=0}^{2n}\int_0^1 g(t)P_i(t)u^i\,dt = \sum_{i=0}^{2n} a_i u^i,$$

where each $a_i = \int_0^1 g(t)P_i(t)\,dt$. Finally, put $f_n(x) = g_n[(x-a)/(b-a)]$ and $f_n(x)$ will be the approximating polynomial for $f(x)$ on $[a,b]$. □

In Chapter 7 we exhibited three quadratic approximations for the function $f(x)=2^x+1$ on $[-1,1]$. It should be a pleasant diversion to use the idea introduced in the proof of the Weierstrass approximation theorem to find yet another quadratic which will approximate (rather roughly) the function $f(x)$. If $f(x)=2^x+1$ on $[-1,1]$, we can extend continuously this function to all of $\mathbb{R}$ so that the extension will be uniformly bounded by defining $f(x)=0$ for $|x|\geqslant 1+\delta$, $\delta>0$ arbitrarily small, and then defining f to be linear over the gaps $1<|x|<1+\delta$. By taking $\delta>0$ arbitrarily small, the integrals $\int_{-\infty}^{\infty} f(x)\,dx$ and $\int_{-1}^{1} f(x)\,dx$ are virtually equal.

Let $K(t)$ be the first function of the Dirac sequence defined above:

$$K(t) = \begin{cases} \frac{3}{4}(1-t^2), & |t|\leqslant 1,\\ 0 & \text{otherwise.}\end{cases}$$

We might just as well assume that $f(x)=0$ off $[-1,1]$; transform $f(x)$ to a function $g(u)$ on $[0,1]$ by putting $x=2u-1$:

$$g(u) = \begin{cases} 2^{2u-1}+1 = (\frac{1}{2})4^u+1, & 0\leqslant u\leqslant 1,\\ 0 & \text{otherwise.}\end{cases}$$

Since $|u-t|\leqslant 1$ when $0\leqslant u\leqslant 1$ and $0\leqslant t\leqslant 1$, we can write

$$\begin{aligned} g_1(u)&=(g*K)(u)=\int_{-\infty}^{\infty} g(t)K(u-t)\,dt=\tfrac{3}{4}\int_0^1 g(t)\left(1-[u-t]^2\right)dt\\ &=\tfrac{3}{4}\int_0^1 (4^t/2)(1-u^2+2ut-t^2)\,dt+\tfrac{3}{4}\int_0^1 (1-u^2+2ut-t^2)\,dt\\ &=\frac{9}{8\ln 4}-\frac{9u^2}{8\ln 4}+\frac{3u}{\ln 4}-\frac{9u}{4(\ln 4)^2}-\frac{3}{2\ln 4}+\frac{3}{(\ln 4)^2}\\ &\quad-\frac{9}{4(\ln 4)^3}-\frac{3u^2}{4}+\frac{3u}{4}+\frac{1}{2}\\ &=-\left(\frac{9}{16\ln 2}+\frac{3}{4}\right)u^2+\left(\frac{3}{4}-\frac{9}{16(\ln 2)^2}+\frac{3}{2\ln 2}\right)u\\ &\quad+\left(\frac{1}{2}-\frac{9}{32(\ln 2)^3}+\frac{3}{4(\ln 2)^2}-\frac{3}{16\ln 2}\right)\\ &=-Au^2+Bu+C, \end{aligned}$$

where $A\doteq 1.5616$, $B\doteq 1.7433$, and $C\doteq 0.9461$.

Transforming $g_1(u)$ back to a polynomial in x by replacing u with $(x+1)/2$, we get

$$P(x)=-\tfrac{1}{4}Ax^2+\tfrac{1}{2}(B-A)x+\left(C+\tfrac{1}{2}B-\tfrac{1}{4}A\right),$$

or

$$P(x)\doteq -0.3904x^2+0.0909x+1.4274.$$

As we said, this is a rather rough approximation; had we used $K_3(t)$ or $K_7(t)$, the polynomials of higher degree would of course have been much better approximations.

Theorem 11.4.3 ensures that we can approximate $f(x)$ by functions $f_n(x)=(f*K_n)(x)$ on compact sets. Since we can consider $f(x)$ to be continuous and bounded on $\mathbb{R}$ and equal to 0 for $|x|\geqslant 1+\delta$, $\delta>0$ arbitrarily small, we can write

$$\int_{-\infty}^{\infty} f(t)K(x-t)\,dt=\int_{-1}^{1} f(t)K(x-t)\,dt,$$

and

$$\int_{-\infty}^{\infty} K(t)f(x-t)\,dt=\int_{-1}^{1} K(t)f(x-t)\,dt$$

since both $K(t)$ and $f(t)$ vanish off $[-1,1]$. Since convolution is commutative, these four integrals are equal. To integrate $\int_{-1}^{1} f(t)K(x-t)\,dt$ is not simple, because the function $1-(x-t)^2$ has to be redefined as 0 when

$t<x-1$ or $t>x+1$ and $f(t)$ must be redefined as 0 when $t<-1$ or $t>1$. However, we can integrate $\int_{-1}^{1} f(x-t)K(t)\,dt$ without dealing with the problem of redefining $f(x-t)$ whenever $|x-t|>1$ since $f(x-t)$ may be written $2^x 2^{-t}+1=2^x f(t)2^{-t}-(2^x-1)$, and this is a continuous function for all $t\in[-1,1]$ and each fixed x. Thus

$$\tfrac{3}{4}\int_{-1}^{1} (2^{x-t}+1)(1-t^2)\,dt=(1.049)2^x+1$$

by elementary integration by parts, and this function is clearly a fairly good approximation for 2^x+1. Unfortunately, it is not a polynomial; but we already have obtained a "best approximation" for 2^x+1 on the span of $\{1, x, x^2\}$, namely,

$$2^x+1\approx 0.249x^2+0.727x+1.999,$$

so we can write

$$\begin{aligned}(K*f)x &\approx (1.049)(0.249x^2+0.727x+0.999)+1\\ &=0.261x^2+0.763x+2.048.\end{aligned}$$

This concludes our relationship with the function 2^x+1 over $[-1,1]$; may it rest in peace. If we denote the five approximating polynomials by $\{P_i\}_{i=1}^{5}$, these quadratics and $f(x)=2^x+1$ all belong to the linear space of continuous functions on $[-1,1]$. Norm this space with sup norm, and find a beginning calculus student who will tell you for which $i=1,2,\dots,5$ $\|f-P_i\|$ is greatest.

If $K(x,t)$ is a function such that $K(x,t)\to 0$ as $|t|\to\infty$ and if the approach to zero is rapid enough, then for a whole class of functions $\{f(t)\}$ the improper integral $\int_{-\infty}^{\infty} f(t)K(x,t)\,dt$ will converge. This integral, if it converges, of course defines a function of x, and we write

$$\hat{f}(x)=\int_{-\infty}^{\infty} f(t)K(x,t)\,dt.$$

$\hat{f}$ is called a *transform* of f, and the function K is called the *kernel* of the transformation. For example, the functions K_n that make up the Dirac sequences we have talked about are kernels of the transformations which send a function $f(t)$ into the functions $f_n(x)$, via the convolution we defined. The particular K_ns used in the proof of the Weierstrass theorem are called the Landau kernels.

We want to talk now about some particular transforms, the Fourier transform and the Laplace transform.

11.5 THE FOURIER TRANSFORM

We begin by considering a special class of continuous functions $\mathfrak{R}$, complex valued and defined on all of $\mathbb{R}$, characterized by their "rapid decrease." This is to say, $f\in\mathfrak{R}$ if and only if for each $m\geq 0$, $|x|^m|f(x)|$ is bounded. As a consequence of this condition, we have that for every m, $\lim_{|x|\to\infty} x^m f(x) = 0$, and this is why we call $\mathfrak{R}$ the class of "rapidly decreasing" functions. We can further restrict the class by insisting that only those functions which are C^∞ on $\mathbb{R}$ and whose derivatives themselves are "rapidly decreasing" be allowed to remain; this restricted class we denote by $\mathbb{S}$. $\mathbb{S}$ is not empty, for it contains the function e^{-x^2}. $\mathbb{S}$ is indeed a vector space over the field of complex numbers, and is referred to as the Schwarz space. It is not hard to see that if $f \in \mathbb{S}$, then $|f|$ is bounded, and $\int_{-\infty}^{\infty} f(x)\,dx$ is absolutely convergent since $|f(x)|$ is dominated by $M(1+x^2)^{-1}$ for some $M>0$, and $\int_{-\infty}^{\infty} M(1+x^2)^{-1}\,dx$ converges.

Now suppose $f(t)$ is a C^1-function such that $\lim_{|t|\to\infty} f(t) = 0$ and that both the integrals $\int_{-\infty}^{\infty} f(t)\,dt$ and $\int_{-\infty}^{\infty} f'(t)\,dt$ converge absolutely. By applying the Lebesgue dominated convergence theorem, we can conclude that for an arbitrary $x \in \mathbb{R}$ the integrals $\int_{-\infty}^{\infty} f(t)\cos xt\,dt$ and $\int_{-\infty}^{\infty} f(t)\sin xt\,dt$ converge, and in this case absolutely, since $\int_{-\infty}^{\infty} f(t)\,dt$ does so. This leads us to the following definitions.

Definition 11.5.1. If f is an odd function, the *Fourier sine transform* of f is

$$\hat{f}(x) = \sqrt{2/\pi}\int_0^\infty f(t)\sin xt\,dt.$$

If f is an even function, the *Fourier cosine transform* of f is

$$\hat{f}(x) = \sqrt{2/\pi}\int_0^\infty f(t)\cos xt\,dt.$$

If f is neither odd or even, the *Fourier transform* of f is given by

$$\hat{f}(x) = \left(1/\sqrt{2\pi}\right)\int_{-\infty}^{\infty} f(t)e^{-ixt}\,dt.$$

It is clear that the mappings $f \to \hat{f}$ are linear mappings. Furthermore, the convergence of these integrals is uniform on every compact x interval, and we can differentiate under the integral sign and get

$$D\hat{f}(x) = \frac{1}{\sqrt{2\pi}}\int_{-\infty}^{\infty} -itf(t)e^{-ixt}\,dt = [-itf(t)]\hat{\ },$$

as long as $\int_{-\infty}^{\infty} tf(t)\,dt$ converges absolutely. This is indeed the case if $f \in \mathbb{S}$, and for such fs we can continue differentiating

$$D^2\hat{f}(x) = \frac{1}{\sqrt{2\pi}}\int_{-\infty}^{\infty} -t^2f(t)e^{-ixt}\,dt = [-t^2f(t)]\hat{\ },$$

and, in general, $D^n\hat{f}(x) = [(-it)^n f(t)]\hat{\ }$.

Analogously, if $f \in \mathbb{S}$, we can obtain

$$\widehat{Df}(x) = \frac{1}{\sqrt{2\pi}} \int_{-\infty}^{\infty} f'(t) e^{-ixt}\, dt$$

$$= \frac{1}{\sqrt{2\pi}} \left(f(t) e^{-ixt} \right) \Bigg|_{-\infty}^{\infty} + \frac{ix}{\sqrt{2\pi}} \int_{-\infty}^{\infty} f(t) e^{-ixt}\, dt = ix\hat{f}(x),$$

and, in general, $\widehat{D^n f}(x) = (ix)^n \hat{f}(x)$.

Finally, we have that $\hat{f} \in \mathbb{S}$, for $|x|^n |\hat{f}(x)|$ is bounded for all n since $D^n f \in \mathbb{S}$ and $\widehat{D^n f}$ is bounded by $\int_{-\infty}^{\infty} |D^n f(t)|\, dt$.

We state these results in the following theorem.

Theorem 11.5.2. *If $f \in \mathbb{S}$, then so is $\hat{f}$, the Fourier transform of f. Moreover, for each positive integer n,*

$$D^n \hat{f}(x) = \left[(-it)^n f(t) \right]^{\wedge} \qquad \textit{and} \qquad \widehat{D^n f}(x) = (ix)^n \hat{f}(x).$$

If $f, g \in \mathbb{S}$ we define the convolution of f and g to be

$$(f * g)(x) = \left(1/\sqrt{2\pi} \right) \int_{-\infty}^{\infty} f(t) g(x-t)\, dt.$$

Note the additional factor of $1/\sqrt{2\pi}$. This integral is absolutely convergent. We have already shown that $f * g = g * f$, and that convolution is associative and distributive with respect to addition.

Theorem 11.5.3. *If $f, g \in \mathbb{S}$, so does $f * g$. Furthermore, for each positive integer n, we have*

$$D^n (f * g) = (D^n f) * g = f * (D^n g).$$

Finally,

$$(f * g)^{\wedge} = \hat{f}\hat{g}.$$

PROOF: Let m be any fixed positive integer. Then

$$|x|^m \leqslant (|x-t| + |t|)^m = \sum_{i=0}^{m} \binom{m}{i} |x-t|^{m-i} |t|^i,$$

and

$$|x|^m |(f * g)(x)| \leqslant \sum_{i=0}^{m} \binom{m}{i} \int_{-\infty}^{\infty} |f(t)|\, |t|^i |g(x-t)|\, |x-t|^{m-i}\, dt < \infty$$

since f and $g \in \mathbb{S}$. This means $f * g \in \mathbb{S}$.

Now, consider $D(f*g)(x)$. We can differentiate under the integral sign to get

$$D(f*g)(x)=\frac{1}{\sqrt{2\pi}}\int_{-\infty}^{\infty}D_x[f(t)g(x-t)]\,dt$$
$$=\frac{1}{\sqrt{2\pi}}\int_{-\infty}^{\infty}f(t)D_xg(x-t)\,dt=(f*Dg)(x),$$

and continuing by induction, $D^n(f*g)=f*D^ng$ for any n. By commutativity,

$$D^n(f*g)=D^n(g*f)=g*D^nf=D^nf=D^nf*g.$$

To prove the last statement, we see that

$$(f*g)\hat{\ }(y)=\frac{1}{2\pi}\int_{-\infty}^{\infty}\int_{-\infty}^{\infty}f(t)g(x-t)e^{-ixy}\,dt\,dx$$

is absolutely integrable, so we can change the order of integration with no problem to get

$$(f*g)\hat{\ }(y)=\frac{1}{2\pi}\int_{-\infty}^{\infty}f(t)\left[\int_{-\infty}^{\infty}g(x-t)e^{-i(x-t)y}\,dx\right]e^{-ity}\,dt$$
$$=\frac{1}{\sqrt{2\pi}}\int_{-\infty}^{\infty}\hat{g}(y)f(t)e^{-ity}\,dt=\hat{g}(y)\hat{f}(y).\quad\square$$

Theorem 11.5.4. *If* $\hat{f}(x)=1/\sqrt{2\pi}\int_{-\infty}^{\infty}f(t)e^{-ixt}\,dt$, *then* $\hat{\hat{f}}(-t)=1/\sqrt{2\pi}\cdot\int_{-\infty}^{\infty}\hat{f}(x)e^{ixt}\,dx=f(t)$.

PROOF: To prove this theorem, we shall use a lemma.

Lemma. *If* $f(x)=e^{-x^2/2}$, *then* $\hat{f}=f$. *This is to say, there is a function equal to its own Fourier transform, indeed, a function in* $\mathcal{S}$.

$$\hat{f}(y)=\frac{1}{\sqrt{2\pi}}\int_{-\infty}^{\infty}e^{-x^2/2}e^{-ixy}\,dx.$$

$$D\hat{f}(y)=\frac{1}{\sqrt{2\pi}}\int_{-\infty}^{\infty}-ixe^{-x^2/2}e^{-ixy}\,dx$$
$$=\frac{1}{\sqrt{2\pi}}\left[ie^{-x^2/2}e^{-ixy}\Big|_{-\infty}^{\infty}-\int_{-\infty}^{\infty}ye^{-x^2/2}e^{-ixy}\,dx\right]=-y\hat{f}(y).$$

Notice—and one might never notice this—that

$$D\left(\frac{\hat{f}(y)}{e^{-y^2/2}}\right)=\frac{e^{-y^2/2}[-y\hat{f}(y)]-\hat{f}(y)\left[-ye^{-y^2/2}\right]}{e^{-y^2}}\equiv 0,$$

so $\hat{f}(y) \equiv ce^{-y^2/2}$, where c is a constant. Since $\hat{f}(0) = (1/\sqrt{2\pi})\int_{-\infty}^{\infty} e^{-x^2/2}\,dx = 1$, $c = 1$, and we have $\hat{f}(y) \equiv f(y)$. This proves the lemma. □

Let $g \in \mathbb{S}$, and write

$$\int_{-\infty}^{\infty} \hat{f}(x)e^{-ixy}g(x)\,dx = \frac{1}{\sqrt{2\pi}}\int_{-\infty}^{\infty}\int_{-\infty}^{\infty} f(t)e^{-itx}e^{-ity}g(x)\,dt\,dx$$

$$= \frac{1}{\sqrt{2\pi}}\int_{-\infty}^{\infty}\int_{-\infty}^{\infty} f(t)g(x)e^{-ix(t+y)}\,dx\,dt$$

$$= \int_{-\infty}^{\infty} f(t)\hat{g}(t+y)\,dt.$$

Note that putting $y = 0$ in this equation shows

$$\int_{-\infty}^{\infty} \hat{f}(x)g(x)\,dx = \int_{-\infty}^{\infty} f(x)\hat{g}(x)\,dx.$$

Now, let $h \in \mathbb{S}$, and let $g(u) = h(au)$, where $a > 0$, so that

$$\hat{h}(u/a) = \int_{-\infty}^{\infty} h(t)e^{-itu/a}\,dt = \int_{-\infty}^{\infty} h(at)e^{-itu}\,d(at) = a\hat{g}(u).$$

Therefore

$$\int_{-\infty}^{\infty} \hat{f}(x)e^{-ixy}h(ax)\,dx = \int_{-\infty}^{\infty} f(t)\hat{h}\left(\frac{t+y}{a}\right)\frac{1}{a}\,dt.$$

Putting $u = (t+y)/a$, and $du = dt/a$, we get

$$\int_{-\infty}^{\infty} \hat{f}(x)e^{-ixy}h(ax)\,dx = \int_{-\infty}^{\infty} f(au-y)\hat{h}(u)\,du.$$

Both integrands are continuous in the parameter a, so let $a \to 0^+$, and at the limit we have

$$h(0)\int_{-\infty}^{\infty} \hat{f}(x)e^{-ixy}\,dx = f(-y)\int_{-\infty}^{\infty} \hat{h}(u)\,du,$$

$$h(0)\hat{\hat{f}}(y) = \frac{f(-y)}{\sqrt{2\pi}}\int_{-\infty}^{\infty} \hat{h}(u)\,du.$$

Let $h(u)$ be the function $e^{-u^2/2}$. By the lemma $\hat{h}(u) = e^{-u^2/2}$, and we have finally

$$\hat{\hat{f}}(y) \equiv f(-y).$$

This is equivalent to writing $\hat{\hat{f}}(-t) \equiv f(t)$, and the theorem is proved. □

This theorem tells us that there is an inverse Fourier transform, at least for functions in $\mathbb{S}$. However, it seems apparent that if a function f is such that it is C^1, and that both f and its derivative are both absolutely integrable over $(-\infty, \infty)$, and similarly for the transforms, then the inversion formula works.

Corollary 11.5.5. *If $f \in \mathbb{S}$, then on transforming f four times we arrive back at f.*

Corollary 11.5.6. *If $f, g \in \mathbb{S}$, then $\int_{-\infty}^{\infty} f\bar{g} = \int_{-\infty}^{\infty} \hat{f}\,\bar{\hat{g}}$. In particular, $\| f \|_2 = \| \hat{f} \|_2$.*

PROOF: $\int_{-\infty}^{\infty} f(t)\bar{g}(t)\,dt$ is the standard inner product for the Schwarz space, where $\bar{g}(t)$ is the complex conjugate of $g(t)$. Refer to the remark made in the proof of Theorem 11.5.4 that

$$\int_{-\infty}^{\infty} \hat{f}(x)g(x)\,dx = \int_{-\infty}^{\infty} f(x)\hat{g}(x)\,dx.$$

This equation, with the result of that theorem, allows us to write

$$\int_{-\infty}^{\infty} \hat{f}(x)g(x)\,dx = \int_{-\infty}^{\infty} \hat{\hat{f}}(-x)\hat{g}(x)\,dx.$$

Therefore

$$\int_{-\infty}^{\infty} f(t)\bar{g}(t)\,dt = \int_{-\infty}^{\infty} \hat{f}(-t)\hat{\bar{g}}(t)\,dt.$$

Replace t by $-x$ to get

$$\int_{-\infty}^{\infty} f(t)\bar{g}(t)\,dt = \int_{-\infty}^{\infty} \hat{f}(x)\hat{\bar{g}}(-x)\,dx.$$

But if $g(x) = a(x) + ib(x)$,

$$\begin{aligned}
\hat{\bar{g}}(-x) &= [a(-x) - ib(-x)]^{\wedge} = \hat{a}(-x) - i\hat{b}(-x) \\
&= \int_{-\infty}^{\infty} a(-x)e^{-it(-x)}\,d(-x) - i\int_{-\infty}^{\infty} b(-x)e^{-it(-x)}\,d(-x) \\
&= \int_{-\infty}^{\infty} a(y)e^{-ity}\,dy - i\int_{-\infty}^{\infty} b(y)e^{-ity}\,dy \\
&= \hat{a}(t) - i\hat{b}(t) = \bar{\hat{g}}(t) = \bar{\hat{g}}(x),
\end{aligned}$$

so we finally have

$$\int_{-\infty}^{\infty} f(t)\bar{g}(t)\,dt = \int_{-\infty}^{\infty} \hat{f}(x)\bar{\hat{g}}(x)\,dx,$$

and in particular

$$\| f \|_2 = \| \hat{f} \|_2. \quad \square$$

We have obtained the result that for $f, g \in \mathbb{S}$, $(f * g)^{\wedge} = \hat{f}\hat{g}$. We can now, by the Fourier inversion theorem just proved, obtain an analogous expression.

Theorem 11.5.7. *If $f, g \in \mathbb{S}$, $(fg)\hat{} = \hat{f} * \hat{g}$.*

PROOF: We introduce the symbol ^{-}f to denote the function

$$^{-}f(x) \equiv f(-x).$$

Hence $\hat{\hat{f}} = {}^{-}f$. Let $\varphi = ({}^{-}f)\hat{}$ and $\psi = ({}^{-}g)\hat{}$, so that $\hat{\varphi} = f$ and $\hat{\psi} = g$. Then

$$(fg)\hat{} = (\hat{\varphi}\hat{\psi})\hat{} = (\varphi * \psi)\hat{\hat{}} = {}^{-}(\varphi * \psi) = {}^{-}\varphi * {}^{-}\psi = \hat{f} * \hat{g}.$$

We leave the details that ${}^{-}(\varphi * \psi) = {}^{-}\varphi * {}^{-}\psi$ and ${}^{-}\varphi = \hat{f}$, ${}^{-}\psi = \hat{g}$ to the reader. □

Recall that we remarked that the map T: $f \to \hat{f}$ is a linear mapping and invertible, certainly, if $f \in \mathbb{S}$. Since $T^4 f = f$, if $g = f + Tf + T^2 f + T^3 f$, then $Tg = g$. Hence there are lots of functions equal to their Fourier transforms.

If f is a function which vanishes outside some compact set, we say f has compact support. If f is C^∞ and has compact support, then $f \in \mathbb{S}$. Let K be the support of f. Then every derivative of f vanishes outside K, so for each n, $D^n f$ has its support contained in K. Since f and all its derivatives are continuous, and hence uniformly continuous on K, they are all bounded on K. Clearly, for each integer $n \geqslant 0$ and each integer $m > 0$, $|x|^m |D^n f(x)|$ is also bounded on K and vanishes outside K. We summarize in the following theorem.

Theorem 11.5.8. *If f is C^∞ on $(-\infty, \infty)$ and has compact support, then $f \in \mathbb{S}$.*

One such function might be

$$f(x) = \begin{cases} 17e^{-1/(1-x^2)}, & \text{if } x \in (-1,1), \\ 0, & \text{otherwise.} \end{cases}$$

We leave it to the reader to show that at the critical points ± 1, the derivatives all approach zero, so that f is indeed C^∞ on $\mathbb{R}$.

Refer back to Chapter 7, if necessary, and recall that if $f(t)$ was a periodic function, with period 2π, we could express $f(t)$ in terms of a Fourier series, under certain conditions; i.e.,

$$f(t) = \frac{\alpha_0}{2} + \sum_{n=1}^{\infty} [\alpha_n \cos nt + \beta_n \sin nt],$$

where

$$\alpha_n = \frac{1}{\pi}\int_0^{2\pi} f(t)\cos nt\, dt, \quad \beta_n = \frac{1}{\pi}\int_0^{2\pi} f(t)\sin nt\, dt, \qquad n = 0, 1, 2, \ldots.$$

Thinking of n now as a continuous variable on $[0, \infty)$, and replacing n by u, we look for an integral analog of this infinite series. It would be

$$\int_0^\infty [\alpha(u)\cos ux + \beta(u)\sin ux]\, du,$$

where

$$\alpha(u)=\frac{1}{\pi}\int_{-\infty}^{\infty} f(t)\cos ut\,dt, \qquad \beta(u)=\frac{1}{\pi}\int_{-\infty}^{\infty} f(t)\sin ut\,dt.$$

Putting it all together, we have an integral analog to a Fourier series which just might be a valid representation for a function $f(x)$, provided $f(x)$ is suitably restricted.

Definition 11.5.9. The integral

$$\frac{1}{\pi}\int_0^{\infty}\int_{-\infty}^{\infty} f(t)[\cos ut\cos ux+\sin ut\sin ux]\,dt\,du$$
$$=\frac{1}{\pi}\int_0^{\infty}\int_{-\infty}^{\infty} f(t)\cos[u(t-x)]\,dt\,du$$

is called the *Fourier integral* of $f(x)$.

We state, but do not prove, the following theorem, not in its most general form, but in a reasonable form.

Theorem 11.5.10. *Suppose f is defined and bounded on $\mathbb{R}$ and absolutely integrable on $\mathbb{R}$ and f is piecewise C^1. Moreover, if one of the (at most finitely many) jump discontinuities occurs at y, the limits*

$$\lim_{h\to 0^+}\frac{f(y+h)-f(y^+)}{h} \qquad \textit{and} \qquad \lim_{h\to 0^+}\frac{f(y-h)-f(y^-)}{h}$$

exist; moreover, the value of f at y is understood to be

$$f(y)=\tfrac{1}{2}[f(y^-)+f(y^+)].$$

Then for each x we have

$$f(x)=\frac{1}{\pi}\int_0^{\infty}\int_{-\infty}^{\infty} f(t)\cos[u(t-x)]\,dt\,du.$$

We remark that if f is even, then

$$f(x)=\frac{2}{\pi}\int_0^{\infty}\int_0^{\infty} f(t)\cos ux\cos ut\,dt\,du,$$

and if f is an odd function,

$$f(x)=\frac{2}{\pi}\int_0^{\infty}\int_0^{\infty} f(t)\sin ux\sin ut\,dt\,du.$$

The first of these is called the Fourier cosine integral, the last the Fourier sine integral. Furthermore, these are valid representations, even if f is neither odd or even, for $x>0$.

As an example of how we might apply this theorem, suppose $a>0$, and $f(x)=e^{-ax}$ if $x>0$, and $f(x)=-e^{ax}$ if $x<0$, and $f(0)=0$. Then f is an odd function;

$$f(x)=\frac{2}{\pi}\int_0^\infty \sin ux\left[\int_0^\infty e^{-at}\sin ut\,dt\right]du.$$

Integration by parts yields $\int_0^\infty e^{-at}\sin ut\,dt=u/(u^2+a^2)$, so

$$f(x)=\frac{2}{\pi}\int_0^\infty \frac{u\sin ux}{u^2+a^2}\,du.$$

This tells us that

$$\int_0^\infty \frac{u\sin ux}{u^2+a^2}\,du=\frac{\pi}{2}e^{-ax},\qquad \text{for}\quad x>0,$$

a formula perhaps worth filing away somewhere.

We give another example. Suppose

$$f(x)=\begin{cases}0, & x<0,\\ \frac{1}{2}, & x=0,\\ 1, & 0<x<a,\\ \frac{1}{2}, & x=a,\\ 0, & a<x.\end{cases}$$

If we set $f(x)$ equal to its Fourier integral or, alternatively, set

$$f(x)=\frac{1}{\sqrt{2\pi}}\int_{-\infty}^\infty \hat{f}(t)e^{ixt}=\frac{1}{2\pi}\int_{-\infty}^\infty e^{ixt}\int_{-\infty}^\infty e^{-iut}f(u)\,du\,dt,$$

we get

$$f(x)=\frac{1}{2\pi}\int_{-\infty}^\infty e^{ixt}\int_0^a e^{-iut}\,du\,dt=\frac{1}{2\pi}\int_{-\infty}^\infty e^{ixt}\left[\frac{e^{-itu}}{-it}\right]\Bigg|_0^a dt$$

$$=\frac{1}{2\pi i}\int_{-\infty}^\infty e^{ixt}\left(\frac{1-e^{-iat}}{t}\right)dt.$$

Letting first $x=0$ and then $x=a$,

$$\frac{\pi}{2}=\int_{-\infty}^\infty \frac{1-e^{-iat}}{2it}\,dt=\int_{-\infty}^\infty \frac{e^{iat}-1}{2it}\,dt.$$

Now, using the Fourier integral,

$$f(x)=\frac{1}{\pi}\int_0^\infty\int_0^a \cos[u(t-x)]\,dt\,du=\frac{1}{\pi}\int_0^\infty \frac{\sin[u(t-x)]}{u}\Bigg|_0^a du$$

$$=\frac{1}{\pi}\int_0^\infty\left[\frac{\sin[u(a-x)]}{u}+\frac{\sin ux}{u}\right]du,$$

and putting $x=0$, or $x=a$, we have

$$\frac{\pi}{2}=\int_0^\infty \frac{\sin au}{u}\,du,\qquad a>0.$$

Suppose now that $f(x)=ax$ if $x\in(-1,1)$, $f(x)=0$ if $x\notin[-1,1]$, and $f(1)=a/2=-f(-1)$. f is an odd function, and we can write

$$\begin{aligned}
f(x)&=\frac{2}{\pi}\int_0^\infty\int_0^1 at\sin ux\sin ut\,dt\,du\\
&=\frac{2a}{\pi}\int_0^\infty \sin ux\left[\int_0^1 t\sin ut\,dt\right]du\\
&=\frac{2a}{\pi}\int_0^\infty \sin ux\left[\frac{-t\cos ut}{u}\Big|_0^1+\int_0^1\frac{\cos ut}{u}\,dt\right]du\\
&=-\frac{2a}{\pi}\int_0^\infty\frac{\sin ux\cos u}{u}\,du+\frac{2a}{\pi}\int_0^\infty\frac{\sin ux\sin u}{u^2}\,du.
\end{aligned}$$

Put $x=1$ to get

$$f(1)=\frac{a}{2}=\frac{2a}{\pi}\int_0^\infty\frac{\sin^2u}{u^2}\,du-\frac{a}{\pi}\int_0^\infty\frac{\sin 2u}{u}\,du=\frac{2a}{\pi}\int_0^\infty\left[\frac{\sin u}{u}\right]^2du-\frac{a}{\pi}\frac{\pi}{2},$$

and, finally, when $a\neq 0$,

$$\frac{\pi}{2}=\int_0^\infty\left[\frac{\sin u}{u}\right]^2du.$$

11.6 THE LAPLACE TRANSFORM

Suppose f is defined on $(0,\infty)$, piecewise continuous there, and bounded on any compact set. We can consider the improper integral

$$F(s)=\int_0^\infty e^{-st}f(t)\,dt.$$

This integral will certainly converge if $f(t)$ is bounded near $t=0$, or even if $t^\varepsilon f(t)$ is bounded near $t=0$ for some $\varepsilon<1$ and if also there is a fixed constant $s_0<s$ such that $|f(t)|\leqslant Me^{s_0t}$ for all sufficiently large t and some constant $M>0$.

Definition 11.6.1. The function $F(s)$ defined above is called the *Laplace transform* of $f(t)$.

We remark that s is assumed to be real; note that the Laplace transform and the Fourier transform are remarkably similar.

In Table 11.1 we list a few functions f and their Laplace transforms $\mathfrak{L}f$.

We state a basic property of the Laplace transform which is not hard to verify: $\mathfrak{L}$ is a linear operator.

Definition 11.6.2. A function $f(t)$ is said to be *of exponential order* $e^{\alpha t}$ if there is a constant α and positive constants M, t_0 such that for all $t\geqslant t_0$, $|f(t)|\leqslant Me^{\alpha t}$.

TABLE 11.1

$f(t)$	$(\mathcal{L}f)(s)$
1	$1/s,\ s>0$
e^{at}	$1/(s-a),\ s>a$
$\sin bt$	$b/(s^2+b^2),\ s>0$
$\cos bt$	$s/(s^2+b^2),\ s>0$
t^n	$n!/s^{n+1},\ s>0$
$t^n e^{at}$	$n!/(s-a)^{n+1},\ s>a$
$t\sin bt$	$2bs/(s^2+b^2)^2,\ s>0$
$t\cos bt$	$(s^2-b^2)/(s^2+b^2)^2,\ s>0$
$e^{-at}\sin bt$	$b/[(s+a)^2+b^2],\ s>-a$
$e^{-at}\cos bt$	$(s+a)/[(s+a)^2+b^2],\ s>-a$

Theorem 11.6.3. *If f is piecewise continuous, bounded in every compact interval $[0, b]$, and of exponential order $e^{\alpha t}$, then $\mathcal{L}f(s)$ exists for all $s>\alpha$.*

PROOF:

$$\int_0^\infty |f(t)|e^{-st}\,dt \leqslant M\int_0^\infty e^{-(s-\alpha)t}\,dt<\infty.$$

Hence the integral converges absolutely. Moreover, the convergence is uniform on every s interval $[b,\infty)$ with $b>\alpha$. □

Theorem 11.6.4. *Suppose $f(t)$ and its first n derivatives are piecewise continuous (assume without loss of too much generality that f is of class C^n). Suppose further that f and its first $n-1$ derivatives are of exponential order $e^{\alpha t}$. Then for all $s>\alpha$, $\mathcal{L}f^{(n)}(s)$ exists and is given by*

$$\mathcal{L}f^{(n)}(s)=s^n\mathcal{L}f(s)-s^{n-1}f(0)-s^{n-2}f'(0)-\cdots-sf^{(n-2)}(0)-f^{(n-1)}(0).$$

PROOF: We proceed by induction.

$$\begin{aligned}\mathcal{L}f'(s)&=\int_0^\infty e^{-st}f'(t)\,dt=e^{-st}f(t)\big|_0^\infty+s\int_0^\infty e^{-st}f(t)\,dt\\&=s\mathcal{L}f(s)-f(0).\end{aligned}$$

Assume the formula holds for $\mathcal{L}f^{(n-1)}(s)$. Then $\mathcal{L}f^{(n)}(s)=s\mathcal{L}f^{(n-1)}(s)-f^{(n-1)}(0)$, and the formula for $\mathcal{L}f^{n}(s)$ follows. □

Theorem 11.6.5 (Translation Property). *Suppose f is such that $\mathcal{L}f$ exists for $s>\alpha$. For any constant a,*

$$\mathcal{L}\{e^{at}f(t)\}=F(s-a),\qquad s>a+\alpha,$$

where $F=\mathcal{L}f$.

PROOF: $F(s)=\int_0^\infty e^{-st}f(t)\,dt$, so

$$F(s-a)=\int_0^\infty e^{-(s-a)t}f(t)\,dt=\int_0^\infty e^{-st}e^{at}f(t)\,dt=\mathfrak{L}\{e^{at}f(t)\},$$

valid as long as $s-a>\alpha$. □

Theorem 11.6.6. *Suppose $f(t)$ satisfies the conditions of Theorem* 11.6.3. *Let $g(t)$ be defined by*

$$g(t)=\begin{cases}0, & 0<t<a,\\ f(t-a), & a<t.\end{cases}$$

Then $\mathfrak{L}\{g(t)\}=e^{-as}F(s)$, where $F(s)=\mathfrak{L}\{f(t)\}$.

PROOF:

$$\int_0^\infty e^{-st}g(t)dt=\int_a^\infty e^{-st}f(t-a)\,dt=\int_0^\infty e^{-s(\tau+a)}f(\tau)\,d\tau$$

$$=\int_0^\infty e^{-s\tau}e^{-as}f(\tau)\,d\tau=\mathfrak{L}\{e^{-as}f(t)\}=e^{-as}F(s).\quad □$$

Theorem 11.6.7. *Let $f(t)$ satisfy the conditions of Theorem* 11.6.3. *Then*

$$\mathfrak{L}\{t^nf(t)\}=(-1)^n\frac{d^n}{ds^n}F(s),\qquad s\geqslant b>\alpha.$$

PROOF:

$$F(s)=\int_0^\infty e^{-st}f(t)\,dt.$$

Since the convergence is uniform for $\alpha<b\leqslant s$, we can differentiate under the integral sign and get $F'(s)=\int_0^\infty -te^{-st}f(t)\,dt$, and, in general, $F^{(n)}(s)=\int_0^\infty(-1)^nt^ne^{-st}f(t)\,dt$, for note that $t^{n+1}e^{-st}|f(t)|\leqslant Mt^{n+1}e^{-(s-\alpha)t}$, and this tends to zero as $t\to\infty$, so the integral $F^{(n)}(s)$ converges absolutely for each n, and the convergence is uniform for $s\geqslant b>\alpha$. We conclude that $(-1)^nD^nF(s)=\mathfrak{L}\{t^nf(t)\}$. □

We shall not say very much about the inverse Laplace transform, except to state without proof that if $F=\mathfrak{L}f$, then f is the inverse transform of the transform F. Moreover, if f and g are two continuous functions having the same Laplace transform F, then $f=g$. How one finds $\mathfrak{L}^{-1}F$, given F, is to enlist the aid of Theorem 11.5.4. If $F(s)=\int_0^\infty e^{-st}f(t)\,dt$, for $s>\alpha$, then $f(t)$ is of exponential order $e^{\alpha t}$, or the integral will not converge absolutely for all $s>\alpha$. Let us assume that $f(t)=0$ for all $t<0$ and $f(0)=\frac{1}{2}\lim_{t\to0^+}f(t)$. Let $\varphi(t)=e^{-at}f(t)$, where $a>\alpha$. Then write

$$F(s+a)=\int_{-\infty}^\infty e^{-(s+a)t}f(t)\,dt=\int_{-\infty}^\infty e^{-st}\varphi(t)\,dt,$$

and

$$\frac{1}{\sqrt{2\pi}} F(is+a) = \frac{1}{\sqrt{2\pi}} \int_{-\infty}^{\infty} e^{-ist} \varphi(t)\, dt = \hat{\varphi}(s).$$

The integral converges since $a > \alpha$, so $\hat{\varphi}(s)$ exists. But now we can write

$$\varphi(t) = \frac{1}{\sqrt{2\pi}} \int_{-\infty}^{\infty} \hat{\varphi}(s) e^{ist}\, ds = \frac{1}{2\pi} \int_{-\infty}^{\infty} F(is+a) e^{ist}\, ds.$$

Put $x = is + a$, and we have

$$\varphi(t) = \frac{1}{2\pi i} \int_{a-i\infty}^{a+i\infty} F(x) e^{xt} e^{-at}\, dx.$$

Multiply by e^{at}, and we have, finally,

$$f(t) = \frac{1}{2\pi i} \int_{a-i\infty}^{a+i\infty} F(x) e^{xt}\, dx, \qquad a > \alpha.$$

We conclude that if $F = \mathfrak{L}f$, then f is given by the integral immediately above.

Theorem 11.6.8. *If $F = \mathfrak{L}f$, then $\mathfrak{L}^{-1}F = f$, given by*

$$f(t) = \frac{1}{2\pi i} \int_{\alpha-i\infty}^{\alpha+i\infty} F(x) e^{tx}\, dx,$$

where α is suitably large.

Naturally, from a table of Laplace transforms one might work backward to try to find the inverse transforms, using all the tricks of manipulating that one can think of.

Theorem 11.6.9. *Let f, g satisfy the conditions of Theorem 11.6.3. Then $\mathfrak{L}\{f*g\} = FG$, where $F = \mathfrak{L}f$, $G = \mathfrak{L}g$, and $(f*g)(t) = \int_0^t f(\tau) g(t-\tau)\, d\tau$, the convolution of f and g.*

PROOF: $\int_0^\infty (f*g)(t) e^{-st}\, dt = \int_0^\infty e^{-st} \int_0^t f(\tau) g(t-\tau)\, d\tau\, dt$. Making the transformation $v = \tau$, $u = t - \tau$, these integrals equal

$$\int_0^\infty \int_0^\infty e^{-s(u+v)} f(v) g(u)\, dv\, du$$

since the Jacobian of this transformation equals 1, or if you prefer, $d\tau\, dt = dv\, du$, by calculating the pullback. But this last integral can be written

$$\int_0^\infty e^{-su} g(u)\, du \int_0^\infty e^{-sv} f(v)\, dv = G(s) F(s). \quad \square$$

Corollary 11.6.10. *If we have $F(s), G(s)$ and we know their inverse transforms $f(t), g(t)$, then*

$$\mathfrak{L}^{-1}\{F(s)G(s)\} = \int_0^t f(\tau)g(t-\tau)\,d\tau = \int_0^t g(\tau)f(t-\tau)\,d\tau$$

since $(f*g)(t) = (g*f)(t)$.

Theorem 11.6.11

$$\mathfrak{L}\left\{\int_0^t f(u)\,du\right\} = F(s)/s.$$

PROOF:

$$\int_0^\infty e^{-st}\int_0^t f(u)\,du\,dt = \left[-\frac{1}{s}e^{-st}\int_0^t f(u)\,du\right]\Bigg|_0^\infty + \frac{1}{s}\int_0^\infty e^{-st}f(t)\,dt$$
$$= 0 + s^{-1}F(s). \quad \square$$

An example of an application of Laplace transforms follows. Suppose we have the equation

$$y''(t) + y(t) = e^{-2t}\sin t$$

with $y(0) = 0 = y'(0)$. Using a number of things we know about Laplace transforms, we start by applying the operation to both sides.

$$\mathfrak{L}\{e^{-2t}\sin t\} = \mathfrak{L}\{y''(t)\} + \mathfrak{L}\{y(t)\} = s^2Y(s) - sy(0) - y'(0) + Y(s)$$
$$= 1/\left[(s+2)^2+1\right].$$

Therefore $Y(s) = [s^2+1]^{-1}[(s+2)^2+1]^{-1}$,

$$\mathfrak{L}^{-1}\{Y(s)\} = y(t) = \mathfrak{L}^{-1}\left\{\left[(s+2)^2+1\right]^{-1}\right\} * \mathfrak{L}^{-1}\left\{[s^2+1]^{-1}\right\}$$
$$= \int_0^t e^{-2\tau}\sin\tau\sin(t-\tau)\,d\tau$$
$$= \sin t\int_0^t e^{-2\tau}\sin\tau\cos\tau\,d\tau - \cos t\int_0^t e^{-2\tau}\sin^2\tau\,d\tau$$
$$= \frac{\sin t}{2}\int_0^t e^{-2\tau}\sin 2\tau\,d\tau - \frac{\cos t}{2}\int_0^t e^{-2\tau}\,d\tau + \frac{\cos t}{2}\int_0^t e^{-2\tau}\cos 2\tau\,d\tau$$
$$= -\frac{e^{-2t}}{8}(\sin t\sin 2t + \sin t\cos 2t) + \frac{\sin t}{8} + \frac{e^{-2t}\cos t}{4} - \frac{\cos t}{4}$$
$$+ \frac{e^{-2t}}{8}(\cos t\sin 2t - \cos t\cos 2t) + \frac{\cos t}{8}$$
$$= \frac{1}{8}(\sin t - \cos t) + \frac{e^{-2t}}{8}(\sin t + \cos t).$$

This then is the solution to the given initial-value problem.

We bring this section to a close by reminding the reader of a couple of fundamental facts. First, if $f(t)$ is a function and $\hat{f}(x)$ is its Fourier transform, then Theorem 11.5.2 shows that

1. the nth derivative of the transform $\hat{f}(x)$ is the transform of the product of $f(t)$ with the nth power of $-it$, and
2. the transform of the nth derivative of $f(t)$ is simply the product of $\hat{f}(x)$ and the nth power of ix.

Hence, given $f(t)$, we can reduce the problem of finding the nth derivative of $f(t)$ to the problem of finding $\hat{f}(x)$, multiplying by $(ix)^n$, and then finding the inverse Fourier transform of the product.

If $f(t)$ is a function, and $F(s)$ its Laplace transform, then the nth derivative of $F(s)$ is the Laplace transform of $(-t)^n f(t)$. On the other hand, the transform of the nth derivative of $f(t)$ is "almost" a polynomial in s of degree n. It misses being such a polynomial only in that the leading coefficient, instead of being a constant, is the transform of $f(t)$, namely, $F(s)$.

11.7 GENERALIZED FUNCTIONS

A synonym for generalized function is distribution. Since the notion of a distribution is important in higher analysis, it is fitting that we at least introduce the reader to the definition. You will then know that such things exist. We start with a preliminary definition.

Definition 11.7.1. By a *test function* φ we mean a real-valued C^∞-function on E^n having compact support.

Let $\mathfrak{T}$ be the set of all test functions on E^n.

Definition 11.7.2. A *distribution* F is a linear functional on $\mathfrak{T}$, continuous in the following sense: If a sequence $\{\varphi_n\}$ of test functions is such that each φ_n vanishes outside some common compact set $K \subset E^n$ and for each $k = 0, 1, 2, \ldots$ the sequences of kth-order derivatives $\{D^k \varphi_n\}_{n=1}^\infty$ converge uniformly to zero, then $\lim_{n\to\infty} F(\varphi_n) = 0$. If $f, |f|$ are integrable (Lebesgue or Darboux) on every compact box in E^n, f determines a distribution F defined by

$$F(\varphi) = \int_{E^n} f\varphi \, d\mu \qquad \forall \varphi \in \mathfrak{T}.$$

A distribution determined by a function f is called *regular*; others are referred to as *singular* distributions.

We simply state the following.

Theorem 11.7.3. *Suppose $\mathfrak{T}$ is the set of test functions on E^1. If F is a distribution on $\mathfrak{T}$, then F has a derivative F', where*

$$F'(\varphi) = -F(\varphi') \qquad \forall \varphi \in \mathfrak{T}.$$

Moreover, if F is regular, determined by f, and f' is integrable, and absolutely integrable as well, on all compact intervals of E^1, then F' is determined by f'; i.e.,

$$F'(\varphi) = \int_{-\infty}^{\infty} f'(x)\varphi(x)\,dx.$$

An example of a singular distribution is the Dirac delta distribution, defined by

$$\delta(\varphi) = \varphi(0) \qquad \forall \varphi \in \mathfrak{T}.$$

PROBLEMS

1. Evaluate

$$\int_0^{\infty} e^{-t^2}\,dt.$$

[*Hint*: $\int_0^\infty e^{-t^2}\,dt = [(\int_0^\infty e^{-x^2}\,dx)(\int_0^\infty e^{-y^2}\,dy)]^{1/2} = [\int_0^\infty \int_0^\infty e^{-(x^2+y^2)}\,dx\,dy]^{1/2}$.] Next, evaluate $\int_0^\infty e^{-t^2 x}\,dt$ and show that

$$\frac{1}{\sqrt{x}} = \frac{2}{\sqrt{\pi}} \int_0^{\infty} e^{-t^2 x}\,dt, \qquad x > 0.$$

If $\int_{-\infty}^{\infty} \cos u^2\,du$ is to be evaluated, let $x = u^2$, and then $\int_{-\infty}^{\infty} \cos u^2\,du = \int_0^\infty x^{-1/2} \cos x\,dx = 2/\sqrt{\pi} \int_0^\infty \int_0^\infty \cos x\, e^{-t^2 x}\,dx\,dt$. Show that this last integral reduces to $(2/\sqrt{\pi})\int_0^\infty t^2(t^4+1)^{-1}\,dt$, and, finally,

$$\int_{-\infty}^{\infty} \cos[u^2]du = \frac{1}{\sqrt{2\pi}} \left[\tfrac{1}{2} \ln \frac{t^2 - \sqrt{2}\,t + 1}{t^2 + \sqrt{2}\,t + 1} + \tan^{-1}[\sqrt{2}t - 1] + \tan^{-1}[\sqrt{2}t + 1] \right]\Bigg|_0^{\infty}$$

$$= \sqrt{\frac{\pi}{2}}\,.$$

Now, show that $\int_{-\infty}^{\infty} \sin u^2\,du = \sqrt{\pi/2}$.

2. If $\varphi(x) = (2/\sqrt{\pi})\int_0^x e^{-t^2}\,dt$, prove that

$$\int_0^x \varphi(at)\,dt = \frac{1}{a\sqrt{\pi}}\left(e^{-a^2x^2} - 1\right) + x\varphi(ax),$$

and

$$\int_0^{\infty} [1 - \varphi(t)]\,dt = \frac{1}{\sqrt{\pi}}.$$

$\varphi(x)$ is known as the *error function* and is denoted erf(x).

3. If $a > -1$ is a parameter, use differentiation with respect to a to evaluate

$$\int_0^\infty \frac{1-e^{-ax}}{xe^x}\,dx.$$

4. Define $\mathrm{Si}(x) = -\int_x^\infty (\sin t)/t\,dt$ and $\mathrm{Ci}(x) = -\int_x^\infty (\cos t)/t\,dt$ (the sine integral and cosine integral, respectively). Show that

$$\int_0^\infty \sin x\,\mathrm{Si}(x)\,dx = \int_0^\infty \cos x\,\mathrm{Ci}(x)\,dx = -\pi/4.$$

5. Compute

$$\int_{-\infty}^\infty\int_{-\infty}^\infty\int_{-\infty}^\infty e^{-x^2-y^2-z^2}\,dz\,dy\,dx.$$

6. Compute

$$\int_0^\infty\int_0^\infty\int_0^\infty (1+x+y+z)^{-7/2}\,dx\,dy\,dz.$$

7. Prove that $\int_0^\infty t^{-1}\sin t\,dt$ does not converge absolutely.

8. Show that

$$\Gamma\left(n+\frac{1}{2}\right) = \frac{(2n)!\sqrt{\pi}}{4^n n!}, \qquad n \in \mathbb{Z}^+.$$

9. Try to show that $\Gamma(x)\Gamma(1-x) = \pi(\sin \pi x)^{-1}$ for x not an integer. That is, try to show how one would integrate $\int_0^\infty [u^{x-1}/(1+u)]\,du$ to get $\pi/\sin \pi x$.

10. Let $K(t) = \frac{1}{2}\cos t$ on $[-\pi/2, \pi/2]$ and zero elsewhere. Show that if $K_n(t) = nK(nt)$, $\{K_n(t)\}$ is a Dirac sequence. If $D_n(t) = [K(t)]^n$, is $\{D_n(t)\}$ a Dirac sequence? How would you normalize each D_n to make $\{D_n\}$ a Dirac sequence?

11. Refer to the proof of Theorem 11.5.7, and complete the proof by showing that ${}^{-}(\varphi * \psi) = {}^{-}\varphi * {}^{-}\psi$, ${}^{-}\varphi = \hat{f}$, and ${}^{-}\psi = \hat{g}$.

12. Show that $f(x) = e^{-(1-x^2)^{-1}}$ is C^∞ at $x = \pm 1$, if $f(\pm 1) = 0$, by showing that
(a) $\lim_{x\to\pm 1} f(x) = 0$,
(b) $\lim_{x\to 1}\{[f(x) - f(1)]/(x-1)\}$ exists,
(c) $\forall n$, $\lim_{x\to 1}\{[f^{(n)}(x) - f^{(n)}(1)]/(x-1)\}$ exists.
Then show that $f(x)$ is C^∞ on $\mathbb{R}$. Finally, show that $g(x) = f(x)$, $-1 < x < 1$, and 0 elsewhere is also C^∞ on $\mathbb{R}$. This will prove that the example given just after Theorem 11.5.8 actually belongs to $\mathbb{S}$ and is a test function on E^1.

13. Let $f(z) = z^2 + 1$ be a complex-valued function of a complex variable. Evaluate $\int_a^b f(z)\,dz$, where $a = 1+i$, $b = 2+3i$, and the path of integration is along the oriented line segment from (1,1) to (2,3). [Do the problem like this:

$$\int_a^b f(z)\,dz = \int_\gamma (x^2 - y^2 + 2xyi + 1)(dx + i\,dy)$$

$$= \int_\gamma \left[(x^2 - y^2 + 1)\,dx - 2xy\,dy\right] + i\int_\gamma \left[(x^2 - y^2 + 1)\,dy + 2xy\,dx\right],$$

and compare the result with $\int_a^b (z^2+1)\,dz = [z^3/3 + z]|_a^b$.]

14. Show that $\int_{E^2} |\sin(x^2+y^2)|\,d\mu$ diverges to ∞.

15. If $g(y) = \int_0^{2\pi} x^3 \sin(x^2+y^2)\,dx$, find $g'(y)$.

16. Consider the integral $\int_0^\infty x(x^2+t^2)^{-1}\,dt$. For what values of x does it converge? For what values of x does

$$\frac{d}{dx}\int_0^\infty \frac{x}{x^2+t^2}\,dt$$

exist? Evaluate this derivative at $x = \pi^{1/\pi}$. Next, what is the value of

$$\int_0^\infty \frac{t^2 - x^2}{(x^2+t^2)^2}\,dt, \qquad \text{for} \quad x \geqslant \delta > 0?$$

17. Discuss the convergence of $\int_2^\infty \sin x / \ln x\,dx$.

18. Evaluate $\Gamma(-\frac{5}{2})$. Can you find a general formula for $\Gamma(\frac{1}{2} - n)$, $n = 1, 2, 3, \ldots$?

19. Express $\Gamma'(x)$. Show for each positive integer $n > 1$, and $0 < x \leqslant 1$, that

$$\frac{\ln\Gamma(n-1) - \ln\Gamma(n)}{(n-1)-n} \leqslant \frac{\ln\Gamma(x+n) - \ln\Gamma(n)}{x} \leqslant \frac{\ln\Gamma(n+1) - \ln\Gamma(n)}{n+1-n},$$

and thus

$$\ln\frac{\Gamma(n)}{\Gamma(n-1)} = \ln(n-1) \leqslant \frac{\ln\Gamma(x+n) - \ln(n-1)!}{x} \leqslant \ln n,$$

$$\ln(n-1)^x(n-1)! \leqslant \ln\Gamma(x+n) \leqslant \ln n^x(n-1)!,$$

$$(n-1)^x(n-1)! \leqslant \Gamma(x+n) = (x+n-1)\cdots(x+1)\Gamma(x) \leqslant n^x(n-1)!,$$

and so finally,

$$\frac{(n-1)^x(n-1)!}{x(x+1)\cdots(x+n-1)} \leqslant \Gamma(x) \leqslant \frac{n^x(n-1)!}{x(x+1)\cdots(x+n-1)}.$$

Since these inequalities hold for all $n \geqslant 2$, replace n by $n+1$ on the left-hand sides, and rewrite the right-hand term as indicated:

$$\frac{n^x n!}{x(x+1)\cdots(x+n)} \leqslant \Gamma(x) \leqslant \frac{n^x(n!)}{x(x+1)\cdots(x+n)}\,\frac{x+n}{n}.$$

This implies that

$$\Gamma(x)\frac{n}{x+n} \leqslant \frac{n^x n!}{x(x+1)\cdots(x+n)} \leqslant \Gamma(x),$$

so what is the limit as $n \to \infty$ of the middle term?

This result holds for $0 < x \leqslant 1$. Try to show that if it holds for such an x, it holds for $x+1$. Now you have a representation for $\Gamma(x)$ for all $x > 0$.

Denote by $\Gamma_n(x)$ the term $n^x n!/x(x+1)\cdots(x+n)$. Show that

$$\Gamma_n(x) = e^{x(\ln n - 1 - 1/2 - \cdots - 1/n)}\frac{1}{x}\frac{e^x}{1+x}\frac{e^{x/2}}{1+x/2}\cdots\frac{e^{x/n}}{1+x/n}.$$

What is $\lim_{n\to\infty}[\ln n - (1+\frac{1}{2}+\cdots+1/n)]$ (see the end of Section 10.2)?

Write a series for $\ln\Gamma(x)$. Is this series uniformly convergent? Now differentiate term by term and check the convergence of the new series.

You should end up with the result

$$\frac{\Gamma'(x)}{\Gamma(x)} = -\gamma - \frac{1}{x} + \sum_{n=1}^{\infty}\left[\frac{1}{n} - \frac{1}{x+n}\right], \qquad x > 0.$$

20. Show that $\Gamma(1+1/x) = \int_0^\infty e^{-t^x}\,dt$ $(x>0)$.

21. Evaluate $\int_0^1 t^{m-1}(1-t^n)^{-1/2}\,dt$.
[*Hint*: Let $x = t^n$.]

22. Solve, using Laplace transforms,

$$y'(t) - 2y(t) = e^{5t}, \qquad y(0) = 3.$$

23. Solve by Laplace transforms

$$y'''(t) + 4y''(t) + 5y'(t) + 2y(t) = 10\cos t,$$
$$y(0) = y'(0) = 0, \qquad y''(0) = 3.$$

24. Find $\mathfrak{L}^{-1}\{1/s(s^2+1)\}$, the inverse Laplace transform.

25. Suppose we have the differential equation $y''(x) - k^2y(x) + f(x) = 0$, $x > 0$, $y(0) = 0$, $y(x)$ bounded on $(0,\infty)$. Put

$$y(x) = \frac{2}{\pi}\int_0^\infty Y(u)\sin ux\,du = \frac{2}{\pi}\int_0^\infty\int_0^\infty y(t)\sin ut\sin ux\,dt\,du,$$

and

$$f(x) = \frac{2}{\pi}\int_0^\infty F(u)\sin ux\,du = \frac{2}{\pi}\int_0^\infty\int_0^\infty f(t)\sin ut\sin ux\,dt\,du.$$

Substitute into the given equation, and obtain

$$\int_0^\infty\left[(u^2+k^2)Y(u) - F(u)\right]\sin ux\,du = 0.$$

Then deduce that $Y(u)=F(u)/(u^2+k^2)$. This is to say,

$$y(x)=\frac{2}{\pi}\int_0^\infty \frac{F(u)}{u^2+k^2}\sin ux\,du, \qquad \text{where} \quad F(u)=\int_0^\infty f(t)\sin ut\,dt.$$

Solve the original equation if $k=1$ and $f(x)=1/x$. Does

$$y(x)=\int_0^\infty \frac{\sin ux}{u^2+1}du$$

satisfy the given differential equation when you put

$$\frac{1}{x}=\frac{2}{\pi}\int_0^\infty\int_0^\infty \frac{1}{t}\sin ut\sin ux\,dt\,du?$$

Now instead of letting $k=1$ and $f(x)=1/x$, use $k=k>0$ and $f(x)=e^{-kx}$. Solve the equation using the same method. Do you get

$$y(x)=\frac{2}{\pi}\int_0^\infty \frac{u\sin ux}{(u^2+k^2)^2}du?$$

If you solve by elementary methods $y''(x)-k^2y(x)+e^{-kx}=0$, do you get

$$y(x)=\frac{x}{2k}e^{-kx}+c_1e^{kx}+c_2e^{-kx}$$

as a general solution? Can you conclude that

$$\int_0^\infty \frac{u\sin ux}{(u^2+k^2)^2}du=\frac{\pi x}{4k}e^{-kx}?$$

Note that what we did previously would lead to the conclusion that

$$\frac{1}{x}=\int_0^\infty \sin ux\,du.$$

26. Define a function $f(x)$ on $[1,\infty)$ as follows. $\forall n\in\mathbb{Z}^+$, $n>1$, $f(n)=1$, and on each interval of the form $[n-1/n^2,n]$ or $[n,n+1/n^2]$, $f(x)$ is linear, equal to zero at the nonintegral endpoints, and equal to 1 at the integral endpoints. For all other $x\geqslant 1$ let $f(x)=0$. Show that $f(x)$ is continuous, $\int_1^\infty f(x)\,dx$ converges, but that $\lim_{x\to\infty}f(x)\neq 0$.

27. Compute $\int_0^\infty\int_0^\infty e^{-(x^2+2xy\cos\alpha+y^2)}\,dx\,dy$.

28. Compute $\int_0^\infty\int_x^\infty e^{-y^2}\,dy\,dx$ (change the order of integration).

29. Compute $\int_0^\infty\int_{2x}^\infty xe^{-y}\sin y/y^2\,dy\,dx$.

30. Compute $\int_0^\infty\int_0^\infty\int_0^\infty xy(1+x^2+y^2+z^2)^{-3}\,dx\,dy\,dz$.

31. If B is the ball centered at the origin with radius R, compute

$$\int_B \frac{1}{(x^2+y^2+z^2)^{3/2}\ln(x^2+y^2+z^2)^{3/2}}d\mu$$

and

$$\int_B \ln(x^2+y^2+z^2)\,d\mu.$$

32. Let R be the region bounded by $z=0$ and $z=(x^2+y^2)\exp(-x^2-y^2)$. What is the measure of R?

33. Evaluate $\int_0^\infty e^{-ax}x^{n-1}\,dx$, where $a>0$, $n\in\mathbf{Z}^+$.

34. Compute

$$\int_0^\infty \frac{1-e^{-ax^2}}{xe^{x^2}}\,dx, \qquad a>-1.$$

[*Hint*: Differentiate with respect to a.]

35. Compute

$$\int_0^\pi \frac{\ln(1+a\cos x)}{\cos x}\,dx, \qquad a^2<1.$$

36. Compute

$$\int_0^\infty e^{-ax}\frac{\sin bx-\sin cx}{x}\,dx, \qquad a>0.$$

(Differentiate with respect to b or c.)

37. Compute

$$\int_0^1 \frac{x^b-x^a}{\ln x}\,dx, \qquad a>-1,\quad b>-1.$$

[*Hint*: What is $\int_a^b x^y\,dy$ equal to?] Write the given integral as a double integral and change the order of integration. You should get $\ln[(1+b)/(1+a)]$ as the answer.

38. Compute

$$\int_0^\infty \frac{e^{-at}-e^{-bt}}{t}\,dt, \qquad 0<a<b.$$

39. Compute

$$\int_0^\infty \frac{e^{-at}\sin xt}{t}\,dt, \qquad a>0.$$

40. Let $y(x)=\int_{-1}^{1}(t^2-1)^{n-1}e^{xt}\,dt$. Show that $xy''+2ny'-xy=0$.

41. Let

$$f(x)=\begin{cases} x, & 0<x<1,\\ \frac{3}{4}, & x=1,\\ x/2, & 1<x<2,\\ \frac{1}{2}, & x=2,\\ 0, & x>2, \end{cases}$$

and $\forall x, f(x)=f(-x)$. Write the Fourier cosine integral for f, and evaluate it if you can.

42. Show that $\Gamma(2p)\Gamma(\frac{1}{2})=2^{2p-1}\Gamma(p)\Gamma(p+\frac{1}{2})$ by making a suitable change of variable in the expression for the beta function.

43. Show that

$$\frac{2}{\pi}\int_0^\infty \frac{\sin t\cos tx}{x}\,dt=\begin{cases} 1, & \text{if } |x|<1,\\ \frac{1}{2}, & \text{if } |x|=1,\\ 0, & \text{if } |x|>1. \end{cases}$$

44. Show that

$$\int_0^\infty \frac{\cos ax}{b^2+x^2}\,dx=\frac{\pi}{2b}e^{-|a|b}, \qquad b>0.$$

45. Show that

$$\int_0^\infty \frac{x\sin ax}{1+x^2}\,dx=(\operatorname{sgn} a)\frac{\pi}{2}e^{-|a|},$$

where $(\operatorname{sgn} a)=\text{signum } a=1$ if $a>0$, 0 if $a=0$, and -1 if $a<0$.

46. Show that

$$\int_0^\infty \frac{\sin^{2n+1}x}{x}\,dx=\frac{\pi(2n)!}{2^{2n+1}(n!)^2}, \qquad n\geqslant 0;$$

$$\int_1^\infty \frac{\ln x}{x^{n+1}}\,dx=n^{-2}, \qquad n\geqslant 1;$$

$$\int_0^\infty x^n(1+x)^{-n-m-1}\,dx=\frac{n!(m-1)!}{(m+n)!}, \qquad m,n\geqslant 1.$$

47. Show that $\int_1^\infty dx/(1+x^4\sin^2 x)$ converges, but that $\int_1^\infty dx/(1+x^2\sin^2 x)$ diverges.

48. Test $\int_1^\infty\int_1^\infty f(x,y)\,dx\,dy=\int_1^\infty\int_1^\infty f(x,y)\,dy\,dx$ when $f(x,y)=(x-y)/(x+y)^3$ and $f(x,y)=(x^2-y^2)/(x^2+y^2)^2$.

49. Evaluate $\int_0^\infty [\tan^{-1}x/x]^2\,dx$.
[*Hint*: If $f(x,y)=\int_0^\infty (1+x^2t^2)^{-1}(1+y^2t^2)^{-1}\,dt$, $(x,y)\neq(0,0)$, can you, by partial fractions, determine that $f(x,y)=\pi/2(x+y)$? Then $\int_0^1\int_0^1 \pi/2(x+y)\,dx\,dy=\pi\ln 2$ by elementary calculus. Now, consider $\partial^2F(x,y)/\partial y\,\partial x$, where

$$F(x,y)=\int_0^\infty \frac{(\tan^{-1}xt)(\tan^{-1}yt)}{t^2}\,dt.$$

What is $F(1,1)$?]

50. Knowing that $\int_0^\infty (\sin x/x)^2\,dx=\pi/2$ and that $\sin^2 x+\cos^2 x=1$, evaluate

$$\int_0^\infty \frac{\sin^4 x}{x^2}\,dx \quad \text{and} \quad \int_0^\infty \frac{\sin^4 x}{x^4}\,dx$$

to get $\pi/4$ and $\pi/3$, respectively.

51. Find a proof for Theorem 11.5.10, or make up your own.

52. Let $f=1/(1+x^2)$ and $S=[-1,1]$. Find a suitable Dirac sequence, and following Theorem 11.4.3, construct a sequence $\{f_n\}$ which will converge uniformly to f on S.

53. Show that

$$\int_0^\infty \frac{\cos ux}{u^2+1}\,du=\int_0^\infty \frac{u\sin ux}{u^2+1}\,du=\frac{\pi e^{-x}}{2}.$$

54. Work on this one for a while:

$$f(x)=\int_0^\infty \frac{\sin xt}{t^2+a^2}\,dt, \qquad a^2>0.$$

Can you express $f(x)$ in closed form? In series form? Can you find this in a table of integrals? For which integers n is $f(x)$ a C^n-function?

55. Evaluate $\int_0^{\pi/2}(\sin\theta)^{2x-1}(\cos\theta)^{2y-1}\,d\theta$.

56. Does $\int_0^\infty(\cos xt)/t\,dt$ converge? How about $\int_{-\infty}^1(\cos xt)/t\,dt$? Can you find a series representation for the latter integral?

57. Show that

$$\int_0^1 \frac{\ln(1+x)}{x}\,dx=\frac{\pi^2}{12}=\sum_{n=1}^\infty \frac{(-1)^{n+1}}{n^2}.$$

58. Let $f(x)=e^{-1/(1-x^2)}$ on $(-1,1)$ and 0 elsewhere. Define $f_n(x)=nf(nx)$. Is $\{f_n\}$ a Dirac sequence? If not, can you find constants k_n so that $\{k_nf_n\}$ is a Dirac sequence? Certainly, $\{f_n\}$ is a sequence of test functions. If δ is the Dirac delta distribution, what is the sequence of numbers $\{\delta(f_n)\}_{n=1}^\infty$?

59. Evaluate by differentiating several times

$$f(x)=\int_0^\infty \frac{\sin xt}{t(1+t^2)}\,dt, \qquad x>0.$$

[*Hint*: Show that $f''(x)-f(x)=-\pi/2$.]

60. Evaluate

$$\int_0^\infty \frac{e^{-ax^2}-e^{-bx^2}}{x^2}\,dx, \qquad a,b>0.$$

61. Evaluate

$$\int_0^{\pi/2} \ln(a^2\cos^2 x+b^2\sin^2 x)\,dx, \qquad a^2+b^2\neq 0.$$

62. Solve $y''(x)+y(x)=1/x$, $y(0)=\pi/2$.

63. Show that

$$y(x)=\int_0^\infty \frac{e^{-xt}}{(1+t^2)^{n+1}}\,dt$$

satisfies $xy''-2ny'+xy=1$.

64. Show that $J_0(x)=(2/\pi)\int_0^{\pi/2}\cos(x\sin t)\,dt$ satisfies

$$xJ_0''(x)+J_0'(x)+xJ_0(x)=0.$$

65. Find the area of the surface:

$$x=u^2+v^2,\ y=uv,\ z=u+v,\ 0\leqslant v\leqslant u\leqslant 1.$$

CHAPTER

12

THE GENERALIZED STOKES THEOREM

12.1 MANIFOLDS AND PARTITIONS OF UNITY

We have avoided as long as possible confronting the fundamental theorem of the calculus of forms. The time for confrontation has come. The first thing we must do is to make some formal definitions in order that we may describe rigorously what kind of set will be the domain of integration for the integral of a form.

Suppose ω is a continuous k-form defined on a set $S \subset E^n$. In order to evaluate the integral of ω over S, we are going to assume at the outset that S is a compact differentiable k-dimensional manifold, with or without a $(k-1)$-dimensional boundary. This is to say, S is a compact set in E^n, and the following hold:

1. We have a finite number of C^1-mappings $\mathbf{F}_1,\ldots,\mathbf{F}_m$ from E^k into E^n such that each $\mathbf{F}_i$ is a diffeomorphism from the k-dimensional cube $K \subset E^k$, $K=\{u_1,\ldots,u_k): |u_j| \leqslant 1\}$, to its image $\mathbf{F}_i(K) \subset E^n$. The maps $\mathbf{F}_i$ are called *charts*, and $\{\mathbf{F}_1,\ldots,\mathbf{F}_m\}$ is called an *atlas*.
2. For each chart $\mathbf{F}_i$ there is a specific closed k-dimensional box $R_i \subset K$ such that the image of a point $\mathbf{u} \in K$ under $\mathbf{F}_i$ is in S if and only if $\mathbf{u} \in R_i$.
3. For each $\mathbf{p} \in S$ there is an i, $1 \leqslant i \leqslant m$, such that $\mathbf{p}=\mathbf{F}_i(\bar{u}_1,\ldots,\bar{u}_k)$, where $\max_j |\bar{u}_j| \leqslant 1$, and there exists an $\varepsilon > 0$ such that every $\mathbf{q} \in E^n$ which is within ε of $\mathbf{p}$ is in S if and only if $\mathbf{q} \in \mathbf{F}_i(R_i)$.
4. The orientations of the images of the charts $\mathbf{F}_i$ agree in the sense that if $\mathbf{p} \in S$ is in the image of two charts $\mathbf{F}_i$ and $\mathbf{F}_j$ and ω is a k-form in E^n defined at $\mathbf{p}$, then the pullback of ω under $\mathbf{F}_i$ evaluated at $\mathbf{F}_i^{-1}(\mathbf{p})$ is a positive multiple of the pullback of ω under $\mathbf{F}_j$ evaluated at $\mathbf{F}_j^{-1}(\mathbf{p})$.
5. If S has a nonempty boundary ∂S, then ∂S consists of those points of S each of which is the image under some $\mathbf{F}_i$ of a point of K which lies in the boundary of the associated box R_i.

We remark that such a set S as we have described may have a boundary ∂S, or no boundary, which would be to say that $\partial S=\varnothing$. Note the distinction between boundary, as we use the term here, and frontier, a topological term. When $k<n$, every point of S is a frontier point, but S might not have a boundary. Moreover, S is bounded since it is compact, even when it has no boundary.

If the compact differentiable manifold S over which we want to integrate ω were simply the image of K under a single chart $\mathbf{F}$, we would have fairly smooth sailing, but we cannot in general make this assumption. Hence we must devise a complicated procedure to handle a complicated situation. Our plan is to break ω up into a finite sum of continuous k-forms, $\omega=\Sigma_{i=1}^{r}\omega_i$, where each ω_i is zero on S except for those points of S which are in the image under $\mathbf{F}_i$ of the interior of the corresponding R_i.

For each point $\mathbf{p}\in S$ there is a continuous real-valued function $c_{\mathbf{p}}$ on E^n such that

i. $c_{\mathbf{p}}\geqslant 0$ on E^n, and, in particular, $c_{\mathbf{p}}(\mathbf{p})>0$;
ii. all points $\mathbf{x}\in S$ for which $c_{\mathbf{p}}(\mathbf{x})>0$ are in the interior of the image $\mathbf{F}_i(R_i)$ of any chart $\mathbf{F}_i$ if $\mathbf{p}$ is an interior point of $\mathbf{F}_i(R_i)$.

The proof of this may be obtained by using Urysohn's lemma (Corollary 6.8.6). The point $\mathbf{p}$ is closed in E^n, as is the set $A=S\setminus F_i(\mathring{R}_i)$, and the lemma ensures the existence of a positive function f which is continuous, vanishes on A, and is equal to 1 at $\mathbf{p}$.

Next, if such a function $c_{\mathbf{p}}$ is chosen for each $\mathbf{p}\in S$, then it is possible to select a finite number of these functions, say $c_1,\ldots,c_r$, such that $\Sigma_{i=1}^{r}c_i(\mathbf{x})>0$, $\forall\mathbf{x}\in S$. This is a consequence of the compactness of S and the fact that S is covered by a finite number of charts $\{\mathbf{F}_i(R_i)\}_{i=1}^{m}$. Suppose the statement were false. Then for at least one chart $\mathbf{F}_i$ no finite sum of the cs would be everywhere positive on $\mathbf{F}_i(R_i)$. Partition R_i into "orthants"; for a least one of these orthants $R_{i'}$, no finite sum of the cs would be everywhere positive on $\mathbf{F}_i(R_{i'})$. Continuing in this way we would come to the conclusion that in an arbitrarily small neighborhood of a point $\mathbf{p}\in S$ no finite sum of the cs would be everywhere positive on that neighborhood yet $c_{\mathbf{p}}$ itself would be strictly positive on that neighborhood, a contradiction.

Now that we have a finite collection $c_1, c_2,\ldots,c_r$ such that for each $\mathbf{x}\in S$, $\Sigma_{i=1}^{r}c_i(\mathbf{x})>0$, put

$$a_j=c_j\Big/\sum_{i=1}^{r}c_i,\qquad j=1,\ldots,r.$$

Note that for each $j=1,\ldots,r$, a_j is a continuous function having the properties of c_j, but $\Sigma_{j=1}^{r}a_j=1$. It is for this reason that we call the collection $\{a_j\}_{j=1}^{r}$ a partition of unity.

The next step is to put

$$\omega_i=\left(\sum a_j\right)\omega,$$

this sum consisting of only those a_j which are positive somewhere in the interior of $\mathbf{F}_i(R_i)$. Put $A_i = \Sigma a_j$, the particular sum just mentioned, and we have $\omega_i = A_i\omega$. Each ω_i is zero on S except for points of S in the interior of $\mathbf{F}_i(R_i)$.

Having broken ω down into a sum of m k-forms $\omega_1,\ldots,\omega_m$, we assume, without loss of generality, that these ω_i are very simple. That is, we assume that, say, ω_i is simply the k-form $G_i(x_1,\ldots,x_n)\,dx_1\cdots dx_k$. Now that part of S on which ω_i does not vanish is given by the chart $\mathbf{F}_i$, and we put $\mathbf{F}_i(R_i) = S_i$. Since $\mathbf{F}_i$ is a diffeomorphism, we can be assured that $x_{k+1},\ldots,x_n$ can all be solved for in terms of $x_1,\ldots,x_k$, thanks to the implicit function theorem. Hence the integral

$$\int_{S_i} G_i(x_1,\ldots,x_n)\,dx_1\cdots dx_k = \int_{\pi S_i} B_i(x_1,\ldots,x_k)\,dx_1\cdots dx_k,$$

where πS_i is the projection of $S_i = \mathbf{F}_i(R_i)$ to $\mathrm{sp}(x_1,\ldots,x_k)$ and B_i is the function obtained from G_i by substituting the expressions in $x_1,\ldots,x_k$ for the terms $x_{k+1},\ldots,x_n$. Since ω_i is continuous, this integral exists, and its value may be obtained by integrating the pullback.

This last statement means that the integral of the k-form ω_i over S_i is really independent of the parametrization $\mathbf{F}_i$ since the solutions for $x_{k+1},\ldots,x_n$ in terms of $x_1,\ldots,x_k$ depends only upon the algebraic relationship existing among the xs and not upon the particular parametrization used to describe S_i. Hence, if ω_i^* is the pullback of ω_i under $\mathbf{F}_i$, we can write

$$\int_{S_i} \omega_i = \int_{R_i} \omega_i^*,$$

and finally,

$$\int_S \omega = \sum_{i=1}^{m} \int_{R_i} \omega_i^*.$$

What we have shown so far is that what we talked about in Chapter 9 as the integral of a continuous k-form over a "k-dimensional surface" in E^n is really an integral of a continuous function over a bounded measurable set in E^k, or a sum of such things, and hence it exists and can be evaluated, either directly or by integrating a pullback. We could go through an analogous argument to show that if $\boldsymbol{\beta}$ is a continuous $(k-1)$-form and the manifold under consideration is now the boundary ∂S, then $\int_{\partial S}\boldsymbol{\beta}$ exists and can be evaluated by integrating pullbacks.

12.2 THE STOKES THEOREM

Recall that if ω is a C^1 k-form in E^n, $k<n$, we described in Chapter 5 what we mean by $d\omega$. Moreover, as a consequence of the "equality of mixed partials" theorem, if ω is C^2 then $d(d\omega)\equiv 0$. In case $k=n$, $d\omega$ itself turns out to be zero.

Theorem 12.2.1 (Stokes). *Let S be a compact, orientable, differentiable k-dimensional manifold with boundary in E^n and ω a $(k-1)$-form in E^n, defined, and C^1 at all points of S. Then*

$$\int_{\partial S} \omega = \int_S d\omega.$$

PROOF: We use the "partitioning of unity" idea described above to partition $d\omega$. Since

$$\int_{\partial S_i} \omega_i = \int_{\partial R_i} \omega_i^*, \qquad \int_{S_i} d\omega_i = \int_{R_i} (d\omega_i)^*,$$

we now observe that $d(\omega_i^*) = (d\omega_i)^*$, a consequence of the chain rule for differentiation (however, the details of this observation should be worked out by the reader, and we leave this as an exercise), so we need first to show that

$$\int_{\partial R_i} \omega_i^* = \int_{R_i} (d\omega_i)^*$$

for each i. Let $i = 1, \ldots, m$ be arbitrary, and put $\boldsymbol{\sigma}_i = \omega_i^*$, so $d\boldsymbol{\sigma}_i = d(\omega_i^*) = (d\omega_i)^*$. Suppose $\boldsymbol{\sigma}_i = A(u_1, \ldots, u_k)\, du_2 \cdots du_k$. Then

$$\int_{\partial R_i} \boldsymbol{\sigma}_i = \int_{\pi R_i} [A(b_1, u_2, \ldots, u_k) - A(a_1, u_2, \ldots, u_k)]\, du_2 \cdots du_k,$$

where πR_i is the projection of R_i to the $(u_2, \ldots, u_k)$-coordinate hyperplane, and $a_1 \leqslant u_1 \leqslant b_1$. This is the case because $\boldsymbol{\sigma}_i$ vanishes on all the "faces" of R_i on which $u_2, \ldots, u_k$ are constant. The orientation of R_i accounts for the minus sign.

But the integral on the right-hand side is

$$\int_{\pi R_i} \left[\int_{a_1}^{b_1} \frac{\partial A}{\partial u_1}(u_1, \ldots, u_k)\, du_1 \right] du_2 \cdots du_k,$$

which is precisely $\int_{R_i} d\boldsymbol{\sigma}_i$. This suffices to prove that for a more general $\boldsymbol{\sigma}_i$ the same result will be obtained, and we conclude that for each i

$$\int_{\partial S_i} \omega_i = \int_{S_i} d\omega_i.$$

Now, $d\omega = \Sigma\, d\omega_i$, where each $d\omega_i$ vanishes outside of $\mathring{S}_i$. Hence, if we sum on i the integrals $\int_{S_i} d\omega_i$, we get

$$\sum_i \int_{S_i} d\omega_i = \sum_i \int_S d\omega_i = \int_S d\omega.$$

Put $\boldsymbol{\sigma} = \omega^* = (\Sigma\, \omega_i)^*$. Now the pullback of this sum is just the sum of the pullbacks, so we can write $\boldsymbol{\sigma} = \Sigma\, \boldsymbol{\sigma}_i$ and $d\boldsymbol{\sigma} = \Sigma\, d\boldsymbol{\sigma}_i$, where $d\boldsymbol{\sigma}_i = d(\omega_i^*) = (d\omega_i)^*$, and note that $d\boldsymbol{\sigma}_i$ vanishes outside the interior of R_i. Moreover, we have shown actually that for this $\boldsymbol{\sigma}$ (and not just $\boldsymbol{\sigma}_i$)

$$\int_{\partial R_i} \boldsymbol{\sigma} = \int_{R_i} d\boldsymbol{\sigma}.$$

Hence

$$\int_{\partial R_i} \boldsymbol{\sigma} = \int_{R_i} d\boldsymbol{\sigma} = \int_{R_i} \Sigma\, d\boldsymbol{\sigma}_i = \int_{R_i} d\boldsymbol{\sigma}_i = \int_{R_i} (d\omega_i)^*$$

for each i, so

$$\int_{\partial R_i} \omega^* = \int_{R_i} d(\omega^*).$$

Thus

$$\int_{\partial S_i} \omega = \int_{S_i} d\omega.$$

The boundary of S consists of points which belong to the sets ∂S_i, so for each i we can split ∂S_i into two disjoint sets

$$C_i = (\partial S) \cap (\partial S_i) \qquad \text{and} \qquad D_i = \partial S_i \setminus C_i.$$

We can now write

$$\int_{\partial S_i} \omega = \int_{C_i} \omega + \int_{D_i} \omega.$$

If we sum on i, we get

$$\sum_i \int_{C_i} \omega = \int_{\partial S} \omega.$$

What remains to be shown is that

$$\sum_i \int_{D_i} \omega = 0.$$

To do this without first making a great simplification is quite messy, so we make the simplification and leave it to the reader to justify what we have done. Assume now that the regions $S_i \subset S$ are such that for each i, D_i can be partitioned into parts D_{ij} such that the part D_{ij} is the common boundary shared by S_i and S_j; i.e., $D_{ij} = D_i \cap D_j$, and that $D_{ij} \cap D_k = \varnothing$ for any $k \neq i, j$. Under this assumption it is easy to see that, since the orientation of D_{ij} is opposite to the orientation of D_{ji},

$$\sum_{i,j} \int_{D_{ij}} \omega = 0,$$

and we can conclude that

$$\sum_i \int_{\partial S_i} \omega = \int_{\partial S} \omega.$$

Having already shown that

$$\sum_i \int_{S_i} d\omega_i = \sum_i \int_{S_i} d\omega = \int_S d\omega,$$

we have

$$\int_{\partial S} \omega = \int_S d\omega,$$

and the proof of Stokes's theorem is complete. □

It is well to make the remark that we have essentially proved that for a suitable manifold S and a continuous form ω of the right order, the integral $\int_S \omega$ behaves like an ordinary integral in that it is linear, additive, independent of parametrization, approximatable for sufficiently small S by an integral of a constant form. Moreover, there is an "integration-by-parts" formula

$$\int_S (d\omega_1)\omega_2 = \int_{\partial S} \omega_1\omega_2 - (-1)^{k_1} \int_S \omega_1(d\omega_2),$$

where S is a C^1 (k_1+k_2+1)-dimensional oriented manifold with boundary, ω_1 is a C^1 k_1-form on S, and ω_2 is a C^1 k_2-form on S. We leave it to the reader to verify that $d(\omega_1\omega_2) = (d\omega_1)\omega_2 + (-1)^{k_1}\omega_1(d\omega_2)$.

We will close this section by applying the generalized Stokes theorem to some special situations.

Theorem 12.2.2 (Green). *Let R be a positively oriented region in the plane E^2 whose boundary C is a simple rectifiable closed curve. Let ω be the C^1 1-form in E^2: $\omega = A\,dx + B\,dy$. Then*

$$\int_C \omega = \int_C A\,dx + B\,dy = \int\int_R \left(\frac{\partial B}{\partial x} - \frac{\partial A}{\partial y} \right) dx\,dy = \int_R d\omega.$$

Theorem 12.2.3 (Divergence Theorem). *Let R be a positively oriented region in E^3 with a regular C^1-boundary S, and ω a C^1 2-form in E^3:*

$$\omega = A\,dy\,dz + B\,dz\,dx + C\,dx\,dy.$$

Then

$$\int_R d\omega = \int\int\int_R \left(\frac{\partial A}{\partial x} + \frac{\partial B}{\partial y} + \frac{\partial C}{\partial z} \right) dx\,dy\,dz = \int\int\int_R \operatorname{div} \mathbf{F}\, dV = \int\int_S \mathbf{F}\cdot\mathbf{n}\, d\Sigma$$

$$= \int\int_S (A\cos\alpha + B\cos\beta + C\cos\gamma)\, d\Sigma = \int_S \omega,$$

where dV is measure in E^3, $d\Sigma$ is measure on S, $\mathbf{F}=(A,B,C)$, $\operatorname{div}\mathbf{F}=\partial A/\partial x+\partial B/\partial y+\partial C/\partial z=\nabla\cdot\mathbf{F}$, *and* $\mathbf{n}$ *is the "unit outer normal" to S. If the surface S is described by an equation $\varphi(x,y,z)=0$, then* $\mathbf{n}=\mathbf{grad}\,\varphi/\|\mathbf{grad}\,\varphi\|=(\cos\alpha,\cos\beta,\cos\gamma)$. *If S is described parametrically by the mapping ψ,*

$$x=x(u,v),\qquad y=y(u,v),\qquad z=z(u,v),$$

then

$$\mathbf{n}=\frac{1}{g_\psi}\left(\det\frac{\partial(y,z)}{\partial(u,v)},\det\frac{\partial(z,x)}{\partial(u,v)},\det\frac{\partial(x,y)}{\partial(u,v)}\right),$$

where g_ψ is the normalizing factor. α,β,γ are the direction angles of $\mathbf{n}$.

Theorem 12.2.4 (Stokes). *Let S be an oriented 2-manifold with boundary C in E^3, and $\omega=A\,dx+B\,dy+C\,dz$ a C^1 1-form in E^3. Then*

$$\int_C\omega=\int_C A\,dx+B\,dy+C\,dz=\int_C\mathbf{F}\cdot\mathbf{T}\,d\sigma=\int\!\!\int_S(\mathbf{curl}\,\mathbf{F})\cdot\mathbf{n}\,d\Sigma$$

$$=\int\!\!\int_S\left[\left(\frac{\partial C}{\partial y}-\frac{\partial B}{\partial z}\right)\cos\alpha+\left(\frac{\partial A}{\partial z}-\frac{\partial C}{\partial x}\right)\cos\beta+\left(\frac{\partial B}{\partial x}-\frac{\partial A}{\partial y}\right)\cos\gamma\right]d\Sigma$$

$$=\int_S d\omega,$$

where $\mathbf{T}$ *is the unit tangent vector to C,* $\mathbf{F}=(A,B,C)$, *$d\sigma$ is measure along C,* $\mathbf{curl}\,\mathbf{F}=\nabla\times\mathbf{F}$, *$d\Sigma$ is measure on S, and* $\mathbf{n}$ *is the unit outer normal to S.*

Theorem 12.2.5. *Let ω be the C^1 zero-form F in E^3 and C a connected compact oriented 1-manifold in E^3 with boundary $\{a,b\}$. Then*

$$\int_{\partial C}\omega=\int_{\{a,b\}}F=F(b)-F(a)=\int_C dF=\int_C\frac{\partial F}{\partial x}dx+\frac{\partial F}{\partial y}dy+\frac{\partial F}{\partial z}dz$$

$$=\int_C(\mathbf{grad}\,F)\cdot\mathbf{T}\,d\sigma=\int_C d\omega,$$

where $d\sigma$ is measure along the arc C and $\mathbf{T}$ *is the unit tangent vector to C.*

Corollary 12.2.6. *If f is a C^1 zero-form in E^1 and I an oriented compact interval with boundary $\{a,b\}$, then*

$$\int_{\partial I}\omega=\int_{\{a,b\}}f=f(b)-f(a)=\int_a^b f'(t)\,dt=\int_I d\omega.$$

Since all of these theorems are special cases of our Theorem 12.2.1, we need not prove them. We refer the reader to the discussion of curl and divergence in Section 9.2.

We think it is time to bring this course to a close. Much has been left unsaid, of course, and the student who continues on in mathematics will find many, many topics which should have been included in this book. On the other hand, remember that we really didn't try to tell you all about calculus, or mathematical analysis; just enough to pique your curiosity. We hope we have succeeded.

PROBLEMS

1. Compute the following integrals:
   (a) $\int_C -y\,dx + x\,dy$ along $y^2 = 4x$ from (1,2) to (4,4);
   (b) $\int_C (x^2 - y^2)\,dx + x\,dy$ along $x^2 + y^2 = 9$ from (0,3) to (3,0), in a clockwise direction;
   (c) $\int_C -3y\,dx + 2x\,dy + 4z\,dz$, where C is the arc of the curve given by $x^2 + y^2 = 3 - z^2$ and $x + y + z = 1$ from $(-1,1,1)$ to $(0,(1-\sqrt{5})/2, (1+\sqrt{5})/2)$;
   (d) $\int_C (y+z)\,dx + (x+z)\,dy + (x+y)\,dz$, where C is the path from (0,0,0) to (1,1,1) given by

$$x = e^t - t(e^t - 1), \qquad y = \sin^{5/3}(\pi t/2), \qquad z = t^7.$$

2. A particle of weight w moves along a path $y = \frac{1}{8}(x-4)^2$, acted upon by gravity in the $-y$ direction and a strange force field F whose components are $(y,0)$ at each point $(x,y) \in E^2$. Find the total work done by these forces in moving the particle from (0,2) to (4,0).

3. Evaluate $\iiint_B (x^2 + y^2 + z^2)^{1/2}\,dx\,dy\,dz$, where B is the ball $x^2 + y^2 + z^2 \leqslant b^2$ in E^3.

4. Evaluate $\iint_S x\,dy\,dz + y\,dx\,dz + z\,dx\,dy$, where S is the surface bounding the solid $x^2/3 + y^2/4 + z^2/5 = 1$.

5. Let R be the region of the first octant bounded by $x^2 + y^2 = 1$, $z = x^2 + y^2$, $z = 0$. If $\omega = y^2 z\,dx\,dy + x^2 y\,dx\,dz + xy\,dy\,dz$, evaluate $\int_{\partial R} \omega$.

6. Let L be a simple closed oriented curve in E^2 and $d\sigma$ measure along L. (By a simple curve we mean one that does not intersect itself, and by a closed curve we mean an arc whose endpoints coincide.) Let α be the angle between the outer normal to L and the positive x axis. Prove that $\int_L (x\cos\alpha + y\sin\alpha)\,d\sigma$, where the integration is in the direction of the positive orientation of L, is equal to twice the area of the region bounded by L.

7. Suppose S is a 2-manifold in E^4 described parametrically by φ, given by

$$x_1 = x_1(u_1, u_2), \qquad x_2 = x_2(u_1, u_2),$$

$$x_3 = x_3(u_1, u_2), \qquad x_4 = x_4(u_1, u_2).$$

Then

$$(dx_i)^2 = \left[\frac{\partial x_i}{\partial u_1} du_1 + \frac{\partial x_i}{\partial u_2} du_2\right]^2.$$

An increment of length is given by the fundamental form $ds^2 = \Sigma_{i=1}^4 dx_i^2$. Express ds^2 in terms of the partial derivatives to get

$$ds^2 = \sum_{\alpha,\beta=1}^{2} \sum_{i=1}^{4} \frac{\partial x_i}{\partial u_\alpha} \frac{\partial x_i}{\partial u_\beta} du_\alpha du_\beta.$$

Just because it is done this way in geometry, put the indices associated with the xs and us up where exponents usually are; they are still indices, but now upper indices. Now

$$ds^2 = \sum_{\alpha,\beta=1}^{2} \sum_{i=1}^{4} \frac{\partial x^i}{\partial u^\alpha} \frac{\partial x^i}{\partial u^\beta} du^\alpha du^\beta.$$

Denote by $g_{\alpha\beta}$ the expression

$$\sum_{i=1}^{4} \frac{\partial x^i}{\partial u^\alpha} \frac{\partial x^i}{\partial u^\beta},$$

so that $ds^2 = \Sigma_{\alpha,\beta=1}^2 g_{\alpha\beta} du^\alpha du^\beta$. Note that $g_{12} = g_{21}$. We call $g_{\alpha\beta}$ a *tensor*; in this case, $g_{\alpha\beta}$ is a *symmetric tensor*. It has four components, and this particular tensor is called the fundamental metric tensor of the surface S. Now let T be the surface given by ψ:

$$x_1 = u_1 \cos u_2, \qquad x_2 = u_1 \sin u_2 \cos u_2$$

$$x_3 = u_1 \sin^2 u_2, \qquad x_4 = u_1^2,$$

$$0 \leqslant u_1 \leqslant R, \qquad 0 \leqslant u_2 \leqslant \pi.$$

Find the fundamental metric tensor for this surface. Next show that the g factor for the first map φ satisfies the equation $g = \sqrt{g_{11}g_{22} - (g_{12})^2}$. Hence the area of S can be expressed in terms of the components of the fundamental metric tensor. Often the symbols E, F, and G are used for the components g_{11}, $g_{12} = g_{21}$, and g_{22}, respectively. Find

$$\int_T \sqrt{EG - F^2}\, d\mu = \int\int_T \sqrt{EG - F^2}\, du_1\, du_2.$$

8. Verify that $d(\omega_1\omega_2) = (d\omega_1)\omega_2 + (-1)^{k_1}\omega_1(d\omega_2)$, where ω_1 is a C^1 k_1-form and ω_2 is a C^1 k_2-form, and hence

$$\int_S (d\omega_1)\omega_2 = \int_{\partial S} \omega_1\omega_2 - (-1)^{k_1} \int_S \omega_1(d\omega_2),$$

where S is a $(k_1 + k_2 + 1)$-dimensional smooth manifold.

9. If γ is the curve $x = 3\cos t$, $y = 2\sin t$, $0 \leqslant t < 2\pi$, can you find the measure of γ? If a surface S is given by $x = 3\cos u$, $y = 2\sin u \cos v$, $z = \sin u \sin v$, $0 \leqslant u < \pi$, $0 \leqslant v < 2\pi$, can you find the measure of S? If not, why not? You might find out what an elliptic integral is, and how to evaluate one.

10. If L is the boundary of the region R bounded by $y=x^2$, $y=0$, $x=2$, evaluate

$$\int_L (1-x^2)y\,dx + x(1+y^2)\,dy.$$

11. Evaluate $\int_S (z+2x+\frac{4}{3}y)\,d\mu$, where S is given by $x/2+y/3+z/4=1$, x, y, $z \geqslant 0$.

12. Evaluate $\int_S x^2y^2z\,dx\,dy$, where S is the positively oriented lower hemisphere $x^2+y^2+z^2=R^2$.

13. Integrate $\int_\gamma [(x\,dx+y\,dy)/(x^2+y^2)]$, where γ is an arc, not containing the origin, initiating at $(5,12)$ and terminating at $(3,4)$.

14. Find u if

(a) $du = x^2\,dx + y^2\,dy$,

(b) $du = (2x\cos y - y^2\sin x)\,dx + (2y\cos x - x^2\sin y)\,dy$,

(c) $du = (x+y+z)^{-1}(dx+dy+dz)$,

(d) $du = e^{y/z}\,dx + (z^{-1}e^{y/z}(x+1) + ze^{yz})\,dy$
$+(e^{-z} + ye^{yz} - z^{-2}e^{y/z}(x+1)y)\,dz$,

(e) $du = z^{-1}(dx - 3\,dy) + z^{-2}(3y - x + z^3)\,dz$.

15. Find the area of the region R enclosed by the curve $x^3+y^3-3axy=0$ by integrating a 1-form around the boundary.
[*Hint*: $\int_R dx\,dy = \int_R d\omega = \int_{\partial R}\omega$, so $\omega = \frac{1}{2}(x\,dy - y\,dx)$. Try letting $y=tx$ to get a parametrization for ∂R. This should lead to

$$x = \frac{3at}{1+t^3}, \qquad y = \frac{3at^2}{1+t^3}.$$

$(x,y)=(0,0)$ when $t=0$ and when $t=\infty$.]

16. Find the area enclosed by $(x^2+y^2)^2 = 2a^2(x^2-y^2)$.

17. Let S be the sphere $x^2+y^2+z^2=R^2$ and r the distance of a point on the sphere and the point $(0,0,c)$, where $c>R$. Evaluate $\iint_S r^{-n}\,d\mu$, $n=0,1,2,3,\ldots$.

18. Find the mass of S in the preceding problem if the density at a point on S is equal to the distance of the point from a fixed diameter.

19. Suppose C is the closed curve

$$x=\cos t,\quad y=\sin t,\quad z=\sin 2t,\qquad 0\leqslant t\leqslant 2\pi,$$

and let S be a surface having C as boundary. Evaluate $\int_C \omega = \int_C (y^2+z^2)\,dx + (x^2+z^2)\,dy + (x^2+y^2)\,dz$. What is the corresponding 2-form $d\omega$ you would integrate over S to get the same result?

Suppose S is given by

$$x = r\cos t,\quad y=r\sin t,\quad z=r^2\sin 2t,\qquad (r,t)\in[0,1]\times[0,2\pi).$$

Evaluate $\int_S d\omega$ and verify that it equals $\int_C \omega$.

Finally, find the measure of S, and if you feel so inclined, see if you can approximate the length of C. (This length will be the value of an elliptic integral.)

20. Find a zero-form ω if

$$d\omega = 2xz^{-1}\,dx + 2yz^{-1}\,dy + (z^2 - x^2 - y^2)z^{-2}\,dz.$$

Integrate $d\omega$ along a sensuous path from $(1,1,1)$ to (a,b,c).

21. Let S be the positively oriented surface in the first octant formed by

$$z = x^2 + y^2, \qquad 1 = x^2 + y^2, \qquad x = 0, \qquad y = 0, \qquad z = 0.$$

Integrate ω over S, where

$$\omega = y^2 z\,dx\,dy + xz\,dy\,dz + x^2 y\,dx\,dz.$$

[*Hint*: Try using the divergence theorem 12.2.3.]

22. If S is the positively oriented sphere $x^2 + y^2 + z^2 = \pi^2$ and ω is the 2-form $x^2\,dy\,dz + y^2\,dx\,dz + z^2\,dx\,dy$, compute $\int_S \omega$.

If σ is the 2-form $(z/3)\,dx\,dy - (y/3)\,dx\,dz + (x/3)\,dy\,dz$, jot down the value of $\int_S \sigma$.

23. Let $f(z)$ be a complex-valued function of a complex variable, given by $f(z) = z(1+z^2)^{1/2}$. Let γ be the path on the complex plane given by $z(t) = x(t) + iy(t)$, where $x(t) = 1 + t^2$, $y(t) = t^{3/2}$, $0 \leqslant t \leqslant 2$. Evaluate

$$\int_\gamma f(z)\,dz \qquad \text{and} \qquad \int_1^{5+2i\sqrt{2}} f(z)\,dz.$$

24. Let B be the closed ball in E^5 with radius 2, centered at the origin, and let ω be the 4-form

$$\omega = x^3\,dx\,dy\,dz\,du - y\,dx\,dz\,du\,dv + z^2\,dx\,dy\,dz\,dv + x\,dy\,dz\,du\,dv$$
$$+ \cos(x^2 + y^2 + v)\,dx\,dy\,du\,dv.$$

Calculate $\int_{\partial B} \omega$.

25. Let R be a solid whose base is the circular disk $x^2 + y^2 \leqslant a^2$. The right-sections obtained by intersecting R by planes perpendicular to the x axis are parabolic segments of constant altitude H. Find the volume of R. Can you also find the total surface area of R?

26. Evaluate $\int_C x^2 y^3\,dx + dy + z\,dz$, where C is the positively oriented circle $x^2 + y^2 = R^2$, $z = 0$. Evaluate $-\int_S 3x^2 y^2\,dx\,dy$, where S is the positively oriented hemisphere $z = \sqrt{R^2 - x^2 - y^2}$.

27. Find the area of a torus whose outer diameter is $2(a+b)$ and inner diameter is $2(b-a)$. Find its volume.

28. Let S be the surface $x^2+y^2+z^2=2az$, and α, β, γ the direction angles of the outer normal. If $d\Sigma$ is surface measure, show that

$$\iint_S (x^2\cos\alpha + y^2\cos\beta + z^2\cos\gamma)\, d\Sigma = \tfrac{8}{3}\pi a^4.$$

29. Suppose $\mathbf{F}$ is a vector field in E^3, R is a region having volume V, and we imagine R to be shrinking continuously to a fixed point $\mathbf{p}\in E^3$. Show that $\operatorname{div}\mathbf{F}$ at $\mathbf{p}$ is given by

$$\operatorname{div}\mathbf{F} = \lim \frac{1}{V}\iint_S \mathbf{F}\cdot\mathbf{n}\, d\Sigma,$$

where $\mathbf{n}$ is the unit outer normal to $S=\partial R$ and $d\Sigma$ is surface measure.

30. Suppose $f\colon E^3\to\mathbb{R}$. Recall that the directional derivative of f in the direction $\mathbf{n}$, where $\mathbf{n}$ is a unit vector, is $\nabla f\cdot\mathbf{n}$. Let $g\colon E^3\to\mathbb{R}$. Show that if T is a "suitable" region in E^3 with $S=\partial T$

$$\int_T (g\nabla^2 f + \nabla g\cdot\nabla f)\, dV = \int_S g\nabla f\cdot\mathbf{n}\, dA,$$

where $\mathbf{n}$ is the outer normal to S.
[*Hint*: $\nabla^2 f = D_{11}f + D_{22}f + D_{33}f$, the Laplacian of f. Put $P=gD_1f$, $Q=gD_2f$, $R=gD_3f$; then (P,Q,R) is a vector. What is $\operatorname{div}(P,Q,R)$? Is it not the left-hand integrand? Now what is $(P,Q,R)\cdot\mathbf{n}$? Apply Theorem 12.2.3.] This formula you have verified is known as the first of Green's identities.

31. Verify the second Green identity:

$$\int_T (g\nabla^2 f - f\nabla^2 g)\, dV = \int_S (g\nabla f - f\nabla g)\cdot\mathbf{n}\, d\Sigma.$$

32. Show that $\int_T \nabla^2 f\, dV = \int_S \nabla f\cdot\mathbf{n}\, d\Sigma$ and that if $\nabla^2 f = 0$ on T (i.e., that f is harmonic on T), then

$$\int_T \left[\left(\frac{\partial f}{\partial x}\right)^2 + \left(\frac{\partial f}{\partial y}\right)^2 + \left(\frac{\partial f}{\partial z}\right)^2\right] dV = \int_S f\nabla f\cdot\mathbf{n}\, d\Sigma.$$

33. Consider the transformation that maps "spherical" to "Cartesian coordinates." Show that if $f\colon E^3\to\mathbb{R}$,

$$\nabla^2 f = \frac{1}{\rho^2\sin\varphi}\left[\sin\varphi\frac{\partial}{\partial\rho}\left(\rho^2\frac{\partial f}{\partial\rho}\right) + \frac{1}{\sin\varphi}\frac{\partial^2 f}{\partial\theta^2} + \frac{\partial}{\partial\varphi}\left(\sin\varphi\frac{\partial f}{\partial\varphi}\right)\right].$$

34. Refer to Problem 29, and show that at a point $\mathbf{p}$ to which a region shrinks is

$$\nabla^2 f = \lim\left[\frac{1}{V}\iint_S \nabla f\cdot\mathbf{n}\, d\Sigma\right],$$

where $f\colon E^3\to\mathbb{R}$.

TIPS AND SOLUTIONS FOR SELECTED PROBLEMS

CHAPTER 1

1. $x \in A \cap (B \,\Delta\, C) \Leftrightarrow x \in A$ or $x \in$ exactly one of B or $C \Leftrightarrow x \in (A \cap B$ or $A \cap C$, but not both) $\Leftrightarrow x \in (A \cap B) \,\Delta\, (A \cap C)$.

2. $x \in \bigcup_\alpha \tilde{A}_\alpha \Leftrightarrow \exists_\alpha$ s.t. $x \notin A_\alpha \Leftrightarrow x \notin \bigcap_\alpha A_\alpha \Leftrightarrow x \notin \widetilde{\bigcap_\alpha} A_\alpha$.

3. $x \in \overline{S}$ implies that x must be a member of infinitely many of the sets A_n, but could also not belong to infinitely many of the A's. For example, if x is in each A with an even subscript but fails to be in each A with an odd subscript, then x is in infinitely many of the A's but fails to be in infinitely many as well. If x is in $\underline{S}$, then x must be in each A_n from some n_0 on.

4. If $q = a + bi + cj + dk$ is an arbitrary quaternion, associate it with the 2×2 matrix M_q whose entries are the complex numbers M_{11}, M_{12}, M_{21}, M_{22} which are, respectively, the complex numbers $a + bi$, $c + di$, $-c + di$, $a - bi$. Let Q be another arbitrary quaternion, with its corresponding matrix M_Q. Show that the collection of all such matrices M_q comprise a division ring, that the zero and identity matrices correspond to the zero and unit of the division ring of quaternions, respectively, and that sums, products, quotients and conjugates correspond accordingly.

6. $a/b < (a + 2b)/(a + b)$ iff $a^2 + ab < a^2b + 2b$ iff $a/b < \sqrt{2}$. $\sqrt{2} < (a + 2b)/(a + b)$ iff $(2a^2 + 4ab + 2b^2) < (a^2 + 4ab + 4b^2)$ iff $a/b < \sqrt{2}$. Therefore, $a/b < \sqrt{2}$ iff $a/b < \sqrt{2} < (a + 2b)/(a + b)$ and $a/b > \sqrt{2}$ iff $(a+2b)/(a+b) < \sqrt{2} < a/b$.

7. $x = \{x_i\} \in P \Leftrightarrow \forall i,\ x_i \geqslant 0$, and for at least one i, $x_i \neq 0$.

12. S is a commutative ring of sets with $\cap$ as multiplication and Δ as addition. If $A, B \in S$, then $A \,\Delta\, B = (A \setminus B) \cup (B \setminus A) = \varnothing \Rightarrow A = B$.

14. Use finite induction. Any point $\mathbf{x}$ on a line segment $[\mathbf{x}_1,\mathbf{x}_2]$ can be represented by: $\mathbf{x} = t_1\mathbf{x}_1 + (1 - t_1)\mathbf{x}_2$, $0 \leqslant t_1 \leqslant 1$. Any point $\mathbf{x}$ in the triangular region $[\mathbf{x}_1,\mathbf{x}_2,\mathbf{x}_3]$ can be represented by: $\mathbf{x} = t_2[t_1\mathbf{x}_1 + (1 - t_1)\mathbf{x}_2 + (1 - t_2)\mathbf{x}_3]$, where $(t_1t_2) + (t_2 - t_2t_1) + (1 - t_2) = 1$. It follows by induction that if the point $\mathbf{x}_n$ is not already in the linear span of $\{\mathbf{x}_1, ---, \mathbf{x}_{n-1}\}$, then any point $\mathbf{x}$ on the segment initiating at any point of the polytope determined by the first n−1 points and terminating at $\mathbf{x}_n$ is of the form $\mathbf{x}_n = t_1\mathbf{x}_1 + t_2\mathbf{x}_2 + --- + t_{n-1}\mathbf{x}_{n-1} + t_n\mathbf{x}_n$, with $\sum_{i=1}^{n} t_i = 1$, and $\forall i$, $0 \leqslant t_i \leqslant 1$. The set of all such points $\mathbf{x}$ is clearly convex and contains the initial n points. That it is the smallest such convex set is clear.

16. For any two maps f and g mapping $\mathbb{R} \to \mathbb{R}$ we define $(f \wedge g)x = f[x] \wedge g[x]$ and $(f \vee g)x$ accordingly. For each real number x, an arbitrary collection of such maps f_α gives rise to a set of numbers $\{f_\alpha[x]\}$. If this set is bounded, the numbers $\inf_\alpha\{f_\alpha[x]\}$ and $\sup_\alpha\{f_\alpha[x]\}$ exist, since $\mathbb{R}$ is complete. Define $\sup_\alpha\{f_\alpha\} = f$ by $f[x] = \sup_\alpha\{f_\alpha[x]\}$ for each x. Define $g = \inf_\alpha\{g_\alpha\}$ accordingly. Thus, the set of such maps comprise a conditionally complete lattice.

17. That Card $\mathbb{P} \leqslant$ Card $\mathbb{Z}$ is clear. Suppose $\mathbb{P}$ is finite, with P the largest of the primes. Form the product of all the primes and add 1 to that product, obtaining the rather large number $M = 2*3*5*\ldots*P + 1$. Since M is not divisible by any prime less than M, M itself must be prime. Contradiction; M is clearly larger than P. Hence, the assumption that $\mathbb{P}$ is finite leads to a contradiction, so the set of all primes is an infinite set, but its cardinality cannot be larger than that of the set of all integers, $\mathbb{Z}$.

18. A positive real number y is equal to e^x for a unique real number x. $1 = e^0$ corresponds to $x = 0$, $y = e^x$ corresponds to x, $e^xe^z = e^{x+z}$ corresponds to $x+z$. Since $y = e^x$ is strictly increasing over the reals, we have a group-isomorphism between the additive group of real numbers and the multiplicative group of positive reals. One possible way to obtain a set-isomorphism between the real line and the plane is this: let each point $(.a_1a_2a_3\ldots, .b_1b_2b_3\ldots)$ in the open unit square of quadrant 1 of the plane correspond to the real number $.a_1b_1a_2b_2\ldots$ in the interval (0,1), and conversely. There is an isomorphism between the interval (0,1) and the interval $(-\pi/2, \pi/2)$, and another one, $\theta \leftrightarrow \tan[\theta]$, mapping that interval to the whole real line. All that remains is to get a 1-1 correspondence between the open first quadrant and the plane. Better still, find a neater way to do the problem.

19. $0 < i < 1$, but for no n is $ni > 1$.

23. The polynomial $p[x] = x^6 - 9x^4 - 4x^3 + 27x^2 - 36x - 23$ vanishes if $x = \sqrt{3} + \sqrt[3]{2}$. There is one other zero for $p[x]$, $\sqrt[3]{2} - \sqrt{3}$, and no others. You might try various schemes to see if you can build the polynomial $p[x]$.

32. $\{r/40\}$ and $\{s/40\}$, where r is an integer of the form $4a + 5b$, $a, b \in \mathbb{Z}$ and $s \in \mathbb{Z}$.

34. If A_1 is the singleton $\{1\}$ and the sets A_n are the sets $\{2, 3, ---, n\}$, $n > 2$, it is not hard to see that the union of the power sets of A_1 and A_n will not contain the set $\{1,2\}$, although this union will contain $\{\{1\},\{2\}\}$, whereas the set $\{1, 2\}$ is certainly one of the subsets of the union of the A's.

36. Consider any prime p. Then 1 divides p, but no other integer divides p. Thus, every chain has an upper bound 1, and 1 is the only maximal element. Thus the ordering $m < n$ if n divides m is a partial order and every chain has a maximal element; 1 is the only such element. If $m < n$ when m divides n, then no chain has an upper bound. This **P.O.** has a unique minimal element, 1.

37. The axiom P3 does not hold unless we modify our notion of equality of functions.

46. The set of positive integers is countable; thus the collection S_r of points of the cartesian plane having coordinates (n, r), where n is a natural number and r is a fixed real number, is a countable collection. It follows that if we consider the collection $\{S_r, r \in \mathbb{R}\}$ of all these countable subsets, it is most certainly uncountable. These are but a few of the uncountable subsets of $\mathbb{R}^2$.

48. $\mathbb{Q}$ is itself a field. It is easy to show that all the field axioms hold for the set $\mathbb{A} = \{a + b\sqrt{3}\}$. We call $\mathbb{A}$ a **field extension** of $\mathbb{Q}$.

54. The relation is an equivalence relation; we say a is **congruent** to b modulo m.

CHAPTER 2

1. Let $\{x_n\}$ be Cauchy. Then $\{x_n\} \to x_0 \in E^n$, so for some n_0 we have $\|x_n - x_0\| < 1$ for all $n > n_0$. Let M be the maximum of the values $\|x_k - x_0\|$, $k = 1, \ldots, n_0$. Then all the terms $\{x_n\}$ lie in a ball of radius $M + 1$ centered at x_0.

4. If tan[x] is not equal to tan[y]:

$$\text{write } f[x, y] = \frac{(\sin[x]-\sin[y])\cos[x]\cos[y]}{\sin[x]\cos[y]-\sin[y]\cos[x]} = \frac{2\sin\frac{x-y}{2}\cos\frac{x+y}{2}\cos[x]\cos[y]}{2\sin\frac{x-y}{2}\cos\frac{x-y}{2}} \to \cos^3[x]$$

as $y \to x$. (Those old trig identities are useful upon occasions)

6. $[e^{-e}, e^{1/e}]$

7. (i) $(x^2 - 1)/(x - 1)$, removable; (ii) $\sqrt{x^2 - 1}$, gap; (iii) $\begin{cases} x - 123, \text{ if } x < 0 \\ x^{10} + 1, \text{ if } x > 0 \end{cases}$, jump; (iv) x^{-2}, infinite.

10. Fix n and let $m \to \infty$. If x is rational, $f[x] \to 0$ if $n!x$ is not an integral multiple of $\pi/2$. Then let $n \to \infty$ and 0 is the iterated limit. If x is irrational, then as $m \to \infty$, $f[x] \to 0$. Now fix m. If x is rational, as $n \to \infty$ the limit of $\cos^{2m}(n!\pi x) = 1$, so the iterated limit is 1. If x is irrational, the cosine is less than 1 in absolute value, so the limit as $m \to \infty$ is 0.

15. Let x_0 be arbitrary and $\varepsilon > 0$. Let $b = f[x_0] + \varepsilon$. If $f^{-1}[(-\infty,b)]$ is "open" (i.e., an interval not containing its endpoints) and contains x_0, we must have that for all x sufficiently close to x_0, $f[x] < f[x_0] + \varepsilon$. If x_0, or any x sufficiently close to x_0, is not in this preimage, then $f^{-1}[(-\infty,b)]$ is not "open".

18. If $x = g(f[x])$, then $1 = \frac{dg}{df} * \frac{df}{dx}$, or $1 \equiv \frac{dx}{dy} * \frac{dy}{dx}$.

21. 6

22. There does not exist such an M; consider $f[x] = \sqrt{x}$ on [0,1].

32. π

33. Since $f[x]$ is strictly positive over $[a,b]$ and continuous, we can be sure that f does not vanish in the interval or at the endpoints. Let c be an arbitrary point of the interval and $\varepsilon > 0$. For all x sufficiently close to c $|f[x] - f[c]| < \varepsilon$. There is a positive number M_c such that for all those x sufficiently close to c $f[x] > M_c$. Thus, for all such x sufficiently close to c, it follows that $|1/f[c] - 1/f[x]| = |f[x] - f[c]|/(f[x]f[c]) < \varepsilon/M^2$.

36. Suppose $\alpha > 0$. Let $\varepsilon > 0$. Let I be an interval centered at x_0. Choose I so that for all $x \in$ I, $|x - x_0| < \varepsilon$. If, for some $M > 0$, $|f[x] - f[x_0]| \leqslant M|x - x_0|^\alpha < \varepsilon$ for all $x \varepsilon$ I, this means that $|x - x_0| < (\varepsilon/M)^{1/\alpha} = \delta$, ensuring that $|f[x] - f[x_0]| < \varepsilon$ for all $x \in$ I; i.e., that f is continuous at x_0. If $\alpha > 1$, we can write $\varphi[x] = (f[x] - f[x_0])/(x - x_0)$. We have $|\varphi[x]| \leqslant M|x - x_0|^{\alpha-1} < \varepsilon$, for all $x \in$ I, and this is to say, the limit of $\varphi[x]$ as x approaches x_0 exists. Thus, f is differentiable at x_0.

37. **(c)** Let $y \to 0$. $f[x,y]$ has no limit since $(x\ (\sin[1/x])(\lim_0 \sin[1/y])$ does not exist. Similarly, the other iterated limit does not exist. However, for all (x,y) sufficiently close to (0,0) we have $|(x+y)\sin[1/x]\sin[1/y]|$ necessarily close to 0.

(f) The two iterated limits exist; the first is 1, the second, 0. However, the limit of f at (0,0) does not exist. If $(x,y) \to (0,0)$ along the path $y = x$, then $f[x,y] \to 1/2$, but if $(x,y) \to (0,0)$ along the path $y = 3x$, then $f[x,y] \to 1/4$.

38. **(d)** $f[x,y] = \frac{1}{y^2} \sum_{n=1}^{[y]} \frac{n \sin[{}^n/_x]}{{}^n/_x}$. As $y \to \infty, f \to 0$ and thus the first iterated limited exists. On the other hand, as $x \to \infty$, $f \to \frac{1}{(2y)} + \frac{1}{2}$, and the second iterated limit exists and equals $1/2 \neq 0$.

39. $f[x]$ is u.s.c. if x is rational since $\forall x \in [0,1], f[x] < 1 + \varepsilon, \forall \varepsilon > 0$. If x is irrational, f is not u.s.c. $f[x]$ is l.s.c. at each irrational x, since $\sin[1/x]$ is continuous except at $x = 0$.

44. Since $a_n = 2/(\sqrt{n+1} + \sqrt{n-1})$, $\lim\{a_n\} = 0$. Assume a_n is rational for some n and equal to j/k for some relatively prime integers j and k. Just as in the standard proof of the irrationality of the square root of 2 you will arrive at a contradiction. Apparently $\{a_n\} \downarrow 0$, so the limit is rational.

46. If $F[x] = \int_0^x f[t]dt$, then $\frac{dF}{dx} = \frac{d}{dx}(x\sin[\frac{1}{x}]) = \sin[\frac{1}{x}] - \frac{1}{x}\cos[\frac{1}{x}] = f[x]$. $F[x] = x\sin[1/x]$ is differentiable and bounded on [0,10] but its derivative is not bounded on [0,10] or defined on [0,10].

CHAPTER 3

4. Form the product of the two binomials, which must be equal to 1, and obtain the equation $a^2c^2 + b^2d^2 = 1 - (b^2c^2 + a^2d^2)$. Add $2abcd$ to both sides of this and arrive at the desired result.

6. This is a consequence of the GAM inequality. The inequality holds if the corresponding inequality of the logarithms of each side holds. You should be able to reduce this to:

$$n\log\sum k^r > n\log[n] + r(\log[n!]) \Leftrightarrow \log\sum k^r > \log[n] + \frac{r}{n}\sum_1^n \log[k] \Leftrightarrow$$

$$\log\left[\frac{1}{n}\sum k^r\right] > \frac{1}{n}\sum\log[k^r] \Leftrightarrow \text{AM}[\{k^r\}] > \text{GM}[\{k^r\}].$$

10. This is a bit tricky. If $0 < m < 1$, use Proposition 3.1.3, the generalized Hölder inequality: $\sum_1^n a_i^m 1_i^{1-m} \leqslant \left(\sum_1^n a_i\right)^m (\Sigma 1)^{1-m} = (\Sigma a_i)^m \frac{n}{n^m}$. Thus, $\frac{1}{n}\Sigma a_i^m \leqslant \left(\frac{\Sigma a_i}{n}\right)^m$. If $m > 1$, use similar tactics.

11. This one is again a direct application of the GAM inequality. Write the inequalities in terms of logarithms, and make use of the fact that the log function is "concave down". The left-hand inequality can be reduced to the equivalent inequality: $\frac{1}{n}\sum_1^n \log[k] < \log\left[\frac{n+2}{2}\right]$. The left side is the average of the n values of the logs of the first n integers, whereas the right side is the log of the average of the first and last integers. Since the log function is concave down, the inequality follows. A similar argument can be used to prove the right-hand inequality.

12. $1*3*\ldots*(2n-1) = (2n)!/(2^n n!) < n^n$ iff $(2n)!/n! < (2n)^n$ iff $(2n)(2n-1)*\ldots*(2n-[n+1]) < (2n)^n$ iff $(2n)*\ldots*(n+1)/(2n)^n < 1$.

14. The GAM inequality ensures that $(ab)^{(a+b)/2} < [(a+b)/2]^{(a+b)}$, assuming a and b are positive unequal numbers. Assume $0 < a < b$. Since $b*\log[a] + a*\log[b] < [(a+b)/2][\log[a] + \log[b]]$ iff $(b-a)\log[a] < (b-a)\log[b]$, we can conclude that $a^b b^a < (ab)^{(a+b)/2}$.

23. Note that $x = \lim\{x_n\} = \sqrt{2 + \sqrt{2 + \sqrt{2 + \ldots}}}$, so $x^2 = 2 + x$ and $x > 0$. Hence, $x = 2$. In general, if c is any positive real number, and $x_1 = \sqrt{c}$, then $x = \dfrac{1+\sqrt{1+4c}}{2}$.

CHAPTER 4

5. Let A be the matrix that maps the vectors $\mathbf{b}_1(1,1,1)$, $\mathbf{b}_2(4,-2,-1)$ and $\mathbf{b}_3(5,3,1)$ to (1,0,0), (0,1,0) and (0,0,1), respectively. Calculate the inverse matrix. You should get the matrix M the columns of which are {1,2,1}/14, {−9,−4,5}/14 and {22,2,−6}/14. Then $M \cdot \mathbf{x}$ is (−89,4,23)/14.

6. Det[A] = 112. A^{-1} = {{8,−40,40,16}, {12,24,−38,10}, {−16,24,4,−4}, {−36,40,−54,26}}/112. The inside bracket enclosures are the respective rows of the matrix.

7. $\mathbf{z} = 0$ is the invariant 2-dimensional subspace; the 1-dimensional ones are sp(1,0,0) and sp(1,1,0).

16. A = sp{(0,1,2,3), (1,2,1,0)} + (1,1,1,1). Yes: (0,0,2,4) = (0,1,2,3) − (1,2,1,0) + (1,1,1,1).

24. The affine space contains the collection of points of the form: $\alpha(1,3,4,0,0) + \beta(-1,3,0,1,0) + (1-\alpha-\beta)(4,0,1,1,1)$. The square of the distance of such a point is a quadratic in the variables. Take the two partial derivatives of this quadratic, set them equal to 0, and solve the system to obtain $\alpha = -2/140$, $\beta = 99/140$, $\gamma = 63/140$. The length of this vector $\frac{1}{140}\langle 131, 231, -25, 162, 63\rangle$ is approximately equal to 2.274077521.

26. 394

27. $X = \pm 1, \pm 2$

29. $p[x] = 7/18\, x^2 - 5/18\, x + 1/18$, $q[x] = 7/18\, x + 16/18$. There are other pairs of higher degree; for example, $p[x] + x^3 + 1$, $q[x] + x^2 + 3x + 1$.

30. A matrix the rows of which are {−2, 0, 2}, {−1/4, 0, −1/4} and {−2, 0, 2}.

31. Assume A is non-scalar and let $B = (e_{ji})$.

34. ker T can be described as the set of vectors in E^3 of the form $(3s, 2s + t, 3t)$, s and t arbitrary real numbers, and $T^{-1}(1)$ those vectors of the form $(s, t, 3t - 2s + 1)$.

CHAPTER 5

1. 28/3

2. 28/3; the pullback of $x^2dx + y^2dy + z^2dz$ is $(52t^5 + 44t^3 - 64t^2 + 94t - 36)dt$.

3. $2\pi^2R^3$, $\pi^2R^4/2$

4. $2\sqrt{21311}$

11. −2.2479...

12. 24, 24

16. $\mu(P) = 5\sqrt{10}$. $\begin{pmatrix} 3 & 2 & 1 & 0 \\ 1 & 1 & -1 & 4 \end{pmatrix}^t$ maps the unit square of E^2 to P. The g-factor is $5\sqrt{10}$. The pullback of $xyzw$ is $f^*[u,v] = 24u^3v - 4u^2v^2 - 16uv^3 - 4v^4$. $\int f^*g = -\frac{11}{9}\sqrt{10}$.

17. (5/2, 3/2, 1, 2)

18. $-8/5$; $\omega = -\frac{1}{2}\left[\frac{1+y^2}{x^2} + y^2\right]$

19. $2\ln[5/3]$; $\omega = \ln[y + \sqrt{y^2 - x^2}]$

20. NO, YES

22. To show what is asked, proceed like this:

$$\omega = \ln[x^2 + y^2],\ d\omega = \frac{2x}{x^2+y^2}dx + \frac{2y}{x^2+y^2}dy.\quad \int_{\partial\gamma}\omega = \int_{\gamma}d\omega.$$

$$\int_{\partial\gamma}\omega = \ln[c^2 + d^2] - \ln[a^2 + b^2] = \int_{\gamma}d\omega = \int_0^1 \frac{2f[t]f'[t]+2g[t]g'[t]}{(f[t])^2+(g[t])^2}dt = \ln[(f[t])^2 + (g[t])^2] = \ln\left[\frac{c^2+d^2}{a^2+b^2}\right].$$

CHAPTER 6

3. A open $\Rightarrow$ for all $x \in A$, there exists $N_x \subset A$, and $\bar{B}$ closed $\Rightarrow \forall y \notin \bar{B}\ \exists N_y$, s. t. $N_y \cap B = \phi$. Thus, $x \in A \cap \bar{B} \Rightarrow \exists N_x$, s.t. $N_x \cap A \neq \varnothing$ and $N_x \cap B \neq \varnothing \Rightarrow x \in \overline{A \cap B}$.

4. Suppose $\exists\gamma$, $\alpha < \gamma < \beta$, s.t. for no $\mathbf{x} \in [\mathbf{x}_0, \mathbf{x}_1]$ is $f'[\mathbf{x}] = \gamma$. Let $g[\mathbf{x}] = f[\mathbf{x}] - \gamma\mathbf{x}$. $g'[\mathbf{x}] = f'[\mathbf{x}] - \gamma$ for all $\mathbf{x} \in [\mathbf{x}_0, \mathbf{x}_1]$, and therefore $g'[\mathbf{x}] < 0$ at $\mathbf{x}_0$, and $g'[\mathbf{x}] > 0$ at $\mathbf{x}_1$, implying that $g\downarrow$ at $\mathbf{x}_0$ and $g\uparrow$ at $\mathbf{x}_1$. At some point $\mathbf{p} \in (\mathbf{x}_0, \mathbf{x}_1)$, $g[\mathbf{x}]$ changes from a decreasing function to a non-decreasing one. $\mathbf{p}$ cannot be a cusp because the derivative of g exists at every point and hence must be 0 at $\mathbf{p}$. Hence we have a contradiction. At $\mathbf{p}$, the derivative of f equals γ.

9. If the intersection is empty it is closed. Otherwise, let G_α = the complement of F_α for each α. The union of the G_α is open, and, by the DeMorgan theorem, one can conclude that the intersection of the F_α is closed.

15. Let C be the cofinite topology for $\mathbb{Z}$. Then $\mathbb{Z} \in C$ and $\phi \in C$. If A, B $\in C$, then the complement of the intersection of these two sets is equal to the union of the complements, a finite set. Clearly, the complements of arbitrary unions of sets of C are finite. Therefore, C is a topology.

16. Let $\mathbf{x} \in V$ be arbitrary. For each point $\mathbf{p} \in V$, $\mathbf{p} \neq \mathbf{x}$, there exists an N_p in the topology which does not contain $\mathbf{x}$. The union of all such sets, one for each point of V other than **x**, is an open set. Thus its complement, the singleton {**x**}, is a closed set.

27. Let $\mathfrak{B}$ be the family of closed, connected sets of the compact connected set X which contain the closed set A. Not only does $\mathfrak{B}$ have the *finite intersection property* (see problem 64) but the intersection of all the sets of $\mathfrak{B}$ contains A and is a closed set. However, it may not be connected. To get around this difficulty, let $S \subset \mathfrak{B}$ be a chain of sets, ordered by inclusion. There may be many such chains, and each has a minimal element, the intersection of all the sets in the chain. Now apply Zorn's lemma.

29. If T is norm-reducing, then for all **x**, **y** in the unit ball of V, if $\|\mathbf{x} - \mathbf{y}\| < \epsilon$, then $\|T(\mathbf{x} - \mathbf{y})\| = \|T\mathbf{x} - T\mathbf{y}\| < \epsilon$. If $\|T\mathbf{x} - T\mathbf{y}\| \leqslant \|\mathbf{x} - \mathbf{y}\|$, for all **x**, **y** in the unit ball, then T is bounded and hence continuous.

46. Let **x** be in the complement of the derived set of a non-empty set S. If the derived set is empty, it is by definition a closed set. If every neighborhood of **x** contains a point of S', then each such neighborhood must contain a point of S, so $\mathbf{x} \in S'$. Contradiction. Thus, the derived set of any non-empty set of real numbers (or any Hausdorff space) is closed.

48. Uniformly continuity of f ensures that given an ϵ, there is a δ such that for any two points in the domain of f within δ of each other, $f[\mathbf{x}]$ and $f[\mathbf{y}]$ will be within ϵ of each other. If a sequence is Cauchy, then for all sufficiently large m and n, $f[\mathbf{x}_m]$ and $f[\mathbf{x}_n]$ will necessarily be close together.

64. Suppose X is compact. Let $\mathfrak{F} = \{F_\alpha\}$ be an arbitrary collection of closed sets of X such that the intersection of the sets of any finite subcollection of $\mathfrak{F}$ is non-empty ($\mathfrak{F}$ has the finite intersection property). $\forall F$, the complement is an open set in X, and the empty set is not one of the family of closed sets. The union of the complements of all the sets F_α in $\mathfrak{F}$ is the complement of the intersection of all the sets F_α. If this union covers X, the cover can be reduced to a finite sub-cover, and this implies that the intersection of the complements of this finite number of sets is empty. Impossible, since $\mathfrak{F}$ has the f.i.p. Now suppose X is not compact, and follow the same line of argument to show that there is a collection of closed sets with the f.i.p. for which the intersection of all the sets is empty.

CHAPTER 7

2. **(a)** $\pi^2/3 - 4\cos[x] + \cos[2x] - (4/9)\cos[3x] + (1/4)\cos[4x] - (4/25)\cos[5x] + (1/9)\cos[6x] - \ldots$

(b) $4\pi^2/3 + 4\cos[x] + \cos[2x] + (4/9)\cos[3x] + \ldots - (4\pi)\sin[x] - 2\pi\sin[2x] - (4/3)\pi\sin[3x] - \ldots$

3. Too long to do by hand. Use a computer, go out to 35 or 40 terms and plot.

4. $(T\mathbf{x},\mathbf{y}) = (\mathbf{x},T\mathbf{y})$ and $T\mathbf{x} = \lambda\mathbf{x}$, $T\mathbf{y} = \mu\mathbf{y}$, $(\lambda \neq \mu)$. $(T\mathbf{x},\mathbf{y}) = (\lambda\mathbf{x},\mathbf{y}) = (\mathbf{x},T\mathbf{y}) = (\mathbf{x},\mu\mathbf{y})$. $\lambda(\mathbf{x},\mathbf{y}) = \mu(\mathbf{x},\mathbf{y}) \Rightarrow (\mathbf{x},\mathbf{y}) = 0 \Rightarrow \mathbf{x} \perp \mathbf{y}$.

5. L symmetric $\Rightarrow \int_a^b y[t][p[t]x'[t]]'dt + \int_a^b q[t]x[t]y[t]dt = \int_a^b x[t][p[t]y'[t]]'dt + \int_a^b q[t]x[t]y[t]dt \Leftrightarrow p[t]x'[t]y[t]|_a^b - \int_a^b p[t]x'[t]y'[t]dt = p[t]y'[t]x[t]|_a^b - \int_a^b p[t]y'[t]x'[t]dt \Leftrightarrow p[t]x'[t]y[t]|_a^b = p[t]y'[t]x[t]|_a^b$. Equality holds; $p[b] = p[a]$, $x'[b] = x'[a]$, $y'[b] = y'[a]$, $x[b] = x[a]$, $y[b] = y[a]$.

To arrive at the second conclusion, you might try induction. Note that $p_3[t]$ is a constant multiple of the polynomial $5t^3 - 3t$ and $L[p_3[t]] = (d/dt)[(t^2 - 1)(15t^2 - 3)] = 12(5t^3 - 3t)$. Thus, $L[p_3[t]] = kp_3[t]$.

7. If you look closely you will see that $\text{sp}\{\mathbf{u}_1,\mathbf{u}_2,\mathbf{u}_3,\mathbf{e}_4,\mathbf{e}_5,\mathbf{e}_6\}$ is **not** all of E^6. Using $\mathbf{e}_3$, $\mathbf{e}_4$, $\mathbf{e}_5$ we get: $\{\mathbf{u}_1,\mathbf{u}_2,\mathbf{u}_3,\mathbf{u}_4,\mathbf{u}_5,\mathbf{u}_6\} = \{(0,2,1,1,2,1)/\sqrt{14}), (4,0,2,2,-1,-1)/\sqrt{26}), (7,-13,-3,-3,8,8)/\sqrt{91}), (-1,-1,3,-1,0,0)/2\sqrt{3}), (-1,-1,0,2,0,0)/\sqrt{6}), (0,0,0,0,1,-1)/\sqrt{2})\}$. The inverse of this matrix is: precisely the transpose of that matrix!

10. (a) $\pi^2/6$ (b) $\pi^2/12$ (c) $\pi^2/8$

11. $\Sigma(-1)^{n-1}/n^4 = 7\pi^4/720$

16. $\ln[z] = \ln[6] - \mathbf{i}\,\text{Arcsin}[\sqrt{3}/2]$

20. The set of all sequences of rational numbers having only finitely many non-zero entries has cardinality $\aleph_0$. Let $\mathbf{x} = \{x_n\}$ be an arbitrary square-summable sequence in l^2. Let $\epsilon > 0$ be given; we can truncate this series $\mathbf{x}$ to obtain a series $\mathbf{x}'$, a series having a tail of zeros and agreeing with $\mathbf{x}$ for sufficiently many terms up to its tail of zeros so that $\|\mathbf{x} - \mathbf{x}'\| < \varepsilon/4$. Let $\mathbf{x}''$ be a series whose terms x_n'' are such that $|x_n' - x_n''| < \varepsilon/2^{n+1}$ and each x_n'' is rational. Certainly, $\|\mathbf{x}' - \mathbf{x}''\| < \varepsilon/2$. It follows that $\|\mathbf{x} - \mathbf{x}''\| < \varepsilon$. Since $\varepsilon > 0$ was arbitrarily small, we can conclude that the countable subset of rational sequences with finitely many non-zero terms is dense in l^2.

25. The determinant of the matrix of the 5 vectors is equal to 2, so they are linearly independent. G.S. will give: $\{(1,1,-2,0,1)/\sqrt{7}, (-19,9,10,7,30)/\sqrt{1491}), (101,210,115,187,-81)/16\sqrt{426}), (51,30,53,-83,25)/48\sqrt{6}), (-3,3,-1,-2,-2)/(3\sqrt{3})\}$.

37. $(11x^5 - 35x^4 - 45x^3 + 155x^2 + 4x - 30)/30$

38. The Gram determinant is $\det[\{(27,2,-13,-11), (2,4,0,0), (-13,0,11,5), (-11,0,5,6)\}] = 604 = \det[\text{diag}\{27,104/27,9/2,151/117\}]$. The measure of the parallelotope is $2\sqrt{151}$.

39. $(-10,-7,0)$

40. $5\sqrt{17}$

CHAPTER 8

2. Measure is continuous from above and below (Theorem 8.1.3).

5. Roughly .7788

7. $-469/90$

9. $(179 - 84\sqrt{2})/12$

18. $15\pi^2$

19. $\pi(R^2 + 1)\ln[R^2 + 1] - \pi R^2$

26. 61/201

32. 1; 1/4

35. $f[\sin[x]]$ is symmetric about $x = \pi/2$.

45. Given that $|f[x] - f[y]| < M|x - y|^\alpha$, $\alpha > 1$, $M > 0$, for all x and y, it follows that $|f[x] - f[y]| < M|b - a|^\alpha$, so f is uniformly continuous. Moreover, if $|(f[x] - f[y])/(x - y)| < M|x - y|^{\alpha-1}$, f is not only continuous at each x but is differentiable at each x, with $f'[x] = 0$. Therefore, f is constant on $[a,b]$.

CHAPTER 9

1. Each component of $\varphi[x,y]$ must be expanded in a Taylor series about the point (1,1). The first component: $\sin[1] + \cos[1](x - 1) + \cos[1](y - 1) - \sin[1](x - 1)^2/2 + (\cos[1] - \sin[1])(x - 1)(y - 1) - \sin[1](y - 1)^2/2 + \ldots$. The second, third and fourth components are found similarly, but the last is just the polynomial $3 + 3(x - 1) + 3(y - 1) + (x - 1)^2 + (x - 1)(y - 1) + (y - 1)^2$.

2. $8\pi^2R^5/15$

5. $g^2 = (J_\varphi)^t(J_\varphi) = a^4$. $\varphi[D] = \int_D g = \pi a^3/12$. $a = \sqrt{x_1^2 + x_2^2 + x_3^2 + x_4^2}$.

6. $y = 4/(5x^2) + x^3/5$

7. $2\sqrt{3} - 1$

8. Max $\prod_{i=1}^{n} x_i^2$, with $\sum_{i=1}^{n} x_i^2 = \mathrm{R}^2$, is $\frac{R^{2n}}{n^n}$; i.e., when each $x_i = \mathrm{R}/\sqrt{n}$.

11. $(a^2 + b^2)\pi/2$

12. $\ln[3] - \frac{1}{2}$

16. The gradient vanishes at $(\pm 1,0,0)$, $(0,\pm 1,0)$, $(0,0,\pm 1)$, $(0,0,0)$; the last is a min, the first a max.

17. Let $G[t] = \int f[x,t]dt$, so $F[x] = G[x, g[x]] - G[x, h[x]]$. Then, $dF = \frac{\partial G}{\partial x}dx + \frac{\partial G}{\partial g}g'(x)dx - \frac{\partial G}{\partial x}dx - \frac{\partial G}{\partial h}h'(x)dx$, and $\frac{dF}{dx} = \frac{\partial G}{\partial g}g'(x) - \frac{\partial G}{\partial h}h'(x)$.

18. $R^2\sqrt{2}$

19. $\pi/6$

20. $(\pi/2)\text{Arcsinh}[a]$.

21. $(3x^2 - 6xy + 3y^2, -3x^2 + 6xy)_{(3,1)} \cdot V = (12,-9) \cdot (.6,.8) = 0$

30. $L \approx 2.04637$; $(\bar{x}, \bar{y}, \bar{z}) = (.407976, .571421, .710276)$

31. (c) $(-1/3, -1/3)$; minimum

32. .768476

33. $f[\pm 2,\pm 3,0,0] = 13$, $f[0,0,\pm\sqrt{3},\pm 1] = 0$, $f[\pm 1,0,0,\pm\sqrt{3}] = 1$, $f[0,\pm 1,\pm 2,0] = 1$

41. (4/3, 1), a maximum

44. $-19/50$

49. curl $\mathbf{F} = (xz - y, x^2 - yz, 0)$, and div $\mathbf{F} = x^2 + z + yz$.

50. $(-1/3)(14\sqrt{14} - 5\sqrt{5})$ and 7

58. 13.346 and -4.346

59. $(0, \pi/4, \pi/2)$

CHAPTER 10

3. $1/x$ is not bounded.

5. A^3 is the 3×3 zero matrix, so the four functions of the matrix A are quite easy to calculate.

6. **(c)** Show that for all $x > 0$, $|\sin[x]| < x$. Then it is clear that the convergence of $\Sigma\pi/2^n$ ensures the convergence of the series in question. **(d)** Compare with $\Sigma 1/(n + 1)$ and conclude divergence. **(f)** Try the integral test and conclude convergence.

16. See problem 36 of Chapter 7.

24. **(a)** LimSup = 1, LimInf = −1. **(b)** Note that the only values the sine function can assume are 0 and $\pm\sqrt{3}/2$, so no limits. **(c)** The limit is 0. **(d)** The limit is $(\sqrt{5} + 1)2$. **(e)** 4/5 and 0.

31. $\Sigma nu_n = \ln \dfrac{2\cdot 2}{3}\,\dfrac{3\cdot 3\cdot 4\cdot 4\cdot 5\cdot 5\cdot 6\cdot 6\cdot 7\cdot 7\cdots}{2\cdot 4\cdot 3\cdot 5\cdot 4\cdot 6\cdot 5\cdot 7\cdot 6\cdot 8\cdots} \to \ln[2]$, so Σu_n converges (to .397127...).

33. Try letting $x = y^2$, use the Maclaurin series for $\tan^{-1}[y]$, and do some long division cleverly.

53. $xy' = y(x + 1) \quad y = kxe^x = k\sum_{1}^{\infty} \dfrac{x^n}{(n-1)!}$

CHAPTER 11

3. As suggested, differentiating the integrand with respect to the parameter a leaves one with an integrand of $\exp[-(a+1)x]$ which after integration and evaluation reduces to $1/(a+1)$. Now integrate wrt a to get $\ln[1+a] + C$. Set $a = 0$ as a trial value and conclude that C must be zero.

5. $\pi^{(3/2)}$

6. 8/15

15. Can you differentiate under the integral sign and then integrate, or must you integrate first? (Either way; $dg/dy = -y\cos[y^2] + y\cos[4\pi^2 + y^2] + 4\pi^2 y \sin[4\pi^2 + y^2]$.)

18. $\dfrac{\pi}{\sin[n\pi + \frac{\pi}{2}]\Gamma[n + \frac{1}{2}]}$, where $\Gamma[n + \frac{1}{2}] = \dfrac{2n!\sqrt{\pi}}{4^n n!}$

23. $y[t] = 2\sin[t] - \cos[t] + 2e^{-t} - 2te^{-t} - e^{-2t}$

34. Compare with Problem #3; $(1/2)\ln[1 + a]$

35. $\pi\sin^{-1}[a]$

36. $\tan^{-1}[|b|/a] - \tan^{-1}[|c|/a]$

CHAPTER 12

1. **(a)** −14/3 **(b)** $-9\pi/4$ **(c)** $\left(\dfrac{10\pi}{3\sqrt{3}} + \dfrac{7\sqrt{5}}{12} - \dfrac{3}{4} - \dfrac{20}{3\sqrt{3}}\sin^{-1}[1/4]\right)$ **(d)** 3

2. $2w + 8/3$

3. πb^4

4. $\frac{8\sqrt{15}}{3}\pi$

5. 1/5

10. 1312/105; $\iint\left(\frac{\partial B}{\partial x} - \frac{\partial A}{\partial y}\right)dxdy = \int_{(0,0)}^{(0,2)} 0 + \int_{(2,0)}^{(2,4)} B + \int_{(t,t^2)} \omega$

22. You might want to integrate $d\omega$ over the whole ball first; that is an elementary integration. $d\omega = (2x - 2y + 2z)dx\,dy\,dz$. The result is 0. If you integrate ω over the whole sphere you will get 0. However, try integrating ω over one half the sphere and see if you get $\pi^5/2$. If you integrate $d\omega$ over the top half of the ball, or the first octant, and multiply by 2, or 8, you will get π^5. If $2(x - y + z)$ expresses a strange density of the solid ball, then π^5 is the result you would choose for the mass of the ball.

24. $512\pi^2/15$

BIBLIOGRAPHY

Ahlfors, L. V. (1966). *Complex Analysis*, 2nd ed., McGraw-Hill, New York.

Albert, A. A. (1937). *Modern Higher Algebra*, University of Chicago Press, Chicago, Illinois.

Alexandrov, P. S., and Hopf, H. (1935). *Topologie I*, Berlin.

Apostol, T. M. (1957). *Mathematical Analysis*, Addison-Wesley, Reading, Massachusetts.

Apostol, T. M. (1967). *Calculus*, 2nd ed., Ginn-Blaisdell, Waltham, Massachusetts.

Artin, E. (1964). *The Gamma Function*, Holt, Rinehart & Winston, New York.

Bohnenblust, H. F. (1939). "Real variables," unpublished lecture notes, Princeton University.

Buck, R. C. (1965). *Advanced Calculus*, 2nd ed., McGraw-Hill, New York.

Cullen, C. (1966). *Matrices and Linear Transformations*, Addison-Wesley, Reading, Massachusetts.

Dixmier, J. (1953). "Sur les bases orthonormales dans les espaces pre-Hilbertiens," *Acta Sci. Math. (Szeged)* 15, 29–30.

Dugundji, J. (1968). *Topology*, Allyn and Bacon, Boston.

Eisenhart, L. P. (1947). *An introduction to Differentiable Geometry*, Princeton University Press, Princeton, New Jersey.

Edwards, H. M. (1969). *Advanced Calculus*, Houghton-Miflin, Boston.

Fine, H. B. (1936). *Calculus*, Macmillan, New York.

Fine, H. B. (1901). *A College Algebra*, Ginn, Lexington, Massachusetts.

Gelbaum, B. R., and Olmsted, J. M. H. (1963). *Counterexamples in Analysis*, Holden-Day, San Francisco.

Gemignani, M. C. (1972). *Elementary Topology*, 2nd ed., Addison-Wesley, Reading, Massachusetts.

Goffman, C. (1960). *Real Functions*, Rinehart, New York.

Halmos, P. R. (1961). *Measure Theory*, Van Nostrand, New York.

Hewitt, E., and Stromberg, K. (1965). *Real and Abstract Analysis*, Springer-Verlag, New York.

Hogg, R. V., and Craig, A. T. (1978). *Introduction to Mathematical Statistics*, 4th ed., Macmillan, New York.

James, G., and James, R. C. (1976). *Mathematics Dictionary*, 4th ed., Van Nostrand, Reinhold, New York.

Knopp, K. (1948). *Theory and Application of Infinite Series*, Hafer, New York.

Lang, S. (1968). *Analysis I*, Addison-Wesley, Reading, Massachusetts.

Lorch, E. R. (1962). *Spectral Theory*, Oxford University Press, New York.

McShane, E. J. (1957). *Integration*, Princeton University Press, Princeton, New Jersey.

Munroe, M. E. (1959). *Introduction to Measure and Integration*, Addison-Wesley, Reading, Massachusetts.

Nickerson, M. K., Spencer, D. C., and Steenrod, N. E. (1959). *Advanced Calculus*, Van Nostrand, New York.

Randol, B. (1969). *An Introduction to Real Analysis*, Harcourt, Brace and World, New York.

Schneider, D. M., Steeg, M., and Young, F. H. (1982). *A Concrete Introduction to Linear Algebra*, Macmillan, New York.

Schurle, A. W. (1979). *Topics in Topology*, North Holland, New York.

Shilov, G. E. (1961). *Theory of Linear Spaces*, Prentice-Hall, Englewood Cliffs, New Jersey.

Simmons, G. F. (1963). *Topology and Modern Analysis*, McGraw-Hill, New York.

Spivak, M. (1965). *Calculus on Manifolds*, Benjamin, New York.

Sz.-Nagy, B. (1965). *Introduction to Real Functions and Orthogonal Expansions*, Oxford University Press, New York.

Taylor, A. E. (1955). *Advanced Calculus*, Ginn, Lexington, Massachusetts.

Taylor, A. E. (1958). *Functional Analysis*, Wiley, New York.

Widder, D. V. (1961). *Advanced Calculus*, 2nd ed., Prentice-Hall, Englewood Cliffs, New Jersey.

Wilansky, A. (1964). *Functional Analysis*, Blaisdell, New York.

This is not intended as a list of references, alternative treatments, or suggestions for further study. Rather, it is a listing of works to which I have referred over the years. There is no question but that these authors have had a considerable influence on what I have been teaching. It seems only fitting that I acknowledge their contributions.

INDEX

A CATALOG OF SELECTED

DOVER BOOKS

IN SCIENCE AND MATHEMATICS

A CATALOG OF SELECTED

DOVER BOOKS

IN SCIENCE AND MATHEMATICS

QUALITATIVE THEORY OF DIFFERENTIAL EQUATIONS, V.V. Nemytskii and V.V. Stepanov. Classic graduate-level text by two prominent Soviet mathematicians covers classical differential equations as well as topological dynamics and ergodic theory. Bibliographies. 523pp. 5⅜ x 8½. 65954-2 Pa. $14.95

MATRICES AND LINEAR ALGEBRA, Hans Schneider and George Phillip Barker. Basic textbook covers theory of matrices and its applications to systems of linear equations and related topics such as determinants, eigenvalues and differential equations. Numerous exercises. 432pp. 5⅜ x 8½. 66014-1 Pa. $10.95

QUANTUM THEORY, David Bohm. This advanced undergraduate-level text presents the quantum theory in terms of qualitative and imaginative concepts, followed by specific applications worked out in mathematical detail. Preface. Index. 655pp. 5⅜ x 8½. 65969-0 Pa. $14.95

ATOMIC PHYSICS (8th edition), Max Born. Nobel laureate's lucid treatment of kinetic theory of gases, elementary particles, nuclear atom, wave-corpuscles, atomic structure and spectral lines, much more. Over 40 appendices, bibliography. 495pp. 5⅜ x 8½. 65984-4 Pa. $13.95

ELECTRONIC STRUCTURE AND THE PROPERTIES OF SOLIDS: The Physics of the Chemical Bond, Walter A. Harrison. Innovative text offers basic understanding of the electronic structure of covalent and ionic solids, simple metals, transition metals and their compounds. Problems. 1980 edition. 582pp. 6⅛ x 9¼. 66021-4 Pa. $16.95

BOUNDARY VALUE PROBLEMS OF HEAT CONDUCTION, M. Necati Özisik. Systematic, comprehensive treatment of modern mathematical methods of solving problems in heat conduction and diffusion. Numerous examples and problems. Selected references. Appendices. 505pp. 5⅜ x 8½. 65990-9 Pa. $12.95

A SHORT HISTORY OF CHEMISTRY (3rd edition), J.R. Partington. Classic exposition explores origins of chemistry, alchemy, early medical chemistry, nature of atmosphere, theory of valency, laws and structure of atomic theory, much more. 428pp. 5⅜ x 8½. (Available in U.S. only) 65977-1 Pa. $11.95

A HISTORY OF ASTRONOMY, A. Pannekoek. Well-balanced, carefully reasoned study covers such topics as Ptolemaic theory, work of Copernicus, Kepler, Newton, Eddington's work on stars, much more. Illustrated. References. 521pp. 5⅜ x 8½. 65994-1 Pa. $12.95

PRINCIPLES OF METEOROLOGICAL ANALYSIS, Walter J. Saucier. Highly respected, abundantly illustrated classic reviews atmospheric variables, hydrostatics, static stability, various analyses (scalar, cross-section, isobaric, isentropic, more). For intermediate meteorology students. 454pp. 6⅛ x 9¼. 65979-8 Pa. $14.95

RELATIVITY, THERMODYNAMICS AND COSMOLOGY, Richard C. Tolman. Landmark study extends thermodynamics to special, general relativity; also applications of relativistic mechanics, thermodynamics to cosmological models. 501pp. 5⅜ x 8½. 65383-8 Pa. $13.95

APPLIED ANALYSIS, Cornelius Lanczos. Classic work on analysis and design of finite processes for approximating solution of analytical problems. Algebraic equations, matrices, harmonic analysis, quadrature methods, much more. 559pp. 5⅜ x 8½. 65656-X Pa. $13.95

INTRODUCTION TO ANALYSIS, Maxwell Rosenlicht. Unusually clear, accessible coverage of set theory, real number system, metric spaces, continuous functions, Riemann integration, multiple integrals, more. Wide range of problems. Undergraduate level. Bibliography. 254pp. 5⅜ x 8½. 65038-3 Pa. $8.95

INTRODUCTION TO QUANTUM MECHANICS With Applications to Chemistry, Linus Pauling & E. Bright Wilson, Jr. Classic undergraduate text by Nobel Prize winner applies quantum mechanics to chemical and physical problems. Numerous tables and figures enhance the text. Chapter bibliographies. Appendices. Index. 468pp. 5⅜ x 8½. 64871-0 Pa. $12.95

ASYMPTOTIC EXPANSIONS OF INTEGRALS, Norman Bleistein & Richard A. Handelsman. Best introduction to important field with applications in a variety of scientific disciplines. New preface. Problems. Diagrams. Tables. Bibliography. Index. 448pp. 5⅜ x 8½. 65082-0 Pa. $12.95

MATHEMATICS APPLIED TO CONTINUUM MECHANICS, Lee A. Segel. Analyzes models of fluid flow and solid deformation. For upper-level math, science and engineering students. 608pp. 5⅜ x 8½. 65369-2 Pa. $14.95

ELEMENTS OF REAL ANALYSIS, David A. Sprecher. Classic text covers fundamental concepts, real number system, point sets, functions of a real variable, Fourier series, much more. Over 500 exercises. 352pp. 5⅜ x 8½. 65385-4 Pa. $11.95

PHYSICAL PRINCIPLES OF THE QUANTUM THEORY, Werner Heisenberg. Nobel Laureate discusses quantum theory, uncertainty, wave mechanics, work of Dirac, Schroedinger, Compton, Wilson, Einstein, etc. 184pp. 5⅜ x 8½. 60113-7 Pa. $6.95

INTRODUCTORY REAL ANALYSIS, A.N. Kolmogorov, S.V. Fomin. Translated by Richard A. Silverman. Self-contained, evenly paced introduction to real and functional analysis. Some 350 problems. 403pp. 5⅜ x 8½. 61226-0 Pa. $10.95

PROBLEMS AND SOLUTIONS IN QUANTUM CHEMISTRY AND PHYSICS, Charles S. Johnson, Jr. and Lee G. Pedersen. Unusually varied problems, detailed solutions in coverage of quantum mechanics, wave mechanics, angular momentum, molecular spectroscopy, scattering theory, more. 280 problems plus 139 supplementary exercises. 430pp. 6½ x 9¼. 65236-X Pa. $13.95

SPECIAL FUNCTIONS, N.N. Lebedev. Translated by Richard Silverman. Famous Russian work treating more important special functions, with applications to specific problems of physics and engineering. 38 figures. 308pp. 5⅜ x 8½. 60624-4 Pa. $9.95

OBSERVATIONAL ASTRONOMY FOR AMATEURS, J.B. Sidgwick. Mine of useful data for observation of sun, moon, planets, asteroids, aurorae, meteors, comets, variables, binaries, etc. 39 illustrations. 384pp. 5⅜ x 8¼. (Available in U.S. only) 24033-9 Pa. $8.95

INTEGRAL EQUATIONS, F.G. Tricomi. Authoritative, well-written treatment of extremely useful mathematical tool with wide applications. Volterra Equations, Fredholm Equations, much more. Advanced undergraduate to graduate level. Exercises. Bibliography. 238pp. 5⅜ x 8½. 64828-1 Pa. $8.95

POPULAR LECTURES ON MATHEMATICAL LOGIC, Hao Wang. Noted logician's lucid treatment of historical developments, set theory, model theory, recursion theory and constructivism, proof theory, more. 3 appendixes. Bibliography. 1981 edition. ix + 283pp. 5⅜ x 8½. 67632-3 Pa. $8.95

MODERN NONLINEAR EQUATIONS, Thomas L. Saaty. Emphasizes practical solution of problems; covers seven types of equations. ". . . a welcome contribution to the existing literature...."–*Math Reviews*. 490pp. 5⅜ x 8½. 64232-1 Pa. $13.95

FUNDAMENTALS OF ASTRODYNAMICS, Roger Bate et al. Modern approach developed by U.S. Air Force Academy. Designed as a first course. Problems, exercises. Numerous illustrations. 455pp. 5⅜ x 8½. 60061-0 Pa. $10.95

INTRODUCTION TO LINEAR ALGEBRA AND DIFFERENTIAL EQUATIONS, John W. Dettman. Excellent text covers complex numbers, determinants, orthonormal bases, Laplace transforms, much more. Exercises with solutions. Undergraduate level. 416pp. 5⅜ x 8½. 65191-6 Pa. $11.95

INCOMPRESSIBLE AERODYNAMICS, edited by Bryan Thwaites. Covers theoretical and experimental treatment of the uniform flow of air and viscous fluids past two-dimensional aerofoils and three-dimensional wings; many other topics. 654pp. 5⅜ x 8½. 65465-6 Pa. $16.95

INTRODUCTION TO DIFFERENCE EQUATIONS, Samuel Goldberg. Exceptionally clear exposition of important discipline with applications to sociology, psychology, economics. Many illustrative examples; over 250 problems. 260pp. 5⅜ x 8½. 65084-7 Pa. $8.95

LAMINAR BOUNDARY LAYERS, edited by L. Rosenhead. Engineering classic covers steady boundary layers in two- and three- dimensional flow, unsteady boundary layers, stability, observational techniques, much more. 708pp. 5⅜ x 8½. 65646-2 Pa. $18 95

LECTURES ON CLASSICAL DIFFERENTIAL GEOMETRY, Second Edition, Dirk J. Struik. Excellent brief introduction covers curves, theory of surfaces, fundamental equations, geometry on a surface, conformal mapping, other topics. Problems. 240pp. 5⅜ x 8½. 65609-8 Pa. $8.95

GEOMETRY OF COMPLEX NUMBERS, Hans Schwerdtfeger. Illuminating, widely praised book on analytic geometry of circles, the Moebius transformation, and two-dimensional non-Euclidean geometries. 200pp. 5⅝ x 8¼. 63830-8 Pa. $8.95

MECHANICS, J.P. Den Hartog. A classic introductory text or refresher. Hundreds of applications and design problems illuminate fundamentals of trusses, loaded beams and cables, etc. 334 answered problems. 462pp. 5⅜ x 8½. 60754-2 Pa. $11.95

TOPOLOGY, John G. Hocking and Gail S. Young. Superb one-year course in classical topology. Topological spaces and functions, point-set topology, much more. Examples and problems. Bibliography. Index. 384pp. 5⅝ x 8¼. 65676-4 Pa. $10.95

STRENGTH OF MATERIALS, J.P. Den Hartog. Full, clear treatment of basic material (tension, torsion, bending, etc.) plus advanced material on engineering methods, applications. 350 answered problems. 323pp. 5⅜ x 8½. 60755-0 Pa. $9.95

ELEMENTARY CONCEPTS OF TOPOLOGY, Paul Alexandroff. Elegant, intuitive approach to topology from set-theoretic topology to Betti groups; how concepts of topology are useful in math and physics. 25 figures. 57pp. 5⅜ x 8½. 60747-X Pa. $3.95

ADVANCED STRENGTH OF MATERIALS, J.P. Den Hartog. Superbly written advanced text covers torsion, rotating disks, membrane stresses in shells, much more. Many problems and answers. 388pp. 5⅜ x 8½. 65407-9 Pa. $10.95

COMPUTABILITY AND UNSOLVABILITY, Martin Davis. Classic graduate-level introduction to theory of computability, usually referred to as theory of recurrent functions. New preface and appendix. 288pp. 5⅜ x 8½. 61471-9 Pa. $8.95

GENERAL CHEMISTRY, Linus Pauling. Revised 3rd edition of classic first-year text by Nobel laureate. Atomic and molecular structure, quantum mechanics, statistical mechanics, thermodynamics correlated with descriptive chemistry. Problems. 992pp. 5⅜ x 8½. 65622-5 Pa. $19.95

AN INTRODUCTION TO MATRICES, SETS AND GROUPS FOR SCIENCE STUDENTS, G. Stephenson. Concise, readable text introduces sets, groups, and most importantly, matrices to undergraduate students of physics, chemistry, and engineering. Problems. 164pp. 5⅜ x 8½. 65077-4 Pa. $7.95

THE HISTORICAL BACKGROUND OF CHEMISTRY, Henry M. Leicester. Evolution of ideas, not individual biography. Concentrates on formulation of a coherent set of chemical laws. 260pp. 5⅜ x 8½. 61053-5 Pa. $8.95

THE PHILOSOPHY OF MATHEMATICS: An Introductory Essay, Stephan Körner. Surveys the views of Plato, Aristotle, Leibniz & Kant concerning propositions and theories of applied and pure mathematics. Introduction. Two appendices. Index. 198pp. 5⅜ x 8½. 25048-2 Pa. $8.95

THE DEVELOPMENT OF MODERN CHEMISTRY, Aaron J. Ihde. Authoritative history of chemistry from ancient Greek theory to 20th-century innovation. Covers major chemists and their discoveries. 209 illustrations. 14 tables. Bibliographies. Indices. Appendices. 851pp. 5⅜ x 8½. 64235-6 Pa. $18.95